T0340184

Agricultural Water Management

Theories and Practices

Agricultural Water Management

Theories and Practices

Edited by

Prashant K. Srivastava
Remote Sensing Laboratory, Institute of Environment and Sustainable Development, Banaras Hindu University, Varanasi, India

Manika Gupta
Department of Geology, University of Delhi, Delhi, India

George Tsakiris
Laboratory of Reclamation Works and Water Resources Management, National Technical University of Athens, Greece

Nevil Wyndham Quinn
Faculty of Environment & Technology, University of the West of England, Bristol, United Kingdom

Academic Press is an imprint of Elsevier
125 London Wall, London EC2Y 5AS, United Kingdom
525 B Street, Suite 1650, San Diego, CA 92101, United States
50 Hampshire Street, 5th Floor, Cambridge, MA 02139, United States
The Boulevard, Langford Lane, Kidlington, Oxford OX5 1GB, United Kingdom

Library of Congress Cataloging-in-Publication Data
A catalog record for this book is available from the Library of Congress

British Library Cataloguing-in-Publication Data
A catalogue record for this book is available from the British Library

ISBN: 978-0-12-812362-1

For information on all Academic Press publications visit our website at https://www.elsevier.com/books-and-journals

Publisher: Charlotte Cockle
Acquisitions Editor: Nancy Maragioglio
Editorial Project Manager: Timothy Bennett
Production Project Manager: Sruthi Satheesh
Cover Designer: Miles Hitchen

Typeset by TNQ Technologies

I would like to dedicate this book to my wife Mary – **George Tsakiris**

I dedicate this book to my parents, husband and children Prakhar and Kaashvi for their continuing love and support in all the endeavours of my life – **Manika Gupta**

I would like to dedicate this book to my parents, wife, children and to all my laboratory group members for their significant efforts and contributions – **Prashant K. Srivastava**

Contents

7. Irrigation water quality appraisal using statistical methods and WATEQ4F geochemical model

Sudhir Kumar Singh, Ram Bharose, Jasna Nemčić-Jurec, Kishan S. Rawat and Dharmveer Singh

Part III
Earth observation techniques

8. Estimation of potential evapotranspiration using INSAT-3D satellite data over an agriculture area

Prachi Singh, Prashant K. Srivastava and R.K. Mall

9. Estimation of evapotranspiration using surface energy balance system and satellite datasets

Garima Shukla, Prasoon Tiwari, Vikas Dugesar and Prashant K. Srivastava

Part IV
Computational intelligence techniques

14. Bistatic scatterometer for the retrieval of soil moisture

Dileep Kumar Gupta, Rajendra Prasad and Prashant K. Srivastava

15. Soil water content influence on pesticide persistence and mobility

Manika Gupta, N.K. Garg and Prashant K. Srivastava

Part V
Geospatial techniques

16. Irrigation water demand estimation in Bundelkhand region using the variable infiltration capacity model

Varsha Pandey, Prashant K. Srivastava, Pulakesh Das and Mukunda Dev Behera

17. Drainage network analysis to understand the morphotectonic significance in upper Tuirial watershed, Aizawl, Mizoram

Binoy Kumar Barman, Gautam Raj Bawri, K. Srinivasa Rao, Sudhir Kumar Singh and Dhruvesh Patel

18. Development of android application for visualisation of soil water demand

Prashant K. Srivastava, Prachi Singh, Varsha Pandey and Manika Gupta

19. GIS-based analysis for soil moisture estimation via kriging with external drift

Akash Anand, Prachi Singh, Prashant K. Srivastava and Manika Gupta

Contributors

Dleen Al-Sharafany, Geomatics Department, College of Engineering, University of Salahaddin, Erbil, Kurdistan Region, Iraq

Akash Anand, Remote Sensing Laboratory, Institute of Environment and Sustainable Development, Banaras Hindu University, Varanasi, Uttar Pradesh, India

Binoy Kumar Barman, Department of Geology, Mizoram University, Aizawl, Mizoram, India

Gautam Raj Bawri, Department of Geology, Mizoram University, Aizawl, Mizoram, India

Mukunda Dev Behera, Centre for Oceans, Rivers, Atmosphere and Land Sciences, Indian Institute of Technology Kharagpur, Kharagpur, West Bengal, India

Shaily Bhardwaj, St. Xavier's, Balrampur, Uttar Pradesh, India

Ram Bharose, Department of Environmental Sciences & NRM, College of Forestry, SHUATS, Prayagraj, Uttar Pradesh, India

Toby N. Carlson, Pennsylvania State University, Department of Meteorology, University Park, State college, PA, United States

Shivani Chaudhary, Institute of Environment and Sustainable Development, Banaras Hindu University, Varanasi, Uttar Pradesh, India

Sumit Kumar Chaudhary, Remote Sensing Laboratory, Institute of Environment and Sustainable Development, Banaras Hindu University, Varanasi, Uttar Pradesh, India

Pulakesh Das, Centre for Oceans, Rivers, Atmosphere and Land Sciences, Indian Institute of Technology Kharagpur, Kharagpur, West Bengal, India; Remote Sensing and GIS Department, Vidyasagar University, Midnapore, West Bengal, India

Vikas Dugesar, Department of Geography, Banaras Hindu University, Varanasi, Uttar Pradesh, India

Daniela Silva Fuzzo, Pennsylvania State University, Department of Meteorology, University Park, State college, PA, United States

N.K. Garg, Department of Civil Engineering, Indian Institute of Technology, Delhi, New Delhi, India

Sandeep Kumar Gautam, School of Environmental Sciences, Jawaharlal Nehru University, New Delhi, India; Department of Geology, University of Lucknow, Lucknow, Uttar Pradesh, India

Ayushi Gupta, Remote Sensing Laboratory, Institute of Environment and Sustainable Development, Banaras Hindu University, Varanasi, Uttar Pradesh, India

Dileep Kumar Gupta, Remote Sensing Laboratory, Institute of Environment and Sustainable Development, Banaras Hindu University, Varanasi, Uttar Pradesh, India

Manika Gupta, Department of Geology, University of Delhi, New Delhi, India

Dionissios T. Hristopulos, Geostatistics Laboratory, School of Mineral Resources Engineering, Technical University of Crete, Chania, Greece

Pavan Kumar, College of Horticulture and Forestry, Rani Lakshmi Bai Central Agricultural University, Jhansi, Uttar Pradesh, India

R.K. Mall, DST-Mahamana Centre of Excellence in Climate Change Research, Institute of Environment and Sustainable Development, Banaras Hindu University, Varanasi, Uttar Pradesh, India

Arundhati Misra, Microwave Techniques Development Division (MTDD), Advanced Microwave and Hyperspectral Techniques Development Group (AMHTDG), Earth, Ocean, Atmosphere and Planetary Science Applications Area (EPSA), Space Applications Centre (SAC), Indian Space Research Organization (ISRO), Ahmedabad, Gujarat, India

M. Mruthyunjaya, Department of Agronomy, Institute of Agricultural Sciences, Banaras Hindu University, Varanasi, Uttar Pradesh, India

Semonti Mukherjee, Haryana Space Application Centre, Hisar, Haryana, India

Jasna Nemčić-Jurec, Institute of Public Health of Koprivnica-Križevci County, Koprivnica, Croatia

Matthew R. North, Geostatistics Laboratory, School of Mineral Resources Engineering, Technical University of Crete, Chania, Greece

Dharmendra Kumar Pandey, Microwave Techniques Development Division (MTDD), Advanced Microwave and Hyperspectral Techniques Development Group (AMHTDG), Earth, Ocean, Atmosphere and Planetary Science Applications Area (EPSA), Space Applications Centre (SAC), Indian Space Research Organization (ISRO), Ahmedabad, Gujarat, India

Varsha Pandey, Remote Sensing Laboratory, Institute of Environment and Sustainable Development, Banaras Hindu University, Varanasi, Uttar Pradesh, India

Dhruvesh Patel, Department of Civil Engineering, School of Technology, PDPU, Raisan, Gandhinagar, Gujarat, India

Andrew Pavlides, Geostatistics Laboratory, School of Mineral Resources Engineering, Technical University of Crete, Chania, Greece

George P. Petropoulos, Geostatistics Laboratory, School of Mineral Resources Engineering, Technical University of Crete, Chania, Greece

Rajendra Prasad, Remote Sensing Laboratory, Department of Physics, Indian Institute of Technology (BHU), Varanasi, Uttar Pradesh, India

Vishal Prasad, Institute of Environment and Sustainable Development, Banaras Hindu University, Varanasi, Uttar Pradesh, India

Deepak Putrevu, Microwave Techniques Development Division (MTDD), Advanced Microwave and Hyperspectral Techniques Development Group (AMHTDG), Earth, Ocean, Atmosphere and Planetary Science Applications Area (EPSA), Space Applications Centre (SAC), Indian Space Research Organization (ISRO), Ahmedabad, Gujarat, India

K. Srinivasa Rao, Department of Geology, Mizoram University, Aizawl, Mizoram, India

Kishan S. Rawat, Centre for Remote Sensing and Geo-Informatics, Sathyabama University, Chennai, Tamilnadu, India

Ajay Shankar, Institute of Environment and Sustainable Development, Banaras Hindu University, Varanasi, Uttar Pradesh, India

Jyoti Sharma, Remote Sensing Laboratory, Department of Physics, Indian Institute of Technology (BHU), Varanasi, Uttar Pradesh, India

Nitin Kumar Sharma, St. Xavier's, Balrampur, Uttar Pradesh, India

Garima Shukla, Department of Geography, Banaras Hindu University, Varanasi, Uttar Pradesh, India

Anjali Singh, Institute of Environment and Sustainable Development, Banaras Hindu University, Varanasi, Uttar Pradesh, India

Dharmveer Singh, Department of Chemistry, University of Allahabad, Prayagraj, Uttar Pradesh, India

Prachi Singh, Remote Sensing Laboratory, Institute of Environment and Sustainable Development, Banaras Hindu University, Varanasi, Uttar Pradesh, India

R.K. Singh, Department of Agronomy, Institute of Agricultural Sciences, Banaras Hindu University, Varanasi, Uttar Pradesh, India

Ram Kumar Singh, Department of Natural Resources, TERI School of Advanced Studies, New Delhi, Delhi, India

Sudhir Kumar Singh, K. Banerjee Centre of Atmospheric and Ocean Studies, IIDS, Nehru Science Centre, University of Allahabad, Prayagraj, Uttar Pradesh, India

Ujjwal Singh, Remote Sensing Laboratory, Institute of Environment and Sustainable Development, Banaras Hindu University, Varanasi, Uttar Pradesh, India

Prashant K. Srivastava, Remote Sensing Laboratory, Institute of Environment and Sustainable Development, Banaras Hindu University, Varanasi, Uttar Pradesh, India; DST-Mahamana Centre of Excellence in Climate Change Research, Institute of Environment and Sustainable Development, Banaras Hindu University, Varanasi, Uttar Pradesh, India

Swati Suman, Remote Sensing Laboratory, Institute of Environment and Sustainable Development, Banaras Hindu University, Varanasi, Uttar Pradesh, India

Ionuţ Şandric, Faculty of Geography, University of Bucharest, Bucharest, Romania

Prasoon Tiwari, Department of Geography, Banaras Hindu University, Varanasi, Uttar Pradesh, India

Jayant K. Tripathi, School of Environmental Sciences, Jawaharlal Nehru University, New Delhi, India

Preface

Information of our planet's water resources is indispensable for a number of practical applications related to both the society and ecosystems. Agricultural water management has gained considerable momentum in recent decades amongst the Earth and environmental science community for solving and understanding its concept. It plays an important role in the plant growth, yield and promotion of sustainable agriculture practices. Moreover, agricultural water is a key variable in the water and energy exchanges that occur at the land–surface/atmosphere interface and conditions the evolution of weather and climate over continental regions. Globally, the monitoring of the agricultural water has developed into a very important and urgent research direction, especially towards water resource management, irrigation scheduling, improved weather forecasts, natural hazards mitigation analysis, predictions of agricultural productivity, crop insurance, climate predictions, ecological health and services, improved trafficability, groundwater recharge, water quality and quantity, etc. In this context, there is a growing need of monitoring and understanding of the theory and practices of agricultural water management. This can be of crucial importance because of fast declining groundwater and drying up of river water.

As compared to other natural resources, agricultural water management is complicated in practice. Previously, it was handled by conventional techniques, but growing population has led to an increase in utilisation of both land and surface water by agricultural practitioners. Agricultural water demand can be also predicted from in-situ probes; however, in-situ measurements of agricultural water are currently limited to discrete measurements at specific locations, and such point-based measurements do not represent the spatial distribution accurately as soil water is highly variable both spatially and temporally. The numerous Earth Observation satellites, computational intelligence techniques, geospatial technology and sophisticated models have the potential to reshape the agricultural water management practices in an efficient way.

Therefore, the primary aim of this book is to advance the scientific understanding, development and application of agricultural water management. In this essence, a book needs to be compiled, devoted to putting together a collection of the recent developments and rigorous applications of the agricultural water management practices. By linking all the systems, this book will be a first handbook promoting the synergistic and multidisciplinary activities among scientists and users working in the field hydrometeorological and agricultural sciences.

This has only become possible because of extensive and valuable contributions from interdisciplinary experts from all over the world.

This book has been divided into six sections. Section I contains Introduction, Section II details the Conventional Techniques, Section III provides Earth Observation-Based Techniques and Section IV provides Computational Intelligence Techniques, while Sections V and VI deal with Geospatial Techniques and Future Challenges in Agricultural Water Management, respectively.

Chapter 1 in introduction section has been written by Srivastava et al. and provides an introduction to concept and methodologies of agricultural water management, Chapter 2 by Singh et al. gives a glimpse of traditional water management techniques and Chapter 3 by Singh et al. provides several approaches to understand the role of geospatial technology in agriculture water management.

Section II starts with Chapter 4 by Gupta et al., which gives an insight into analysis of wastewater and its potential impact on crop production. Chapter 5 deals with water quality assessment of the river Ghaghra for irrigation purpose (Gautam et al.). Chapter 6 is about screening and characterisation of drought tolerance ability of rhizobacteria by Shankar et al. Chapter 7 is written by Singh et al. and provides a detailed study of irrigation water by using WATEQ4F and geochemical models.

Section III focuses on evapotranspiration fluxes at agricultural sites and begins with Chapter 8 written by Singh et al. The chapter is about assessing evapotranspiration derived from satellite and station-based observations over agricultural landscape. Chapter 9 is about estimating the evapotranspiration using surface balance system and satellite datasets by Shukla et al. Chapter 10 is given by Pandey et al., on Large Scale Soil Moisture Mapping using Earth Observation Data and its Validation at selected Agricultural Sites over Indian Region. Chapter 11, by Petropoulos et al., provides simplified triangle method for mapping surface soil moisture and evaporative fluxes using Sentinel-3.

Section IV contains Chapter 12 by Chaudhary et al., on the role of Artificial Neural Network for estimating soil moisture using satellite datasets. Chapter 13 by Al-Sharafany et al. is a study about soil moisture retrieval using AMSR-E datasets. Chapter 14 deals with soil moisture retrieval using Bistatic scatterometer over agricultural field by Gupta et al. Chapter 15 is provided by Gupta et al. for understanding pesticide persistence and mobility with different irrigation treatments in field conditions through experiment and numerical model.

Section V contains Chapter 16, which is about estimating irrigation water demand using a land surface model, variable infiltration capacity model by Pandey et al. Chapter 17 by Barman et al., provides an understanding of the morphotectonic significance in drainage network analysis in upper Tuirial watershed for water resource management. Chapter 18 by Srivastava et al., shows an implementation of android application for visualising soil water demand in agricultural field, and Chapter 19 by Anand et al. provides a detailed analysis of cokriging, a GIS-based technique with external drift for estimation of soil moisture. Chapter 20 is written by Suman et al., which discusses the key parameters characterising land surface in one-dimensional SimSphere SVAT model in European ecosystem.

Section VI is the last section of this book by Chaudhary and Srivastava that summarises about the Future Challenges in agricultural water management.

We believe that this book would be read by the people with a common interest in agriculture as well as geospatial techniques, remote sensing, sustainable water resource development, applications and other diverse backgrounds within earth and environment, meteorology and hydrological sciences field.

Editors
Prashant K. Srivastava
Manika Gupta
George Tsakiris
Nevil Wyndham Quinn

Part I

Introduction

Chapter 1

Concepts and methodologies for agricultural water management

Prashant K. Srivastava[1,2], Swati Suman[1], Varsha Pandey[1], Manika Gupta[3], Ayushi Gupta[1], Dileep Kumar Gupta[1], Sumit Kumar Chaudhary[1], Ujjwal Singh[1]
[1]*Remote Sensing Laboratory, Institute of Environment and Sustainable Development, Banaras Hindu University, Varanasi, Uttar Pradesh, India;* [2]*DST-Mahamana Centre of Excellence in Climate Change Research, Institute of Environment and Sustainable Development, Banaras Hindu University, Varanasi, Uttar Pradesh, India;* [3]*Department of Geology, University of Delhi, New Delhi, India*

1. Introduction

The world's population is growing faster and estimates show that one-fifth of the total population lives in water-scarce regions and will face a 40% shortfall between forecast demand and available supply of water by 2030 (World Health Organization, 2009). Rapid population growth, increase in per capita food consumption, change in diet composition, chronic water scarcity, hydrological uncertainties and extreme weather events are believed to be the greatest threats to global prosperity and stability (Wallace, 2000). According to statistics, by the year 2050 world's total population is estimated to reach 9 billion, which will require 60% increase in food production than today and 15% increase in water withdrawal will be needed to irrigate this production (Nelson et al., 2010). Climate change is believed to worsen the situation by altering the hydrological cycle, changing rainfall pattern and increasing the intensity and duration of hydrological extreme event such as floods and droughts affecting one billion people living in the monsoonal basins and 500 million in deltas (Pandey et al., 2020; Singh and Kumar, 2018; Snyder et al., 2019; Trenberth, 1999).

To strengthen water security against this backdrop of increasing demand, water scarcity, growing uncertainty, greater extremes and fragmentation challenges, we will need to make efforts to improve water usage in irrigation in order to have marked effect on sustainable agriculture and on soil and water

Agricultural Water Management. https://doi.org/10.1016/B978-0-12-812362-1.00001-1

resource conservation (Zilberman et al., 1997). As such efficient management of irrigation water can play a crucial role in reducing wastage and conservation of water resources.

Irrigation water management is the practice to monitor and manage rate, volume and timing of water application in the field based on the crop conditions such as growing and non-growing seasons, growth stage, weed and pest resistance and soil characteristics such as water holding capacity under prevailing financial and technological limitations to get optimum yield (Garg and Gupta, 2015; Majumdar, 2001). Effective agronomic practices are also considered essential components for effective irrigation water management. Approaches such as proper management of soil fertility, crop selection and rotation based on local conditions, selection of irrigation technique based on drainage of the area, soil reclamation and erosion control, use of organic manures and mulching are some of the steps which can make incremental differences (FAO, UN).

To combat water quality deterioration due to irrigation primarily due to discharges of salts, pesticides and nutrients to the ground water and discharges, Irrigation Scheduling is considered as the most practical approach. Irrigation scheduling is a practice to provide exact amount of water in fields at appropriate timings through natural and artificial sources based on criteria such as soil moisture content and soil moisture tension with the main purpose to increase water-use efficiency, minimise wastage, reduce nonpoint sources of pollution of ground and surface water and maximise production (Gupta et al., 2012, 2014; Howell and Meron, 2007). A number of methods for scheduling irrigation have been developed and are used with varying degrees of success (Dudley et al., 1971; Paul et al., 2000). Some of these notable decision-support systems include HYDRA, DSSAT (Decision Support System for Agrotechnology Transfer), HYDRUS, Linear Network Flow Model, Bayesian Data-Driven Model, CropSyst Models for irrigation water management, water balance methods, irrigation techniques such as drift and sprinkler and mulching. In the following section of this chapter, we present and discuss in detail these agriculture water management approaches.

2. Irrigation water management techniques

2.1 HYDRA

HYDRA Decision Support System (DSS) is a strategic model developed to help farmers and water managing authorities for improved and efficient irrigation water use and practices (Jacucci et al., 1995). Hydra is the result of a European community project started to help the farmers in making efficient and profitable irrigation water consumption decisions, as well as the planners in policy framing and implementation. The main aim of the HYDRA DSS is to use various public domain software tools in developing a knowledge-based, graphically oriented, user-friendly water distribution portal.

Main components of Hydra DSS are a hierarchical set of crop growth simulation models, a database and data entry system, a Graphical Information System (GIS), an expert system module and a Graphical User Interface (GUI) and DSS scenario controller system. Briefing the role of the various components of the Hydra DSS, the first component, the crop growth simulation model is its main component. The unit comprises of a single to complex crop growth and a water balance model such as ISAREG (Teixeira and Pereira, 1991) the UN FAO Model (Santos and de Camargo, 2006), WOFOST (Van Diepen et al., 1989), SWATRE (Belmans et al., 1983) which require daily inputs of weather information of the location of simulation such as maximum and minimum temperatures, precipitation, pan evapotranspiration or grass reference evapotranspiration and soil hydraulic information such as soil type, irrigation records to predict the crop phenology and water consumption rate. The second component or the data entry system is used for viewing and modifying the database of the system. The GIS part of the Hydra DSS is used to provide information and displays the physical features of the location such as the land use of the area, soil map, soil type, etc. Based on the management choices made by the user, the fourth component, the Expert System Module, helps the users to select and project the best water practice for the current agricultural season. The expert system module uses the long-term historic weather records for decision-making. The GUI provides a user-friendly environment to the operators to use HYDRA DSS. This component of the model helps the end-user to feed, update and retrieve the input parameters into the Hydra DSS. Hydra DSS has been very useful in improving water management and irrigation practices in various parts of European Mediterranean agricultural ecosystems and can be a useful aid for efficient water management for developing countries and in tackling water-scarce situations.

2.2 Decision support system for agrotechnology transfer (DSSAT)

Water is a very valuable commodity. Shortage of water and its irregular consumption need better management of water resources (Heeren et al., 2006). Since irrigation is the largest consumptive use of water in agriculture field, it will be necessary for the farmer to optimise the use of water in irrigation. Science-based knowledge, improving our irrigation water management for conservation of water, along with role of modelling approach can help in understanding the interrelation in the atmosphere-soil-plant processes. In this context, soil water balance—based decision support system for agrotechnology transfer (DSSAT) model as given below proves to be an excellent tool (Lascano, 1991; Ritchie, 1998).

$$SW_i = (ETa_i + RO_i + OP_i) - (EP_i + G_i + IRR_i) + SW_{i-1} \qquad (1.1)$$

where SW_i is soil water depletion in the root zone at end of the day i [mm], Eta_i is actual crop evapotranspiration on day i [mm], RO_i is runoff from the soil surface on day i [mm], OP_i is deep percolation on day i [mm], Epi is effective precipitation on day i [mm], G_i is capillary rise from the groundwater table on day i [mm], IRR_i is net irrigation on day i [mm], SW_{i-1} is soil water depletion in the root zone at end of the previous day, i−1 [mm].

The DSSAT is based on tipping bucket approach explained by Ritchie which simulates the soil-water flow and root uptake of each soil layer (Soler et al., 2013). Also DSSAT has a variety of modules involved in evaluating the soil water content in root zone, daily crop water requirement and soil moisture data, weather data for quantifying the relationships between crop yield and water-use for optimising irrigation management across various seasons over long periods (Thorp et al., 2008). Further, DSSAT has been utilised worldwide in recent years because it is applicable in large variety of crops such as CERES for cereals and CROPGRO for legumes, root crop model, validated with regional crop also (Jones et al., 2003).

2.3 Hydrus

Hydrus-1D (Šimůnek and Hopmans, 2009) is a unidimensional numeric model for simulating water flow and solute transport in variably saturated porous media. Water flow can be simulated using a wide range of governing equations including van Genuchten-Mualem, Brooks-Corey and Kosugi equations. It can account for macropore flow using mobile-immobile dual porosity equations, and simulate subsurface tile drainage using either the Hooghoudt equation or the Ernest equation (Simunek et al., 2006). Hydrus-1D has a versatile solute transport module as it can simulate a general solute with its adsorption, transport and reactions, as well as simulate major ions such as CO_2, Ca, Mg, Na, K, SO_4, Cl, NO_3, H_4SiO_4 (Simunek et al., 2006) Other functions of the Hydrus-1D model include heat transport, bacterial transport, 2-site adsorption and inverse solution of problems. This model requires a large number of input parameters such as climatological data, soil properties, solute transport and reaction parameters, root depth and crop properties. Limitations of the model include its inability to address uneven tile drainage spacing, controlled tile drainage, subirrigation, solid and manure fertilisers, as well as agricultural practices such as cropping and tillage. In the meantime, one major advantage is that both the model and its source code are both freely available, which enables model modifications/coupling work, such as the Hydrus-NICA.

It evaluates the nonsteady unsaturated water flow through Richard's equation.

$$\frac{\partial \theta}{\partial t} = \frac{\partial}{\partial x}\left[K(h)\left(\frac{\partial h}{\partial x} + \cos\alpha\right)\right] - D \tag{1.2}$$

where θ is the volumetric soil water content (cm^3/cm^3), h is the water pressure head (Koopal et al., 1994), and time (days), respectively, x is the spatial coordinate (positive upward), D and α are the sink/drop term ($cm^3/cm^3 day^{-1}$)

and the angle between the flow direction, respectively). K is used for unsaturated soil hydraulic conductivity function (cm/day). The vertical axis (i.e., = 0° for vertical flow, 90° for horizontal flow, and $0° < \alpha < 90°$ for inclined flow.

Soil water hydraulic properties are the model inputs and are calculated using Van Genuchten (1980) and in correspondence with Mualem pore size distribution (Mualem, 1976) functions.

$$\theta(h) = \begin{cases} \theta_r + \dfrac{\theta_s - \theta_r}{\left[1 + |ah|^n\right]^m} & h < 0 \\ \theta_s & h \geq 0 \end{cases}$$

$$K(h) = K_s S_e^l \left[1 - \left(1 - S_e^{1/m}\right)^m\right]^2 \tag{1.3}$$

$$m = 1 - 1/n, \quad n > 1$$

where θ = water content (−), h = hydraulic head, θ_r = residual water content, θ_s = saturated water content, a = inverse of air entry value or bubbling pressure (1/cm), m = coefficient related to n, the pore size distribution index (−), n = pore size distribution index (−), k_s = hydraulic conductivity, s_e = effective water content (−), l = pore connectivity, usually 0.5 is the average value for soil (−).

Tim and Mostaghimi (1989) developed a two-dimensional model that describes phosphorous movement and its transformation in vadose zone. This model was able to simulate phosphorous in soil column and the data was found well with only 7% deviation in it. This study was more focused on phosphorous movement not on its loss pathways for tile drainage and surface runoff. In a study, Hydrus (2D/3D) model was tested for simulation of water flow and phosphorus transport in southern Ontario having a clay loam soil type. The model was calibrated and validated between 2008 and 2011 using field data from 1000 × 2 sq. m test plots. These plots had corn-soybean crop rotation with controlled tile drainage. The surface and subsurface flow of water in test plots were accounted and monitored. The samples were collected continuously year-round using an auto-sampling system. The simulation of water flow and phosphorous came out relatively well by the model. The weekly modelling efficiency observed was 513 to 0.738. Validation of this weekly modelling for water flow and dissolved phosphorous loss in tile drainage came out to be 0.587 to 0.768. Some of the errors in simulation were attributed to soil cracking in the summer which consequently enhanced macropore flow. Limitations of the model include lack of simulation of surface runoff phosphorous loss and particulate phosphorous loss (Qiao, 2014).

2.4 MOPECO model

The MOPECO model is one of the popular tools used for the water irrigation management planning. This model can find crop production, crop yield using the irrigation plan and suitable crop pattern (Jacucci et al., 1995). The initial use of MOPECO was by Stewart (1977) as a function of actual and maximum evapotranspiration ratio (ET_a/ET_m) during the different crop stages for the finding of crop yield. If $ET_a < ET_m$ after which crop suffers from stress that causes a lower crop yield (actual yield (Y_a)< potential yield (Y_m)) and is expressed as (Teixeira and Pereira, 1991):

$$\frac{Y_a}{Y_M} = \prod_{n=1}^{n=4}\left(1 - K_{YJ}\left(1 - \frac{ET_{aj}}{ET_{mj}}\right)\right) \quad (1.4)$$

where Y_M and Y_a represent the potential and actual crop yields (kg/ha); n represents the growing stages (Santos and de Camargo, 2006); ET_m and ET_a represent crop evapotranspiration for maximum yield (mm) and actual crop evapotranspiration (mm), respectively; while *j* represents for growing stages.

Doorenbos and Pruitt (1977) proposed the calculation of daily ET_m (Van Diepen et al., 1989). For daily ET_m required the value of kc for each growth stage including daily reference evapotranspiration (ET_o in mm) (Allen et al., 1998; Belmans et al., 1983).

$$ET_m = KcET_O \quad (1.5)$$

The values of K_{YJ} were proposed by Doorenbos and Kassam (1979) (Heeren et al., 2006) and Kc values were proposed by Allen et al. (1998). Domínguez et al. (2011) and Dominguez et al. (2008) vigorously used MOPECO model for the studies to find different suitable irrigation methods under water stress and saline conditions in arid and semi-arid environments.

2.5 Linear optimisation model for efficient use of irrigation water

The linear optimisation model for irrigation was developed by Difallah et al. (2017) to assess the effect of precipitation to determine the actual amount of irrigation water required to optimise the water use. The concept of model was outlined using the Knapsack Optimisation problem, which uses right amount of water and replaces the traditional strategies of irregular irrigation based on plant needs. To formulate this model, field experiments were conducted at the meteorological station of Ain Skouna, municipality of Zarma, situated 23 km east of the city Batna in Algeria. Winter wheat crop was selected for the experiments over a time period of November to June months of the year. Timing and amount of irrigation water were decided considering factors such as (1) prevailing climatic conditions of the region, (2) plant type and growth stage, (3) soil properties and (4) existing irrigation methods.

The model was designed considering three case scenarios which may exist in natural field conditions which include: (Case 1) When there is no rainfall and the soil moisture is insufficient to cover the lost water, the irrigation water provided will be equal to the difference between plant water requirement and soil water content. (Case 2) If the soil moisture is equivalent to the plant water needs either with or without rainfall, in this case, the irrigation water will take the value 0. (Case 3) If rain and soil moisture are not effective, and there is a lack of coverage of the plant water needs, the value of irrigation water will be equal to the difference needed to cover this lack.

Important stage in the formulation process of the linear optimisation model contains

(i) the identification of the decision variables,
(ii) the formulation of the objective function,
(iii) the identification and the formulation of the constraints.

A generalised layout of the model is presented in Fig. 1.1 below.

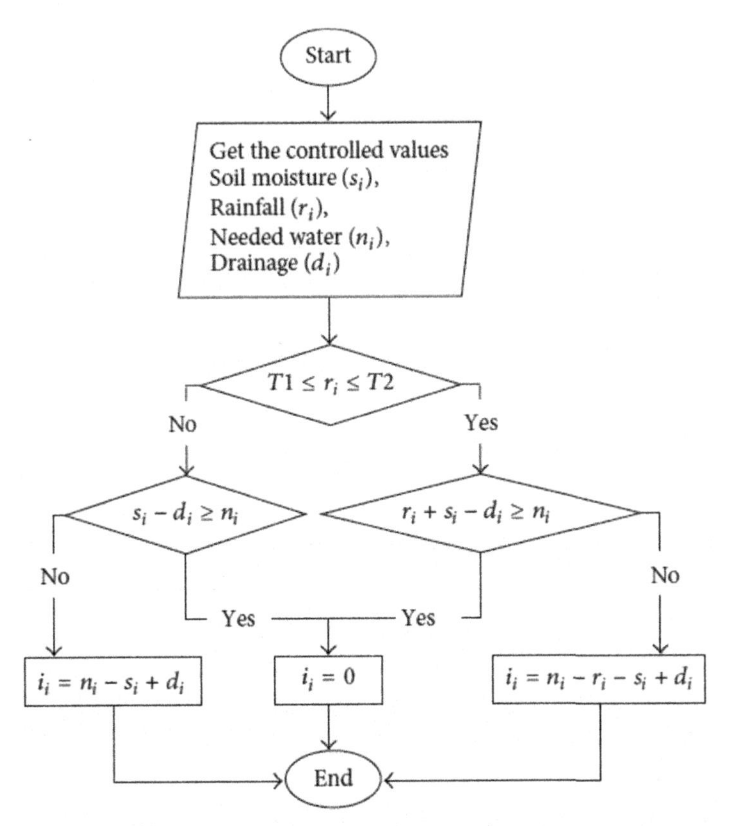

FIGURE 1.1 Flowchart of linear optimisation model of irrigation water management.

2.6 A network flow model for irrigation water management

The network flow approach is a very adaptive tool and widely used in many areas, including transportation, transshipment, communication networks, image segmentation, oil and natural gas production and distribution system (Evans, 2017; Magnanti and Orlin, 1993). This approach was used by Murthy and Murthy (2012) in developing a user-friendly network-based optimisation computer software as a decision support system for irrigation water management. They have used the primal network simple algorithm that is one of the popular and competent algorithms of network flow in their software.

There are two main components of this software. The first one is code for network solver utilised for solving general minimum cost network flow-related inadequacy. The second component is a code that acts as a software interface between the network solvers and the users in irrigation management (Murthy and Murthy, 2012). The latter one is user-friendly and formulates the problems by taking minimal inputs from the user; it then prepares the input file for the problem solver, again it follows the solver's solution and finally gives the results in user's specified format. Research is needed in improving this software by adding new modules for other applications as well. The formulation of this model is explained as below.

An irrigation system generally consists of a main reservoir, main channels, branch channels, distributary channels and several storage and/or regulatory points. These channels are used to carry water to the agricultural fields (AFs). An important feature of the irrigation system under consideration is that every AF is connected to one and only one channel in the network. Technically speaking, the irrigation network is a tree in Graph Theory terminology (Wilson and Watkins, 1990). Based on this nature of irrigation network, minimum cost flow model is used to optimise the water supply. The inputs for this model will be the total amount of water available at various storage points during the period and the total requirement of water of the AFs.

2.6.1 Minimum cost network flow problem

A directed network consists of a set of nodes N nd a set $A = \{(i_1, j_1),\ldots, (i_m, j_m)\}$ of m directed arcs. Each arc (i, j) is obtained by joining two nodes i and j from N. Here, i and j are called the tail and head nodes of the arc (i, j). Two subsets S and T of N are specified, S is called the set of source nodes and T is called the set of sink nodes. A certain commodity is to be supplied from nodes of S to nodes of T through the arcs of the network. Each node i in S is assigned an integer a_i which means that i can supply at most a_i units of the commodity. Similarly, each node j in T is assigned an integer d_j which means that j requires d_j units of the commodity. Since the network is a directed network, arc (j, i), if exists in A, is treated as different from (i, j). Three numbers, l_{ij} , k_{ij} and c_{ij} are associated with each arc (i, j) in A. Here l_{ij} and k_{ij} are called the lower and upper bounds of the arc (i, j) and the quantity of the commodity transferred through the arc, called the flow in the arc, should be within the limits l_{ij} and k_{ij}. Next, c_{ij} is the cost of flow per one unit of commodity (Busacker and Gowen, 1960). Given these inputs, the minimum cost network flow problem is to determine the flows in the network arcs so that the total cost of transportation is a minimum and that the requirements of T are met, availability constraints of S are not violated, and the flow restriction on the arcs are not violated. A mathematical statement of this problem can be stated as follows (Bertsekas and Tseng, 1988): Find the flows f_{ij}, $(i, j) \in A$ so as to minimise

$$\sum_{(i,j) \in A} c_{ij} \cdot f_{ij} \tag{1.6}$$

subject to $f(N,j) - f(j,N) = d_j$ for each $j \in T$.
$f(N, j) - f(j,N) \leq a_j$ for each $j \in S$,
$f(N, j) = f(j,N)$ for each $j \in N/(S \cup T)$,
$l_{ij} \leq f_{ij} \leq k_{ij}$ for each $(i, j) \in A$,
and $fij \geq 0$ for all $(i, j) \in A$.
where, $f(N, j) = \sum_{i \in N:(i,j) \in A} f_{ij}$ and $(i, N) = \sum_{j \in N:(i,j) \in A} f_{ij}$. The total availability of the material is equal to $\sum a_i$ and the total demand is equal to $\sum d_i$. If $\sum a_i = \sum d_i$, then we call the network problem as a balanced network problem.

2.7 Bayesian data-driven models for irrigation water management

The real-time management for today's irrigation systems involves the coordination of diversions and delivery of water to croplands. Most of the irrigation systems fail to address lags between when water is diverted and when it is delivered. For a model to be used as irrigation management tool three processes should be precise. These are (1) potential evapotranspiration forecasting, (2) hydraulic model error correction and (3) estimation of aggregate water demands (Mamitimin et al., 2015; Xue et al., 2016).

These tools are usually based on statistical machine learning or data-driven modelling and have a wide range of applications from engineering to irrigation management for better, improved and periodic information to water managers. The development of such models requires development based on Bayesian data-driven algorithm called the Relevance Vector Machine (RVM) and its extension, known as Multivariate Relevance Vector Machine (MVRVM). The use of these learning machines has the advantage of avoidance of model overfitting, high robustness in the presence of unknown data and accurate uncertainty estimation in the results (error bars) (Torres-Rua, 2011).

Tipping (2001) introduced the Relevance Vector Machine (RVM). The RVM concept is based on giving a training dataset of input-target vector pairs (x_n, t_n) n=1, where n is the number of observations that the model has to learn the dependency between input and output target with the purpose of making accurate predictions at t for previously unknown values of x.

$$t = \varnothing(x)w + \varepsilon \tag{1.7}$$

where w is a vector of weight parameters and $\varnothing$ (x) = [1, K(x,x_1),... K(x, x_n)] is a design matrix where K (x, x_n) is a fixed kernel function. The error ε is conventionally assumed to be zero-mean Gaussian with variance σ^2. A Gaussian likelihood distribution for the target vector can be written as:

$$p\left(t|w, \sigma^2\right) = (2\pi)^{-\frac{n}{2}}\, \sigma^{-N} \exp\left\{-\frac{t - y^2}{2\sigma}\right\} \tag{1.8}$$

Tipping (2001) proposed an additional prior term to the likelihood or error function to avoid the maximum likelihood estimation of w and σ^2 which suffers from severe overfitting from Eq. (3.22). This term is added by applying a Bayesian perspective, and thereby constraining the selection of parameters by defining an explicit zero-mean Gaussian prior probability distribution over them:

$$p(w|\alpha) = \left|(2\pi)^{-\frac{m}{2}} \prod_{m=1}^{m} \alpha_m^{\frac{1}{2}} \exp\left(-\frac{\alpha_m W_m^2}{2}\right)\right| \tag{1.9}$$

where m is the number of independent hyperparameters $\alpha = (\alpha, \ldots, \alpha_m)^T$. Each α is associated independently with every weight to moderate the strength of the term and to control the generalisation ability of the model (Khalil et al., 2006). Bayesian inference considers the posterior distribution of the model parameters, which is given by the combination of the likelihood and prior distributions:

$$p\left(w|t, a, \sigma^2\right) = \frac{p(t|w, \sigma^2)\, p(w|\alpha)}{p(t|\alpha, \sigma^2)} \tag{1.10}$$

This model has been widely used in several studies for effective irrigation planning at various sites by a number of users. The study of Torres-Rua (2011) performed across model attributes including input set, kernel scale parameter and model update scheme for checking whether the model provides superior prediction capability using RVM for simulating Richfield Canal flow and irrigation is prominent among other studies as it employed a data-driven approach for simulating canal flow using the relevance vector machine (RVM). The results obtained from (Torres-Rua, 2011) study indicate that the combination of the hydraulic simulation machine learning approach was capable of minimising the aggregate error adequately, capturing its behaviour pattern along the irrigation season. This provides means to reduce the aggregate error that ultimately improves the performance of the hydraulic simulation model.

2.8 CropSyst model

CropSyst (Cropping Systems Simulation Model) can be used for multiple crop and years with daily time step crop growth simulation. It has a user-friendly interface, a link to GIS software and weather generator (Stockle et al., 1996). It is a risk analysis model which serves as a link between agricultural production and economics. This model is an analytical tool for studying the effect of cropping systems on crop productivity and environment. CropSyst in case of soil can simulate the soil water budget, soil-plant nitrogen budget, decomposition and soil erosion by water. In case of crop it can simulate crop phenology, crop canopy, root growth, biomass production, crop yield, residue production and pesticide fate. These parameters are affected by various weather conditions, crop characteristics and soil characteristics. It is also dependent on user-dependent cropping system management options like irrigation, cultivar selection, crop rotation, nitrogen fertilisation, pesticide applications, tillage operations, soil and irrigation water salinity and residue management.

The model codes are written in Pascal and C++. It can run on DOS and Windows operating system. Its interface allows users to easily change and verify input parameters for checking range errors and cross-compatibility. It can execute single and batch run simulations. The output can be customised, as well as generated, in text or graphical format linked to spreadsheet programs.

The water and nitrogen budgets is parameterised by the user to produce a simulation of nitrogen transport within the soil. The same is true with chemical budgets whether dealing with pesticides (decay and absorption) or salinity. All balances are made sure at each time step and errors are reported in case of departures within the set threshold ranges (Pala et al., 1996). Integration of these time steps is performed using the Euler's method.

The model accounts for four factors of crop growth: water, nitrogen, light and temperature. The relationship between crop transpiration and biomass production is given by Tanner and Sinclair (1983) in the equation

$$B_{AT} = K_{BT}\, AT/VPD_m \tag{1.11}$$

where B_{AT} is the biomass production dependent on actual transpiration (kg/m^2 day^{-1}), T is actual transpiration (kg/m^2 day^{-1}) and VPD_m is the mean daily vapour pressure deficit of the air (kPa).

CropSyst presents three options to calculate grass reference ET dependent on decreasing order of required weather data input. These options are: The Penman-Monteith model, Priestley-Taylor model and a simpler implementation of the Priestley-Taylor model which only requires air temperature. Crop ET is determined from a crop coefficient at full canopy and ground coverage determined by canopy leaf area index. This relationship becomes unstable at low VPD; hence, it would predict infinite growth at near zero VPD.

To overcome this problem, a second estimate of biomass production is calculated following Monteith (1977):

$$B_L = e\, I_{PAR} \quad (1.12)$$

where B_L is the light-dependent biomass production (kg/m^2 day^{-1}), *e* is the light-use efficiency (kg/MJ) and I_{PAR} is the daily amount of crop-intercepted photosynthetically active radiation (MJ^{-1} m^{-2} day^{-1}). The simulation for each day, the minimum of B $_T$ and B_L is taken as the biomass production for that day.

The nitrogen effects on biomass production are accounted using the minimum B_T and B_L as base to determine the nitrogen-dependent biomass production (B_N):

$$B_N = \mathrm{Min}\{B_T,\ B_L\}\ [1 - (N_{pcrit} - N_p) / (N_{pcrit} - N_{pmin})] \quad (1.13)$$

where B_N is in kg/m^2 day^{-1}, N_p is plant nitrogen concentration (kg/kg), N_{pcrit} is the critical plant N concentration (kg/kg) below which growth is limited and N_{pmin} is the minimum plant nitrogen concentration (kg/kg) at which growth stops. The values of N_{pcrit} and N_{pmin} fluctuates as a function of accumulated biomass, following the concept of growth dilution.

A study checked the ability of the CropSyst model (Suite-4) to simulate actual evapotranspiration (ET_a), biomass and grain yield, and to identify major evaporation (E) losses in case of winter wheat (WW) and summer maize (SM) crop rotations. Field experiments were conducted at the Luancheng Agro-ecosystem station, NCP, in 2010–11 to 2012–13. The CropSyst model was calibrated for wheat/maize (weekly leaf area/biomass data available for 2012–13) and validated onto measured experimental station from 2010 to 11 to 2011–12. The results showed that CropSyst is a valid model that can be used with a reliable degree of accuracy for optimising WW and SM grain yield production and water requirement on the NCP (Umair et al., 2017).

2.9 Water balance method

The water balance approach is a dynamic process that requires new insight to improve water resource and policy-making understanding by using enhanced algorithms and techniques in hydrological science. The principle of water balance method is to maintain a balance between incoming (e.g., rainfall and irrigation) and outgoing (e.g., evapotranspiration) soil water for sufficient water availability to the plants. The principle involved in water balance is utilized in a number of the hydrological process more importantly in the irrigation process by accounting the addition and subtraction of the soil root zone water to know soil water deficit for plants. In irrigation scheduling, the water balance method is used for the estimation of the required amount and duration of irrigation for the crops (Harrison and Tyson, 2002).

The water balance method can be applied if initial soil water content in the root zone, crop evapotranspiration (ET_c), precipitation and irrigation amounts are known. Eq. (1.15) can be used to calculate water balance in the soil for irrigation scheduling (Andales et al., 2011; Broner, 1989);

$$D_c = D_p + ET_c - EP - Irr - U - SRO - DP \quad (1.14)$$

here D_c and D_p denote the soil water deficit (or irrigation demand) in the root zone in the current and previous day, respectively, ET_c is the crop evapotranspiration, EP is the effective precipitation, I_{rr} is the net irrigation amount, U is the upflux of shallow groundwater into root zone, SRO is surface runoff and DP is deep percolation.

Generally, the water balance equation is not much impacted by the last three variables (U, SRO and DP) in the equation. As the estimation of these variables is difficult in field conditions, we can eliminate it for irrigation management. Moreover, mostly the water table is significantly deeper than the root zone and, in that case, U will be zero. Also, in most cases, runoff and deep percolation are accounted to be zero or minimal if good irrigation scheduling is followed. Using these assumptions, we can simplify Eq. 1.14 as follows;

$$D_c = D_p + ET_c - EP - Irr \quad (1.15)$$

Here, it is noticeable that D_c is zero in case of its negative value. The reason behind this is the excessive amount of effective precipitation and irrigation water than crop evapotranspiration and previous day soil water deficit (i.e., $EP + Irr > D_p + ET_c$). It indicates water added to the root zone is exceeded to field capacity and assumed to be lost through the process of SRO and DP (Andales et al., 2011; Nyvall, 2004).

2.10 Irrigation practices

Nowadays the problem of water storage has become a major socioeconomic challenge for most of the developing countries. According to a study by United States (2012), around 70% of the world's annual fresh water is used by irrigation in the agricultural field. To reduce the excessive loss of irrigation water, the prime focus of irrigation water conservation is technological advancement such as altered tillage practices, microirrigation (drip and sprinkler), mulch, etc., in arid and semi-arid lands. These irrigation practices have become popular because they require low labour and energy; on the other hand, they offer weed suppression, increased yield, enhanced germination, rain shedding, minimising soil erosion, reducing crop cycle and others (Ingman et al., 2015). In the following segments of this subsection, we have discussed some of the most water-efficient irrigation techniques, including drip sprinkler and mulching techniques for sustainable water-use practices and their suitability for developing countries like India.

The drip or trickle irrigation is the most common form of microirrigation that reduces around 47%–62% water requirement of the crop compared to the flash and furrow irrigation (Berihun, 2011). In this system, water is allowed to drip directly to the roots of the plants through the small diameter pipes fitted with small outlets called drippers with a very slow rate (2–20 L/h). Drip system of irrigation provides a favourable high moisture condition to the soil for the plants to flourish. It is suitable for most of the soils and row crops like vegetables and soft fruits, trees and vine crops (http://agroman.in/drip-irrigation-cost-benefits-future-india/); whereas, in sprinkler irrigation method, water is supplied over the entire crop field by sprinklers which have nozzles that break water into small water droplets, similar to rainfall.

Though both drip and sprinkler irrigation techniques are categorised as microirrigation methods, they differ from each other in terms of flow rate, the pressure applied, mobility and wetted area. Sprinkler irrigation is not much popular like drip as it shows relatively low irrigation water-saving capacity (Kulkarni, 2011). The relatively very slow rate of drip irrigation regulates water stress in the root zone by supplying water continuously and in frequent quantities compared to the other pressurized irrigation system (Hargreaves and Merkley, 1998).

Mulching is another essential process of water conservation practice that involves covering of top soil by organic (hay and straw, seaweed, ground corn cobs, pine needles, etc.) and inorganic (plastic, shingle, slate, stone chippings, etc.) protecting materials. It reduces the water loss by preventing evaporation by bare soil, suppresses weed germination, improves soil condition, reduces soil erosion, maintains soil temperature by working as an insulator (Ramakrishna et al., 2006; Tiwari et al., 1998). Nowadays plastic mulch is extensively used in large-scale in different parts of the world ranging from fully mechanised installation to hand-tool installation. China accounts for around 40% of the world's plastic mulch.

According to Berihun (2011), Ingman et al. (2015) and Amare and Simane (2017), drip irrigation more significantly improves the water management practices when it is combined with plastic mulch method. The related research shows these combined practices are highly related to water management efficiency and increased yield production. Though we expected that both plastic mulch and drip irrigation would share an increasing trend in irrigation practices, more research is needed for cost-effective installation of this technology which is still hindered in developing countries. Another future research and innovation challenge is disposal alternative for such practices. In aggregate, there is need to makes these technologies economically and socially feasible and sustainable in developing and poor countries.

3. Conclusion

Water is the most important agricultural resource. Due to climate change, the scarcity of water is the global concern for agricultural industries. Therefore, management of irrigation water is the prime requirement for sustainable agricultural growth. The management of water supply from source to the end node of irrigation system needs to be planned accurately using technical applications to solve this problem. This scenario of the global rising issue needs to be solved through optimum management of hydrological irrigation management and adapting user-friendly, cost-efficient agricultural practices. Therefore, in this chapter we present a detailed report on some of these techniques, models and practices for efficient water resource management.

References

Allen, R.G., Pereira, L.S., Raes, D., Smith, M., 1998. Crop evapotranspiration. Guidelines for Computing Crop Water Requirements.

Amare, A., Simane, B., 2017. Determinants of smallholder farmers' decision to adopt adaptation options to climate change and variability in the Muger Sub basin of the Upper Blue Nile basin of Ethiopia. Agric. Food Secur. 6 (1), 64.

Andales, A.A., Chávez, J.L., Bauder, T.A., Broner, I., 2011. Irrigation Scheduling: the Water Balance Approach. Service in action; no. 4.707.

Belmans, C., Wesseling, J.G., Feddes, R.A., 1983. Simulation model of the water balance of a cropped soil: SWATRE. J. Hydrol. 63, 271–286.

Berihun, B.J., 2011. Effect of mulching and amount of water on the yield of tomato under drip irrigation. J. Hortic. For. 3 (7), 200–206.

Bertsekas, D.P., Tseng, P., 1988. The relax codes for linear minimum cost network flow problems. Ann. Oper. Res. 13 (1), 125–190.

Broner, I., 1989. Irrigation Scheduling: The Water-Balance Approach. Colorado State University Cooperative Extension.

Busacker, R.G., Gowen, P.J., 1960. A Procedure for Determining a Family of Minimum-Cost Network Flow Patterns. Research Analysis Corp Mclean Va. Operations Research Office, Bethesda, MD.

Difallah, W., Benahmed, K., Draoui, B., Bounaama, F., 2017. Linear optimization model for efficient use of irrigation water. Int. J. Agron. 2017.

Dominguez, A., Lopez-Mata, E., Juan, A.d., Artigao, A., Tarjuelo, J.M., 2008. Deficit irrigation under water stress and salinity conditions. The use of MOPECO model. In: Central Theme, Technology for All: Sharing the Knowledge for Development. Proceedings of the International Conference of Agricultural Engineering, XXXVII Brazilian Congress of Agricultural Engineering, International Livestock Environment Symposium-ILES VIII, Iguassu Falls City, Brazil, 31st August to 4th September, 2008.

Domínguez, A., Tarjuelo, J., De Juan, J., López-Mata, E., Breidy, J., Karam, F., 2011. Deficit irrigation under water stress and salinity conditions: the MOPECO-Salt Model. Agric. Water Manag. 98, 1451–1461.

Doorenbos, J., Pruitt, W.J.F., 1977. Crop water requirements. FAO Irrigation and drainage paper No. 24 34–37. Rome.

Doorenbos, J., Kassam, A.H., 1979. Yield response to water. Irrigation and drainage paper No. 33 257.

Dudley, N.J., Howell, D.T., Musgrave, W.F., 1971. Optimal intraseasonal irrigation water allocation. Water Resour. Res. 7 (4), 770–788.

Garg, N.K., Gupta, M., 2015. Assessment of improved soil hydraulic parameters for soil water content simulation and irrigation scheduling. Irrig. Sci. 33 (4), 247–264.

Gupta, M., Garg, N.K., Joshi, H., Sharma, M.P., 2012. Persistence and mobility of 2, 4-D in unsaturated soil zone under winter wheat crop in sub-tropical region of India. Agric. Ecosyst. Environ. 146 (1), 60–72.

Gupta, M., Garg, N.K., Joshi, H., Sharma, M.P., 2014. Assessing the impact of irrigation treatments on thiram residual trends: correspondence with numerical modelling and field-scale experiments. Environ. Monit. Assess. 186 (3), 1639–1654.

Evans, J., 2017. Optimization Algorithms for Networks and Graphs. Routledge.

Hargreaves, G.H., Merkley, G.P., 1998. Irrigation Fundamentals: An Applied Technology Text for Teaching Irrigation at the Intermediate Level. Water Resources Publication, Colorado, USA.

Harrison, K., Tyson, A., 2002. Irrigation Scheduling Methods.

Heeren, D.M., Werner, H.D., Trooien, T.P., 2006. Evaluation of irrigation strategies with the DSSAT cropping system model. In: ASABE/CSBE North Central Intersectional Meeting, p. 1.

Howell, T.A., Meron, M., 2007. Irrigation scheduling. In: Developments in Agricultural Engineering, Vol. 13. Elsevier, pp. 61–130.

Ingman, M., Santelmann, M.V., Tilt, B., 2015. Agricultural water conservation in China: plastic mulch and traditional irrigation. Ecosyst. Health Sust. 1 (4), 1–11.

Jacucci, G., Kabat, P., Verrier, P., Teixeira, J., Steduto, P., Bertanzon, G., et al., 1995. HYDRA: a decision support model for irrigation water management. In: Crop-water-simulation Models in Practice; Selected Papers of the 2nd Workshop on Crop-Water-Models Held at the Occasion of the 15th Congress of the International Commission on Irrigation and Drainage (ICID) at the Hague, The Netherlands in 1993. Wageningen, Wageningen Pers, 1995, pp. 315–332.

Jones, J.W., Hoogenboom, G., Porter, C.H., Boote, K.J., Batchelor, W.D., Hunt, L., et al., 2003. The DSSAT cropping system model. Eur. J. Agron. 18, 235–265.

Khalil, A.F., McKee, M., Kemblowski, M., Asefa, T., Bastidas, L., 2006. Multiobjective analysis of chaotic dynamic systems with sparse learning machines. Adv. Water Resour. 29, 72–88.

Koopal, L., Van Riemsdijk, W., De Wit, J., Benedetti, M., 1994. Analytical isotherm equations for multicomponent adsorption to heterogeneous surfaces. J. Colloid Interface Sci. 166, 51–60.

Kulkarni, S., 2011. Innovative technologies for water saving in irrigated agriculture. Int. J. Water Resour. Arid Environ. 1 (3), 226–231.

Lascano, R., 1991. Review of Models for Predicting Soil Water Balance, vol. 199. IAHS Publ, pp. 443–458.

Magnanti, T.L., Orlin, J.B., 1993. Network Flows. PHI, Englewood Cliffs NJ.

Majumdar, D.K., 2001. Irrigation Water Management: Principles and Practice. PHI Learning Pvt. Ltd, New Delhi, India.

Mamitimin, Y., Feike, T., Doluschitz, R., 2015. Bayesian network modeling to improve water pricing practices in northwest China. Water 7, 5617–5637.

Monteith, J.L., 1977. Climate and the efficiency of crop production in Britain. Philos. Trans. R. Soc. Lond. B Biol. Sci. 281, 277–294.

Mualem, Y., 1976. A new model for predicting the hydraulic conductivity of unsaturated porous media. Water Resour. Res. 12, 513–522.

Murthy, A., Murthy, G., 2012. A network flow model for irrigation water management. Algorithmic Oper. Res. 7.

Nelson, G.C., Rosegrant, M.W., Palazzo, A., Gray, I., Ingersoll, C., Robertson, R., Msangi, S., 2010. In: Food Security, Farming, and Climate Change to 2050: Scenarios, Results, Policy Options, Vol. 172. International Food Policy Research Institute, Washington, DC.

Nyvall, J., 2004. Sprinkler Irrigation Scheduling Using a Water Budget Method, Resource Management Branch, BC Ministry of Agriculture, Foods and Fisheries, Tech. Rep.

Pala, M., Stockle, C., Harris, H., 1996. Simulation of durum wheat (*Triticum turgidum* ssp. durum) growth under different water and nitrogen regimes in a Mediterranean environment using CropSyst. Agric. Syst. 51, 147–163.

Paul, S., Panda, S.N., Kumar, D.N., 2000. Optimal irrigation allocation: a multilevel approach. J. Irrig. Drain. Eng. 126 (3), 149–156.

Pandey, V., Srivastava, P.K., Petropoulos, G.P., 2020. The contribution of earth observation in disaster prediction, management, and mitigation: a holistic view. In: Techniques for Disaster Risk Management and Mitigation. John Wiley & Sons, Hoboken, NJ, pp. 47–62.

Qiao, S.Y., 2014. Modeling Water Flow and Phosphorus Fate and Transport in a Tile-Drained Clay Loam Soil Using HYDRUS (2D/3D). McGill University Libraries.

Ramakrishna, A., Tam, H.M., Wani, S.P., Long, T.D., 2006. Effect of mulch on soil temperature, moisture, weed infestation and yield of groundnut in northern Vietnam. Field Crops Res 95 (2-3), 115–125.

Ritchie, J., 1998. Soil water balance and plant water stress. In: Understanding Options for Agricultural Production. Springer, pp. 41–54.

Santos, M.A.D., de Camargo, M.B.P., 2006. Parametrização de modelo agrometeorológico de estimativa de produtividade do cafeeiro nas condições do Estado de São Paulo. Bragantia 65.

Simunek, J., Jacques, D., van Genuchten, M.T., Mallants, D., 2006. Multicomponent geochemical transport modeling using HYDRUS-1 D and HP 1. J. Am. Water Resour. Assoc. 42, 1537–1547.

Šimůnek, J., Hopmans, J.W., 2009. Modeling compensated root water and nutrient uptake. Ecol. Model. 220, 505–521.

Singh, U.K., Kumar, B., 2018. Climate change impacts on hydrology and water resources of Indian River Basins. Curr. World Environ. 13 (1), 32.

Snyder, K.A., Evers, L., Chambers, J.C., Dunham, J., Bradford, J.B., Loik, M.E., 2019. Effects of changing climate on the hydrological cycle in cold desert ecosystems of the Great Basin and Columbia Plateau. Rangeland Ecol. Manag. 72 (1), 1–12.

Soler, C.M.T., Suleiman, A., Anothai, J., Flitcroft, I., Hoogenboom, G., 2013. Scheduling irrigation with a dynamic crop growth model and determining the relation between simulated drought stress and yield for peanut. Irrigat. Sci. 31, 889–901.

Stewart, J.B., 1977. Evaporation from the wet canopy of a pine forest. Water Resour. Res. 13 (6), 915–921.

Stockle, C., Cabelguenne, M., Debaeke, P., 1996. Validation of CropSyst for water management at a site in southwestern France. In: Proc. 4th European Society of Agronomy Congress, Wageningen.

Tanner, C., Sinclair, T., 1983. Efficient Water Use in Crop Production: Research or Re-search? 1, Limitations to efficient water use in crop production, pp. 1–27.

Teixeira, J., Pereira, L., 1991. ISAREG, Programma Para Simular a Rega, Gufa Do Utilizador. ISAREG. A Program for Irrigation Simulation. User's Guide. Instituto Superior de Agronomfa, Lisbon, Portugal.

Thorp, K.R., DeJonge, K.C., Kaleita, A.L., Batchelor, W.D., Paz, J.O., 2008. Methodology for the use of DSSAT models for precision agriculture decision support. Comput. & Electron. Agric. 64, 276–285.

Tim, U.S., Mostaghimi, S., 1989. Modeling transport of a degradable chemical and its metabolites in the unsaturated zone. Gr. Water 27, 672–681.

Tipping, M.E., 2001. Sparse Bayesian learning and the relevance vector machine. J. Mach. Learn. Res. 1, 211–244.

Tiwari, K.N., Mal, P.K., Singh, R.M., Chattopadhyay, A., 1998. Response of okra (Abelmoschus esculentus (L.) Moench.) to drip irrigation under mulch and non-mulch conditions. Agric. Water Manag. 38 (2), 91–102.

Torres-Rua, A.F., 2011. Bayesian Data-Driven Models for Irrigation Water Management.

Trenberth, K.E., 1999. Conceptual framework for changes of extremes of the hydrological cycle with climate change. In: Weather and Climate Extremes. Springer, Dordrecht, pp. 327–339.

Umair, M., Shen, Y., Qi, Y., Zhang, Y., Ahmad, A., Pei, H., et al., 2017. Evaluation of the CropSyst model during wheat-maize rotations on the North China Plain for identifying soil evaporation losses. Front. Plant Sci. 8, 1667.

Van Diepen, C.v., Wolf, J., Van Keulen, H., Rappoldt, C., 1989. WOFOST: a simulation model of crop production. Soil Use & Manag. 5, 16–24.

Van Genuchten, M.T., 1980. A closed-form equation for predicting the hydraulic conductivity of unsaturated soils 1. Soil Sci. Soc. Am. J. 44, 892–898.

Wilson, R.J., Watkins, J.J., 1990. Graphs: An Introductory Approach: A First Course in Discrete Mathematics. Wiley, New York.

Wallace, J.S., 2000. Increasing agricultural water use efficiency to meet future food production. Agr. Ecosyst. Environ. 82 (1-3), 105–119.

World Health Organization, 2009. Summary and policy implications Vision 2030: the resilience of water supply and sanitation in the face of climate change.

Xue, J., Gui, D., Lei, J., Mao, D., 2016. Development of a participatory Bayesian network model for integrating ecosystem services into catchment-scale water resources management. Hydrol. Earth Syst. Sci. 2016, 1–37.

Zilberman, D., Chakravorty, U., Shah, F., 1997. Efficient management of water in agriculture. In: Decentralization and Coordination of Water Resource Management. Springer, New York City, US, pp. 221–246.

Chapter 2

Traditional water management in India

R.K. Singh[1], M. Mruthyunjaya[1], Prashant K. Srivastava[2,3]

[1]*Department of Agronomy, Institute of Agricultural Sciences, Banaras Hindu University, Varanasi, Uttar Pradesh, India;* [2]*Remote Sensing Laboratory, Institute of Environment and Sustainable Development, Banaras Hindu University, Varanasi, Uttar Pradesh, India;* [3]*DST-Mahamana Centre of Excellence in Climate Change Research, Institute of Environment and Sustainable Development, Banaras Hindu University, Varanasi, Uttar Pradesh, India*

1. Introduction

Water, an indispensable resource, forms the basis for life on earth. The ever-increasing human population, technological advances, and changing life patterns have summoned calamities in the form of declining water quantity (Kala and Kala, 2006). There is a problem of water scarcity occurring and it is paramount to learn what our ancestors and elders were doing in the past to mitigate water scarcity problems. Throughout India, several indigenous ways have been devised to conserve water. They are called as traditional water conservation systems. History tells us that both floods and droughts were regular occurrences in ancient India. The Indian system has witnessed many indigenous systems of water harvesting for surface water or ground water. Possibly this is why every region in the country has its own traditional water harvesting techniques that reflect the geographical distinctiveness and cultural uniqueness of the regions. The basic thought underlying all these techniques is that rain should be harvested whenever and wherever it falls. Evidences revealed our civilization had excellent systems of water harvesting and drainage. The settlement of Dholavira, laid out on a slope between two storm water channels, is a great example of water engineering. Chanakya's *Artha-shashtra* mentions irrigation using water harvesting systems. Sringaverapura, near Allahabad, had a sophisticated water harvesting system that used the natural slope of the land to store the flood waters of the river Ganga. Chola King Karikala built the Grand Anicut or Kallanai across the river Cauvery to divert water for irrigation (it is still functional), while King Bhoja of Bhopal built the largest artificial lake in India. In hilly and mountainous terrains, the communities have devised certain means of water tapping and transfer systems

Agricultural Water Management. https://doi.org/10.1016/B978-0-12-812362-1.00002-3

that have been used for generations to provide water for drinking and agriculture. In fact, even today most of these hilly tracts do not enjoy the benefits of modern irrigation systems and still depend on their age-old practices. People living in hilly areas still construct structures to catch, hold and store monsoon rainwater. Since ages, people across different regions of India, have experienced either excess or scarce water due to varied rainfall and land topography. Yet, they have managed to irrigate their agricultural fields using localized water harvesting methods. Their traditional ways, though less popular, are still in use and efficient. They are enriched with the knowledge to manage water in collective ways.

2. The role of traditional knowledge

Indigenous knowledge is regarded as the sum of experience and knowledge for the given ethnic group on specific aspects, which form the basis for decision making (Gupta et al., 2015). This indigenous knowledge is learnt from nature since time immemorial (Baul and McDonald, 2015). People make use of their own sagacity and accumulated knowledge from their forerunners (Sarkar et al., 2015). One of the salient features of indigenous knowledge is that it is highly adaptable (Negi, 2015). Sustainable natural resource management is driven by the beliefs and behaviour of the human communities and local cultures (Shah et al., 2015). People are the sole custodians of the valuable but disappearing traditional knowledge, which remained deeply rooted and very popular among their ancestors (Maiti et al., 2014). Local people hold good knowledge base regarding changing scenario and adopt suitable strategies to cope with it. Traditional knowledge is vital for the sustainability of natural resources including water, soil, forests etc. Only recently has there been a revived interest in evaluating traditional water management techniques as viable alternatives alongside new technologies and systems. The old technology is regaining importance. Different regions in India have developed state-of-the-art technologies that have employed various means for accessing underground water sources, which are the main sources of water in these regions.

3. Revisiting some traditional practices

3.1 Bamboo drip irrigation system

The spring water storage technique is widely used by the inhabitants of north-eastern states for the prudent use of water mostly for irrigation purposes during non-rainy days. It is mainly constructed in sloppy and undulating hilly terrain where construction of ground channel is not feasible. It involves a network of channels in which bamboo pipes of different diameters are used to control the flow by means of gravity to undertake downward flow. The main bamboo channel usually run hundreds of meters or sometimes a few kilometres to tap water from inaccessible uphill source and enables irrigation far away from the water sources. To regulate the irrigation in agricultural field, a complex network of secondary and tertiary bamboo channels is created to encompass the entire plantation area and for effective irrigation. The construction materials include bamboo and fibre which are available locally in adequate amount. The cost-effectiveness, low maintenance, and minimum manpower makes it more effective and useful technique in such sloppy terrain. To create a network for irrigating 1 ha of agricultural land requires one to two labourers for 15 days. The average life of this system is around 3 years after which rotten bamboo and wood decomposed and increases soil fertility. This unique technique is successfully used by Khasi and Jaintia tribes' of Meghalaya in irrigating agricultural lands of black pepper, betel, etc. (Anonymous, 2017). It is also popular in urban areas, where water stored on roof-top water tanks for irrigating kitchen gardens and local fields through bamboo channels. The advantages of this irrigation system include it doesn't need any fuel or power, made-up of biodegradable substances, economical, and easy to install. One can think through implementing it in regions where bamboo is accessible for free or at a low cost. As bamboo starts to rot in rain, so the entire network needs to be restored after 2–3 years. Overall, it is economical and sustainable irrigation technique which is applicable in agricultural farm as well as in urban permaculture garden (Singh and Gupta, 2002).

3.2 *Zabo* system

Another popular irrigation practice of north-eastern state is *zabo* (Nagamese word which means 'impounding run-off') which is commonly found in Nagaland. It consists of a protected forestland towards the top of the hill, near water-harvesting tanks in the middle, and besides cattle yard and paddy fields at the bottom. In the catchment, silt retention tank and water harvesting tank are created near the mid-hill region with the development of earthen embankments. Silt retention tanks are constructed at two or more points, where water is kept for two or three days before transferring to the main tank. The tanks are cleaned annually and the separated impurities are used as a compost in the terrace fields due to rich organic matter and nutrients. However during water-harvesting tank construction, the bottom surface is properly rammed and sidewalls are mortared with paddy husk to curtail the loss of water through seepage. The water from the pond is passed through the cattle yard before taking it to the paddy field for irrigation. The water brings with it the dung and urine of the animals to the fields via split bamboo conduits, which include good source of nutrition for the crops. Paddy fields, which are generally 0.2–0.8 ha in size, are located at the lower elevations. The fields are thoroughly rammed at the time of puddling through human treading, cattle in-group and wooden sticks to create a hard pan in order to avoid percolation of water. Using of paddy husk checks seepage losses from shoulder bunds. Two additional irrigations are given from the water harvesting tank. Majority of the farmers practice fish culture in the paddy field. They create a small pit in the middle of the rice field and keep fish fingerlings in the field. At the time of harvesting and intercultural operation, when the water is drained out from the paddy field, the fishes remain in the pit. The zabo farming is an effective and sustainable practice that contributes in well-being of water harvesting and recycling systems and thus in management of water resources by checking soil erosion, increasing soil fertility, and insuring water availability (Pal, 2016).

3.3 Katta

Katta is a transitory structure made by binding mud and loose stones available locally along small streams and rivers. It is cost-effective, simple, and viable method, commonly found in rural areas. This stone bund slowdowns the flow of water, and stores a large quantity (depending upon its height) during the dry months (Laxmi et al., 2018). The collected water gradually trickles into ground and increases the water levels of proximate wells and also abate the fresh water flow into the sea. It is constructed by double layer arrangement of stone bunds that makes it more effective structure than the concrete structures. Though, its construction requires several skilled labourers, the total cost is generally shared by all the villagers. With the increase in the numbers of personal borewells and handpumps, the subsurface water level experiencing significant reduction and limiting the water availability for the marginal villages. Thus, the spring water storage, Zabo and Katta can be alternative water harvesting structures for sustainable water management (Anonymous, 2017).

3.4 *Madakas/Johads/Pemghara*

These water soak pits called by several names, *Madakas* in Karnataka, *Pemghara* in Odisha and *Johads* in Rajasthan, are one of the age-old systems used to conserve and recharge groundwater. Constructed on an area with naturally high elevation on three sides, the soil is excavated to create a storage area and used to create a wall on the fourth side to hold water. *Johads* gather monsoon water, which slowly soaks into recharge groundwater and sustains soil moisture. Sometimes, several *Johads* are interconnected with a gulley or deep channels with a single outlet in a river or stream nearby to check structural mutilation. This cost-efficient and simple structure necessitates annual upkeep of desilting and cleaning the storage area of weed growth. Water from *Johads* is still been extensively used by farmers to irrigate fields in various parts of India. What needs to be done today is the revitalization of old *Johads*, many of which have fallen into disrepair due to the growth of weed plants and dumping of waste (Verma, 2017).

3.5 Eri

Eris are commonly found in the irrigated areas of Tamil Nadu. *Eris* has played several important roles in maintaining ecological harmony such as flood-control systems, preventing soil erosion, and wastage of runoff during periods of heavy rainfall, and recharging the groundwater in the surrounding areas. The presence of *eris* provided an appropriate microclimate for the local areas. Data indicates that in the 18th century about 4%–5% of the gross produce of each village was allocated to maintain *eris* and other irrigation structures. Assignments of revenue-free lands, called *manyams*, were made to support village functionaries who undertook to maintain and manage *eris*. These allocations ensured *eri* upkeep through regular desilting and maintenance of sluices, inlets, and irrigation channels (Gopalakrishna, 2014).

3.6 Kuhl

In Himachal Pradesh, a traditional irrigation system where surface channels diverting water from natural flowing streams were used (*khuds*). A typical public *kuhl* services 6–30 farmers, irrigating an area of about 20 ha. In this system a temporary headwall (usually with river boulders) was built across a *khud* (ravine) for storage and diversion of the flow through a canal to the fields. The water would flow from one field to the other and the surplus water, if any, would drain back to the khud. The kuhls that were found in the region were potentially built and preserved by the local village community. At the time when irrigation starts, the Kohli (traditional water masters of the region) would take care of the small constructions including the headwall, repairing of the Kuhl etc. (Pal, 2016). to make sure that the system is operational. Being responsible for the local engineering needs, any person who refuses to take part these activities without any valid explanations would be left without water for that season. Since individual rejection of water for a particular season was a sacred punishment, but it ensures the community participation and ground level solidarity of traditional water management of India. An individual was always free to participate in different tasks as per their skills and a fixed labour was also provided in return (Pal, 2016).

3.7 Surangam

Another region in Surangam Kasaragod district in Malabar, Kerala, where the people were not totally dependent on the direct use of surface water. In this region the altitudinal variation is such that the local river receives very high discharge in the monsoon season and low discharge in the dry seasons. As there is not a natural way to store the water above the surface, therefore, people mostly depend on the groundwater and in the irrigation season they follow the harvesting structure known as suangam. The Surangam word is originated from a Kannada which means tunnel and locally in Kasaragod it is also called thurangam, thorapu, mala, etc. It is basically a horizontal well like structure, due to the hard laterite rock formations within the ground the water is found little deeper than usual. The water goes down in the monsoon season through the seepage in hard rock formations and gets stored in the bottom of the tunnel. And the stored water through an open pit which is constructed outside the surangam. The geometry of the surangam is about 0.45–0.70 m wide, 1.8–2.0 m high and 3 to 300 m in length. Usually there are several subsidiary surangams inside the main surangam due to the presence of several outlet created by the local villagers. In case of a very long surangam, several vertical air shafts also provided to guarantee the required atmospheric pressure inside. The distance between consecutive air shafts vary from 50–60 m, whereas the dimension of the shafts is normally 2 x 2 m and the depth varies depending on the local subsurface geology (Padre, 2006).

3.8 Jackwells

The variation in the physiography, topography, rock types and rainfall meant that the tribes in the different islands followed different approaches for harvesting rain and groundwater. For illustration, the southern part of the Great Nicobar Island near Shastri Nagar has a comparatively rugged topography in contrast to the northern part of the islands. The Shompen tribals here made full use of the topography to reap water. In lower parts of the rolling terrain, bunds were made using logs of hard bullet wood, and water would gather in the pits so formed. They make wide use of split bamboos in their water harvesting systems. A full length of bamboo is cut longitudinally and retained along a moderate slope with the lower end leading into a shallow pit. These serve as ditches for rainwater, which is collected drop by drop in pits called jackwells. Repeatedly, these split bamboos are placed beneath trees to harvest the through falls (of rain) through the leaves. A series of progressively bigger jackwells are built, connected by split bamboos so that overflows from one lead to the other, finally leading to the principal jackwell, with an rough diameter of 6 m and depth of 7 m, so that overflows from one lead to the other (Jain, 2007).

3.9 Ramtek model

The Ramtek model has been named after the water harvesting structures in the town of Ramtek in Maharashtra. The elaborate network of groundwater and surface water bodies were often built and upheld by the malguzars (landowners) of the particular region. It comprised tanks associated with surface and underground canals forming a chain that ranged from the foothills to the plains. As soon as tanks located in the hills get filled, the water flows down as per elevation to fill succeeding reservoirs, commonly ending in a small waterhole. This system ensures proper distribution and conserves about 60%–70% of the total runoff of the area (Pal, 2016).

3.10 Pat system

The Pat system utilizes the uniqueness of the terrain to divert water from hills streams into irrigation networks, was developed in the Bhitada village in Jhabua district of Madhya Pradesh. Diversion embankments are built across a stream close to the village by piling up stones and then coating them with teak leaves and mud to make them leak-proof. The Pat channel channelizes the diverted water through deep ditches and stone aqueducts that are expertly cut into stone cliffs to create an irrigation system that for the villagers use (Pal, 2016).

3.11 Zings

Zings, found in Ladakh, are small tanks that collect the melting glacier water through a network of guiding channels. Trickle from the morning becomes melting waters of the glacier till it reaches zings from where it gets to turn into a flowing stream by the afternoon. On the next day, water is used in the fields, collected by the evening of the previous day. A water official called chirpun responsible for the reasonable distribution of water in this dry region that solely relies on melting glacial water to meet their farming needs (Pal, 2016).

3.12 Khadins

Khadins are native constructions first constructed by the Paliwal Brahmins of Jaisalmer in the 15th century which is very similar to the irrigation methods of the people of ancient Iraq. This structure harvested surface overflow water for agriculture. Its unique feature involves khadin, also known as dhora, which is an extensive earthen embankment that is constructed across the hill slopes of rough, gravely highlands. Sluices and spillways permit the surplus water to drain off, and the water-saturated lands are then employed for crop production (Pal, 2016).

3.13 Swing basket

Swing baskets are popular in India and Bangladesh. The device is constructed from economical raw materials like iron sheets, knitted bamboo strips or leather and four supports are attached to this. Containers filled with water from a source in double quantity facing one another are held by two people and are lifted to the course of water that pass to the fields by swinging. The effectiveness of this device is up to a depth of 0.75 m. The frequency of operation is about 15–20 swings/minute discharge may vary from 3500 to 5000 L/ha (Olley, 2008).

3.14 Moat (rope-and-bucket lift)

It is also known as *charsa* or *pur*. It is a manual irrigation method. The primary application of this device is in lifting the water from wells bordered in a line to a depth of 30. The design of this device consists of a bucket, basket, container or bag that is made up of either leather or GI sheet and attached to a pulley-based system through a rope. This rope is finally is attached to the yoke or harness that is put to the animals like bull or horse. These animals would lift the water by walking down on a ramp made of clay that is sloped at an angle of 5–10 degrees. Anonymous, (2004) says that up to 9000 Liters of water could be lifted utilizing two pairs of bulls with this device. The traditional pulley based or channel-based irrigation is a time consuming and inefficient irrigation system that works by pulling water from a water source for facilitating irrigation. On the other hand, it does not cost a lot of money to install a moat or pulley irrigation system, as it does not require vast technology or machinery invested in it.

3.15 Dhekli

It is also known as counterpoise lift or picottach. It is a system of drawing water from unlined wells, stream or ponds for irrigating small fields. Here it consists of a rope and bucket which is tied to a pole. At the other end, a heavy stick or any other object is tied as a counterbalance. And this pole is used to draw up water. About 2000 L of water can be lifted from depth of 2–3 m in 1 h (Anonymous, 2004).

3.16 Persian wheel

It is also known as *Rahat* (Urdu). It is believed that the technology originated in Egypt and as world shrunk through extensive trading, it spread to India and China. It is used to lift water from a depth up to 20 m. The efficiency of the device is considerably reduced after 7.5 m. The device consists of circular chain of buckets made of GI sheet having capacity of 8–15 L. The chain of bucket is mounted on a drum and is submerged in the water to sufficient depth. In the abovementioned device, the drum is coupled to a vertical wheel having tooth-like projections via a long shaft and is kept below the ground level. A horizontal beam is used to gear up the vertical wheel to a large horizontal wheel having tooth-like projections. This beam is then harnessed to a pair of ox or horse. A rotatory movement of animals rotates the containers to carry out the water via the gear system. The water is released when the bucket reaches the top. Average discharge of Persian wheel is about 10,000 L per hour from a depth of 9 m with one pair of bullocks (Anonymous, 2004).

3.17 Don

The principle of operation of don is similar to counterpoise lift. The don consists of trough made from wooden leg or iron sheet, closed at one end and open at the other. A manger is attached to a hinged pole that is balanced by a counterweight via a rope. This manger is then lowered through the application of pressure either by drawing the rope or by application of the pressure using feet of the worker till the end is merged in the water. The manger could be brought back to its original spot by releasing the applied pressure. Water could be lifted at the rate of 80–160 Liters per minute from a depth of 0.8–1.2 m (Anonymous, 2004).

3.18 Archimedean screw

An Archimedean Screw is a machine utilized in transferring water from a lowland source to irrigation trenches. This is also known as a screw pump and water is pumped by turning a bolt attached inside a pipe. It consists of a helical screw mounted on a spindle, which is rotated inside a wooden or metallic cylinder by windmill, manual labor, cattle, or by modern means, such as a motor. One end of the cylinder remains submerged in water and is placed in inclined position at an angle of 30 degrees. As the shaft turns, the bottom end scoops up a volume of water. This water is then pushed up the tube by the rotating helicoid until it pours out from the top of the tube. It is used for lifting of water from a depth of 0.6–1.2 m and may discharge 1600 L per hour (Rorres, 2000).

3.19 Paddle wheel

It is also known as *chakram* and is mostly used in coastal regions for irrigating paddy fields. It consists of small paddles mounted radially to a horizontal shaft, which moves in close fitting concave trough, thereby pushing water ahead of them. The number of blades depends on the size of wheel, which may be 8 for 1.2 m and up to 24 for 3–3.6 m diameters. The wheel having 12 blades may lift about 18,000 L per hour from a depth of 0.45–0.6 m (Anonymous, 2004).

4. Conclusion

It is being realized, albeit slowly, that the time-tested techniques, developed by ancient people, for sustainable management of limited available water resources, are more efficient and economical than the new systems that have been introduced in the 20th century. But unfortunately traditional knowledge is already becoming extinct. On the other hand, increased demand on water resources cannot be met by existing traditional systems alone. It can be concluded that indigenous practices are dynamic mechanisms of creativity and innovativeness, which support the lives of the communities making use of their own technological knowhow, do-how and inputs. Identification and integration of traditional science and knowledge with modern technology would be an effective and efficient way forward for sustainability, thereby ensuring the conservation of natural resources, especially soil and water. National and international research organizations need to allocate higher priority to the study and investigation of long-established traditional knowhow.

References

Anonymous, November 30, 2004. Directory of Agricultural Machinery and Manufacturers, pp. 316–325. https://farmer.gov.in/dacdivision/Machinery1/.

Anonymous, April 1, 2017. Traditional and innovative water conservation methods in India. Bhoomi Mag. http://bhoomimagazine.org/2017/04/01/traditional-and-innovative-water-conservation-methods-in-india/.

Baul, T.K., McDonald, M., 2015. Integration of Indigenous Knowledge in Addressing Climate Change.

Gopalakrishna, S., 2014. Ancient Engineering Marvels of Tamil Nadu. http://www.indiawaterportal.org/articles/ancient-engineering-marvels-tamil-nadu.

Gupta, L., Tiwari, G., Garg, R., 2015. Documentation of Ethnoveterinary Remedies of Camel Diseases in Rajasthan (India).

Jain, A.K., 2007. Water: A Manual for Engineers, Architects, Planners and Managers. Daya publishers.

Kala, R., Kala, C.P., 2006. Indigenous Water Conservation Technology of Sumari Village (Uttaranchal).

Laxmi, B., Bopanna, P.K., Bhavish, S.B., Bhat, A., 2018. Improvement of ground water by replacement of earthern Kattas by Bison panel sheet katta system. Int. J. Adv. Mech. Civil Eng. 5 (2), 26–28.

Maiti, S., Jha, S.K., Garai, S., Nag, A., Chakravarty, R., Kadian, K.S., Upadhayay, R.C., 2014. Adapting to Climate Change: Traditional Coping Mechanism Followed by the Brokpa Pastoral Nomads of Arunachal Pradesh (India).

Negi, C.S., 2015. Developing Sacred Forests into Biodiversity Heritage Sites-Experiences from the State of Uttarakhand. Central Himalaya, India.

Olley, J., November 2008. Human & Animal Powered Water-Lifting Devices for Irrigation. Practical Action.

Padre, S., March 2, 2006. Surangas, Manmade Caves to Tap Underground Water. India Together.

Pal, S., July 15, 2016. Modern India Can Learn a Lot from These 20 Traditional Water Conservation Systems. https://www.thebetterindia.com/61757/traditional-water-conservation-systems-india/.

Rorres, C., 2000. The turn of the screw: optimal design of an Archimedes screw. J. Hydraul. Eng. 126 (1), 72–80.

Sarkar, S., Padaria, R.N., Vijayragavan, K., Pathak, H., Kumar, P., Jha, G.K., 2015. Assessing the potential of Indigenous Technological Knowledge (ITK) for adaptation to climate change in the Himalayan and arid ecosystems. Indian J. Trad. Know. 14 (2), 251–257.

Shah, A., Ahmad, J., Sharma, M.P., 2015. Medicinal shrubs used by Gujjar-Bakerwal tribes against various non-communicable diseases in Rajouri district, (J&K), India. Indian J. Trad. Know. 14 (3), 466–473.

Singh, R.A., Gupta, R.C., 2002. Traditional Land and Water Management Systems of North-East Hill Region.

Verma, S., 2017. Traditional Water Conservation System of India that Are Still in Use and Worked Perfectly. Anonymous, March 22, 2018, from: https://www.awaaznation.com/social-issues/traditional-water-conservation-system-of-india/.

Chapter 3

Application of geospatial technology in agricultural water management

Ram Kumar Singh[1], Pavan Kumar[2], Semonti Mukherjee[3], Swati Suman[4], Varsha Pandey[4], Prashant K. Srivastava[4,5]

[1]*Department of Natural Resources, TERI School of Advanced Studies, New Delhi, Delhi, India;* [2]*College of Horticulture and Forestry, Rani Lakshmi Bai Central Agricultural University, Jhansi, Uttar Pradesh, India;* [3]*Haryana Space Application Centre, Hisar, Haryana, India;* [4]*Remote Sensing Laboratory, Institute of Environment and Sustainable Development, Banaras Hindu University, Varanasi, Uttar Pradesh, India;* [5]*DST-Mahamana Centre of Excellence in Climate Change Research, Institute of Environment and Sustainable Development, Banaras Hindu University, Varanasi, Uttar Pradesh, India*

1. Introduction

Developing countries are mostly dependent on agricultural production for their livelihood (Singh et al., 2020), so they are called as agrarian nations. Agriculture is the main source of food supply for about two-thirds of the population in the South Asian countries (Singh et al., 2020) (IMF, 2014; FAO, 2006). However, agriculture is also the biggest consumer of water for irrigation purposes. This demand is increasing day by day due to growing population whose food security needs to be guaranteed. The public and tenants know that water is life, but they do not pay attention to its management and proper use. About 97% of the water is in the form of salt water in the oceans, 2% is stored in the form of ice on the mountain ranges and the remaining 1% water is available for use with us. In developing countries, most of the water available is generally used in agriculture followed by industries. Because of extensive extraction of groundwater, today there is a need to do irrigation in the most efficient manner (Galiani et al., 2005; Molden, 2013).

The issues of agricultural water management can be addressed through basic and strategic research activities such as through rainwater harvesting, drainage, groundwater and waterlogged area management (Couceiro et al., 2007; Thornton and Herrero, 2010; Masese et al., 2012). There are several techniques in practice by the farming community for efficient water usage.

Agricultural Water Management. https://doi.org/10.1016/B978-0-12-812362-1.00003-5

Adequate quantity and quality of water are essential for high productivity. Since long, in India agricultural areas were irrigated by farmers using traditional methods. Due to this, irrigation efficiency was reduced considerably, and problems like water sedimentation and soil salinity occurred (Snyder, 1993; Ngoumou et al., 1994; Hill et al., 2000). Though there are initiatives that laid a network of canals in the many regions to provide irrigation facility to the farmers, due to nonavailability of proper amount of water, farmers have to depend on ground water for irrigation of their crops (Varela-Ortega et al., 1998; Rockström et al., 2003; Carr et al., 2011).

To safeguard the irrigation water management, geospatial technologies can play an important role towards sustainable ground water resources, mapping and applications at a larger scale. Following the introduction, this chapter provides several geospatial techniques and approaches towards irrigation water management broadly divided into two categories, i.e., conventional and advanced techniques.

1.1 Conventional geospatial techniques for water resource management

All the large area projects to maintain agriculture land irrigated in various South Asian countries and at other various regional levels are initiated based on various conventional techniques. Even recently new geospatial techniques were used for irrigation in developed and developing countries and are discussed below:

Geographic information systems (GIS) supported Water Use Master Plan is developed by Mahesh Neupane (2016) in Nepal, which is a planning tool for integrated water resources management and discusses an integrated approach for water mapping and master plan. The GIS tools used for water mapping analytically integrate to maintain ground water recharge, retention, and reuse. The integrated analysis assesses the current recent and future possible impacts of climate change on water catchment unit area. This helps to identify the critical watershed area and hardship zones and prioritize the works. The study is based on pilot implementation of GIS-based integrated works for Water Use Master Plan (WUMP) and sustainable development of village committee (Thakur et al., 2012).

Likewise, morphometric analysis is another promising technique for watershed management. It provides quantitative descriptions of river basin and is useful for understanding the behaviour of basin (Maurya et al., 2016). Morphometric parameters such as patterns, shape, and linearity, are relevant and useful to identify hydraulic characteristics of the drainage basin (Pradhan et al., 2019). On the other hand, GIS is useful in positioning the ideal site for water harvesting structures (Khan et al., 2001; Patel et al., 2012a; Makwana and Tiwari, 2016). In a study by Maurya et al. (2016), the AHP-MCE approach has been used for the watershed ranking and prioritization in the Pahuj River

Basin in India. They concluded that the integration of morphometric analysis, satellite precipitation and soil hydraulic parameters can provide useful information for soil and water conservation. The prioritization of watersheds will help reducing the runoff and provide guidance for reducing water logging. The study by Patel et al. (2012b) indicated that the prioritization of mini-watershed area is an important aspect for locating water harvesting structures, as well as for soil conservative practices. They provided a possible GIS-integrated approach for prioritization of sub-watersheds and deduced the location of check dams for water resource management that can be used for irrigation purpose as well. Similarly, in other work of Patel et al. (2015), prioritization of watershed has been carried out by considering various morphometric parameters and then on the basis of priority and weighted sum analysis, possible locations of check dams were proposed. In recent years, the mappings of soil erosion prone areas for conservation of soil and water management studies are in demand. The study carried out by Kumar Pradhan et al. (2020) in the Kosi River Basin provided the high soil erosion prone sites by integrating the erosion model, compound factor and field capacity through the AHP-MCE technique that can be used for soil and water conservation in the area. The combination of morphometric analysis and geovisualization was used for positioning water harvesting structure for checking the excessive water coming from the Varekhadi mini-watersheds. With the structures, water can be conserved and used for irrigation purposes (Patel et al., 2012a).

A range of ground-based instruments are also able to precisely predict soil moisture content at point locations; however these estimations fail to represent the spatial and temporal resolution of soil moisture at larger scale (Srivastava et al., 2013a). As such spatial interpolation and geographic information systems (GIS) play an important role in providing continuous data for environmental variables such as soil moisture, crucial for environmental modelling (Zhou et al., 2007; Srivastava, 2017; Srivastava et al., 2019; Suman et al., 2020). These methods act as powerful tools and solution to get continuous data at unsampled and ungauged locations using information from point measurements within same location (Burrough and McDonnell, 1998). Approaches such as multifarious forms of Kriging, inverse distance weighting (IDW), regularized spline with tension and thin plate spline have proved very useful in researches focusing on interpolation of soil moisture (Lu and Wong, 2008; Ding et al., 2011; Yao et al., 2013; Fang et al., 2016) and can help in estimation of soil water content.

1.2 Advanced techniques

The modelling techniques are used to model various processes of agriculture crops and for yield prediction. Three methods are generally used for the predictions of yield. First deals with methods of regression, correlation, basic estimation, trend line analysis based on historical data. This kind of analysis is very good for short- and long-term estimation and provides the results very fast, and its accuracy is also in-between moderate to good. The second method deals with machine learning techniques. The machine learning techniques are dependent on training the model with various supporting phenomena and providing prediction on given test datasets. The few machine learning methods are Random Forest (Halmy et al., 2015), Cart Tree (Shao and Lunetta, 2012), Support Vector Machine (Surangsrirat et al., 2010), Multinomial logistic regression (Koller, 2016; Koller et al., 2016), K-neighborhood and Artificial Neural Network (Lee and Tong, 2011)-based methods. The ANN-based Multi-Layer Perceptron Neural Network (MLPNN) methods are used to predict the future Agriculture, Forestry and Other Land use (AFOLU) Representative Concentration pathway (RCP) scenarios in year 2050 using various drivers including future RCP climate data (Singh et al., 2020). Third, simulation-based techniques are used for predicting the agriculture production. The methods APSIM, DSSAT, INFOCROP, ORYZA, and WTGROWS are used for yield estimation, growth model, climate variability in production, management and monitoring purposes. It uses climate data, soil parameters, crop cultivars for single agricultural crop, available irrigation information to predict agriculture crop information (Kalra and Kumar, 2018) and is useful for smaller areas. The Dynamic Growth Vegetation Model (DGVM) uses simulation modeling, which is useful to assess the large areas of agriculture and forest land cover for their productivity assessment using soil and climate data for current and long-term predictions, which are required for policy making. Limited work was carried in this field. The DGVM simulation models are Integrated Biospheres Simulator (IBIS) (Kucharik et al., 2000; Kumar et al., 2018), Joint UK Land Environment Simulator (JULES) (Slevin et al., 2017; Van den Hoof et al., 2011) and Lund-Postdam-Jena (LPJ) (Bonan 1996; Sitch et al., 2003). The DGVM simulation models provide Leaf Area Index (LAI), Net Primary Productivity (NPP), Soil Available Water (SAW), Potential Evapo-Transpiration (PET) for various plant functional types (PFT) to map larger areas. It processes very large areas with multilayer gridded data for hourly, daily and annual datasets.

Prem Chandra Pandey (2016) discussed the spatial integration of rice-based cropping systems for soil and water quality assessment using geospatial tools, and agriculture production is being identified as major source of Gross National Product. It provides livelihood to more than three-fourths of the population and associated economic activities. Agricultural production depends on good climate, soil type, management practices and other factors like fertilizers. The digital remote sensing data provides information about mapping and dynamic information about the cropping pattern (Singh et al., 2020). With this approach, the study focuses on Rice Equivalent Yield (REY) and Sodium Absorption Ratio (SAR) with presence of various fertilizers for mapping of rice-based cropping pattern in the districts of Bihar, India. The study illustrated to determine the REY with the availability of fertilizers contains nitrogen, phosphorous, potassium and soil micronutrients. The relational information pattern is determined for rice field, nitrate, and soil micronutrients between rice equivalent yield. Water resource was used for 'a GIS based assessment of Hydropower Potential within the Upper Indus Basin in Pakistan' region by Abdul Moiz (2016). The hydropower is source for sustainable, environment friendly and renewable energy source. This study assessed the hydropower potential for the Upper Indus Basin using Digital Elevation Model (DEM) data and spatial tools. The spatial tools and data identify the location for development of run-of-river hydropower plant. They identified locations and mapped for further decision making (Gergeľová et al., 2013) and could be used for irrigation management.

Vandana Tomar (2016) described about geospatial technology for water resource development for a watershed area, helpful to conserve and manage natural resources. Watershed is a natural unit to provide better management, development and execution of development plans. This study for Water Development Plan (WRDP) was prepared using remote sensing data and various land, topography and soil characteristics for creating decision rules for different conservation structures (Saptarshi and Raghavendra, 2009). The weighted-overlay and spatial query techniques were used for proposal of water harvesting structures in agriculture areas. Suggestions were also provided for urban rural settlement areas for collecting, recharging and storing rainwater. Sandeep Kumar Gautam (2016) explored the appraisal of surface and ground water for river basin using remote sensing irrigation indices and statistical techniques. The specific objective of the study was to find out the drinking and irrigational suitability of surface and ground water for a watershed of Jharkhand, India region. The various water hydro-quality parameters were observed; the result observed was that the ground water and surface water were suitable for irrigation and not for drinking purpose (Merchant, 1994; Maharana et al., 2015).

Panagiota Louka (2016) explored deterministic model to predict frost hazard in agricultural land utilizing remotely sensed imagery and GIS and have depicted frost risk mapping model for agricultural crops in Mediterranean environment by using remote sensing data of MODIS and ASTER. The main focus of his work is to provide time-efficient and cost-effective ways to enhance agricultural management and planning activities. Thus, Earth Observations (EO) and GIS have immensely helped in detecting frost events (Martsole et al., 1985; Gessler et al., 2000). Nearly, 30 years of datasets of 11 climatic stations within and nearby the study area have been taken as an input for frost risk. The result during validation is showing a very high correlation level between frost frequency map and ground observation of frost damages in crops.

Qiang Dai (2016) discussed in spatiotemporal uncertainty model radar rainfall and proposed fully functional and formulated uncertainty models, which statistically quantify the characteristics of the radar rainfall errors and their spatial and temporal structure. It pictured a radiant knowledge of input radar rainfall uncertainty and can support the uncertainty estimation in the hydrological model. Apart from rainfall, soil moisture is an important descriptor of soil water status (Engman and Chauhan, 1995). It not only regulates the evolution of climate and weather over continental regions but also controls the exchange of energy and water at the surface-atmosphere interface (Entekhabi et al., 2010). Global and frequent monitoring of soil moisture plays an important role in water resource management, irrigation scheduling, improved weather forecasts, natural hazards mitigation analysis, predictions of agricultural productivity, crop insurance, climate predictions, ecological health and services, improved trafficability, groundwater recharge, water quality and quantity, etc. (Srivastava et al., 2013b,c). Other than hydrology, precise estimation of soil moisture also plays a crucial role in earth system sciences, ecosystem, biochemistry and climate change studies. Remote sensing techniques, especially microwave approaches, have demonstrated immense potential in surface soil moisture estimation. Both microwave techniques, active (Radar/backscatter) and passive (radiometer/brightness temperature), offer unique ways for measuring soil moisture with each having distinct advantages and limitations over a range of topographic conditions and vegetation conditions up to an effective depth (Jackson and Schmugge, 1995). With the launch of the Tropical Rainfall Measurement Mission (TRMM), a joint space mission between the National Aeronautical and Space Administration (NASA) and the Japan Aerospace Exploration Agency (JAXA), designed to monitor and study tropical rainfall with the daily input of rainfall into the numerical model such as HYDRUS 1D, simulation of soil moisture at certain soil depths is now possible. A similar study was performed by Gupta et al. (2014) for soil moisture simulation in vertical strata, up to a soil depth 0–30 cm under a real-field condition in subtropical climate over a paddy field in Roorkee, Uttarakhand, India, and found satisfactory performance of the datasets when compared with the in situ soil moisture measurements. Land surface models are important elements in climate modelling studies. Surface models such as Noah Land Surface Model (LSM) and Noah Multi Physics in integration with mesoscale modelling systems such as Weather Research and Forecasting (WRF) model have been very useful in simulating the land surface conditions (Skamarock et al., 2001; Zaitchik et al., 2008). These models are often used for prediction of spatially distributed environmental parameters such as fluxes and soil moisture using global precipitation and temperature dataset at much finer spatial scales. WRF's are often simulated at various domains to generate soil moisture at different spatial scales to be compared with coarse resolution satellite operational products (Srivastava et al., 2013c, 2015a,b).

Earth observation system offers an effective platform for observing land surface changes at a global scale at a regular interval that has a great impact in mapping, motoring and future planning (Srivastava et al., 2014, 2016b). In this regard, the satellite remote sensing data–derived estimates are widely used to assess the present status and account irrigation requirement in forecasting crop grain yields (Ines et al., 2013; Singh et al., 2017; Petropoulos et al., 2018). Optical remote sensing typically acquires different responses from the top canopy that exhibits crop growth, health, water, and nutrient stress conditions from various rainfed and water irrigated lands, where the open-source satellite data such as Landsat, Sentinel, MODIS data are widely used to monitor irrigated areas at small to large spatial extent (Vörösmarty et al., 2000; Ambika et al., 2016). Thermal remote sensing with the land surface energy balance model is capable of estimating the ET-based irrigation information (Hain et al., 2015; Rosas et al., 2017). Irrigation water demand estimated by ET only represents the water consumption by plants during transpiration and evaporation. In contrast, the actual amount of irrigation water use can be potentially captured by microwave remote sensing by estimating soil moisture variation in the topsoil layers (Zhang et al., 2018). For irrigation water demand, Soil Moisture and Ocean Salinity (SMOS) data provides a flow of coarse resolution soil moisture data for hydrological applications. The other Soil Moisture Deficit (SMD) indicator represents soil water content changes and other soil-water related applications. The SMOS data is used for estimation of SMD. The statistical techniques, binomial with identity and poisson log functions, show good results for simulation of SMD from SMOS soil moisture (Srivastava et al., 2016a).

Recently, the state-of-the-art long-term record having ESA Climate Change Initiative (CCI) merges the different available microwave satellite soil moisture products into high resolution spatiotemporal consistent datasets, which has been recognized as a more suitable and effective way to verify the irrigation water use (Pandey et al., 2019; Zohaib and Choi, 2020). The ESA CCI soil moisture provides a multidecadal observational data record and is expected to generate more reliable products with NRT capabilities in the near future by incorporating advanced satellite soil moisture products (Dorigo et al., 2015; Enenkel et al., 2016). Qiu et al. (2016) highlighted the usefulness of ESA CCI soil moisture products to examine and manage the irrigated area in Eastern China by cooperative trend analysis of satellite-derived soil moisture and precipitation products.

In another work, the semi-distributed macroscale Variable Infiltration Capacity (VIC) hydrological model, which incorporates the water and surface energy balances within a grid cell was used and includes the subgrid variations in land surface and vegetation classes, topography, soil layers, moisture content, etc., in a heterogeneous landscape (Pandey and Srivastava, 2018). The VIC model uses the Penman-Monteith equation for evapotranspiration (ET) estimation by integrating climate, soil and vegetation parameters (Liang et al., 1994, 1996). The vegetation parameters encompass the leaf area index, albedo, stomatal conductance, architectural resistance, roughness length, the relative fraction of roots in each soil layer and displacement length, and accurately measure the ET. Assuming water availability is not a limiting factor, this model facilitates modelling irrigation water requirements in addition to the precipitation (indicated by the soil moisture content) for optimal plant growth. Various studies verified the suitability of VIC model for agriculture water management practices such as studies by Haddeland et al. (2006a, 2006b, 2007) highlighted the great applicability of VIC model integrating the simulated soil moisture deficit and irrigation scheduling to enhance the ET, which again subsequently reduces the surface temperature during the water scarcity periods. Employing the VIC model, Tatsumi and Yamashiki (2015) studied the effects of irrigation water withdrawals on ET, surface temperature and runoff. They observed significant increase in ET with the relative decrease in surface temperature and runoff compared to 'no irrigation' baseline during the dry season. Using the VIC model, Issac et al. (2017) simulated soil moisture deficit and estimated the field level irrigation water requirement during the wet and dry periods from soil water stress coefficient and corroborated the crop production data during the studied period; they suggested that it can help in generating various scenarios under the project climate conditions.

2. Way forward

Due to increasing industrialization, the requirement of water in newly established industrial units every day is increasing at a high speed. As a result, with increasing dependence on ground water, water is being drawn out in excessive quantity. Along with our tendency to be philanthropic, luxurious, and modernist, we are using more water than necessary due to insensitivity to water shortage. It is very important for us to save water today for tomorrow's golden future. Accumulation of water should be the top priority of the present time. It is also very important to renovate the traditional water harvesting structures. There is not much problem of availability of water in our country, but the proper management of water is a major problem here. We should manage water properly at our own level.

The geospatial technology is based on spatial data, which considers real-earth coordinate and is linked with information, attributes, geo-tag photos and survey devices. The geospatial data is created using recent various datasets and information. The prime challenges are to access the various recent data resources such as from satellite, aerial, Lidar, climate data and derived products, land use and land cover, analyzed reports, vegetation changes, productivity and soil data; this is followed by associated demarcation and updation of changes at different scales of data (Srivastava et al., 2016a). Generally the latest data are available with various global agencies, but its data interoperability, class definition, specialized formats, schema, uniform-scale, acquisition cost, end user agreements and different sources have different API (Application programming interface) as also identified as a challenge (Congalton et al., 2014; Singh et al., 2020). For higher scale data acquisition we may need to use unmanned aerial vehicles, Lidar aerial and terrestrial survey, ground-based measurement, runoff calculation, weather control station data, capturing of any phenomenon, ground survey and truth verification data, which take long duration to get captured and the cost of getting the data also is more. The time-series analysis needs to be captured for continuous and consistent time interval of data, so the consistent data should be acquired for analysis of the dynamic phenomenon. The agriculture land identification and its production mapping are dynamic phenomena; time-series analysis data is properly captured to study its growing stages and monitoring yield and availability of food cereals (Singh et al., 2020), as well as irrigation requirement.

In the changing world all the features in the ecosystem are changing day by day. So even the geospatial processes also need to be modified so that it can capture the changes in the system behaviour. The geospatial data programming at various platforms like Python, R, Java/Java Script, various script languages, Geospatial centralized database, Open Geospatial Consortium (OGC) compliances, Hexagon server platforms and other platforms reduces the processing time-data sharing analytics and decision-making related issues. The geospatial processes should be focus on to map natural features, agriculture production and climatic data changes. Air-water quality changes, environment and impact assessments, and can be used for application development in the field of agricultural water management , so that by knowing the crop type, hydro-meteorological conditions, soil type and texture, etc., proper agricultural water management can be planned.

References

Abdul Moiz, M.S.B., Ali, M., Shamim, M.A., Ali Naeem, U., 2016. A Geographic Information System (GIS) Based Assessment of Hydropower Potential within the Upper Indus Basin Pakistan. Geospatial Technology for Water Resource Applications. CRC Press, pp. 27−50.

Ambika, A.K., Wardlow, B., Mishra, V., 2016. Remotely sensed high resolution irrigated area mapping in India for 2000 to 2015. Scientific data 3, 1−14.

Bonan, G.B., 1996. Land Surface Model (LSM version 1.0) for Ecological, Hydrological, and Atmospheric Studies: Technical Description and Users Guide. Technical Note. National Center for Atmospheric Research, Boulder, CO (United States). Climate and Global Dynamics Div 1−144. https://doi.org/10.5065/D6DF6P5X.

Burrough, P.A., McDonnell, R.A., 1998. Creating Continuous Surfaces from Point Data. Principles of Geographic Information Systems. Oxford University Press, Oxford, UK.

Carr, G., Potter, R.B., Nortcliff, S., 2011. Water reuse for irrigation in Jordan: perceptions of water quality among farmers. Agric. Water Manag. 98, 847−854.

Congalton, R., Gu, J., Yadav, K., Thenkabail, P., Ozdogan, M., 2014. Global land cover mapping: a review and uncertainty analysis. Remote Sens 6 (12), 12070−12093. https://doi.org/10.3390/rs61212070.

Couceiro, S.R., Hamada, N., Luz, S.L., Forsberg, B.R., Pimentel, T.P., 2007. Deforestation and sewage effects on aquatic macroinvertebrates in urban streams in Manaus, Amazonas, Brazil. Hydrobiologia 575, 271−284.

Ding, Y., Wang, Y., Miao, Q., 2011. Research on the spatial interpolation methods of soil moisture based on GIS. In: International Conference on Information Science and Technology. IEEE, pp. 709–711.

Dorigo, W., Gruber, A., De Jeu, R., Wagner, W., Stacke, T., Loew, A., Albergel, C., Brocca, L., Chung, D., Parinussa, R., 2015. Evaluation of the ESA CCI soil moisture product using ground-based observations. Remote Sens. Environ. 162, 380–395.

Enenkel, M., Reimer, C., Dorigo, W., Wagner, W., Pfeil, I., Parinussa, R., De Jeu, R., 2016. Combining satellite observations to develop a global soil moisture product for near-real-time applications. Hydrol. Earth Syst. Sci. 20, 4191.

Engman, E.T., Chauhan, N., 1995. Status of microwave soil moisture measurements with remote sensing. Remote Sens. Environ. 51, 189–198.

Entekhabi, D., Njoku, E.G., O'Neill, P.E., Kellogg, K.H., Crow, W.T., Edelstein, W.N., Entin, J.K., Goodman, S.D., Jackson, T.J., Johnson, J., 2010. The soil moisture active passive (SMAP) mission. Proc. IEEE 98, 704–716.

Fang, K., Li, H., Wang, Z., Du, Y., Wang, J., 2016. Comparative analysis on spatial variability of soil moisture under different land use types in orchard. Sci. Hortic. 207, 65–72.

FAO, 2006. World Agriculture: Towards 2030/2050 - Interim Report:Global Perspective Unit: Food and Agriculture Organization of the United Nations. FAO, Rome, Italy. https://doi.org/10.1007/s10266-011-0028-z.

Galiani, S., Gertler, P., Schargrodsky, E., 2005. Water for life: the impact of the privatization of water services on child mortality. J. Polit. Econ. 113, 83–120.

Gergeľová, M., Kuzevičová, Ž., Kuzevič, Š., 2013. A GIS based assessment of hydropower potential in Hornád basin. Acta Montan. Slovaca 18.

Gessler, P., Chadwick, O., Chamran, F., Althouse, L., Holmes, K., 2000. Modeling soil–landscape and ecosystem properties using terrain attributes. Soil Sci. Soc. Am. J. 64, 2046–2056.

Gupta, M., Srivastava, P.K., Islam, T., Ishak, A.M.B., 2014. Evaluation of TRMM rainfall for soil moisture prediction in a subtropical climate. Environ. Earth Sci. 71, 4421–4431.

Haddeland, I., Lettenmaier, D.P., Skaugen, T., 2006a. Effects of irrigation on the water and energy balances of the Colorado and Mekong river basins. J. Hydrol. 324, 210–223.

Haddeland, I., Skaugen, T., Lettenmaier, D., 2007. Hydrologic Effects of Land and Water Management in North America and Asia: 1700? 1992.

Haddeland, I., Skaugen, T., Lettenmaier, D.P., 2006b. Anthropogenic impacts on continental surface water fluxes. Geophys. Res. Lett. 33.

Hain, C.R., Crow, W.T., Anderson, M.C., Yilmaz, M.T., 2015. Diagnosing neglected soil moisture source–sink processes via a thermal infrared–based two-source energy balance model. J. Hydrometeorol. 16, 1070–1086.

Halmy, M.W.A., Gessler, P.E., Hicke, J.A., Salem, B.B., 2015. Land use/land cover change detection and prediction in the north-western coastal desert of Egypt using Markov-CA. Appl. Geogr. 63, 101–112.

Hill, G., Mitkowski, N., Aldrich-Wolfe, L., Emele, L., Jurkonie, D., Ficke, A., Maldonado-Ramirez, S., Lynch, S., Nelson, E., 2000. Methods for assessing the composition and diversity of soil microbial communities. Appl. Soil Ecol. 15, 25–36.

Hoof, C. Van den., Hanert, E., Vidale, P.L., 2011. Simulating dynamic crop growth with an adapted land surface model - JULES-SUCROS: model development and validation. Agr. Forest Meteorol. 151 (2), 137–153. https://doi.org/10.1016/j.agrformet.2010.09.011.

IMF, 2014. International Monetary Fund. Retrieved December 1, 2014, from. http://www.imf.org/external/datamapper.

Ines, A.V., Das, N.N., Hansen, J.W., Njoku, E.G., 2013. Assimilation of remotely sensed soil moisture and vegetation with a crop simulation model for maize yield prediction. Remote Sens. Environ. 138, 149–164.

Issac, A.M., Raju, P., Joshi, S., Rao, V., 2017. Decadal trends in field level irrigation water requirement estimated by simulation of soil moisture deficit. Proc. Natl. Acad. Sci. India: Phys. Sci. 87, 901–910.

Jackson, T., Schmugge, T., 1995. Surface soil moisture measurement with microwave radiometry. Acta Astronaut. 35, 477–482.

Kalra, N., Kumar, M., 2018. Simulating the impact of climate change and its variability on agriculture. In: Sheraz Mahdi, S. (Ed.), Climate Change and Agriculture in India: Impact and Adaptation. Springer International Publishing, Cham, pp. 21–28. https://doi.org/10.1007/978-3-319-90086-5_3.

Khan, M., Gupta, V., Moharana, P., 2001. Watershed prioritization using remote sensing and geographical information system: a case study from Guhiya, India. J. Arid Environ. 49, 465–475.

Koller, M., 2016. Robustlmm: an R package for robust estimation of linear mixed-effects models. J. Stat. Softw. 75 (6), 1–48. https://doi.org/10.18637/jss.v075.i06.

Koller, I., Carstensen, C.H., 2016. Wiedermann W. and Eye A. von., Granger meets rasch: investigating granger causation with multidimensional longitudinal item response models. In: Wiedermann, W., Eye, A. von. (Eds.), Wiley Series in Probability and Statistics. https://doi.org/10.1002/9781118947074.ch10.

Kucharik, C.J., Foley, J.A., Delire, C., Fisher, V.A., Coe, M.T., Lenters, J.D., Ramankutty, N., 2000. Testing the performance of a dynamic global ecosystem model: water balance, carbon balance, and vegetation structure. Global Biogeochem. Cy. 14 (3), 795–825. https://doi.org/10.1029/1999GB001138.

Kumar, M., Rawat, S.P.S., Singh, H., Ravindranath, N.H., Kalra, N., 2018. Dynamic forest vegetation models for predicting impacts of climate change on forests: an Indian perspective. Indian J. For. 41 (1), 1–12.

Kumar Pradhan, R., Srivastava, P.K., Maurya, S., Kumar Singh, S., Patel, D.P., 2020. Integrated framework for soil and water conservation in Kosi River Basin. Geocarto Int. 35, 391–410.

Lee, Y.S., Tong, L.I., 2011. Forecasting time series using a methodology based on autoregressive integrated moving average and genetic programming. Knowl.-Based Syst. 24 (1), 66–72. https://doi.org/10.1016/j.knosys.2010.07.006.

Liang, X., Lettenmaier, D.P., Wood, E.F., Burges, S.J., 1994. A simple hydrologically based model of land surface water and energy fluxes for general circulation models. J. Geophys. Res.: Atmosphere 99, 14415–14428.

Liang, X., Wood, E.F., Lettenmaier, D.P., 1996. Surface soil moisture parameterization of the VIC-2L model: evaluation and modification. Global Planet. Change 13, 195–206.

Lu, G.Y., Wong, D.W., 2008. An adaptive inverse-distance weighting spatial interpolation technique. Comput. Geosci. 34, 1044–1055.

Maharana, C., Gautam, S.K., Singh, A.K., Tripathi, J.K., 2015. Major ion chemistry of the Son River, India: weathering processes, dissolved fluxes and water quality assessment. J. Earth Syst. Sci. 124, 1293–1309.

Mahesh Neupane, M.R.B., 2016. Rubika Shrestha, Jay Krishna Thakur, Ragindra man Rajbhandari and Bikram Rana Tharu. In: GIS Supported Water Use Master Plan: A Planning Tool for Integrated Water Resources Management in Nepal. Geospatial Technology for Water Resource Applications. CRC Press, pp. 27–50.

Makwana, J., Tiwari, M.K., 2016. Prioritization of agricultural sub-watersheds in semi arid middle region of Gujarat using remote sensing and GIS. Environ. Earth Sci. 75, 137.

Martsole, J., Gerber, J.F., Chen, E.Y., Jackson, J.L., Rose, A.J., 1985. What do satellite and other data suggest about past and future Florida freezes?. In: Proceedings of the Annual Meeting of the Florida State Horticulture Society (USA).

Masese, F.O., Raburu, P.O., Mwasi, B.N., Etiégni, L., 2012. Effects of deforestation on water resources: integrating science and community perspectives in the Sondu-Miriu River Basin, Kenya. New Ad. Contribut. Forest. Res. 268, 1–18. InTech, Rijeka.

Maurya, S., Srivastava, P.K., Gupta, M., Islam, T., Han, D., 2016. Integrating soil hydraulic parameter and microwave precipitation with morphometric analysis for watershed prioritization. Water Resour. Manag. 30, 5385–5405.

Merchant, J.W., 1994. GIS-based groundwater pollution hazard assessment: a critical review of the DRASTIC model. Photogramm. Eng. Rem. Sens. 60, 1117-1117.

Molden, D., 2013. Water for Food Water for Life: A Comprehensive Assessment of Water Management in Agriculture. Routledge.

Ngoumou, P., Walsh, J., Mace, J., 1994. A rapid mapping technique for the prevalence and distribution of onchocerciasis: a Cameroon case study. Ann. Trop. Med. Parasitol. 88, 463–474.

Panagiota Louka, I.P., Petropoulos, G.P., Nikolaos Stathopoulos, 2016. A deterministic model to predict frost hazard in agricultural land utilizing remotely sensed imagery and GIS. In: Geospatial Technology for Water Resource Applications. CRC Press, pp. 27–50.

Pandey, V., Srivastava, P., 2018. Integration of satellite, global reanalysis data and macroscale hydrological model for drought assessment in sub-tropical region of India. Int. Arch. Photogramm. Remote Sens. Spat. Inf. Sci. 42, 1347–1351.

Pandey, V., Maurya, S., Srivastava, P.K., 2019. Assessment of agricultural drought using a climate change initiative (CCI) soil moisture derived/soil moisture deficit: case study from Bundelkhand. In: Wastewater Reuse and Watershed Management: Engineering Implications for Agriculture, Industry, and the Environment, Vol. 63. Apple Academic Press, CRC press, New York, pp. 63–71.

Patel, D.P., Dholakia, M.B., Naresh, N., Srivastava, P.K., 2012a. Water harvesting structure positioning by using geo-visualization concept and prioritization of mini-watersheds through morphometric analysis in the Lower Tapi Basin. J. Indian Soci. Remote Sens. 40, 299–312.

Patel, D.P., Gajjar, C.A., Srivastava, P.K., 2012b. Prioritization of Malesari mini-watersheds through morphometric analysis: a remote sensing and GIS perspective. Environ. Earth Sci. 1–14.

Patel, D.P., Srivastava, P.K., Gupta, M., Nandhakumar, N., 2015. Decision Support System integrated with Geographic Information System to target restoration actions in watersheds of arid environment: a case study of Hathmati watershed, Sabarkantha district, Gujarat. J. Earth Syst. Sci. 124, 71–86.

Petropoulos, G.P., Srivastava, P.K., Piles, M., Pearson, S., 2018. Earth observation-based operational estimation of soil moisture and evapotranspiration for agricultural crops in support of sustainable water management. Sustainability 10, 181.

Pradhan, R.K., Maurya, S., Srivastava, P.K., 2019. Morphometric Analysis and Prioritization of Sub-watersheds in the Kosi River Basin for Soil and Water Conservation. Wastewater Reuse and Watershed Management: Engineering Implications for Agriculture, Industry, and the Environment, p. 353.

Prem Chandra Pandey, A.S.R., Mandal, V., Tomar, V., Katiyar, S., Ravishankar, N., Kumar, P., Nathawat, M.S., 2016. Spatial Integration of Rice-based Cropping Systems for Soil and Water

Quality Assessment Using Geospatial Tools and Techniques. Geospatial Technology for Water Resource Applications. CRC Press, pp. 27–50.

Qiang Dai, D.H., Rico-Ramirez, M.A., Srivastava, P.K., 2016. Spatio-temporal uncertainty model for radar rainfall. In: Geospatial Technology for Water Resource Applications. CRC Press, pp. 27–50.

Qiu, J., Gao, Q., Wang, S., Su, Z., 2016. Comparison of temporal trends from multiple soil moisture data sets and precipitation: the implication of irrigation on regional soil moisture trend. Int. J. Appl. Earth Obs. Geoinf. 48, 17–27.

Rockström, J., Barron, J., Fox, P., 2003. Water productivity in rain-fed agriculture: challenges and opportunities for smallholder farmers in drought-prone tropical agroecosystems. Water product. agric.: Limits opportun. improve. 85199, 8.

Rosas, J., Houborg, R., McCabe, M.F., 2017. Sensitivity of landsat 8 surface temperature estimates to atmospheric profile data: a study using modtran in dryland irrigated systems. Rem. Sens. 9, 988.

Sandeep Kumar Gautam, A.K.S., Tripathi, J.K., Singh, S.K., Srivastava, P.K., Narsimlu, B., Singh, P., 2016. Appraisal of surface and groundwater of the Subarnarekha river basin, Jharkhand, India: using remote sensing, irrigation indices and statistical technique. In: Geospatial Technology for Water Resource Applications. CRC Press, pp. 27–50.

Saptarshi, P.G., Raghavendra, R.K., 2009. GIS-based evaluation of micro-watersheds to ascertain site suitability for water conservation structures. J. Indian Soc. Remote Sens. 37, 693–704.

Shao, Y., Lunetta, R.S., 2012. Comparison of support vector machine, neural network, and CART algorithms for the land-cover classification using limited training data points. ISPRS J. Photogramm. Remote Sens. 70, 78–87. https://doi.org/10.1016/j.isprsjprs.2012.04.001.

Singh, D., Gupta, P., Pradhan, R., Dubey, A., Singh, R., 2017. Discerning shifting irrigation practices from passive microwave radiometry over Punjab and Haryana. J. Water Climate Change 8, 303–319.

Singh, R.K., Sinha, V.S.P., Joshi, P.K., Kumar, M., 2020. Modelling agriculture, Forestry and other land use (AFOLU) in response to climate change scenarios for the SAARC nations. Environ. Monit. Assess. 192, 1–18.

Sitch, S., Smith, B., Prentice, I.C., Arneth, A., Bondeau, A., Cramer, W., Sykes, M.T., 2003. Evaluation of ecosystem dynamics, plant geography and terrestrial carbon cycling in the LPJ dynamic global vegetation model. Glob. Change Biol. 9 (2), 161–185.

Slevin, D., Tett, S.F.B., Exbrayat, J.F., Bloom, A.A., Williams, M., 2017. Global evaluation of gross primary productivity in the JULES land surface model v3.4.1. Geosci. Model Dev. 10 (7), 2651–2670. https://doi.org/10.5194/gmd-10-2651-2017.

Skamarock, W.C., Klemp, J.B., Dudhia, J., 2001. Prototypes for the WRF (weather research and forecasting) model. In: Preprints, Ninth Conf. Mesoscale Processes, J11–J15. Amer. Meteorol. Soc., Fort Lauderdale, FL.

Snyder, F., 1993. The effectiveness of European Community law: institutions, processes, tools and techniques. Mod. Law Rev. 56, 19–54.

Srivastava, P.K., Pandey, P.c., Kumar, P., Raghubanshi, A.S., Han, D., 2016a. Monitoring soil moisture deficit using SMOS satellite soil moisture: correspondence through rainfall-runoff model. In: Geospatial Technology for Water Resource Applications. CRC Press, pp. 27–50.

Srivastava, P.K., Han, D., Ramirez, M.R., Islam, T., 2013a. Machine learning techniques for downscaling SMOS satellite soil moisture using MODIS land surface temperature for hydrological application. Water Resour. Manag. 27 (8), 3127–3144.

Srivastava, P.K., Han, D., Ramirez, M.A.R., Islam, T., 2013b. Appraisal of SMOS soil moisture at a catchment scale in a temperate maritime climate. J. Hydrol. 498, 292–304.

Srivastava, P.K., Han, D., Rico-Ramirez, M.A., Al-Shrafany, D., Islam, T., 2013c. Data fusion techniques for improving soil moisture deficit using SMOS satellite and WRF-NOAH land surface model. Water Resour. Manag. 27 (15), 5069–5087.

Srivastava, P.K., Mukherjee, S., Gupta, M., Islam, T. (Eds.), 2014. Remote Sensing Applications in Environmental Research, Vol. 211. Springer, Basel.

Srivastava, P.K., Islam, T., Gupta, M., Petropoulos, G., Dai, Q., 2015a. WRF dynamical downscaling and bias correction schemes for NCEP estimated hydro-meteorological variables. Water Resour. Manag. 29 (7), 2267–2284.

Srivastava, P.K., Han, D., Rico-Ramirez, M.A., O'Neill, P., Islam, T., Gupta, M., Dai, Q., 2015b. Performance evaluation of WRF-Noah Land surface model estimated soil moisture for hydrological application: synergistic evaluation using SMOS retrieved soil moisture. J. Hydrol. 529, 200–212.

Srivastava, P.K., Petropoulos, G., Kerr, Y.H., 2016b. Satellite Soil Moisture Retrieval: Techniques and Applications. Elsevier.

Srivastava, P.K., 2017. Satellite soil moisture: review of theory and applications in water resources. Water Resour. Manag. 31 (10), 3161–3176.

Srivastava, P.K., Pandey, P.C., Petropoulos, G.P., Kourgialas, N.N., Pandey, V., Singh, U., 2019. GIS and remote sensing aided information for soil moisture estimation: a comparative study of interpolation techniques. Resources 8 (2), 70.

Suman, S., Srivastava, P.K., Petropoulos, G.P., Pandey, D.K., O'Neill, P.E., 2020. Appraisal of SMAP operational soil moisture product from a global perspective. Remote Sens 12 (12), 1977.

Surangsrirat, D., Tapia, M.A., Zhao, W., 2010. Classification of endoscopie images using support vector machines. In: Proceedings of the IEEE SoutheastCon 2010. SoutheastCon), Concord, NC, pp. 436–439. https://doi.org/10.1109/SECON.2010.5453834.

Tatsumi, K., Yamashiki, Y., 2015. Effect of irrigation water withdrawals on water and energy balance in the Mekong River Basin using an improved VIC land surface model with fewer calibration parameters. Agric. Water Manag. 159, 92–106.

Thakur, J.K., Srivastava, P., Singh, S., Vekerdy, Z., 2012. Ecological monitoring of wetlands in semi-arid region of Konya closed Basin, Turkey. Reg. Environ. Change 12, 133–144.

Thornton, P.K., Herrero, M., 2010. The Inter-linkages between Rapid Growth in Livestock Production, Climate Change, and the Impacts on Water Resources, Land Use, and Deforestation. The World Bank.

Vandana Tomar, A.N.K., Indal, K., Ramteke, P.C.P., Kumar, P., 2016. Geospatial technology for water resource development in WGKKC2 watershed. In: Geospatial Technology for Water Resource Applications. CRC Press, pp. 27–50.

Varela-Ortega, C., Sumpsi, J.M., Garrido, A., Blanco, M.a., Iglesias, E., 1998. Water pricing policies, public decision making and farmers' response: implications for water policy. Agric. Econ. 19, 193–202.

Vörösmarty, C.J., Green, P., Salisbury, J., Lammers, R.B., 2000. Global water resources: vulnerability from climate change and population growth. Science 289, 284–288.

Yao, X., Fu, B., Lü, Y., Sun, F., Wang, S., Liu, M., 2013. Comparison of four spatial interpolation methods for estimating soil moisture in a complex terrain catchment. PloS One 8.

Zaitchik, B.F., Rodell, M., Reichle, R.H., 2008. Assimilation of GRACE terrestrial water storage data into a land surface model: results for the Mississippi River basin. J. Hydrometeorol. 9, 535–548.

Zhang, X., Qiu, J., Leng, G., Yang, Y., Gao, Q., Fan, Y., Luo, J., 2018. The potential utility of satellite soil moisture retrievals for detecting irrigation patterns in China. Water 10, 1505.

Zhou, F., Guo, H.-C., Ho, Y.-S., Wu, C.-Z., 2007. Scientometric analysis of geostatistics using multivariate methods. Scientometrics 73, 265–279.

Zohaib, M., Choi, M., 2020. Satellite-based global-scale irrigation water use and its contemporary trends. Sci. Total Environ. 136719.

Part II

Convetional technqiues

Chapter 4

Treated waste water as an alternative to fresh water irrigation with improved crop production

Ayushi Gupta[1], Nitin Kumar Sharma[2], Shaily Bhardwaj[2], Prashant K. Srivastava[1,3]

[1]*Remote Sensing Laboratory, Institute of Environment and Sustainable Development, Banaras Hindu University, Varanasi, Uttar Pradesh, India;* [2]*St. Xavier's, Balrampur, Uttar Pradesh, India;* [3]*DST-Mahamana Centre of Excellence in Climate Change Research, Institute of Environment and Sustainable Development, Banaras Hindu University, Varanasi, Uttar Pradesh, India*

1. Introduction

Industrialisation and urbanisation are the need of the hour for uplifting a country's economy. However, in the era of rapid industrialisation, huge amount of wastes is generated on a daily basis (Phantumvanit and Panayotou, 1990). Wastes are introduced into the environment in the form of sewage, vegetable wastes, crop residues, city garbage and industrial effluents. India generates 62 million tonnes of waste every year out of which less than 60% is collected and around 15% processed. With landfills ranking third in terms of greenhouse gases emissions in India, and increasing pressure from the public, the Government of India is now bringing change in waste perspective by treating it as a resource. There is a constant increase in establishment of waste-to-energy plants for solid waste management. Furthermore, industrial wastes also add to this urban waste in solid, liquid and gaseous forms from various industrial sectors like food processing, oil and gas industry, metal working industry, coal industry, sugar mills, etc. Out of the total effluents generated 60% of industrial waste water, mostly from large-scale industries, is treated. Other effluents are just dumped in to nearby water bodies (Kaur et al., 2012) All kinds of wastes have detrimental effects on the environment, as they cause pollution as well as direct exposure leads to serious health problems. Therefore, proper management and recycling of wastes are required in order to minimise their ill-effects.

Agricultural Water Management. https://doi.org/10.1016/B978-0-12-812362-1.00004-7

The two main sources of water contamination are sewage and industrial waste. With both the population of India and its industrial landscape increasing at a phenomenal speed, the amount of waste water generated is also at an all-time alarming high. One of the productive uses of waste water could be application of industrial effluent in irrigation, as they contain many useful nutrients in appreciable amount; hence, it can be used wisely as substitute of fresh water which is limited in nature. Productive use of waste water will contribute to achieve 'every drop count' mission as we are facing severe water scarcity (Jain, 2011). Also, effluents are considered as liquid organic material with fertiliser value. Efforts were made to know the utility of such effluents in improving soil productivity (Girisha and Raju, 2008). Many workers have tried to assess the profit in agriculture in terms of quantity and quality of effluent used, soil type, crop type, etc. (Ajmal and Khan, 1983; Reyer et al., 1990; Yaduvanshi and Yadav, 1996). To feed the continuously growing population with limited resources is the biggest challenge. Rapid industrialisation and increasing demand of food and fibre are causing exploitation of water resources (Najafi et al., 2002). Hence, many researchers have shown that the best utility of waste water after treatment is in agriculture (Pescod, 1992). In a study the use of treated municipal waste water for irrigation has caused increase in yield as compared to irrigation with well water (Erfani and Haghnia, 2001). Industrial organic wastes can be used effectively to increase soil fertility (Schuler and Larson, 1975). But waste water should be applied only after assuring that it will not cause any ill-effects to soil, crops, as well as on surface and subsurface water. It has been found that drip irrigation is the only method of application which caused least problems in using waste water in irrigation (Pescod, 1992). The response of plants to the application of polluted effluent depends upon several factors like climate, soil type, and nature of pollutant present. Hence, numerous studies have been conducted so far to know the effect of effluents produced by different kinds of industries on the growth phases of plants. Use of dairy effluent in irrigation has given several beneficial results. In sandy soils, NPK value was increased as well as in fodder (Zabek, 1976). It was also found that higher dose (100% waste water) caused reduction in plant height whereas use of 25%−75% effluent waste water increased the plant height in kidney beans and pearl millet (Ajmal and Khan, 1984). Germination of seed was found to be more with diluted dairy effluent in comparison to undiluted one (Gautam and Bishnoi, 1990).

Diluted paper mill effluents were also found to be beneficial. In case of coconut plants, an increase in the concentration of copper, lead, zinc, etc., was observed in roots, leaves and coconut water (Fazeli et al., 1991). However, a reduction in germination and growth was reported in case of rice, black gram and tomato (Oblisami and Rajannan, 1978). In case of soil, nutrient status was enhanced. Chemical factory and tannery effluents were also found useful.

Use of distillery effluents like spent wash resulted in increase in yield in case of sugarcane at lower dose because of its manurial values (Bajpai and Dua, 1972). Several research studies have shown that sewage water is best as a nutrient source in irrigation (El-Bassan et al., 1977; Sommers, 1977). It was also reported that municipal waste water contained more minerals than tube well water, hence found suitable for irrigation (Singh and Kansal, 1985). Likewise, all other effluents were found effective in various aspects if used instead of fresh water. Thus, the main objective of this research is a comparative study of the growth of agricultural plants by using treated effluents and normal tap water. Various aspects of plant growth were analysed to have in-depth understanding of the positive aspects of using waste water in agriculture. Physicochemical properties of soil and water were investigated. Its impact on parts of seeds were also analysed by observing germination and morphology. Since human body derives energy from the nutrients, therefore, nutritional studies were also taken under consideration to account for its positive effect on humans. Growth of roots, epicotyl and hypocotyl were observed through anatomical study. Hence the study highlights the overall perspective of the growth to mark visible difference between the final products produced.

2. Materials and methods

2.1 Experimental sites

The study site is located in Surat, Gujarat, which lies near the mouth of Tapti River. It is situated at 21.1702°N, 72.8311°E coordinates having population of approximately 4.4 millions. This city is famous for textile and diamond industry. Manufacturing units of several recognised sectors are present in Surat like Reliance, Essar Steel, Larsen and Toubro, Torrent Power and ONGC.

Oil and Natural Gas Corporation (ONGC) is one of the most renowned companies, situated at Hazira near Surat. It contributes about 77% of India's gas production. It has various units like LPG, recovery, gas sweetening, gas dehydration, condensation, fractionation and dew point depression and boiler unit. Waste water treatment plant of ONGC is designed to effectively treat waste water to meet the water-quality standards. This plant releases 1500 cubic feet per second of water per day.

Two sites were chosen for the cultivation of plants for comparative study:

1. Veer Narmada South Gujarat University Site (VNSGU).
2. Oil and Natural Gas Corporation Site (ONGC).

2.2 Materials

Well-nourished seeds of eight plants were obtained from Gandhi Agro Pvt. Ltd., Surat. These seeds were sown at different plots at both sites for comparative study. Plant to plant distance was kept 20–25 cm for all plots. Seedlings were collected at different developmental stages for the study. The study was conducted on eight agricultural plants namely, bottle gourd (*Lagenaria scieraria* (Mol.) Standl.), brinjal (*Solanum melongena* L.), cucumber (*Cucumis sativa* L.), guar (*Cyamopsis tetragonaloba* L. Taub.), lady's finger (*Abelmoschus esculentus* L. Moench.), sunflower (*Helianthus annuus* L.), tomato (*Lycopersicon esculentum* Mill.) and marigold (*Tagetes erecta* L.). These selected plants were cultivated using treated effluent at ONGC site and tap water at VNSGU site.

Several components such as physicochemical properties of soil, effluent, germination and seedlings morphology, vegetative growth and nutritional studies of seedlings are provided in this chapter.

2.3 Experimental design

Selected plants were cultivated for 2 years starting from March 2005 to February 2007. Plants were divided into two groups for study and were sown twice in a year. Results were calculated by taking average of two trials. These are listed in Table 4.1 with the trial durations.

The soil and water analysis for the whole study period was split into six groups; each group was of 4 months duration. Samples of water and effluents were analysed at 15 days intervals for each group. Total eight readings were taken and results were calculated as the average of these readings. Soil

TABLE 4.1 Experimental period and crops planted according to groups.

Groups	Sowing time	Crops planted
Group 1 plants	First sowing—March 2005 to August 2005	Brinjal
		Guar
		Sunflower
		Tomato
	Second sowing—September 2005 to February 2006	Brinjal
		Guar
		Sunflower
		Tomato
Group 2 plants	First sowing—March 2006 to August 2006	Lady's finger
		Cucumber
		Bottle gourd
		Marigold
	Second sowing—September 2006 to February 2007	Lady's finger
		Cucumber
		Bottle gourd
		Marigold

samples were analysed at interval of 1 month for each study group and average of four readings was obtained. The water-treated and untreated effluents were analysed twice a month to assure more accuracy.

2.4 Methods

2.4.1 Physicochemical analysis of soil and effluent

Soil samples were collected from both the sites and analysed with the help of referenced methods (Table 4.2).

From VNSGU site normal tap water was collected and from ONGC site both treated and untreated effluents were collected. Treated effluents were collected from Final Effluent Disposal System (FEDS). Certain listed parameters in Table 4.3 were evaluated.

2.4.2 Heavy metal and biochemical analysis

Study was carried out at Pollucon lab Pvt. Ltd., Surat. Heavy metals like cadmium, copper, iron, lead, nickel, zinc were analysed using various specialised kits. Fruits and seeds of the plants grown at both sites were analysed biochemically to know the nutritional content (Table 4.4).

TABLE 4.2 List of parameters analyzed in soil.

Sl No.	Parameters	Method	References
1	pH	Electrometric method by using pH meter	Apha (1998)
2	Electrical conductivity	Electrometric method by using conductivity meter	Apha (1998)
3	Available phosphorus	Stannous chloride method	Apha (1998)
4	Total alkalinity	Titrimetric method	Apha (1998)
5	Organic matter	Walkley and Black method	Trivedy et al. (1998)
6	Calcium	Titrimetric method	Trivedy and Goel (1984)
7	Magnesium	Titrimetric method	Trivedy and Goel (1984)
8	Sodium	Flame photometric method	Trivedy and Goel (1984)
9	Potassium	Flame photometric method	Trivedy and Goel (1984)
10	Sulphate	Turbid metric method	Apha (1998)
11	Nitrate	Cadmium reduction method	Apha (1998)

TABLE 4.3 List of parameters analyzed in effluent.

S. No	Parameters	Method	References
1.	Colours	Visual comparison method	Apha (1998)
2.	Temperature	By thermometer	Trivedy and Goel (1984)
3.	Total solids	Dried at (103–105)°C	Apha (1998)
4.	Total dissolved solids	Dried at 180°C	Apha (1998)
5.	Total suspended solids	Dried at (103–105)°C	Apha (1998)
6.	pH	By electrometric method using pH meter	Apha (1998)
7.	COD	Dichromate oxidation open reflux method followed by titration	Apha (1998)
8.	Total hardness	EDTA titrimetric method	Apha (1998)
9.	Total alkalinity	Titrimetric method	Apha (1998)
10.	Chloride	Titrimetric method	Trivedy and Goel (1984)
11.	Oil and grease	Soxhlet extraction method	Apha (1998)
12.	Phosphorus (P)	Stannous chloride method	Apha (1998)
13.	Silicate	Spectrophotometric method	Apha (1998)
14.	Nitrate	Cadmium reduction method	Apha (1998)
15.	Calcium (Ca)	Titrimetric method	Apha (1998)
16.	Magnesium (Mg)	Titrimetric method	Apha (1998)
17.	Sodium (Na)	Flame photometric method	Trivedy and Goel (1984)
18.	Potassium (K)	Flame photometric method	Trivedy and Goel (1984)
19.	Sulphate	Turbid metric method	Apha (1998)

2.4.3 Germination and vegetative growth study

Germination is the most fundamental and significant factor in measuring plant growth. It is the most sensitive stage and is highly influenced by external environmental conditions. Studies conducted have shown that germination is adversely affected by effluents due to its chemical composition. Thus, the present study included parameters for the same.

TABLE 4.4 List of biochemical parameters and methods.

S. No	Parameters	Method	References
1.	Protein	Lowery et al.	Lowery et al. (1951)
2.	Total sugar	Nelson and somagyi	Nelson (1944)
3.	Ascorbic acid	Chinoy et al.	Chinoy (1969)
4.	Calcium	Thimmaiah, S.K.	Thimmaiah and Thimmaiah (2004)
5.	Magnesium	Thimmaiah, S.K.	Thimmaiah and Thimmaiah (2004)
6.	Sodium	Tondon, H.L.S.	Tondon (1995)
7.	Potassium	Tondon H.L.S.	Tondon (1995)
8.	Phosphorus	Thimmaiah, S.K.	Thimmaiah and Thimmaiah (2004)

TABLE 4.5 List of stage and days for growth measurement.

Sl No.	Stages	Growth
1	First stage	10 days of growth
2	Middle stage	40 days of growth
3	Mature stage	Fully mature with fruits and flowers

The growth period between germination and flowering is commonly known as vegetative growth of plant development. The study for vegetative growth and cambial activity were carried out at three developmental stages, i.e., after 10th, 20th and 30th days of germination (Table 4.5). Length of root, hypocotyl and epicotyl were measured and an average of 20 readings at both the sites.

3. Results and discussion

3.1 Physicochemical properties of soil

Soil samples of VNSGU and ONGC sites where plants were grown by using normal tap water and treated effluent analysed for parameters in Table 4.2. pH of soils at VNSGU and ONGC sites was alkaline. Electrical conductivity of ONGC soil was reported in permissible limit which was suitable for the plant growth (CPCB, standards). The maximum alkalinity was found in the later period of the crop cultivation at both the sites. This indicated that the irrigation of treated effluent maintained the alkalinity of soil. Organic matter of soil was increased after 20 months of irrigation of treated effluent. This was beneficial effect of treated effluent. Calcium content of soil of both sites was almost same in the initial stage. But it was increased later on after the irrigation of treated effluent at ONGC site. P, K, Mg and sulphate content were found higher in the soil of ONGC than the soil of VNSGU site. This was due to irrigation of effluent having the same element content higher in them. Thus, the irrigation of treated effluent was favorable for plant growth. Sodium content was slightly higher in the soil of ONGC but it had no any adverse effect on the growth of experimental plants. The higher nitrate content found in the soil of ONGC site was due to the more nitrate content of effluent which was because of the addition of urea during the purifying treatment of effluent at the ONGC plant.

3.2 Physicochemical properties of water samples

Parameters selected to check the quality of water mentioned in Table 4.3 were analysed. In the treated effluent and control water pH was alkaline. The growth was better in the experimental plants at optimum temperature at both sites. Present investigation showed that the treated effluent of ONGC had higher COD value than control. Total solids, total dissolved solids and total suspended solids were found higher in ONGC treated effluent, which had no inhibitory effect on the growth of plants selected for the study. The alkalinity was 95.0–133.7 mg/L in the treated effluent of ONGC and 58.0–137.0 mg/L in control which was under the permissible limit for irrigation (Joint, 1985). The hardness of treated effluent of ONGC was in the range of 106.5–197.5 mg/L. The oil and grease content was within the permissible limit (CPCB standards) in treated effluent, which had no adverse effect on the growth of agricultural plants. The excess of oil and grease can cause obstacle in respiration in case of plants; hence the above parameter becomes crucial.

Phosphate content was higher in the treated effluent of ONGC site compared to VNSGU site. In the treated effluent of ONGC it was found in the range of 0.23–0.81 mg/L. This was in the limits of irrigation of industrial effluent and useful for the plant growth (Joint, 1985). Treated effluent of ONGC contained 1.4–2.2 mg/L nitrate, which was beneficial for the plant growth as well. Calcium content in ONGC treated effluent were in the range of 20.6–26.8 mg/L which was under permissible limit of irrigation (CPCB standards) and useful in the cultivation of plants. The contents of magnesium, sodium and potassium were relatively higher in treated effluent of ONGC than control water. This has increased the utility of treated effluent for the cultivation of agricultural plants. Similarly, chloride and sulphate contents were found in the required proportion in treated effluent for the cultivation of plants.

3.3 Heavy metal analysis

The heavy metals were absent in the water of VNSGU site. In treated ONGC effluent, cadmium, cobalt, copper, iron, lead, nickel and zinc were observed in

TABLE 4.6 List of heavy metals and their acceptable standard values.

Sl NO.	Metals	Proportion (mg/L)	Standard CPCB (2000)/NAS (1972) (mg/L)
1	Cadmium	0.0–0.02	2.0
2	Cobalt	ND[a]	0.05
3	Copper	0.08–0.09	3.0
4	Iron	0.07–0.09	5.0
5	Lead	0.02–0.06	1.0
6	Nickel	0.0–0.17	5.0
7	Zinc	0.35	15

[a]*ND, Not detected.*

negligible proportion and had no any adverse effect on the plant growth as well as no further concern for human health. The values of heavy metal obtained are given in Table 4.6.

3.4 Germination and vegetative growth of seedlings

Seeds were sown and facilitated as per requirement by control tap water at VNSGU site and treated effluent at ONGC site for germination. Germination percentages were higher in bottle gourd, brinjal and marigold at ONGC site than VNSGU site. The germination started with the emergence of radicle which was pushed out of the seed coat by elongating hypocotyl. The period of germination was variable in different plants. The germination period was similar in most of the plants at VNSGU and ONGC site except bottle gourd, guar and marigold where the germination was earlier at ONGC site. The percentage of seed germination was higher at ONGC (Table 4.7).

In recent studies it was concluded that the type of industry from which effluent is generated and concentration in which it was used has higher impact on growth rate. In the study conducted, it was found that growth of epicotyl, hypocotyl and root was better at ONGC site which is one of the favourable aspects of using effluent in irrigation. The growth rate of the plants of ONGC site was comparatively higher than the plants of VNSGU site. The germination percentage was 100% at ONGC site for bottle gourd, cucumber, guar and sunflower.

3.5 Biochemical/nutritional studies

A comparative study of the nutritional contents mentioned in Section 2.2 was done for fruits of bottle gourd, brinjal, cucumber, guar, lady's finger and

TABLE 4.7 Germination features of the taxa under study

		Period of germination (days)		Stage of fall of cotyledons		Germination (%)	
Plants	Types of germination	VNSGU	ONGC	VNSGU	ONGC	VNSGU	ONGC
Bottle gourd	Macranga	8	6	5 leaf stage	6 leaf stage	70	100
Brinjal	Macranga	3	3	2 leaf stage	4 leaf stage	65	80
Cucumber	Macranga	4	4	4 leaf stage	6 leaf stage	80	100
Guar	Macranga	3	2	4 leaf stage	4 leaf stage	85	100
Lady's finger	Macranga	2	2	4 leaf stage	4 leaf stage	60	80
Sunflower	Sloanea	3	3	5 leaf stage	5 leaf stage	80	100
Tomato	Sloanea	3	3	4 leaf stage	4 leaf stage	60	85
Marigold	Sloanea	8	6	4 leaf stage	4 leaf stage	50	70

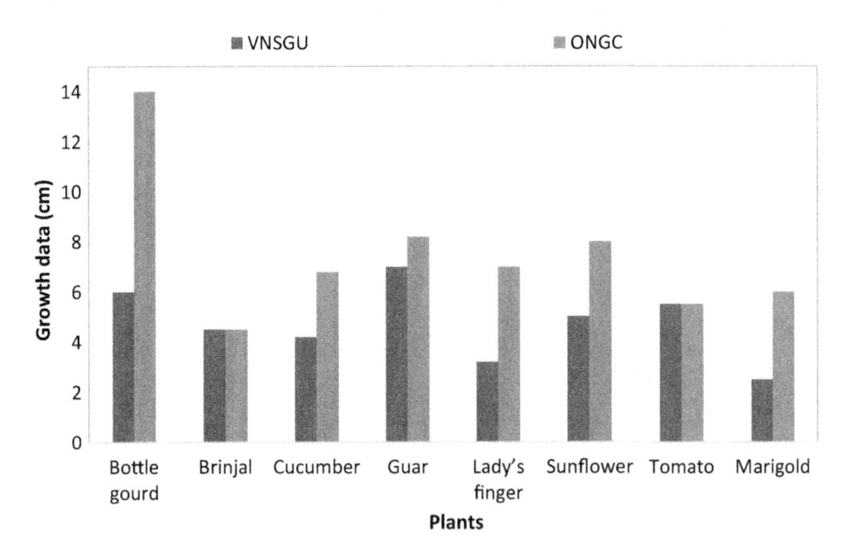

FIGURE 4.1 Comparison of vegetative growth in fruits/seeds of VNSGU and ONGC site in 10-day old plants.

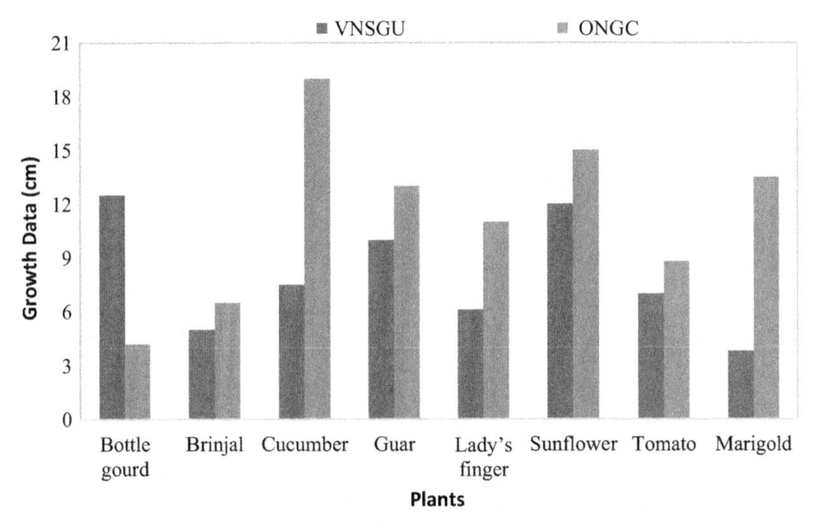

FIGURE 4.2 Comparison of vegetative growth in fruits/seeds of VNSGU and ONGC site in 40-day old plants.

tomato; and for the seeds of sunflower and marigold plants at VNSGU and ONGC sites to find the effect of treated effluent on nutritional status (Figs. 4.1–4.3).

3.5.1 Protein and Ascorbic acid

Protein is very important for all cellular and metabolic activities in plants. These proteins are consumed by humans through various diets and thus serve

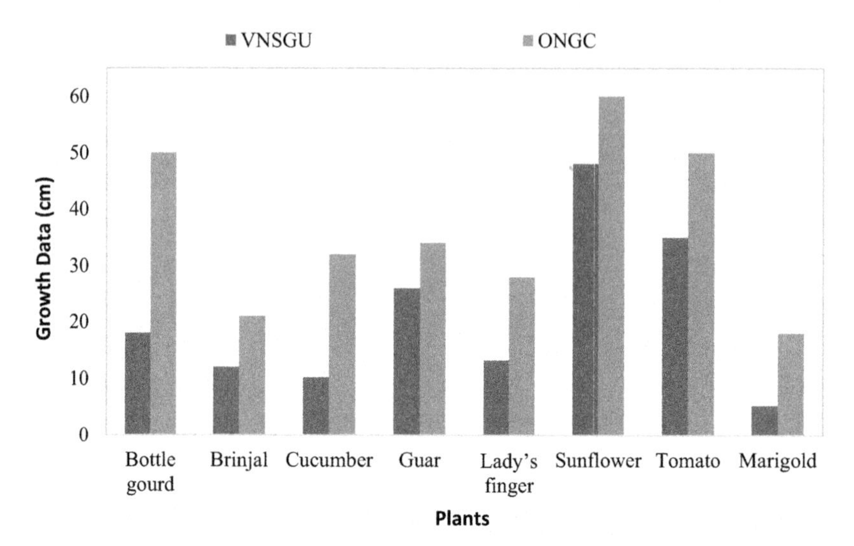

FIGURE 4.3 Comparison of vegetative growth in fruits/seeds of VNSGU and ONGC site in mature plants.

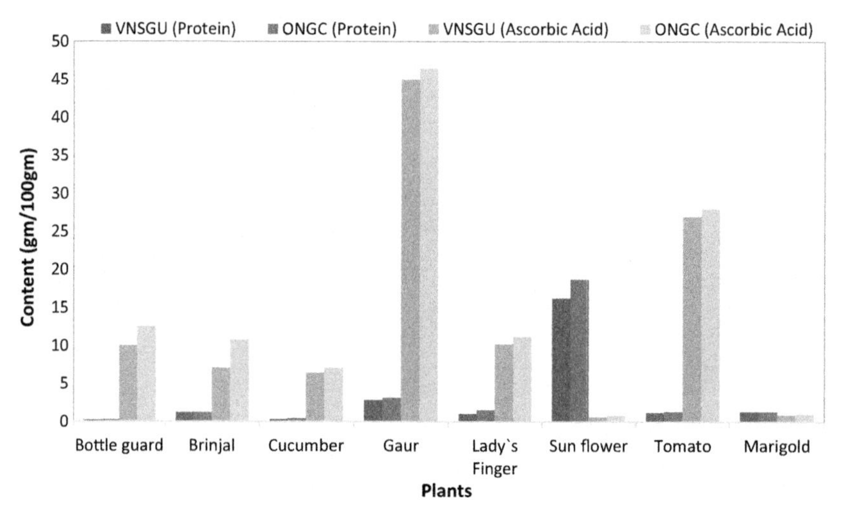

FIGURE 4.4 Comparison of protein and ascorbic acid content in fruits/seeds of VNSGU and ONGC site.

as energy source. It was observed for crops like wheat, maize and rice. Protein content was increased due to use of industrial effluent after dilution (Singh and Mishra, 1987). Protein content of fruits of most of the plants of ONGC site was higher than the plants of VNSGU site (Fig. 4.4). It was 19.04% higher in bottle gourd, 33.3% higher in cucumber, 10.7% higher in guar and 8.3% higher in tomato of ONGC site than VNSGU site. The seeds of sunflower were analysed and 15.43% more protein content was found in plants of ONGC site than VNSGU site. The protein content of the seeds of marigold of both the sites was almost same. In the study conducted at ONGC site, protein content was found increased and thus proves beneficial effects of sewage treated water.

Ascorbic acid is essential to plants as it is involved in regulation of photosynthesis, biosynthesis of hormones and generation of antioxidants. The seeds and fruits of plants cultivated at ONGC site was found higher as pH of effluent was not harmful to plants.

The values of ascorbic acid of the fruits/seeds of both sites are presented in Fig. 4.4. Ascorbic acid content was 25% higher in bottle gourd cultivated at ONGC site than VNSGU site. Brinjal showed 54.28% higher ascorbic acid content in ONGC site than VNSGU site. In cucumber, the proportion of ascorbic acid was 9.3% higher at ONGC site than VNSGU site. Guar produced highest ascorbic acid content followed by tomato.

3.5.2 *Phosphorus (P) and Potassium (K)*

Phosphorus is important for the plants for the formation of roots, flowers and seeds. Potassium helps to fight various diseases in plants. Increase in potassium in leaves of sorghum and maize was found by use of textile mill and sugar factory effluent, respectively (Garg and Kaushik, 2008; Parameswari and Udayasoorian, 2013). Both potassium and phosphorus content were found higher at ONGC site and no adverse effect of effluent was observed (Fig. 4.5).

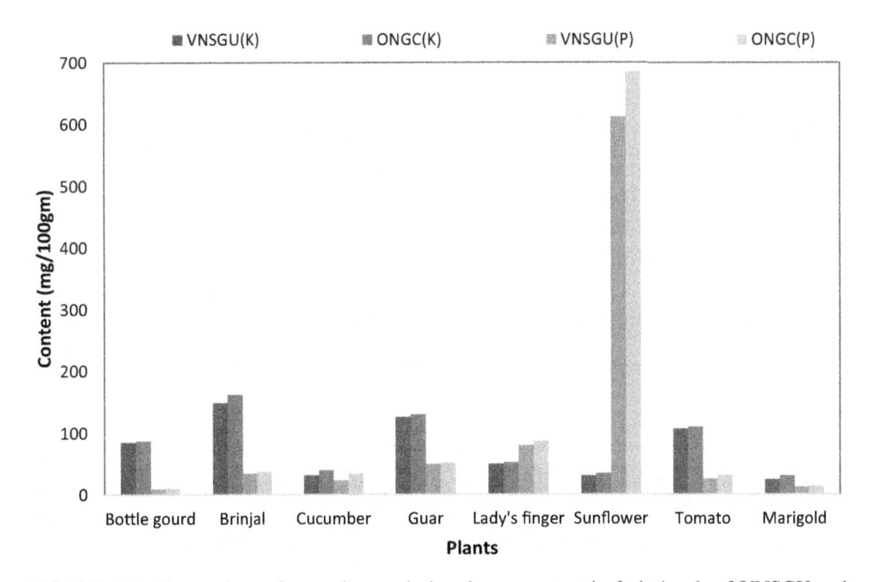

FIGURE 4.5 Comparison of potassium and phosphorus content in fruits/seeds of VNSGU and ONGC site.

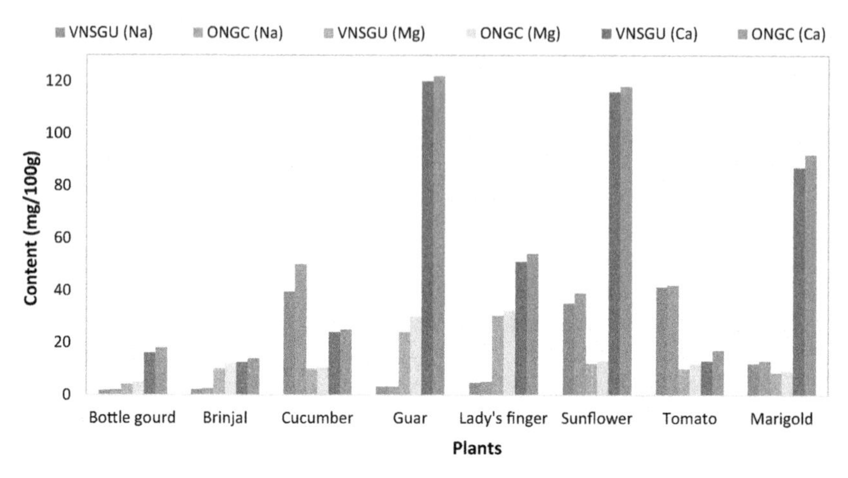

FIGURE 4.6 Comparison of sodium, magnesium and calcium content in fruits/seeds of VNSGU and ONGC site.

3.5.3 Sodium (Na), Magnesium (Mg) and Calcium (Ca)

Sodium is important for photosynthesis and helps in opening and closing of stomata. Sodium contents are presented in Fig. 4.6. The amount of sodium was 11.76% higher in bottle gourd, 19.0% higher in brinjal and 6.5% higher in cucumber fruits of ONGC site than VNSGU site. Sodium content in the fruits/ seeds was adequate in the plants cultivated at ONGC site and had no adverse effect on the plant growth.

Magnesium has various key roles in plants and it is building block of chlorophyll and thus responsible for green colour of plants. The proportion of magnesium was 17.7% higher in bottle gourd, 20.0% higher in brinjal and guar, 3.96% higher in cucumber, 5.60% higher in lady's finger and 18.0% higher in tomato fruits of plants cultivated at ONGC site than VNSGU site.

Calcium is responsible for binding of cell wall. Calcium content of the fruits of ONGC site was higher than the fruits of VNSGU site. It was 12.5% higher in bottle gourd, 12.0% higher in brinjal, 4.16% higher in cucumber and 5.8% higher in lady's finger of ONGC site. The maximum difference in calcium content was found in tomato where calcium content was 30% higher in fruits of ONGC site. In guar, the difference in the calcium content in the fruits of both the sites was less. Sunflower seeds had 1.73% and marigold seeds had 5.7% higher calcium content in the plants of ONGC site than of VNSGU site.

Sodium, magnesium and calcium were found more at the ONGC site which is appreciable. Thus, the cultivation of plants by using treated effluent of ONGC site had beneficial effect on mineral content, ultimately leading to a better nutritional status of the crop cultivated with ONGC effluent.

4. Conclusion

From the above research it was found that plants grown at ONGC site gave better response in almost every aspect in comparison to plants grown at VNSGU site. Hence, treated industrial effluent can be effectively used for irrigation in agriculture. It will be a better option for the areas which are near the industries and are facing water scarcity. Beneficial nutrients were observed in treated effluent with higher COD values. Nutritional value of the plants grown at ONGC site was higher as fruits contained appreciable amount of protein, ascorbic acid, calcium, magnesium, sodium, potassium and phosphorus, which is capable to fulfilling the nutritional demand of human bodies. Also, the seed germination percentage were found higher in case of plants grown at ONGC site.

The study strongly suggests that above research will help to understand the utility and benefits of treated effluents use in agriculture for all the crops included in the study. In future, more advance management and regulation can be developed in application of effluents on agricultural land for achieving higher efficiency in irrigation without causing adverse effects to plants, soil and environment.

Acknowledgements

The authors acknowledge the cooperation of Veer Narmada South Gujarat University Site (VNSGU) and Oil and Natural Gas Corporation Site (ONGC) for necessary infrastructure and financial support for the study.

References

Ajmal, M., Khan, A.U., 1983. Effects of sugar factory effluent on soil and crop plants. Environ. Pollut. Ecol. Biol. 30, 135–141.

Ajmal, M., Khan, A.U., 1984. Effects of vegetable ghee manufacturing effluent on soil and crop plants. Environ. Pollut. Ecol. Biol. 34, 367–379.

Apha, A., 1998. Wef. Standard Methods for the Examination of Water and Wastewater.

Bajpai, P., Dua, S., 1972. Studies on the Utility of Distillery Effluent (Spent Wash) for its Manurial Value and its Effect on Soil Properties. Indian Sugar Calcutta.

CPCB: *Effluent standards*. New Delhi: Central Pollution Control Board, Ministry of Environment and Forest, Government of India. (http://www.cpcbenvis.nic.in/scanned%20reports/PCL%204%20Environmental%20Standards.pdf).

Chinoy, N., 1969. On the specificity of the alcoholic, acidic silver nitrate reagent for the histochemical localization of ascorbic acid. Histochemie 20, 105–107.

El-Bassan, N., Kepper, H., Tietjen, C., 1977. Effect of wastewater, sewage sludge, and refuse compost on accumulation and movement of arsenic and selenium in cultivated soils. Landforsch Voelkenrode 27, 105–110.

Erfani, A.A., Haghnia, G.H., 2001. Effect of Irrigation by Treated Wastewater on the Yield and Quality of Tomato.

Fazeli, M.S., Sathyanarayan, S., Satish, P., Muthanna, L., 1991. Effect of paper mill effluents on accumulation of heavy metals in coconut trees near Nanjangud, Mysore District, Karnataka, India. Environ. Geol. Water Sci. 17, 47–50.

Garg, V., Kaushik, P., 2008. Influence of textile mill wastewater irrigation on the growth of sorghum cultivars. Appl. Ecol. Environ. Res. 6, 1–12.

Gautam, D., Bishnoi, S., 1990. Effect of diurnal dairy effluent on soil characteristics and plant growth in Avena sativa Linn. Geobios 17, 91–94.

Girisha, S., Raju, N., 2008. Effect of sewage water on seed germination and vigour index of different varieties of groundnut (Arachis Hypogaea L.). J. Environ. Biol. 29, 937–939.

Jain, S.K., 2011. Population rise and growing water scarcity in India–revised estimates and required initiatives. Curr. Sci. 101, 271–276.

Joint, F., 1985. Energy and Protein Requirements: Report of a Joint FAO/WHO/UNU Expert Consultation. World Health Organization.

Kaur, R., Wani, S., Singh, A., Lal, K., 2012. Wastewater production, treatment and use in India. In: National Report Presented at the 2 Nd Regional Workshop on Safe Use of Wastewater in Agriculture.

Lowery, O.H., Rosebrough, N.J., Farr, A.L., Randall, R.J., 1951. Protein estimation with the folin phenol method. Biol. Chem. 193 (265).

Najafi, B., Aminian, K., Loew, F., Blanc, Y., Robert, P.A., 2002. Measurement of stand-sit and sit-stand transitions using a miniature gyroscope and its application in fall risk evaluation in the elderly. IEEE Trans. Biomed. Eng. 49, 843–851.

Nelson, N., 1944. A photometric adaptation of the Somogyi method for the determination of glucose. J. Biol. Chem 153, 375–380.

Oblisami, G., Rajannan, G., 1978. Effect of industrial effluents on soil and crop plants. In: Int. Symposium on Environmental Agents and Their Biological Effects. Hyderabad, India.

Parameswari, M., Udayasoorian, C., 2013. Influence of textile and dye effluent irrigation and amendments on micronutrients Iron and Copper status in soil under Maize crop. Int. J. Cur. Tr. Res 2, 163–167.

Pescod, M., 1992. Wastewater Treatment and Use in Agriculture.

Phantumvanit, D., Panayotou, T., 1990. Industrialization and Environmental Quality: Paying the Price.

Reyer, G., Van Beukering, P., Verma, M., Yadav, P., Pandey, P., 1990. Integrated Modeling of Solid Waste in India. Working paper.

Schuler, V.J., Larson, L.E., 1975. Improved Fish Protection at Intake Systems. Journal of the Environmental Engineering Division, ASCE 101.

Singh, J., Kansal, B., 1985. Amount of Nutrients and Heavy Metals in the Waste Water of Different Towns of Punjab and its Evaluation for Irrigation [India]. Journal of Research Punjab Agricultural University.

Singh, K., Mishra, L., 1987. Effects of fertilizer factory effluent on soil and crop productivity. Water Air Soil Pollut. 33, 309–320.

Sommers, L., 1977. Chemical composition of sewage sludges and analysis of their potential use as fertilizers 1. J. Environ. Qual. 6, 225–232.

Thimmaiah, S., Thimmaiah, S., 2004. Standard Methods of Biochemical Analysis. Kalyani publishers.

Tondon, H., 1995. Sulphur Fertilizers for Indian Agriculture. A Guide Book, second ed. Fertilizer Development and Consultation Organization, New Delhi, pp. 1–23.
Trivedy, R., Goel, P., 1984. Chemical and Biological Methods for Water Pollution Studies. Environmental publications.
Trivedy, R., Goel, P., CL, T., 1998. Practical Methods in Ecological and Environmental Sciences. Media publication, Karad, India.
Yaduvanshi, N., Yadav, D., 1996. Residual effect of pressmud cake with nitrogen on cane yield, juice quality and soil nitrogen in sugarcane ratoon. J. Indian Soc. Soil Sci. 44, 158–160.
Zabek, S., 1976. Content of macro and micro-elements in crops irrigate with dairy plant sewage. Rockizi. Nauk Rolniczych 101, 71–78.

Chapter 5

Assessing the suitability of Ghaghra River water for irrigation purpose in India

Sandeep Kumar Gautam[1,2], Jayant K. Tripathi[1], Sudhir Kumar Singh[3]
[1]*School of Environmental Sciences, Jawaharlal Nehru University, New Delhi, India;*
[2]*Department of Geology, University of Lucknow, Lucknow, Uttar Pradesh, India;* [3]*K. Banerjee Centre of Atmospheric and Ocean Studies, IIDS, Nehru Science Centre, University of Allahabad, Prayagraj, Uttar Pradesh, India*

1. Introduction

Rivers have played an important role in the social development of humans since historical times. The history suggests that many great human civilisations have started near rivers (Singh et al., 2014). Famous river valley civilisations flourished along Indus, Nile, Tigris-Euphrates and Yellow Rivers. Rivers support a vast biodiversity of flora and fauna and provide food and habitat to aquatic lives and humans; and hence, they maintain the ecological balance of the environment. Generally, due to unprecedented population growth, unplanned and unmanaged industrial and agricultural development in the flood plain of rivers, global warming and climate change, which are also expected to disturb the balance of water supply and demand (Haddeland et al., 2014), the ecological sustainability of rivers is in great danger. Humans settled along rivers because they provide drinking, irrigation and industrial water; fertile floodplain deposits for agriculture; and waterways for transportation. As a result of human proximity to rivers, activities of agriculture, flood control and input of human and industrial wastes have considerably affected rivers (Lerman, 1980). In result the riverine environment has heavily suffered due to the increasing urbanization, industrialization, growth in population and increasing pressure on the natural resource (Gautam et al., 2013, 2015, 2018; 2016; Singh et al., 2013; Singh et al., 2014). The hydrological cycle thus got manipulated both quantitatively and qualitatively through addition of solid or liquid materials or through withdrawal of water and sediments, which has resulted in great damages to rivers (Singh et al., 2014).

Agricultural Water Management. https://doi.org/10.1016/B978-0-12-812362-1.00005-9

In today's world the availability of water influences economic, industrial and agricultural growth which is essential to mankind and decides the prosperity of a region or a country. Although water is in plenty on the planet earth (1,386,000,000 km^3) the surface-water resources, e.g., rivers and lakes, constitute only about 2120 km^3 and 176,400 km^3, respectively (Gleick, 1996). Globally, the major use of water (70%) is in the agricultural sector. In developing countries, whose income mainly depends on agricultural products, water use in agriculture can occupy up to 90% of all water withdrawals (FAO, 2010). The distribution of different types of water resources is shown in Table 5.1.

In India river water quality is mainly affected by either point or diffused sources of pollution. Changes in surface and groundwater quality in recent years have been reported by many researchers (Singh et al., 2012, 2015; Gautam et al., 2015; Nemčić-Jurec et al., 2017). Hence, a detailed appraisal of water resources for their better management is necessary to cope with the

TABLE 5.1 **Inventory of water at the earth's surface (Gleick, 1996).**

Estimate of global water distribution		
Water source	**Water volume, in cubic kilometers**	**Percent of-total water**
Oceans, seas and bays	1,338,000,000	96.5
Ice caps, glaciers and permanent snow	24,064,000	1.74
Groundwater	23,400,000	1.7
Fresh	10,530,000	0.76
Saline	12,870,000	0.94
Soil moisture	16,500	0.001
Ground ice and permafrost	300,000	0.022
Lakes	176,400	0.013
Fresh	91,000	0.007
Saline	85,400	0.006
Atmosphere	12,900	0.001
Swamp water	11,470	0.0008
Rivers	2120	0.0002
Biological water	1120	0.0001
Total	1,386,000,000	100

anthropogenic and geogenic impacts that alter the quality and quantity of water, and also to develop appropriate treatment technologies for the removal of pollutants and to rely on alternative water sources. Hence, monitoring and protection of water resources are primary requirements for a sustainable environment and at the same time for satisfying the demand of fresh water (Gupta et al., 2014; Singh et al., 2017a,b; Das et al., 2017).

The quality of irrigation water has a major concern in sustaining the soil fertility and high agricultural output (Rawat et al., 2018); hence, keeping this in view the present study is to evaluate the suitability of the irrigation water quality of the Ghaghra River and its major tributaries in Indian territory. Irrigation water quality is linked to its effects on soils and cultivated crops and its management. The objective of the work to evaluate the water for irrigation purpose.

2. Study area

The Ghaghra River, a significant river of the Ganga Plain, is one of the largest and left bank tributaries of the Ganga River. The Ghaghra River is also known as 'Manchee' or 'Karnali' in Nepal and is an international river flowing through Tibet, Nepal and states of Uttar Pradesh and Bihar in India. It is a perennial river and one of the significant rivers of the northern India. The Ghaghra River system is bounded by longitude 79°30′ to 85°0′E and latitude 25°45′ to 30°15′N. The headwater is sourced from the Himalayan glacier in Tibet at an elevation of 5500 m near Shinglabtsa, nearly 60 km southwest of the lake Manasarowar. The total length of the river before its confluence with the Ganga is 1080 km of which 72 km lies in the Tibetan Himalaya, 513 km in Nepal, 425 km in Uttar Pradesh and remaining 70 km in Bihar. The total catchment area is 127,950 km^2 of which 45% (i.e., 57,647 km^2) is in India (Rao, 1995). The catchment area in Uttar Pradesh is about 54,417 km^2 and in Bihar nearly 3230 km^2. The Rapti, Chhoti Gandak and Sharda are the major tributaries of the Ghaghra. The Ghaghra River basin is one of the worst flood affected basins in the country. Most of the annual discharge of the Ghaghra River occurs during southwest monsoon season. The basin in India is dominated by the agricultural activities and forested land. The majority of the people of the basin are engaged in agricultural activities (Fig. 5.1).

3. Material and methods

For detailed hydrogeochemical analysis of water from the Ghaghra River system, systematic samplings have been carried out from the entire region of the basin in two seasons. Water samples were collected from the mid-stream of the river with the help of rope and plastic bucket from the middle of bridge to avoid local heterogeneity and possible human influence near the riverbanks. A Garmin (GPSMAP-76CSX) global positioning system (GPS) was used for location readings. The water samples have been collected in 1 L narrow-mouth

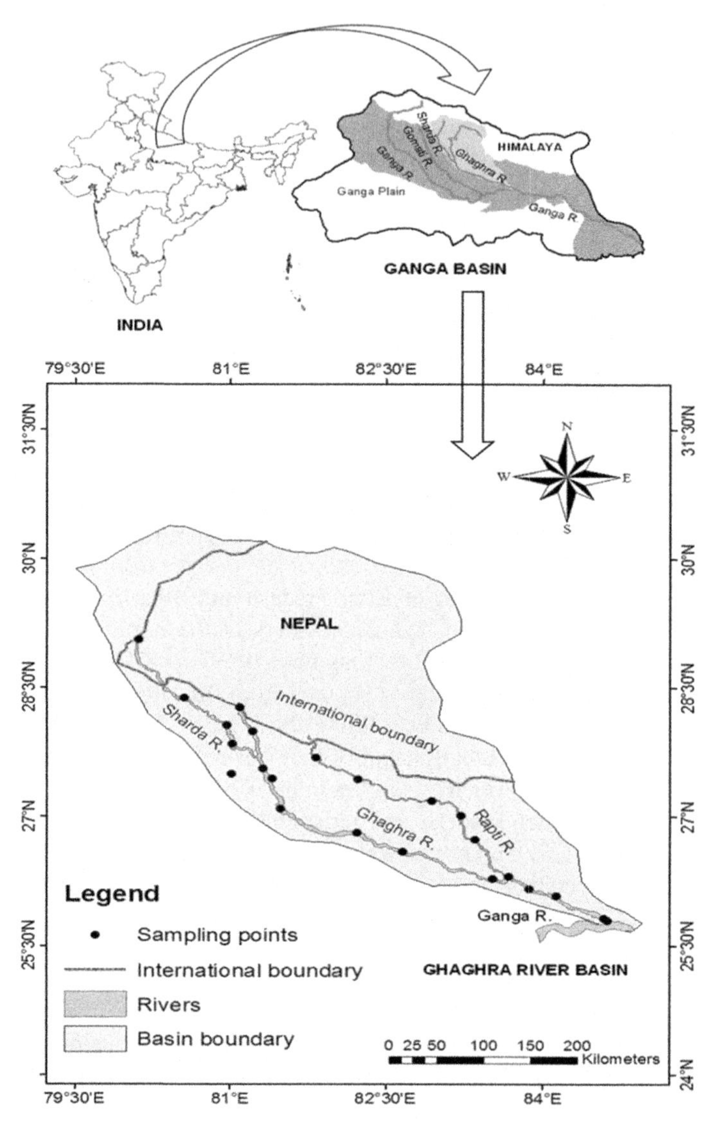

FIGURE 5.1 Sampling location map of Ghaghra River basin and sampling points.

bottles (pre-washed high density polyethylene material) from different parts of the basin in two seasons. For metal analysis 100 mL of water samples have been filtered in the field with help of syringe Millipore filter (0.45 μm) and acidified with 1 mL concentrated HNO_3 and then preserved for the analysis. Water samples collected from field were brought to the laboratory and stored at 4°C temperature in order to avoid any major chemical alteration (APHA, 2005). *In-situ* measurements were performed for temperature, electrical conductivity (EC) and pH using digital probe.

Bicarbonate was determined by potentiometric titration method. Standards of HCO_3 were prepared for required concentration from chemical salt $NaHCO_3$. The 50 mL of each standard and samples were titrated against 0.03N HCl. pH (4.5) was taken as the endpoint of the reaction. A graph was plotted for standard concentrations against the volume of HCl consumed. The concentrations of the samples were determined from the graph plotted for standards.

Major cations (Ca^{2+}, Mg^{2+}, Na^+ and K^+) were measured by ion chromatograph using cation column (CS12A/CS12G) and cation self-regenerating suppressor (CSRS) in recycle mode. The eluent used for cation analysis was prepared by diluting 22 mL of 1N sulphuric acid (H_2SO_4) in 1 L of deionized water. The 1N H_2SO_4 was prepared by mixing of 27.76 mL of concentrated sulphuric acid with 1 L of deionized water. Analysis of major cations of some samples was repeated on ICP-OES to monitor the accuracy of the analysis. The analytical precision was maintained by running the known standard after every 15 samples. An overall precision, expressed as percent relative standard deviation (RSD) was obtained for the entire samples. Analytical precision for cations (Ca^{2+}, Mg^{2+}, Na^+ and K^+) were within 10% (Fig. 5.1).

4. Results and discussion

4.1 Physicochemical characteristics (pH, EC, TDS and TH)

Physicochemical characteristics during the pre-monsoon and post-monsoon seasons are given in Table 5.2. The pH was slightly alkaline and its value varied from 7.64 to 8.32 (average 7.99) in the pre-monsoon and in the post-monsoon 8.00 to 8.52 (average 8.34). During the post-monsoon the pH was relatively high; this may be due to enhanced dissolution caused by higher interaction between rainwater and soil/rocks (Subramanian and Saxena, 1983). The electrical conductivity is a measurement of the ionic strength of solutions, which varies from 198 to 435 μS/cm (average 274 μS/cm) in the pre-monsoon and 244 to 408 μS/cm (average 325 μS/cm) in the post-monsoon. Total dissolved solids (TDS) of river water samples vary from 184 to 420 mg/L with the average being 254 mg/L in the pre-monsoon, and in the post-monsoon values vary from 244 to 376 mg/L (average 393 mg/L).

Total hardness (TH) in water is defined as concentration of multivalent cations. These mainly include divalent Ca^{2+} and Mg^{2+} ions. These ions move into water supply by leaching from minerals in the aquifer. Common calcium-containing minerals are calcite and gypsum. A common magnesium mineral is dolomite which also contains calcium. Rainwater is soft due to presence of fewer ions. According to US Department of Interior and Water Quality Association's water hardness classification, the soft water hardness is in the range of 0–17.1 mg/L, hardness for slightly hard water is in the range 17.1–60 mg/L, moderately hard water has the range of 60–120 mg/L, hard water has the range of 120–180 mg/L and very hard water has values 180 mg/L or above.

TABLE 5.2 The statistics of the physicochemical parameters and irrigation indices in the pre-monsoon and post-monsoon season.

	Pre-monsoon						Post-monsoon					
Variables	Max.	Min.	Ave.	Stdev.	[a]Ganga ave.	[b]World ave.	Max.	Min.	Ave.	Stdev.	[a]Ganga ave.	[b]World ave.
pH	8.3	7.6	8.0	0.2	7.7	–	8.5	8.0	8.3	0.1	7.7	–
EC (μScm^{-1})	435	198	274	69	239	–	408	244	325	39	239	–
TDS (mgL^{-1})	420	184	254	69	197	110	376	204	293	40	197	110
TH (mgL^{-1})	197.4	99.8	132.2	29.0	–	–	211.5	126.9	171.8	21.1	–	–
HCO_3 (mgL^{-1})	305	125	176	53	110	53	260	126	195	32	110	53
Ca (mgL^{-1})	45.3	26.0	31.6	5.7	32	14.7	52.3	31.4	41.8	5.4	32	14.7
Mg (mgL^{-1})	20.5	8.2	12.9	3.8	10	3.7	21.0	11.8	16.4	2.2	10	3.7
Na (mgL^{-1})	20.2	2.0	5.4	4.5	10	7.2	10.8	3.2	6.0	1.7	10	7.2
K (mgL^{-1})	6.3	1.5	3.4	1.5	5	1.4	4.8	2.8	3.9	0.5	5	1.4
%Na	21.2	5.0	10.1	3.8	–	–	12.2	6.7	9.4	1.3	–	–
SAR	0.6	0.1	0.2	0.1	–	–	0.3	0.1	0.2	0.0	–	–
RSC	3.1	1.0	1.6	0.6	–	–	2.1	0.8	1.5	0.3	–	–
PI	74.8	59.6	67.5	4.2	–	–	60.0	51.0	55.5	2.1	–	–
MH	46.7	32.9	39.7	4.0	–	–	48.2	37.1	39.3	2.4	–	–
KI	0.1	0.0	0.1	0.0	–	–	0.2	0.0	0.1	0.0	–	–

EC, Electrical conductivity; *KI*, Kelly's index; *MH*, Magnesium hazard; *PI*, Permeability index; *RSC*, Residual sodium carbonate; *SAR*, Sodium adsorption ratio; *TDS*, Total dissolved solids and *TH*, Total hardness.
[a]*Subramanian, 2011.*
[b]*Berner and Berner, 1987.*

The high TH may cause encrustation on water supply distribution systems. Hardness has no known adverse consequence on health but it can inhibit formation of lather with soap and increases the boiling point of water. TH of river water sample varies from 100 to 197 mg/L with the average being 132 mg/L in the pre-monsoon, and in the post-monsoon hardness varies from 127 to 212 mg/L (average 172 mg/L). On the basis of hardness assessment, we conclude that all the water samples of the Ghaghra, Rapti and Sharda Rivers are in hard group during pre and post-monsoon season.

4.2 Irrigation indices

The suitability of Ghaghra River water for irrigation purpose depends upon the effect of mineral constituents of water on both plants and soils. The general criteria for assessing the irrigation water quality are the total major ions concentration as measured by EC, relative proportions of sodium as expressed by sodium percentage (Na%), Sodium absorption ratio (SAR), and Residual sodium carbonate (RSC). Some other parameters adopted by Laboratory of the United States Department of Agriculture are also evaluated, which are used for the suitability of water for irrigation, like permeability index (PI), magnesium hazard (MH) and Kelly's index (KI), as explained in the following sub-sections.

4.2.1 Sodium percentage (Na%)

The study of sodium concentration is important in classifying irrigation water because excess sodium in water produces undesirable effects of changing soil properties and reducing soil permeability (Kelley, 1951). The Na% is calculated using Eq. (5.1) is given below,

$$Na\% = \frac{(Na^{+} + K^{+})}{(Ca^{2+} + Mg^{2+} + Na^{+} + K^{+})} \times 100 \tag{5.1}$$

where all the ions concentrations are in meq/L.

The Wilcox (1955) diagram uses sodium percentage and electrical conductivity (Fig. 5.2) to classify surface water, which can be divided into five divisions. As per the Bureau of Indian Standard (BIS), maximum sodium of 60% is recommended for irrigation water. In the present study the sodium percentage in the premonsoon of the surface water ranges from 5.01% to 21.18% and from 6.70% to 12.18% in the post-monsoon season. Based on the % Na of surface water samples on the Wilcox diagram are fall in the categories of excellent to good region in both seasons. Thus, the quality of surface water is suitable for irrigation.

4.2.2 United States Salinity Laboratory's (USSL's) diagram

The combined effect of EC and SAR on plant growth is shown graphically by the United States Salinity Laboratory (USSL; Richards, 1954), which is widely used for classification of water quality for irrigation. EC and sodium concentration are very important in classifying irrigation water.

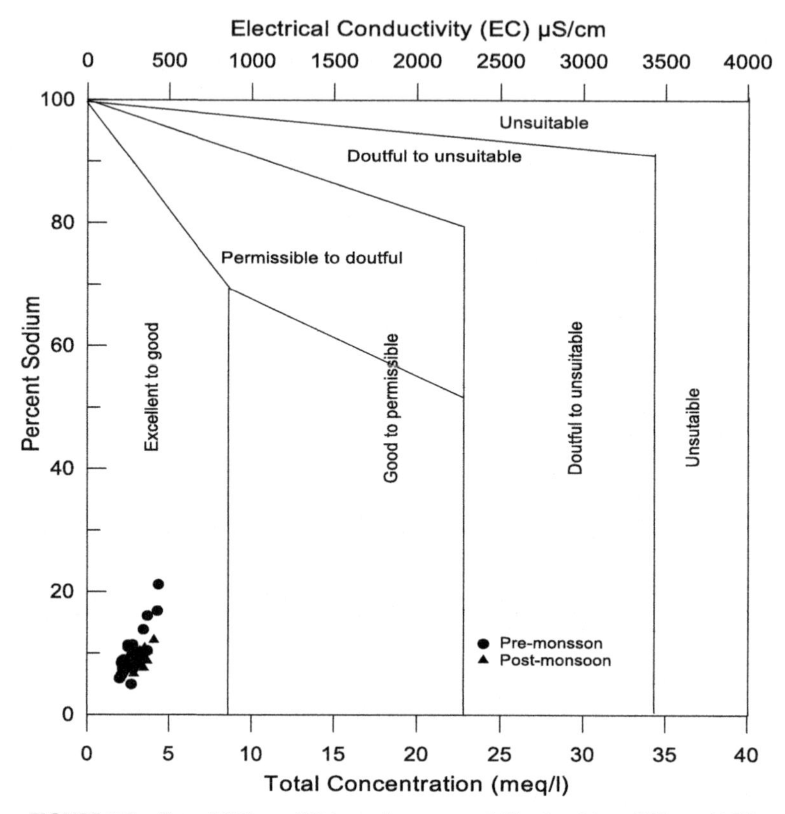

FIGURE 5.2 Plot of %Na vs. EC in surface water of Ghaghra River (Wilcox, 1955).

The total concentration of soluble salts in irrigation water can be expressed for the purpose of classification of irrigation water as low (EC = <250 μS/cm), medium (250–750 μS/cm), high (750–2250 μS/cm) and very high (2250–5000 μS/cm) salinity zones (Richards, 1954). Eq. (5.2) of SAR is given as below:

$$SAR = \frac{Na^+}{\sqrt{\left(\frac{Ca^{2+} + Mg^{2+}}{2}\right)}} \tag{5.2}$$

The calculated value of SAR in the surface water ranges from 0.08 to 0.64 in the premonsoon, in the post-monsoon season it varies from 0.11 to 0.32. The plot of surface water data on the US salinity diagram (Fig. 5.3) shows that the surface water samples fall in the category C1S1 (47.83%), and C2S1 (52.17%) region, indicating low to medium salinity and low alkalinity in the pre-monsoon The post-monsoon samples fall in the region C1S1 (8.70%) and remaining 91.30% in C2S1 region, indicating low to medium salinity and low alkalinity.

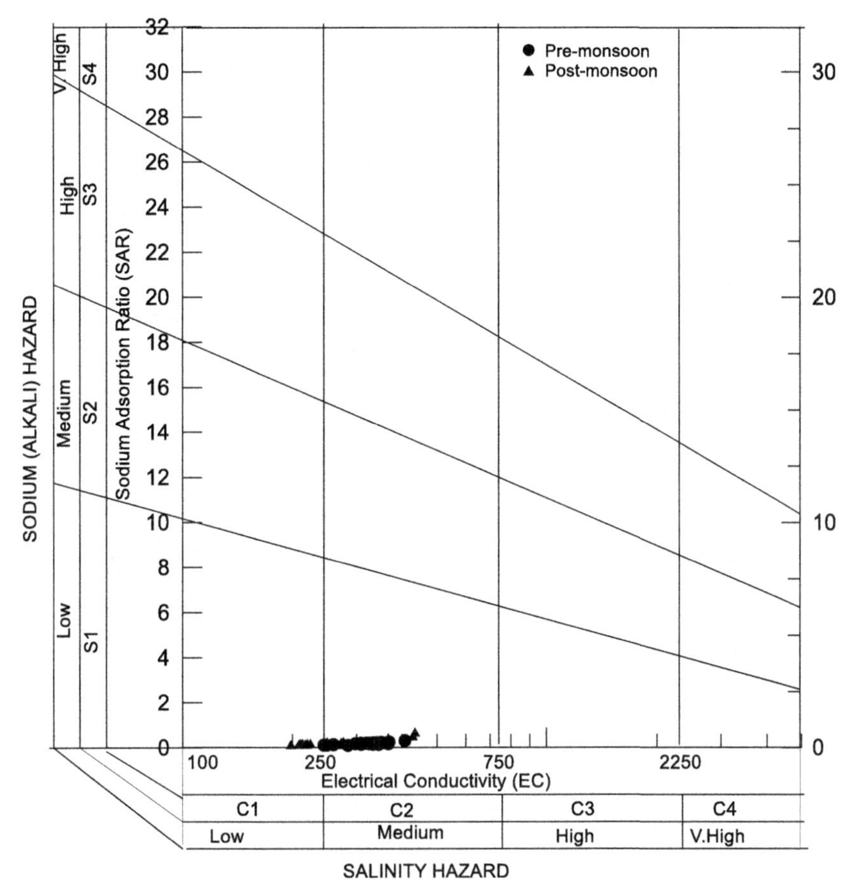

FIGURE 5.3 USSL salinity diagram of surface water of Ghaghra River for classification of irrigation water. *(After Richards, L.A., 1954. Diagnosis and Improvement of Saline and Alkali Soils, vol. 60. US Department of Agriculture Handbook.)*

4.2.3 Residual sodium carbonate (RSC)

The quantity of bicarbonate and carbonate in excess of alkaline earths (Ca + Mg) also influences the suitability of water for irrigation purposes. When the sum of carbonates and bicarbonates is in excess of calcium and magnesium, there may be a possibility of complete precipitation Ca and Mg (Raghunath, 1987). To quantify the effects of carbonate and bicarbonate, residual sodium carbonate (RSC) has been computed by Eq. (5.3):

$$RSC = \left(CO_3^- + HCO_3^-\right) - \left(Ca^{2+} + Mg^{2+}\right) \quad (5.3)$$

Ions values are in meq/L.

TABLE 5.3 Residual sodium carbonate (RSC).

RSC	Water class	Pre-monsoon	Post-monsoon
<1.25	Safe	G1b, G3b, G5b, G6b, G7b, S21b, S22b (30.43%)	G1a, S19a, S21a, S23a, (17.39%)
1.25–2.5	Marginally safe	G2b, G4b, G8b, G9b, G10b, G11b, G12b, S19b, S20b, S23b, R13b, R14b, R15b, R16b (60.87%)	G2a, G3a, G4a, G5a, G6a, G7a, G8a, G9a, G10a, G11a, G12a, R13a, R14a, R15a, R16a,R17a, R18a, S20a, S22a (82.61%)
>2.5	Unsafe	R17b, R18b (8.70%)	

A high value of RSC in water leads to an increase in the adsorption of sodium in soil (Eaton, 1950). Waters having RSC values greater than 5 meq/L are considered harmful to the growth of plants, while waters with RSC values above 2.5 meq/L are not considered suitable for irrigation purpose.

In the present study, the surface waters samples show RSC values range from 0.95 to 3.10 meq/L in the pre-monsoon and 0.80–2.14 meq/L in the post-monsoon season. According to RSC classification 30.43% of surface water is in safe region (RSC < 1.25 meq/L), 60.87% samples lie in marginally safe (RSC range 1.25–2.5 meq/L) and remaining 8.70% samples in the region are unsafe (RSC < 2.5 meq/L) in the pre-monsoon season. During the post-monsoon season about 17.39% water samples are safe and remaining 82.61% lie in marginally safe region (Table 5.3). This indicates that surface water is suitable for irrigation uses.

4.2.4 Permeability index (PI)

Doneen (1964) classified irrigation waters based on the permeability index (PI). PI is defined by Eq. (5.4):

$$PI = \frac{\left(Na^{+} + \sqrt{HCO_3^{-}}\right)}{\left(Ca^{2+} + Mg^{2+} + Na^{+}\right)} \times 100 \tag{5.4}$$

where all ionic concentrations are expressed in meq/L.

Accordingly, the PI is classified under class I (>75%), class II (25%–75%) and class III (<25%) orders. Class I and class II waters are categorized as good for irrigation with 75% or more of maximum permeability. Class III waters are unsuitable with 25% of maximum permeability. The permeability index of river water varied in pre-monsoon from 59.65 to 74.75, and in the post-monsoon

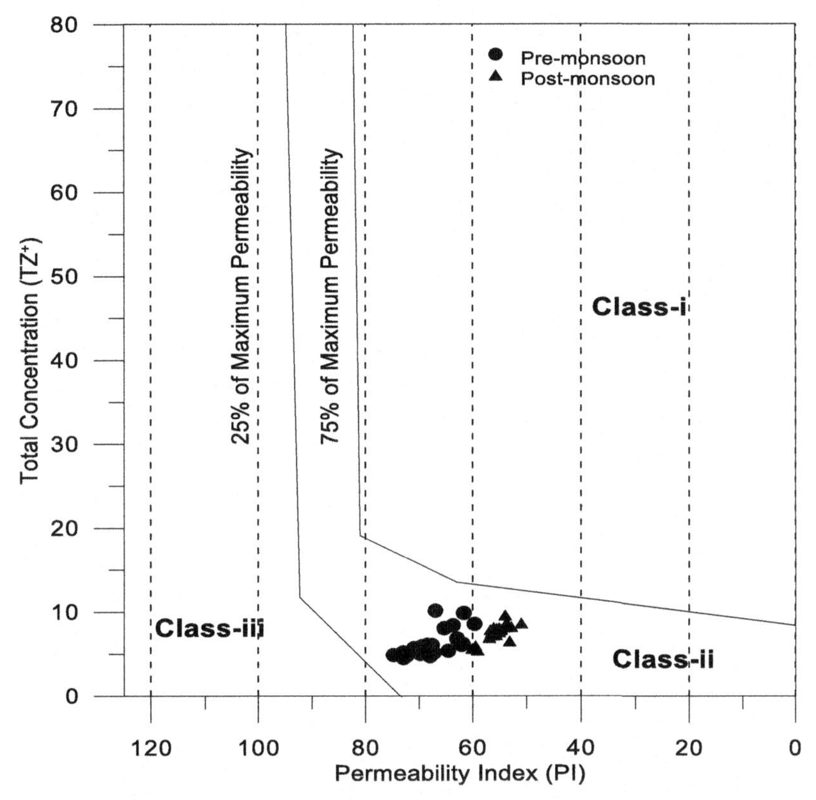

FIGURE 5.4 Plot of TZ^+ vs. Permeability index in surface water of Ghaghra River.

values vary from 50.96 to 60.04. All the samples fall into the Class II (PI % 25–75) category of Doneen's chart (Domenico and Schwartz, 1990) shown in Fig. 5.4. All the surface water samples belong to safe region in the both pre- and post-monsoon seasons.

4.2.5 Magnesium hazard (MH)

Szabolcs and Darab (1964) proposed magnesium hazard (MH) or magnesium ratio (MR) value for irrigation water as given below

$$MH = \frac{Mg^{2+}}{(Ca^{2+} + Mg^{2+})} \times 100 \tag{5.5}$$

where all the ions are expressed in meq/L.

MH < 50 is considered safe and >50 is unsafe for irrigation use. In the analyzed water samples the MH values are <50 in both the pre-monsoon and post-monsoon seasons. So the Ghaghra River water samples are suitable for irrigation uses.

4.2.6 Kelly's index (KI)

Kelly's index is used for the classification of water for irrigation purposes (Kelly, 1957). Sodium measured against calcium and magnesium is considered to calculate Kelly index (5).

$$KI = \frac{Na^{+}}{(Ca^{2+} + Mg^{2+})} \tag{5.6}$$

where all the ions are expressed in meq/L.

KI values greater than 1 (>1) indicate an excess level of sodium in waters, which is unsuitable for irrigation, and KI values less than 1 (<1) are suitable for irrigation purposes. In present studies of the Ghaghra River system, all water samples have KI value < 1 (average 0.07). According to Kelly's index, surface water of Ghaghra River and its tributaries (the Rapti and Sharda) is suitable for irrigation purposes.

5. Management plan

Supply of good quality irrigation water are expected to decrease in the future. New water supplies will not keep pace with the increasing water need of industries and municipalities, rapidly increasing population and change in life style (Oster, 1994). Reducing excessive ions requires multiple solutions ranging from good farm management strategies, innovative farm irrigation schemes, careful design of farms in relation to land geomorphology and geology. Further, to educate the farmers to use biofertilizers instead of synthetic fertilizers. In addition, a selective planting strategy of plant species can also mitigate the negative effects on poor water quality.

Apart from increased levels of nitrogen, phosphorus, and potassium, the salinity (total salt content) and sodicity (sodium content) of these waters will be higher than that of the original source water because of the direct addition of salts to the water through both weather (natural) and anthropogenic activities and the evapo-concentration that occurs as water is used (Oster, 1994). While the use of these waters may require only minor modifications of existing irrigation and agronomic strategies in most cases. There will be some situations that will require major changes in the crops grown, the method of water application and the use of soil amendments. Use of poor-quality water requires three changes from standard irrigation practices: (1) selection of appropriately salt-tolerant crops; (2) improvements in water management, and in some cases, the adoption of advanced irrigation technology; and (3) maintenance of soil-physical properties to assure soil tilth and adequate soil permeability to meet crop water and leaching requirements (LR) (Oster, 1994).

Further, to reduce pollution of water the best solution is to immediately reduce by half the amount of fertilizer in the watershed upstream of the well-field and verify the evolution in time. This is a long operation (several years) depending on the rainfall-runoff characteristics of the watershed. Modelling of flow and transport of ions can be successfully estimated for the rehabilitation of the water quality.

6. Conclusions

For the present study, 24 samples from Ghaghra River, 12 samples from Rapti, and 10 samples from Sharda River were collected from different sampling locations during the post-monsoon and pre-monsoon seasons of year 2010–11. The collected samples were analysed for pH, conductivity, TH, TDS, major cations (Ca^{2+}, Mg^{2+}, Na^{+} and K^{+}), major anions (HCO_3^-, F^-, Cl^-, SO_4^{2-} and NO_3^-), and dissolved silica. The analytical data were evaluated in terms of the chemical characteristics of the water resources of the Ghaghra River basin and its suitability for irrigation use. The parameters like %Na, SAR, RSC, PI, MH and KI were calculated for assessment of water for the irrigation uses. The pH value of the analysed water samples shows that the Ghaghra River system water is alkaline in nature. Sodium concentration is important in classifying irrigation water because sodium reacts with soil to reduce its permeability. On the basis of %Na, SAR, RSC, PI, MH and KI values, water of Ghaghra River basin is found to be suitable for irrigation purpose.

Acknowledgements

First author is grateful to the Council of Scientific and Industrial Research (file no. 09/263 (0703)/2008- EMR-I), and Department of Science & technology [File No: PDF/2017/ 002820] New Delhi, India, for providing the financial support and altogether acknowledges the technical and administrative staff of School of Environmental Sciences, JNU of their technical support. Authors are also astoundingly acknowledged to Director and Dr Abhay K Singh of the Central Institute of Mining and Fuel Research, Dhanbad, for providing research facilities and valuable support throughout the research. Author Sudhir Kumar Singh express his thanks to the Science & Engineering Research Board, New Delhi, India (sanction order no CRG/2019/003551).

References

American Public Health Association (APHA), 2005. Standard Methods for the Examination of Water and Wastewater, twenty-first ed. American Public Health Association, Washington DC.

Berner, E.K., Berner, R.A., 1987. The global water cycle: Geochemistry and Environment. Prentine Hall, Englewood Cliffs, New Jersey, 208.

Das, A., Munoz-Arriola, F., Singh, S.K., Jha, P.K., Kumar, M., 2017. Nutrient dynamics of the Brahmaputra (Tropical River) during the monsoon period. Desalin. Water Treat. Sci. Eng. 76, 212–224.

Domenico, P.A., Schwartz, F.W., 1990. Physical and Chemical Hydrogeology, vol. 824. Wiley, New York.

Doneen, L.D., 1964. Notes on Water Quality in Agriculture. Water Science and Engineering Paper 4001. Department of Water Sciences and Engineering, University of California, California.

Eaton, F.M., 1950. Significance of carbonates in irrigation water. Soil Sci. 69, 123–133.

FAO, 2010. Water at a Glance. Retrieved from. http://www.fao.org/nr/water/docs/waterataglance.pdf.

Gautam, S.K., Sharma, D., Tripathi, J.K., Ahirwar, S., Singh, S.K., 2013. A study of the effectiveness of sewage treatment plants in Delhi region. Appl. Water Sci. 3, 57–65. https://doi.org/10.1007/s13201-012-0059-9.

Gautam, S.K., Evangelos, T., Singh, S.K., Tripathi, J.K., Singh, A.K., 2018. Environmental monitoring of water resources with the use of PoS index: a case study from Subarnarekha River basin, India. Environ. Earth Sci. https://doi.org/10.1007/s12665-018-7245-5, 77-70.

Gautam, S.K., Maharana, C., Sharma, D., Singh, A.K., Tripathi, J.K., Singh, S.K., 2015. Evaluation of groundwater quality in the chotanagpur plateau region of the subarnarekha river basin, Jharkhand state. India Sustain. Water Qual. Ecol. 6, 57–74. https://doi.org/10.1016/j.swaqe.2015.06.001.

Gautam, S.K., Singh, A.K., Tripathi, J.K., Singh, S.K., Srivastava, P.K., Narsimlu, B., Singh, P., 2016. Appraisal of Surface and Groundwater of the Subarnarekha River Basin, Jharkhand, India: Using Remote Sensing, Irrigation Indices and Statistical Techniques. Geospatial Technology for Water Resource Applications. CRC Press, Boca Raton, FL, pp. 144–169.

Gleick, P.H., 1996. Water resources. In: Schneider, S.H. (Ed.), Encyclopedia of Climate and Weather, vol. 2. Oxford University Press, New York, pp. 817–823.

Gupta, L.N., Avtar, R., Kumar, P., Gupta, G.S., Verma, R.L., Sahu, N., Sil, S., Jayaraman, A., Roychowdhury, K., Mutisya, E., Singh, S.K., 2014. A multivariate approach for water quality assessment of river Mandakini in Chitrakoot, India. J. Water Resour. Hydraul Eng. 3, 22–29.

Haddeland, I., Heinke, J., Hester Biemans, H., Eisner, S., et al., 2014. Global water resources affected by human interventions and climate change. Proc. Natl. Acad. Sci. Unit. States Am. 111 (9), 3251–3256. https://doi.org/10.1073/pnas.1222475110.

Kelley, K.R., 1951. Alkali Soils- Their Formation Properties and Reclamation. Reinold Publ. Corp., New York.

Kelly, W.P., 1957. Adsorbed sodium cation exchange capacity and percentage sodium sorption in alkali soils. Science 84, 473–477.

Lerman, A., 1980. Controls on river water composition and the mass balance of river systems. In: Martin, J.M., Burton, J.D., Eisma, D. (Eds.), River Inputs to Ocean Systems. SCOR/UNEP/UNESCO Review and workshop, FAO, Rome, pp. 1–4.

Nemčić-Jurec, J., Singh, S.K., Jazbec, A., Gautam, S.K., Kovač, I., 2017. Hydrochemical investigations of groundwater quality for drinking and irrigational purposes: two case studies of Koprivnica-Križevci County (Croatia) and district Allahabad (India). Sustain. Water Resour. Manag. https://doi.org/10.1007/s40899-017-0200-x.

Oster, J.D., 1994. Irrigation with poor quality water. Agric. Water Manag. 25 (3), 271–297. https://doi.org/10.1016/0378-3774(94)90064-7.

Raghunath, H.M., 1987. Groundwater. Wiley Eastern Ltd, Delhi.

Rao, K.L., 1995. India's Water Wealth: Its Assessment, Uses and Projections. Longman, New Delhi (Reprinted).

Rawat, K.S., Singh, S.K., Gautam, S.K., 2018. Assessment of groundwater quality for irrigation use: a Peninsular case study. Appl. Water Sci. https://doi.org/10.1007/s13201-018-0866-8.

Richards, L.A., 1954. Diagnosis and Improvement of Saline and Alkali Soils, vol. 60. US Department of Agriculture Handbook.

Singh, H., Pandey, R., Singh, S.K., Shukla, D.N., 2017a. Assessment of heavy metal contamination in the sediment of the River Ghaghara, a major tributary of the River Ganga in Northern India. Appl. Water Sci. 7 (7), 4133–4149.

Singh, H., Singh, D., Singh, S.K., Shukla, D.N., 2017b. Assessment of river water quality and ecological diversity through multivariate statistical techniques, and earth observation dataset of rivers Ghaghara and Gandak, India. Int. J. River Basin Manag. 15 (3), 347–360. https://doi.org/10.1080/15715124.2017.1300159.

Singh, S.K., Srivastava, P.K., Gupta, M., Mukherkjee, S., 2012. Modeling mineral phase change chemistry of groundwater in a rural-urban fringe. Water Sci. Technol. 66 (7), 1502−1510. https://doi.org/10.2166/wst.2012.338.

Singh, S.K., Srivastava, P.K., Gautam, S.K., Pandey, A.C., 2013. Integrated assessment of groundwater influenced by a confluence river system: concurrence with remote sensing and geochemical modeling. Water Resour. Manag. 27 (13), 4291−4313. https://doi.org/10.1007/s11269-013-0408-y.

Singh, S.K., Srivastava, P.K., Singh, D., Han, D., Gautam, S.K., Pandey, A.C., 2014. Modeling groundwater quality over a humid subtropical region using numerical indices, earth observation datasets, and X-ray diffraction technique: a case study of Allahabad district, India. Environ. Geochem. Health 37 (1), 157−180. https://doi.org/10.1007/s10653-014-9638-z.

Subramanian, V., 2011. Environmental Chemistry. I.K. International Publishing House Pvt Ltd. New Delhi, 166−168.

Subramanian, V., Saxena, K., 1983. Hydrogeology of ground water in Delhi region of India, Relation of water quality and quantity. In: Proceedings of the Hamberg Symposium IAHS Publication No. 146.

Szabolcs, I., Darab, C., 1964. The influence of irrigation water of high sodium carbonate content of soils. In: Proceedings of 8th International Congress of Isss, Trans, vol. II, pp. 803−812.

Wilcox, L.V., 1955. Classification and use of irrigation waters. USDA Circular No 969, 19.

Chapter 6

Desiccation-tolerant rhizobacteria: a possible approach for managing agricultural water stress

Ajay Shankar, Anjali Singh, Shivani Chaudhary, Vishal Prasad
Institute of Environment and Sustainable Development, Banaras Hindu University, Varanasi, Uttar Pradesh, India

1. Introduction

By 2050 the population of the world will be nine billion, requiring continued increase in crop production to assure food security (Gatehouse et al., 2011; Foley et al., 2011). Therefore it becomes imperative to find solutions for enhancing tolerance of plants against drought and let the crops grow and fulfil the food demands under limited water resource availability (Editorial, 2010; Mancosu et al., 2015; Ngumbi and Kloepper, 2016).

1.1 Drought

In the 21st century climate change is an imminent threat to agriculture sustainability worldwide. It has been found that drastic changes in the different climatic conditions like heat and precipitation are the most influencing factors for global reduction in crop productivity nowadays. Another key factor is limited availability of water leading to water stress, also called as drought, which hinders crop growth and productivity in many regions of the world (Vinocur and Altman, 2005). Drought can be classified into the following types (1) meteorological drought, described as lack of precipitation for a period of time; (2) hydrological drought, described as lack of adequate surface and subsurface water resource for established water use of a given water resources management system; (3) socioeconomic drought, defined as failure of water resource system to meet water demand; and (4) agricultural drought, defined as a period with declining soil moisture resulting in crop failure (Wilhite and Glantz, 1985;American Meteorological Society, 2004; Ngumbi

Agricultural Water Management. https://doi.org/10.1016/B978-0-12-812362-1.00006-0

and Kloepper, 2016). In this chapter the focus is on aspects related to agricultural drought.

1.2 Effect of drought stress on plant growth and development

Drought stress is amongst one of the most severe threats to agriculture nowadays, which is limiting crop productivity in many arid and semi-arid regions of the world (Lesk et al., 2016). Drought is supposed to create serious growth problems for crops on more than 50% of the earth's arable lands by 2050 (Vinocur and Altman, 2005). It affects the plant water potential and turgidity, and also changes the physiological and morphological characteristics of the plants (Hsiao, 2000; Vurukonda et al., 2016). There are several examples describing the effects of drought on crop plants such as maize, rice, wheat and barley (Kamara et al., 2003; Lafitte et al., 2007; Rampino et al., 2006; Samarah, 2005). Drought stress decreases the water content resulting in decrease in fresh weight and also affects the nutrient availability to plants, as nutrients are carried to roots through water. It decreases diffusion and mass-flow of water-soluble nutrients like Ca, Mg, Si, NO_3^- and SO_4^{2-} (Jaleel et al., 2009; Barber, 1995; Selvakumar et al., 2012). The drought stress also enhances the formation of free radicals and reactive oxygen species (ROS) such as superoxide radicals, hydrogen peroxide and hydroxyl free radicals disturbing antioxidant mechanism resulting in oxidative stress. High levels of ROS can degrade the proteins, lipids and nucleic acids in plants. Chlorophyll content also decreases due to drought as seen in case of *Paulownia imperialis* (Astorga and Melendez, 2010), bean (Beinsan et al., 2003) and *Caranthus tinctorius* (Siddiqi et al., 2009). It also heightens the ethylene biosynthesis that inhibits growth of plants (Ali et al., 2014). Drought is a kind of multifaceted stress causing damage to cell organelles, as well as plant as a whole and also decreases the growth and development in plant qualitatively and quantitatively (Choluj et al., 2004).

1.3 Concept behind drought adaptations

Drought resistance can be defined as the capability of plants to sustain growth and survivability when drought occurs (Chaves et al., 2003). Plants have evolved several mechanisms which help them to deal with the drought stress, like morphological and architectural adaptation, osmotic adjustment, water resource optimization, drought-linked ROS scavenging antioxidant system and induction of a variety of stress responsive genes and proteins (Farooq et al., 2009). There are three main categories into which the adaptation of plants to drought can be classified: (1) Drought escape: in this method plant completes its life cycle before beginning of drought and plant is subjected to dormancy before beginning of dry season (Levitt, 1980; Turner et al., 2001; Farooq et al., 2009). (2) Drought avoidance and phenotypic flexibility: this means the plants have the ability to maintain their normal water status under drought conditions

(Blum, 2005). It can be possible when plants acquire more and more water from the soil or reduce water loss through transpiration. (3) Drought tolerance: it is a very important feature for plants, which occurs when normal plant growth and metabolic activities are maintained even under water scarcity or water stress. This includes activities like osmotic adjustment, maintenance of root viability and membrane stability under dehydration, as well as acquisition of proteins and other metabolites that modified the structure directly or indirectly (Nilsen and Orcutt, 1996; Huang et al., 2014; Ngumbi and Kloepper, 2016).

1.4 Bacterial response against water stress in soil

Microbes present in the soil are affected by the drought stress due to osmotic stress, and this results in nucleic acid damages and also accumulation of free radicals, which triggers the protein denaturation and lipid peroxidation and finally death due to cell lysis (Dose et al., 1991; Vriezen et al., 2007; Berard et al., 2015). The tiny soil microorganisms are in close contact with soil water having a semi-permeable membrane, in case of drought the water potential decreases therefore preventing dehydration and death of cells. They accumulate the solutes and decrease their internal water potentials (Schimel et al., 2007). For survivability during the drought the soil microbes have many physiological mechanisms like exo-polysaccharide production, which helps to protect the cell along with the surrounding environment of the cell, spore production and accumulation of compatible solutes like proline, glycine betaine and trehlose to increase thermotolerance of enzymes, prevent enzymes from thermal denaturation and also maintain membrane integrity (Conlin and Nelson, 2007; Allison and Martiny, 2008; Welsh, 2000; Hecker et al., 1996; Feder and Hoffman, 1999; Rossi et al., 2012).

1.5 Bacterial-mediated drought tolerance in crop plants

Till date traditional plant breeding and genetic engineering are the two main approaches to develop the drought-tolerant cultivars for overcoming drought stress in crop plants, but these methods have some disadvantages like traditional plant breeding is time-consuming and labour-intensive, traits are limited to single crop species and also loss of desirable traits from genepool of host species. Although genetic engineering is faster theoretically, but it has its own set of limitations like time and labour and the ethical issues and laws are different in different countries for consumption of genetically modified crops (Barrow et al., 2008; Eisenstein, 2013; Ashraf, 2010; Philippot et al., 2013; Fedoroff et al., 2010). The microbial communities which are chiefly associated with plants like mycorrhizal fungi, nitrogen-fixing and plant growth-promoting rhizobacteria have been in great demand nowadays because of their capabilities to produce a broad range of metabolites and enzymes that help to tolerate the biotic and abiotic stresses in plants (Mayak et al., 2004a,b; Rodriguez and

Redman, 2008; Kloepper et al., 2004; Glick et al., 2007). Along with nutrient management and disease control PGPR also gained importance in managing the abiotic stress like water stress (Yang et al., 2009). To maintain water efficiency during drought the PGPR improves the physiological processes associated with drought resistance such as root characteristics to enhance water uptake, shoot growth, relative water content, osmotic adjustment for drought tolerance, antioxidants metabolism and many plant growth materials like auxins and others.

2. Rhizobacteria

Microorganisms present in the soil play a crucial role in maintaining the quality and health of the soil (Jeffries et al., 2003). Before application of microbial technology, it is important to have the basic knowledge about microbial ecology residing in the rhizosphere and also their function and diversity (Bolton et al., 1992). The rhizosphere is a very thin region surrounding the roots of the plant; it is the physical, chemical and biological environment to microbial community where diversified microbes flourish with their nutrition niche and perform their activities; and in turn they considerably leverage plant health and soil fertility, and also can help the host plant to reciprocate to abiotic stress conditions (Singh et al., 2016). Rhizobacteria are free living tiny microorganisms that have beneficial impact on plants located in 1 mm thin zones surrounding the plant root and struggle vigorously to make a significant impact on plants directly or indirectly (Fig. 6.1) against abiotic stresses (Kloepper et al., 1980; Kohler et al., 2006; Nadeem et al., 2012, 2014; Singh

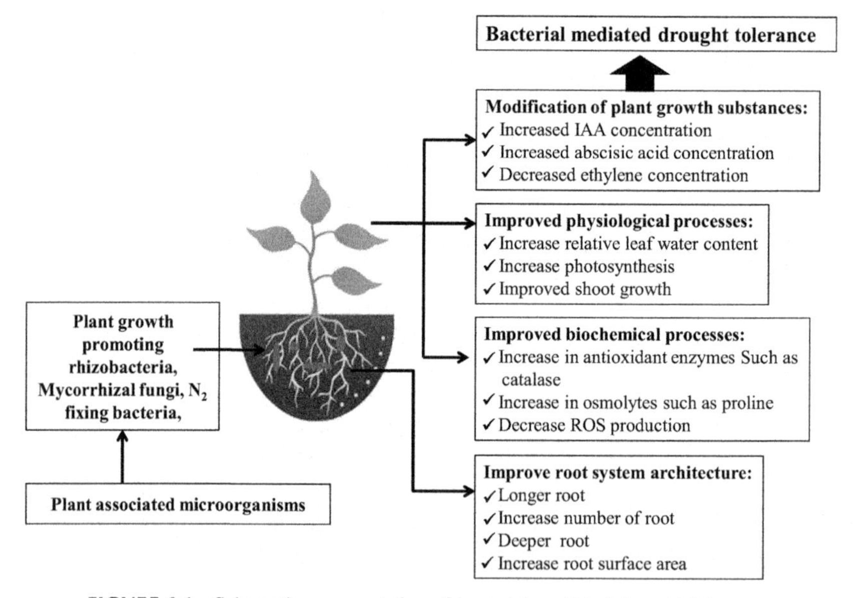

FIGURE 6.1 Schematic representation of bacterial-mediated drought tolerance.

et al., 2016). Therefore these bacteria can be useful for plants and their seeds to promote growth and development and also protects them from harmful impacts of water stress and also other type of abiotic stresses like salinity, temperature and heavy metals (Kloepper et al., 1980; Dimpka et al., 2009; Egamberdieva and Lugtenberg, 2014; Nadeem et al., 2014).

There are many types of rhizobacteria available which belong to genus *Bacillus, Pseudomonas, Serratia, Klebsiella, Azotobacter, Azospirillum, Arthrobacter, Paenibacillus, Burkholderia,* etc. (Zahir et al., 2008; Belimov et al., 2009; Dimpka et al., 2009 and Gupta et al., 2015). The phytohormones play a key role in alleviation of abiotic stresses like drought or water stress. Abscisic acid, auxins and gibberellins are the phytohormones that accelerate the rhizobacterial activity during stress conditions. Along with phytohormones the other compounds/metabolites synthesized by rhizobacteria like certain enzymes, organic acids, siderophore, osmolytes, nitric oxide and antibiotics also help in plant growth and development during normal as well as in stress conditions by various mechanisms (Patten and Glick, 1996; Chakraborty et al., 2006; Dimpka et al., 2009; Singh et al., 2016). Fig. 6.2 presents a schematic representation of various direct and indirect mechanisms used by rhizobacteria for plant growth promotion. One of the major functions that rhizobacteria perform for enhancement of plant growth during stress conditions is the hydrolysis of 1-aminocyclopropane-1-corboxylate (ACC) by production of an enzyme called ACC- deminase; ACC is instant precursor of ethylene. The ACC is used by rhizobacteria as a nitrogen source and decreases the amount of ACC in plant roots resulting in decreased level of ethylene in plants and enhances plant growth under stress conditions (Mayak et al., 2004a,b; Nadeem et al., 2014). In rhizobacteria stress management is a complicated multistep gene regulation process in which several genes are involved (Srivastava et al., 2008).

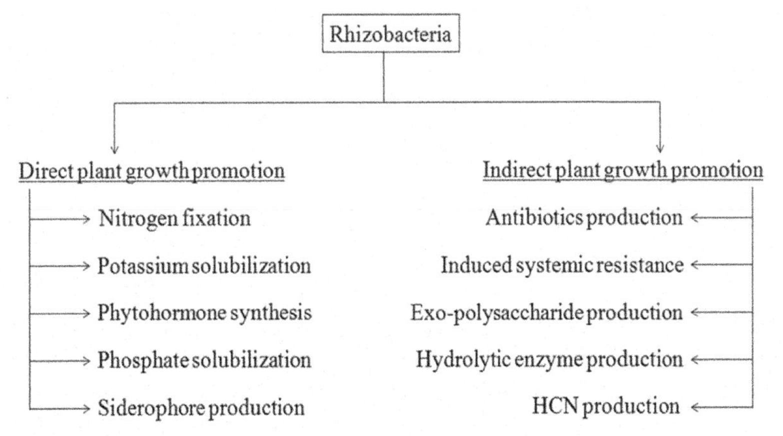

FIGURE 6.2 Schematic representation of various plant growth-promoting mechanisms used by rhizobacteria.

Pseudomonas a gram-negative, rod-shaped bacteria produces exopolysaccharide (EPS), which helps them to survive in the stress condition, EPS protects the microorganism from water stress by improving water retention capability and by regulating the diffusion of organic carbon (Wilkinson, 1958; Chenu and Roberson, 1996). EPS also helps the bacteria to adhere the root surface by forming a network of fibrillar material (Bashan and Holguin, 1997). *Azospirillum* is one of the EPS producing bacteria and has been reported for its role in soil aggregation; under stressful conditions the concentration and composition of microbial EPS could be changed. *Azospirillum brasilense* Sp245 has been reported to have high molecular weight carbohydrate complexes (liposaccharide-protein (LP) complex and polysaccharide-lipid (PL) complex), which could protect under extreme conditions such as desiccation (Bashan et al., 2004). In an experiment, Konnova et al. (2001) found that addition of these complex materials to a suspension of decapsulated cells of *A. brasilense* Sp245 increased the survival significantly under drought conditions. When an EPS producing rhizobial strain YAS34 was inoculated in the rhizosphere of sunflower, then a significant increase in root adhering soil per root tissue (RAS/RT) ratio was observed under drought conditions (Alami et al., 2000). A noteworthy correlation between amounts of EPS produced by cowpea *Bradyrhizobium* strain and its desiccation tolerance has also been observed (Hartel and Alexander, 1986). As the global climate change is increasing continuously, the duration, frequency and severity of drought in soyabean (*Glycine max* L.), cotton (*Gossypium hirsutum* L.) and corn (*Zea mays* L.) in most of the producing areas across the world are also predicted to continue to increase (IPCC, 2007; EEA, 2011). The desiccation-tolerant rhizobacteria may prove beneficial for these crops under such conditions.

3. Plant-microbe interaction

Rhizosphere is the ocean of microorganisms that associate with the plant; the microbes in rhizosphere show manifold profits, such as plant growth promoting rhizobacteria give strength to plants growth and intended increased resistance to abiotic stresses (Lugtenberg and Kamilova, 2009). For fruitful plant-microbe interaction a two-way appreciation and reciprocal reaction is required from both the plant as well as microorganisms. A considerable amount of carbohydrates synthesized during photosynthesis is transferred to microbes present in the roots. As nutrients like nitrate, phosphate, potassium and other minerals present in fixed amounts in soil are not accessible in free form, the rhizospheric microbes are very critical partners of plants that make nutrients accessible to plants. Rhizobacteria helps plants to achieve the induced systemic resistance (ISR) against pathogens, the rhizobacteria also boost the plant-inherited immunity and intended resistance to huge number of pathogens along with little effect on growth and yield (Zelicourt et al., 2013). The rhizospheric bacteria such as *Bacillus subtilis, Bacillus pumilus,*

Pseudomonas fluorescence, Pseudomonas putida, Serratia marcescens and Azospirillum brasiliense colonise the roots and protect the crop plants and trees against different types of biotic stress like foliar disease and abiotic stresses like drought, salinity, heavy metal toxicity and flooding, and thus facilitate plants to endure unfavourable environmental conditions (Van Loon, 2007; Mayak et al., 2004a,b; Glick et al., 2007; Zahir et al., 2008; Sandhya et al., 2009; Grover et al., 2010). The various mechanisms that rhizobacteria use to enhance the growth and development in plants are phosphate solubilisation, siderphore production, nitrogen fixation, production of certain enzymes like chitinase, glucanase, ACC-deaminase and production of plant growth hormones and organic acids (Glick et al., 2007; Berg, 2009; Hayat et al., 2010; Nadeem et al., 2014).

4. Screening and characterization of rhizobacteria for water-stress management in agriculture system

In tropical regions drought is one of the major limiting factors for reduction in crop yield. Drought destabilizes the plant-water communication at cellular level and affects the whole plant. Several morphological and physiological modifications occur in plants under water-scarce conditions such as drought. PGPR can play an important role in such modification to help the plants survive. Plants have their adaptive mechanisms to cope up with the drought stress; they can alter their root architecture. However, plants treated with PGPR could alter the root architecture, which resulted in increased root surface area and enhanced nutrient and water uptake; they also maintained the normal shoot growth and improved crop productivity (Kim et al., 2012; Sarma and Sasikala, 2014; Nagarajan and Nagarajan 2009; Farooq et al., 2009). Casanovas et al. (2002) showed that there is direct correlation between bacterial ABA synthesis and plant-relative water content in maize plants, in which it was found that stomatal closure was induced when maize plant was inoculated with *Azospirillum brasilense* BR11005spp. There are other mechanisms present in the microorganism by which they can ameliorate the drought stress in plants and help the plants to cope up with drought by maintaining normal growth and development during drought-stress condition. The production of ACC deaminase, EPS production, production of endogenous osmolytes such as proline and glycine betaine, capability of osmotolerance, plant-relative water content, as well as various PGPR traits to ameliorate the drought stress are supposed to be of considerable importance for stressed agriculture system. Therefore these characteristic features can be suitable for screening and characterization of drought-tolerant PGPR.

PGPR have the capability to produce various growth hormones in plants like auxins, gibberellins and cytokinins, and also growth inducing materials like siderophore and hydrogen cyanide, which help in growth and development of plants during water stress. In general, ethylene has a beneficial effect on plants; however, a sudden increment in ethylene biosynthesis during stress condition has a negative impact on plants and causes senescence (Bashan and de Bashan, 2005; Glick et al., 1998; Czarny et al., 2006; Saikia et al., 2018). Here an important fact about PGPR is that some strains have ACC deaminase activity, which can cleave the ethylene ACC into α-ketobutyrate and ammonia and overcome the inhibitory effect of ethylene in plants during stress conditions to promote growth and development (Glick et al., 1998). There are a number of scientific reports on plant health improvement by inoculation of ACC deaminase producing rhizobacteria in case of drought stress (Mayak et al., 2004a,b; Sarma et al., 2014). However, in some cases no direct evidence was found in support of the rhizobacterial ACC deaminase in stress tolerance; some other bacterial determinants might be there for alleviation of stress in affected plants. Some plant hormones related to rhizobacterial traits like IAA also have distinct role in evoking stress tolerance in plants (Egamberdieva et al., 2015).

EPS producing microorganisms present in the soil help to mitigate the drought stress in plants. Sandhya et al. (2009) investigated 212 fluorescent pseudomonads from various semiarid regions of India out of which 81 isolates showed mucoid growth in which 26 isolates could grow at water potential of −0.73 MPa and the EPS produced by these selected isolates was greater than the isolates grown under nonstressed conditions and further the EPS production was also increased by increasing the stress level; this inferred that the production of EPS in bacteria occurred due to stress (Roberson and Firestone, 1992). Konnova et al. (2001) suggested the role of *A. brasilense* Sp245 cells in desiccation tolerance. The EPS produced by the bacteria created a microsurrounding which retained water, and loss of moisture was slowed rather than outside the EPS surrounding, hence protecting the rhizobacteria from becoming dry under variations in the water potentials (Hepper, 1975; Wilkinson, 1958). *Pseudomonas putida,* an EPS producing strain, exhibited the production of IAA, P-solubilisation, ammonia production, siderophore production, HCN synthesis and gibberellic acid production during normal conditions as well in stressed conditions; on inoculating this bacteria with sunflower seeds, the results were significant, as the plant not only tolerated drought stress but also showed high root and shoot dry biomass (Sandhya et al., 2009).

Sarma et al. (2014) identified fluorescent pseudomonads from mung bean rhizosphere named GGRJ21 to discuss the drought-stress resistance in host plant. The GGRJ21 showed noticeable growth increment under various osmotic stress conditions such as −0.05, −0.15, −0.30, −0.49 and −0.73 MPa. The considerable production efficiency of endogenous osmolytes like proline and glycine betaine helped GGRJ21 to cope up with stressed conditions. The microbes withstand stress conditions mainly due to the production efficiency of glycine betaine, as well as L-amino acids and their D-isomers (Roberson and Firestone, 1992; Shahjee et al., 2002). During stress conditions a microenvironment is formed inside the bacterial cell mass by production of various osmolytes; the microenvironment protects the cells from further desiccation, the accumulation of glycine betaine and *N*-acetylglutaminylglutamine amide, which is a very important property of *P. aeruginosa* as a cytoplasmic osmoprotectant during osmotic stress (D'Souza-Ault et al., 1993). Therefore, under prolonged osmotic stress conditions heavy accumulation of proline and glycine betaine helps GGRJ21 to survive. Increased production of glycine betaine in plants due to PGPR has been reported by several workers (Yuwono et al., 2005).

Relative water content (RWC) in plant tissues acts as potential indicator to measure plant water status because it takes part in the metabolic activity in tissues. The water stress causes the decrease in RWC and reduction of sunflower shoot dry mass (DM). A decrease in RWC causes loss of turgor resulting in restricted cell expansion, and consequently reduced growth in plants (Ashraf, 2010; Lu et al., 2010; Castillo et al., 2013). It has been reported that the species which are better adjusted to dry environment have higher RWC (Jarvis and Jarvis, 1963; Ngumbi and Kloepper 2016). In most of the studies the ability of PGPR to help plants in drought tolerance has been investigated by measuring the RWC in treated and nontreated plants under drought stress, as a result PGPR treated plants showed comparatively higher RWC than nontreated plants. Therefore, it has been concluded that PGPR which helps the plants to survive during drought usually have higher RWC in plants. For example, sorghum plants treated with *Bacillus* spp. strain KB129 under drought stress showed an increment of 24% in RWC in comparison to nontreated plants (Grover et al., 2014).

5. Rhizobacteria as a tool for water-stress management in agriculture system

Drought stress is one of the most serious threats amongst the abiotic stress associated with plant growth and development, which affects the agricultural demands. Therefore to enhance the productivity under drought stress conditions, use of PGPR can be a beneficial approach, and the PGPR isolated from semi-arid zones having less precipitation with low moisture can be used as bioinoculant to enhance growth of the plants and provide resistance against soil water deficit. These PGPR can make the plants tolerant against drought stress. The PGPR can be used for water stress management in agricultural system by employing various strategies such as the following:

5.1 Improve root system architecture for water uptake

They improve root characteristics for water uptake. Root architecture is one of the most important parameters that decides how plants have to counter drought stress, it combines the spatial distribution of primary and secondary roots along with the diameter, length, number and the topology of root system

(Bacon et al., 2002; Huang et al., 2014; Vacheron et al., 2013). It has been shown that plants having more deeper and prolific roots are more drought tolerant than plants with lesser roots (Gowda et al., 2011; Ngumbi and Kloepper, 2016). According to Naseem and Bano (2014), *Alcaligenes faecalis* (AF3) when applied to maize seeds, showed the 10% increase in root length after 3 weeks in the drought-stressed PGPR treated plants.

5.2 Improve shoot growth

Inhibition of shoot growth is the result of drought stress; it benefits the plants by limiting the leaf area to decrease water loss due to evaporation, and allows them to distract the essential solutes to stress-related house-keeping function rather than growth, so inhibition of shoot growth seems to be an adaptive mechanism which help plants against drought stress and likely to decrease plant size hence inhibition of shoot growth could be a counterproductive response and limit the yield potential (Sinclair and Muchow, 2001; Aachard et al., 2006). Treatment of plants with PGPR usually increases shoot growth; as a result under drought stress PGPR could maintain normal shoot growth resulting in enhanced productivity, for example, Vardharajula et al. (2011) reported that inoculation of *Bacillus* spp. with corn plants improved the shoot growth. Furthermore there are many other crops in which PGPR is reported for enhanced shoot growth under drought stress such as sunflower (*Helianthus annuus* L.), sorghum (*Sorghum bicolor* L.), wheat, mung bean (*Vigna radiata* L.), green gram (*Vigna radiata* L.) and maize (Grover et al., 2014; Castillo et al., 2013; Saravanakumar et al., 2011; Sarma and Saikia, 2014; Sandhya et al., 2010; Ngumbi and Kloepper, 2016).

5.3 Relative water content

Relative water content could be the best standard to measure the plant water status as it is involved in most of the metabolic activities in the plant tissues; it depicts the decreased turgor pressure results in reduced plant growth because of cell shrinkage (Ashraf, 2010; Lu et al., 2010; Castillo et al., 2013). In most of the studies it has been found that PGPR treated plants comparatively had higher RWC than nontreated plants under drought stress; it concludes that PGPR which improves the survivability of the plants under drought conditions usually enhance the RWC of that plant. Several studies have shown that under drought stress, PGPR-treated plants maintained relatively higher RWC compared to nontreated plants, leading to the conclusion that PGPR strains that improve survival of plants under drought stress generally increase RWC in the plants. For example, Grover et al. (2014) reported that sorghum plants treated with PGPR, *Bacillus* spp. strain KB 129 under drought stress showed 24% increase in RWC over plants that were not treated with PGPR. Similar results have been demonstrated in maize (Sandhya et al., 2010; Vardharajula et al., 2011; Bano et al., 2013; Naveed et al., 2014; Naseem and Bano, 2014).

5.4 Osmotic adjustment

Osmotic adjustment operates at cellular level which helps plants to tolerate drought by protecting the enzymes, proteins, membranes and organelles of the cells from oxidative damages (Blum, 2005; Farooq et al., 2009; Huang et al., 2014). During the drought stress the osmotic adjustment constitutes the active accumulation of compatible organic and inorganic solutes like glycine, betaine and proline to maintain turgor pressure at cellular level and help plants to decrease the water potential without loss of actual water content (Kiani et al., 2007; Serraj and Sinclair, 2002). Drought stress is generally countered by proline, which accumulates in plants experiencing drought stress and acts in osmotic adjustment, and it also helps to stabilise the subcellular structure, scavenging free radicals and buffering cellular redox potential (Ashraf and Foolad, 2007; Hayat et al., 2012; Farooq et al., 2009; Huang et al., 2014).

Plants that are treated with PGPR have been found with an increment in proline content; this has been observed in maize, sorghum, potato, mung bean and *Arabidopsis* (Sandhya et al., 2010; Vardharajula et al., 2011; Naseem and Bano, 2014; Grover et al., 2014; Gururani et al., 2013; Sarma and Saikia, 2014; Cohen et al., 2015). A three to four fold increase was seen in leaf proline content when cucumber (*Cucumis sativa* L.) was treated with three PGPR strains (*Bacillus cereus* AR156, *Bacillus subtilis* SM21 and *Serratia* sp. XY21) in comparison to untreated controls (Wang et al., 2012).

5.5 Plant growth regulators

Plant growth regulators play an important role in controlling growth and development of the plants, and various phytohormones like auxins, cytokinins (CKs), ethylene (ET), gibberellins (GAs) and abscisic acid (ABA) are noted for this (Farooq et al., 2009). CKs and GAs promote plant growth while ethylene and abscisic acid inhibit growth (Taiz and Zeiger, 2010). When drought occurs, the secretion of inhibiting growth regulators is increased and inhibits the growth and regulates the water stock (Farooq et al., 2009). When plants are treated with PGPR, it promotes the plant growth even in presence of drought stress, by controlling and modifying the level of phytohormones (Dodd et al., 2010), through IAA signalling and decreasing the ET production (Glick et al., 1998; Contesto et al., 2010). These modifications are associated with bacterial-mediated drought tolerance when PGPR are applied to them.

Indole-3-acetic acid (IAA), a kind of auxin, is a very important factor for plant growth and development. IAA controls various functions in plant like elongation of roots and stems, orientation of root and shoot growth in response to gravity and light, tissue differentiation and initiation of lateral and adventitious roots (Glick, 1995). For example PGPR (*Pseudomonas putida* and *Bacillus megaterium*) used with clover (*Trifolium repens* L.) plants enhanced the root and shoot biomass and also improved water content during drought-stress condition; these increments corresponded with increased IAA production which was elicited by PGPR (Marulanda et al., 2009). Ethylene is synthesized during various types of stresses in plants like wounding, flooding, extreme temperature, pathogens and drought (Johnson and Ecker, 1998). The compound 1-aminocyclopropane-1-carboxylate (ACC) is a direct precursor of ET in plants; the regulation of ACC has been explored as the principle mechanism by which bacteria deploy the beneficial effects on plants during drought stress (Saleem et al., 2007). PGPR have ACC-deaminase activity that hydrolyzes the ACC in to alpha-ketobutyrate and ammonia rather than ethylene. As a result the ET and ACC level decreases, which in turn triggers the reduction of endogenous ET production hence eliminating the inhibitory effect of higher ethylene concentration in plants, and maintain the normal growth of the plants (Glick et al., 1998; Shaharoona et al., 2006). For example, *Achromobcater piechaudii* ARV8 a PGPR when treated with seedlings of tomato (*Solanum lycopersicum* L.) and pepper (*Capsicum annuum* L.) contributed to drought tolerance by reducing the production of ET (Mayak et al., 2004a,b). Therefore, isolation, screening and characterization of stress-tolerant PGPR are important for understanding their ecological behaviour in the rhizosphere and their utilization for sustainable agricultural practices. Thus the use of microbial consortia that can induce drought-stress tolerance and increased growth and development of plants might be beneficial for sustainable agricultural practices (Table 6.1).

6. Conclusion

The PGPR can help in enhancement of water-use efficiency in plants possibly through improved soil structure and plant growth promoting substances. Thus the use of microbial consortia, which can enhance water-use efficiency and also plant growth and development during normal as well as stressed conditions, might be very much beneficial for sustainable agriculture. Thus the isolation, screening and characterization of drought-tolerant rhizobacteria are important for exploring their ecological role in rhizosphere; apart from this, their incorporation is also essential for sustainable and ecofriendly agricultural practices. Moreover, it is necessary that bacterial strains and their consortia formulation must require further field testing and validation before being confirmed as bioinoculants to overcome drought stress and enhance water-use efficiency in the soil-based agro-ecosystems.

Acknowledgement

The author Ajay Shankar is grateful to University grant commission (UGC), New Delhi for providing the financial assistance as RGN-SRF.

TABLE 6.1 List of various rhizobacteria conferring tolerance against drought stress in crop plants.

Microorganisms	Crop	Results of PGPR treatments to plants	References
Pseudomonas fluorescence, Bacillus cereus	Rice (*Oryza sativa* L.)	Improved plant growth and activity of antioxidant defence systems	Gusain et al. (2015)
Bacillus sp. *Klebsiella* sp.	Pepper (*Capsicum annuum* L.)	Enhanced photosynthetic activity and biomass synthesis	Marasco et al. (2012)
Variovorax paradoxus	Pea (*Pisum sativum*)	Synthesis of ACC-deaminase and reduction in ethylene level	Belimov et al. (2009)
Bacillus sp.	Lettuce (*Lactuca sativa*)	Enhanced photosynthetic activity and biomass synthesis	Arkhipova (2007)
Pseudomonas putida, Achromobacter piechaudii	Tomato (*Lycopersicum esculentum*), Pepper (*Capsicum annuum*)	Plant maintained their growth under water-deficit condition, decreased ethylene concentration	Mayak et al. (2004a,b)
Azospirillum lipoferum AZ1,AZ9 and AZ45	Wheat (*Triticum aestivum*)	Enhance drought tolerance of plants by increasing RWC and decreasing leaf water potential	Arzanesh et al. (2011)
Bacillus spp. strains KB122, KB129, KB133 and KB142	Sorghum (*Sorghum bicolour*)	Increased shoot length, root dry biomass, RWC, sugar, chlorophyll and soil moisture	Grover et al. (2014)
Bacillus pumillus strain DH-11, Bacillus firmus strain 40	*Potato (Solanum tuberosum)*	Increased proline content and antioxidant enzymes, enhanced photosynthetic efficiency	Gururani et al. (2013)

References

Aachard, P., Cheng, H., De Grauwe, L., Decat, J., Schoutteten, H., Moritz, T., Van Der Straeten, D., Peng, J., Harberd, N.P., 2006. Integration of plant responses to environmentally activated phytohormonal signals. Science 311, 91–94.

Alami, Y., Achouak, W., Marol, C., Heulin, T., 2000. Rhizosphere soil aggregation and plant growth promotion of sunflowers by exopolysaccharide producing *Rhizobium* sp. strain isolated from sunflower roots. Appl. Environ. Microbiol. 66, 3393–3398.

Ali, S.k.Z., Sandhya, V., Rao, L.V., 2014. Isolation and characterization of drought-tolerant ACC deaminase and exopolysaccharide-producing fluorescent *Pseudomonas* sp. Ann. Microbiol. 64, 493–502.

Allison, S.D., Martiny, J.B.H., 2008. Resistance, resilience and redundancy in microbial communities. Proc. Natl. Acad. Sci. U.S.A. 105, 11512–11519.

American Meteorological Society (AMS), 2004. Statement on meteorological drought. Bull. Am. Meteorol. Soc. 85, 771–773.

Arkhipova, T.N., 2007. Cytokinin producing bacteria enhance plant growth in drying soil. Plant Soil 292, 305–315.

Arzanesh, M.H., Alikhani, H.A., Khavazi, K., Rahimian, H.A., Miransari, M., 2011. Wheat (*Triticum aestivum* L.) growth enhancement by *Azospirillum* sp. under drought stress. World J. Microbiol. Biotechnol. 27, 197–205.

Ashraf, M., 2010. Inducing drought tolerance in plants: some recent advances. Biotechnol. Adv. 28, 169–183.

Ashraf, M., Foolad, M.R., 2007. Roles of glycine betaine and proline in improving plant abiotic stress resistance. Environ. Exp. Bot. 59, 206–216.

Astorga, G.I., Melendez, L.A., 2010. Salinity effects on protein content, lipid peroxidation, pigments and proline in Paulownia imperialis and Paulowina fortune grown in vitro. Electron. J. Biotechnol. 13, 115.

Bacon, M.A., Davies, W.J., Mingo, D., Wilkinson, S., 2002. Root signals. In: Waisel, Y., Eshel, A., Kafkafi, U. (Eds.), Plant Roots: The Hidden Half. Monticello, New York, pp. 461–470.

Bano, Q., Ilyas, N., Bano, A., Zafar, N., Akram, A., Hassan, F., 2013. Effect of Azospirillum inoculation on maize (*Zea mays* L.) under drought stress. Pakistan J. Bot. 45, 13–20.

Barber, S.A., 1995, second ed. Soil Nutrient Bioavailability: A Mechanistic Approach. Wiley, New York.

Barrow, J.R., Lucero, M.E., Reyes-Vera, I., Havstad, K.M., 2008. Do symbiotic microbes have a role in plant evolution, performance and response to stress? Commun. Integr. Biol. 1, 69–73.

Bashan, Y., de-Bashan, L.E., 2005. In: Hillel, D.E. (Ed.), Bacteria/plant Growth-Promotion, pp. 103–115. Oxford.

Bashan, Y., Holguin, G., 1997. Azospirillum-plant relationships: environmental and physiological advances (1990–1996). Can. J. Microbiol. 43, 03–121.

Bashan, Y., Holguin, G., de-Bashan, L.E., 2004. Azospirillum-plant relationships: physiological, molecular, agricultural, and environmental advances. Can. J. Microbiol. 50, 521–577.

Beinsan, C., Camen, D., Sumalan, R., Babau, M., 2003. Study concerning salt stress effect on leaf area dynamics and chlorophyll content in four bean local landraces from Banat areas. Fac. Hortic. 119, 416–419.

Belimov, A.A., Dodd, I.C., Hontzeas, N., Theobald, J.C., Safronova, V.I., Davies, W.J., 2009. Rhizosphere bacteria containing 1-aminocyclopropane- 1-carboxylate deaminase increase yield of plants grown in drying soil via both local and systemic hormone signaling. New Phytol. 181, 413–423.

Bérard, A., Ben Sassi, M., Kaisermann, A., Renault, P., 2015. Soil microbial community responses to heat wave components: drought and high temperature. Clim. Res. 66, 243–264.

Berg, G., 2009. Plant–microbe interactions promoting plant growth and health: perspectives for controlled use of microorganisms in agriculture. Appl. Microbiol. Biotechnol. 84, 11–18.

Blum, A., 2005. Drought resistance, water-use efficiency, and yield potential-are they compatible, dissonant, or mutually exclusive? Aust. J. Agric. Res. 56, 1159–1168.

Bolton, H., Fredrickson, J.K., Eliot, L.F., 1992. Microbial ecology of the rhizosphere. In: Blaine Metting, F. (Ed.), Soil and Microbial Ecology. Marcel Dekker, New York, pp. 27–63.

Casanovas, E.M., Barassi, C.A., Sueldo, R.J., 2002. Azospirillum inoculation mitigates water stress effects in maize seedlings. Cereal Res. Commun. 30, 343–350.

Castillo, P., Escalante, M., Gallardo, M., Alemano, S., Abdala, G., 2013. Effects of bacterial single inoculation and co-inoculation on growth and phytohormone production of sunflower seedlings under water stress. Acta Physiol. Plant. 35, 2299–2309.

Chakraborty, U., Chakraborty, B., Basnet, M., 2006. Plant growth promotion and induction of resistance in *Camellia sinensis* by *Bacillus megaterium*. J. Basic Microbiol. 46, 186–195.

Chaves, M., Maroco, J., Pereira, J., 2003. Understanding plant responses to drought—from genes to whole plant. Funct. Plant Biol. 30, 239–264.

Chenu, C., Roberson, E.B., 1996. Diffusion of glucose in microbial extracellular polysaccharide as affected by water potential. Soil Biol. Biochem. 28, 877–884.

Choluj, D., Karwowska, R., Jasinska, M., Haber, G., 2004. Growth and dry matter partitioning in sugar beet plants (Beta vulgaris L.) under moderate drought. J. Plant Soil Environ. 50, 265–272.

Cohen, A., Bottini, R., Pontin, M., Berli, F., Moreno, D., Boccanlandro, H., Travaglia, C., Picocoli, P., 2015. Azospirillum brasilense ameliorates the response of *Arabidopsis thaliana* to drought mainly via enhancement of ABA levels. Physiol. Plantarum 153, 79–90.

Conlin, L.K., Nelson, H.C.M., 2007. The natural osmolytetrehalose is a positive regulator of the heat-induced activity of yeast heat shock transcription factor. Mol. Cell Biol. 27, 1505–1515.

Contesto, C., Milesi, S., Mantelin, S., Zancarini, A., Desbrosses, G., 2010. The auxin-signaling pathway is required for the lateral root response of Arabidopsis to the rhizobacterium Phyllobacterium brassicacearum. Planta 232, 1455–1470.

Czarny, J.C., Grichko, V.P., Glick, B.R., 2006. Genetic modulation of ethylene biosynthesis and signalling in plants. Biotechnol. Adv. 24, 410–419.

Dimpka, C., Weinand, T., Asch, F., 2009. Plant–rhizobacteria interactions alleviate abiotic stress conditions. Plant Cell Environ. 32, 1682–1694.

Dodd, I.C., Zinovkina, N.Y., Safronova, V.I., Belimov, A.A., 2010. Rhizobacterial mediation of plant hormone status. Ann. Appl. Biol. 157, 361–379.

Dose, K., Biegerdose, A., Kerz, O., Gill, M., 1991. DNA-strand breaks limit survival in extreme dryness. Orig. Life Evol. Biosph. 21, 177–187.

D'Souza-Ault, M.R., Smith, L.T., Smith, G.M., 1993. Roles of N-acetylglutaminylglutamine amide and glycine betaine in adaptation of *Pseudomonas aeruginosa* to osmotic stress. Appl. Environ. Microbiol. 59, 473–478.

Editorial, 2010. How to feed a hungry world. Nature 466, 531–532.

EEA, 2011. Europe's Environment—An Assessment of Assesments. EuropeanEnvironment Agency, Copenhagen.

Egamberdieva, D., Lugtenberg, B., 2014. Use of plant growth promoting rhizobacteria to alleviate salinity stress. In: Miransari, M. (Ed.), Use of Microbes for the Alleviation of Soil Stresses, vol. 1. Springer, New York, pp. 73–96.

Egamberdieva, D., Jabborova, D., Hashem, A., 2015. Pseudomonas induces salinity tolerance in cotton (*Gossypium hirsutum*) and resistance to Fusarium root rot through the modulation of indole-3-acetic acid. Saudi J. Biol. Sci. 522, 773–779.

Eisenstein, M., 2013. Discovery in a dry spell. Nature 501, S7–S9.

Farooq, M., Wahid, A., Kobayashi, N., Fujita, D., Basra, S.M.A., 2009. Plant drought stress: effects, mechanisms and management. Agron. Sustain. Dev. 29, 185–212.

Feder, M.E., Hofmann, G.E., 1999. Heat-shock proteins, molecular chaperones, and stress response: evolutionary and ecological physiology. Annu. Rev. Physiol. 61, 243–282.

Fedoroff, N.V., Battisti, D.S., Beachy, R.N., Cooper, P.J.M., Fischhoff, D.A., Hodges, C.N., Zhu, J.K., 2010. Radically rethinking agriculture for the 21st century. Science 327, 833–834.

Foley, J.A., Ramankutty, N., Brauman, K.A., Cassidy, E.S., Gerber, J.S., Johnston, M., Mueller, N.D., O'Connell, C., Ray, D.K., West, P.C., Balzer, C., Bennett, E.M., Carpenter, S.R., Hill, J., Monfreda, C., Polasky, S., Rockström, J., Sheehan, J., Siebert, S., Tilman, D., Zaks, P.M., 2011. Solutions to a cultivated planet. Nature 478, 337–342.

Gatehouse, A.M.R., Ferry, N., Edwards, M.G., Bell, H.A., 2011. Insect-resistant biotechcrops and their impacts on beneficial arthropods. Philos. Trans. R. Soc. B 366, 1438–1452.

Glick, B., 1995. The enhancement of plant growth by free-living bacteria. Can. J. Microbiol. 41, 109–117.

Glick, B.R., Penrose, D.M., Li, J., 1998. A model for lowering plant ethylene concentrations by plant growth promoting rhizobacteria. J. Theor. Biol. 190, 63–68.

Glick, B.R., Cheng, Z., Czarny, J., Cheng, Z., Duan, J., 2007. Promotion of plant growth by ACC deaminase-producing soil bacteria. In: Bakker, P.A.H.M., Raaijmakers, J.M., Bloemberg, G., Hofte, M., Lemanceau, P., Cooke, B.M. (Eds.), New Perspective and Approaches in Plant Growth Promoting Soil Bacteria. Springer, Netherlands, pp. 329–339.

Gowda, V.R.P., Henry, A., Yamauchi, A., Shashidhar, H.E., Serraj, R., 2011. Root biology and genetic improvement for drought avoidance in rice. Field Crop. Res. 122, 1–13.

Grover, M., Ali, S.Z., Sandhya, V., Rasul, A., Venkateswarlu, B., 2010. Role of microorganisms in adaptation of agriculture crops to abiotic stresses. World J. Microbiol. Biotechnol. 27, 1231–1240.

Grover, M., Madhubala, R., Ali, S.Z., Yadav, S.K., Venkateswarlu, B., 2014. Influence of *Bacillus* spp. strains on seedling growth and physiological parameters of sorghum under moisture stress conditions. J. Basic Microbiol. 54, 951–961.

Gupta, G., Parihar, S.S., Ahirwar, N.K., Snehi, S.K., Gupta, V.S., 2015. Plant growth promoting rhizobacteria (PGPR): current and future prospects for development of sustainable agriculture. J. Microb. Biochem. Technol. 72, 96–102.

Gururani, M.A., Upadhyaya, C.P., Baskar, V., Venkatesh, J., Nookaraju, A., Park, S.W., 2013. Plant growth-promoting rhizobacteria enhance abiotic stress tolerance in Solanum tuberosum through inducing changes in the expression of ROS- Scavenging enzymes and improved photosynthetic performance. J. Plant Growth Regul. 32, 245–258.

Gusain, Y.S., Singh, U.S., Sharma, A.K., 2015. Bacterial mediated amelioration of drought stress in drought tolerant and susceptible cultivars of rice (*Oryza sativa* L.). Afr. J. Biotechnol. 14, 764–773.

Hartel, P.G., Alexander, M., 1986. Role of extracellular polysaccharide production and clays in the desiccation tolerance of cowpea Bradyrhizobia. Soil Sci. Soc. Am. J. 50, 1193–1198.

Hayat, R., Ali, S., Amara, U., Khalid, R., Ahmed, I., 2010. Soil beneficial bacteria and their role in plant growth promotion: a review. Ann. Microbiol. 60, 579–598.

Hayat, S., Hayat, Q., Alyemeni, M.N., Wani, A.S., Pichtel, J., Ahmad, A., 2012. Role of proline under changing environments. Plant Signal. Behav. 7, 1456–1466.

Hecker, M., Schumann, W., Volker, U., 1996. Heat-shock and general stress response in *Bacillus subtilis*. Mol. Microbiol. 19, 417–428.

Hepper, C.M., 1975. Extracellular polysaccharides of soil bacteria. In: Walker, N. (Ed.), Soil Microbiology, a Critical Review. Wiley, New York, pp. 93–111.

Hsiao, A., 2000. Effect of water deficit on morphological and physiological characterizes in Rice (*Oryza sativa*). J. Agric. 3, 93–97.

Huang, B., DaCosta, M., Jiang, Y., 2014. Research advances in mechanisms of turfgrass tolerance to abiotic stresses: from physiology to molecular biology. Crit. Rev. Plant Sci. 33, 141–189.

IPCC, 2007. Climate change 2007: the physical science basis. In: Solomon, S., Qin, D., Manning, M., Chen, Z.S., Marquis, F.M., Averyt, K.B., Miller, H.L. (Eds.), Contribution of Working Group I to the Fourth Assessment Report of the Intergovernmental Panel on Climate Change. Cambridge University Press, Cambridge, p. 996.

Jaleel, C.A., Manivannan, P., Wahid, A., Farooq, M., Al-Juburi, H.J., Somasundaram, R., Vam, R.P., 2009. Drought stress in plants: a review on morphological characteristics and pigments composition. Int. J. Agric. Biol. 11, 100–105.

Jarvis, P.G., Jarvis, M.S., 1963. The water relations of tree seedlings. IV. Some aspects of the tissue water relations and drought resistance. Physiol. Plantarum 16, 501–516.

Jeffries, P., Gianinazzi, S., Perotto, S., Turnau, K., Barea, J.M., 2003. The contribution of arbuscular mycorrhizal fungi in sustainable maintenance of plant health and soil fertility. Biol. Fertil. Soils 37, 1–16.

Johnson, P.R., Ecker, J.R., 1998. The ethylene gas signal transduction pathway: a molecular perspective. Annu. Rev. Genet. 32, 227–254.

Kamara, A.Y., Menkir, A., Badu-Apraku, B., Ibikunle, O., 2003. The influence of drought stress on growth, yield and yield components of selected maize genotypes. J. Agric. Sci. 141, 43–50.

Kiani, S.P., Talia, P., Maury, P., Grieu, P., Heinz, R., Perrault, A., Nishinakamasu, V., Hopp, E., Gentzbittel, L., Paniego, N., Sarrafi, A., 2007. Genetic analysis of plant water status and osmotic adjustment in recombinant inbred lines of sunflower under two water treatments. Plant Sci. 172, 773–787.

Kim, Y.C., Glick, B.R., Bashan, Y., Ryu, C.M., 2012. In: Aroca, R. (Ed.), Enhancement of Plant Drought Tolerance by Microbes. Springer Verlag, pp. 383–413.

Kloepper, J.W., Leong, J., Teintze, M., Schroth, M.N., 1980. Enhanced plant growth by siderophores produced by plant growth promoting rhizobacteria. Nature 286, 885–886.

Kloepper, J.W., Ryu, C.M., Zhang, S., 2004. Induced systemic resistance and promotion of plant growth by Bacillus spp. Phytopathology 94, 1259–1266.

Kohler, J., Caravaca, F., Carrasco, L., Roldan, A., 2006. Contribution of *Pseudomonas mendocina* and *Glomus intraradices* to aggregates stabilisation and promotion of biological properties in rhizosphere soil of lettuce plants under field conditions. Soil Use Manag. 22, 298–304.

Konnova, S.A., Brykova, O.S., Sachkova, O.A., Egorenkova, I.V., Ignatov, V.V., 2001. Protective role of the polysaccharide containing capsular components of Azospirillum brasilense. Microbiology 70, 436–440.

Lafitte, H.R., Yongsheng, G., Yan, S., Lil, Z.K., 2007. Whole plant responses, key processes, and adaptation to drought stress: the case of rice. J. Exp. Bot. 58, 169–175.

Lesk, C., Rowhani, P., Ramankutty, N., 2016. Influence of extreme weather disasters on global crop production. Nature 529, 84–87.

Levitt, J., 1980. Responses of plants to environmental stresses: chilling, freezing, and high temperature stresses. In: Kozlowski, T.T. (Ed.), Water Radiation, Salt and Other Stresses. Academic, New York, pp. 93–186.

Lu, G.H., Ren, D.L., Wang, X.Q., Wu, J.K., Zhao, M.S., 2010. Evaluation on drought tolerance of maize hybrids in China. J. Maize Sci. 3, 20–24.

Lugtenberg, B., Kamilova, F., 2009. Plant growth promoting rhizobacteria. Annu. Rev. Microbiol. 56, 541–556.

Mancosu, N., Snyder, R.L., Kyriakakis, G., Spano, D., 2015. Water scarcity and future challenges for food production. Water 7, 975–992.

Marasco, R., Rolli, E., Ettoumi, B., Vigani, G., Mapelli, F., Borin, S., 2012. A drought resistance-promoting micro biome is selected by root system under desert farming. PLoS One 7, 1–14.

Marulanda, A., Barea, J.-M., Azcón, R., 2009. Stimulation of plant growth and drought tolerance by native microorganisms (AM fungi and bacteria) from dry environments: mechanisms related to bacterial effectiveness. J. Plant Growth Regul. 28, 115–124.

Mayak, S., Tirosh, T., Glick, B.R., 2004a. Plant growth-promoting bacteria confer resistance in tomato plants to salt stress. Plant Physiol. Biochem. 42, 565–572.

Mayak, S., Tirosh, T., Glick, B.R., 2004b. Plant growth-promoting bacteria that confer resistance to water stress in tomatoes and peppers. Plant Sci. 166, 525–530.

Nadeem, S.M., Ahmad, M., Zahir, A., Ashraf, M., 2012. Microbial ACC deaminase biotechnology: perspectives and applications in stress agriculture. In: Maheshwari, A.K. (Ed.), Bacteria in Agrobiology: Stress Management. Springer-Verlag, Berlin, pp. 141–185.

Nadeem, S.M., Ahmad, M., Zahir, Z.A., Javaid, A., Ashraf, M., 2014. The role of mycorrhizae and plant growth promoting rhizobacteria (PGPR) in improving crop productivity under stressful environments. Biotechnol. Adv. 32, 429–448.

Nagarajan, S., Nagarajan, S., 2009. Abiotic tolerance and crop improvement. In: Pareek, A., et al. (Eds.), Abiotic stress adaptation in plants. Springer, pp. 1–11.

Naseem, H., Bano, A., 2014. Role of plant growth-promoting rhizobacteria and their exopolysaccharide in drought tolerance in maize. J. Plant Interact. 9, 689–701.

Naveed, M., Mitter, B., Reichenauer, T.G., Wieczorek, K., Sessitsch, A., 2014. Increased drought stress resilience of maize through endophytic colonization by *Burkholderia phytofirmans* PsJN and *Enterobacter* sp. FD 17. Environ. Exp. Bot. 97, 30–39.

Nilsen, E.T., Orcutt, D.M., 1996. The Physiology of Plants Under Stress. Wiley, NewYork.

Ngumbi, E., Kloepper, J., 2016. Bacterial mediayed drought tolerance: current and future prospects, 105, 109–125.

Patten, C.L., Glick, B.R., 1996. Bacterial biosynthesis of indole-3-acetic acid. Can. J. Microbiol. 42, 207–220.

Philippot, L., Raaijmakers, J.M., Lemanceau, P., van der Putten, W.H., 2013. Going back to the roots: the microbial ecology of the rhizosphere. Nat. Rev. Microbiol. 11, 789–799.

Rampino, P., Pataleo, S., Gerardi, C., Perotta, C., 2006. Drought stress responses in wheat: physiological and molecular analysis of resistant and sensitive genotypes. Plant Cell Environ. 29, 2143–2152.

Roberson, E.B., Firestone, M.K., 1992. Relationship between desiccation and exopolysaccharide production in soil *Pseudomonas* sp. Appl. Environ. Microbiol. 58, 1284–1291.

Rodriguez, R., Redman, R., 2008. More than 400 million years of evolution and some plants still can't make it on their own: plant stress tolerance via fungal symbiosis. J. Exp. Bot. 59, 1109–1114.

Rossi, F., Potrafka, R.M., Pichel, F.G., De Philippis, R., 2012. The role of exopolysaccharides in enhancing hydraulic conductivity of biological soil crusts. Soil Biol. Biochem. 46, 33–40.

Saikia, J., Sarma, R.K., Dhandia, R., Yadav, A., Bharali, R., Gupta, V.K., Saikia, R., 2018. Alleviation of drought stress in pulse crop with ACC deaminase producing rhizobacteria isolated from acidic soil of North East India. Nature 8, 3560.

Saleem, M., Arshad, M., Hussain, S., Bhatti, A., 2007. Perspective of plant growth promoting rhizobacteria (PGPR) containing ACC deaminase in stress agriculture. J. Ind. Microbiol. Biotechnol. 34, 635–648.

Samarah, N.H., 2005. Effects of drought stress on growth and yield of barley. Agron. Sustain. Dev. 25, 145–149.

Sandhya, V., Ali, S.Z., Grover, M., Reddy, G., Venkateswarlu, B., 2009. Alleviation of drought stress effects in sunflower seedlings by the exopolysaccharides producing Pseudomonas putida strain GAP-P45. Biol. Fertil. Soils 46, 17–26.

Sandhya, V., Ali, S.Z., Grover, M., Reddy, G., Venkateswarlu, B., 2010. Effect of plant growth promoting *Pseudomonas* spp. on compatible solutes, antioxidant status and plant growth of maize under drought stress. Plant Growth Regul. 62, 21–30.

Saravanakumar, D., Kavino, M., Raguchander, T., Subbian, P., Samiyappan, R., 2011. Plant growth promoting bacteria enhance water stress resistance in green gram plants. Acta Physiol. Plant. 33, 203–209.

Sarma, R.K., Saikia, R., 2014. Alleviation of drought stress in mung bean by strain *Pseudomonas aeruginosa* GGRJ21. Plant Soil 377, 111–126.

Schimel, J.P., Balser, T.C., Wallenstein, M., 2007. Microbial stress response physiology and its implications for ecosystem function. Ecology 88, 1386–1394.

Selvakumar, G., Panneerselvam, P., Ganeshamurthy, A.N., 2012. Bacterial mediated alleviation of abiotic stress in crops. In: Maheshwari, D.K. (Ed.), Bacteria in Agrobiology: Stress Management. Springer-Verlag, Berlin Heidelberg, pp. 205–224.

Serraj, R., Sinclair, T.R., 2002. Osmolyte accumulation: can it really help increase crop yield under drought condition. Plant Cell Environ. 25, 331–341.

Shaharoona, B., Arshad, M., Zahir, Z.A., 2006. Effect of plant growth promoting rhizobacteria containing ACC-deaminase on maize (*Zea mays* L.) growth under axenic conditions and on nodulation in mung bean (*Vigna radiata* L.). Lett. Appl. Microbiol. 42, 155–159.

Shahjee, H.M., Banerjee, K., Ahmad, F., 2002. Comparative analysis of naturally occurring L-amino acid osmolytes and their disomers on protection of *Escherichia coli* against environmental stresses. J. Biosci. 27, 515–520.

Siddiqi, E.H., Ashraf, M., Hussain, M., Jamil, A., 2009. Assessment of intercultivar variation for salt tolerance in safflower (*Carthamustinctorius* L.) using gas exchange characteristics as selection criteria. Pakistan J. Bot. 41, 2251–2259.

Sinclair, T.R., Muchow, R.C., 2001. System analysis of plant traits to increase grain yield on limited water supplies. Agron. J. 93, 263–270.

Singh, A., Shankar, A., Gupta, V.K., Prasad, V., 2016. Rhizobacteria: Tools for the management of plant abiotic stresses. In: Shukla, P. (Ed.), Microbial Biotechnology: An Interdisciplinary Approach. CRC Press, Taylor & Francis Group, pp. 241–259.

Srivastava, S., Yadav, A., Seem, K., Mishra, S., Choudhary, V., Nautiyal, C.S., 2008. Effect of high temperature on *Pseudomonas putida* NBRI0987 biofilm formation and expression of stress sigma factor RpoS. Curr. Microbiol. 56, 453–457.

Taiz, L., Zeiger, E., 2010. In: Plant Physiology, fifth ed. Sinauer Associates Inc., Massachusetts.

Turner, N.C., Wright, G.C., Siddique, K.H.M., 2001. Adaptation of grain legumes(pulses) to water limited environments. Adv. Agron. 71, 193–271.

Vacheron, J., Desbrosses, G., Bouffaud, M., Touraine, B., Moenne-Loccoz, Y., Muller, D., Legendre, L., Wisniewski-Dye, F., Prigent-Combaret, C., 2013. Plant growth- promoting rhizobacteria and root system functioning. Front. Plant Sci. 4, 356.

Van Loon, L.C., 2007. Plant responses to plant growth-promoting rhizobacteria. Eur. J. Plant Pathol. 119, 243–254.

Vardharajula, S., Ali, S.Z., Grover, M., Reddy, G., Bandi, V., 2011. Drought-tolerant plant growth promoting *Bacillus* spp.: effect on growth osmolytes, and antioxidant status of maize under drought stress. J. Plant Interact. 6, 1−14.

Vinocur, B., Altman, A., 2005. Recent advances in engineering plant tolerance to abiotic stress: achievements and limitations. Curr. Opin. Biotechnol. 16, 123−132.

Vriezen, J.A.C., De Bruijn, F.J., Nüsslein, K., 2007. Responses of rhizobia to dessication in relation to osmotic stress, oxygen, and temperature. Appl. Environ. Microbiol. 73, 3451−3459.

Vurukonda, S.S.K.P., Sandhya, V., Shrivastava, M., Ali, S.k.Z., 2016. Enhancement of drought stress tolerance in crops by plant growth promoting rhizobacteria. Microbiol. Res. 184, 13−24.

Wang, C., Yang, W., Wang, C., Gu, C., Niu, D., Liu, H., Wang, Y., Guo, J., 2012. Induction of drought tolerance in cucumber plants by a consortium of three plant growth- promoting rhizobacterium strains. PLoS One 7 e52565.

Welsh, D.T., 2000. Ecological significance of compatible solute accumulation by microorganisms: from single cells to global climate. FEMS Microbiol. Rev. 24, 263−290.

Wilhite, D.A., Glantz, M.H., 1985. Understanding the drought phenomenon: the role of definitions. Water Int. 10, 111−120.

Wilkinson, J.F., 1958. The extracellular polysaccharides of bacteria. Bacteriol. Rev. 22, 46−73.

Yang, J., Kloepper, J.W., Ryu, C., 2009. Rhizosphere bacteria help plants tolerateabiotic stress. Trends Plant Sci 14, 1−4.

Yuwono, T., Handayani, D., Soedarsono, J., 2005. The role of osmotolerant rhizobacteria in rice growth under different drought conditions. Aust. J. Agric. Res. 56, 715−721.

Zahir, Z.A., Munir, A., Asghar, H.N., Shahroona, B., Arshad, M., 2008. Effectiveness of rhizobacteria containing ACC-deaminase for growth promotion of peas (*Pisum sativum*) under drought conditions. J. Microbiol. Biotechnol. 18, 958−963.

Zelicourt, A.D., Yousif, M.A., Hirta, H., 2013. Rhizosphere microbes as essential partners for plant stress tolerance. Molecular Plant 6, 242−245.

Chapter 7

Irrigation water quality appraisal using statistical methods and WATEQ4F geochemical model

Sudhir Kumar Singh[1], Ram Bharose[2], Jasna Nemčić-Jurec[3], Kishan S. Rawat[4], Dharmveer Singh[5]

[1]*K. Banerjee Centre of Atmospheric and Ocean Studies, IIDS, Nehru Science Centre, University of Allahabad, Prayagraj, Uttar Pradesh, India;* [2]*Department of Environmental Sciences & NRM, College of Forestry, SHUATS, Prayagraj, Uttar Pradesh, India;* [3]*Institute of Public Health of Koprivnica-Križevci County, Koprivnica, Croatia;* [4]*Centre for Remote Sensing and Geo-Informatics, Sathyabama University, Chennai, Tamilnadu, India;* [5]*Department of Chemistry, University of Allahabad, Prayagraj, Uttar Pradesh, India*

1. Introduction

Groundwater has become a major source of water in the agricultural sector in many countries where river and drainage systems are not sufficient. Therefore decline in the quality of groundwater for irrigation purpose is a matter of worry in recent years. Over chemical fertilisation is resulting in groundwater pollution (Ayers and Westcot, 1985; Rowe and Abdel-Magid, 1995; Singh et al., 2013; Gautam et al., 2015; Gautam et al., 2016; Nemčić-Jurec et al., 2017; Rawat et al., 2018a). Groundwater quality depends on the nature of recharging water, precipitation, subsurface, surface water and geo-hydrochemical activities (Keesari et al., 2016a,c), land-use/land-cover change (Amin et al., 2014; Srivastava et al., 2013; Gajbhiye et al., 2015; Nemčić'-Jurec et al., 2017; Rawat et al., 2017; Rawat and Singh, 2018b) and mining activities (Gautam et al., 2016; Keesari et al., 2016b; Gautam et al., 2018). Temporal changes in the constitution and origin of the water recharged and human factors frequently cause periodic changes in groundwater quality (Vasanthavigar et al., 2010; Milovanovic, 2007).

Groundwater quality has been investigated by many researchers (Singh et al., 2012, 2013; Thakur et al., 2015; Gautam et al., 2016; Moharir et al., 2019; Maliqi et al., 2020). Multivariate statistical techniques (MST) help in

Agricultural Water Management. https://doi.org/10.1016/B978-0-12-812362-1.00007-2

effective management of large and complex groundwater data. Principal Component Analysis (PCA) and Cluster Analysis (CA) are widely used statistical techniques for characterisation and evaluation of groundwater quality (Singh et al., 2009; Jacintha et al., 2016). Researchers have used MST to characterise and evaluate surface and groundwater quality and have found them to be very useful for studying the variations caused by different factors (Singh et al., 2009, 2012, 2013, 2015; Maliqi et al., 2020). Astel et al. (2006) have applied MST for the determination of chloro/bromo disinfection byproducts in drinking water at 12 locations in the Gdańsk area (Poland). Ujević Bošnjak et al. (2012) have used cluster factor and discriminant analyses to study the groundwater chemistry of Eastern Croatia and to identify the main geochemical processes responsible for high arsenic concentrations in analysed groundwater.

Generally water quality parameters (major cations as Na^+, Ca^{2+}, Mg^{2+}, K^+ and anions Cl^-, SO_4^{2-}, HCO_3^-, CO_3^{2-}, NO_3^-) are indicators of drinking water use while water quality indices such as Sodium Adsorption Ratio (SAR), Sodium Percentage (SSP; %Na), Residual Sodium Carbonate (RSC), Residual Alkalinity (RA), Kelly's ratio (KR) (or Kelly's index (KI)), Permeability Index (PI), Chloroalkaline indices (CAI1 and CAI2), Potential Salinity (PS), Magnesium hazard (MH) (or Magnesium Adsorption Ratio; MAR), Total Dissolved Salts (TDS) and Total Hardness (TH) based on primary water quality parameters are frequently used to determine quality of water for irrigation (Singh et al., 2015). The aim of the work was to evaluate the groundwater for irrigation uses.

2. Materials and methods

2.1 Study area

The study area of Allahabad district is a part of Quaternary alluvium and Vindhyan Plateau ranges from Proterozoic to recent geological origin. The basement in the area is formed by quartzite of Kaimur group and is unconformably overlain by Quaternary alluvium. The distribution of groundwater depends on the type of formation; in study area it ranges from unconsolidated to alluvial formation and groundwater occurs under unconfined to confined conditions. The groundwater is shallow in northern part of the district and there are deep aquifers in the southern part of the district, whose depth ranges from 2 to 20 m during pre-monsoon period. In the post-monsoon period the groundwater gets recharged and hence the shallow water tables recorded are at a depth of 1–18 m. In the Vindhyans formation (consolidated), the water tables are much dynamic in nature and it ranges from 3 to 10 m below ground level during pre-monsoon period and 2–8 mgbl in post-monsoon period. The seasonal fluctuations range from 1 to 4 m. The lithologs in Trans-Ganga and Trans-Yamuna area suggest that these areas have unique geological and hydrogeological characteristics. Further, the alluvial plain is also classified as younger and older alluvial plains; in the study area older plain has two subdivisons (Fig. 7.1), namely Banda older alluvium and Varanasi alluvium (Singh et al., 2015). The groundwater exploration studies have revealed a three-tier aquifer system in the alluvial area that has unique granular zones as follows: (1) I Aquifer Group (shallow) ranging from 0.0 to 110 mbgl; (2) II Aquifer Group (middle aquifer) ranging from 120 to 250 mbgl; and (3) III Aquifer Group (deeper aquifer) lying below 260 down to depth 400 mbgl. The aquifer material is medium-to-coarse-grained sand admixed with gravel at places. The tube wells which are located in the alluvium plains have high yield as of 2000−3000 Lpm (litre per minute), whereas the tube wells of Vindhyan regions have limited fractured zones, and sustain less water and exist down to 125 m only and yield low, as of 500−1000 Lpm; within this sandstone domain, a silica sand horizon exists having a thickness of 5−40 m which also contains groundwater (Singh et al., 2015).

2.2 Collection of sample and analysis

Water samples were drawn from wells/pumps of mentioned blocks during March 2018 in clean bottles without air bubbles. The sample bottles were rinsed thoroughly with sample water, sealed tightly and labelled in the field after collection. The pumps were continuously pumped prior to avoid contamination from the surface. The depth of sampling location was 140−150 feet deep in these blocks. The samples were collected as per the standard methods of water examination (APHA, 1998). The 90 samples were collected from 10 sites of each block in triplicates systematically and analysed for 10 parameters, pH, EC, total dissolved solid, calcium hardness, magnesium hardness, total hardness, alkalinity, chloride, sulphate and nitrate, as per standards laid by APHA (1998). The detailed methodology showed in Fig. 7.2.

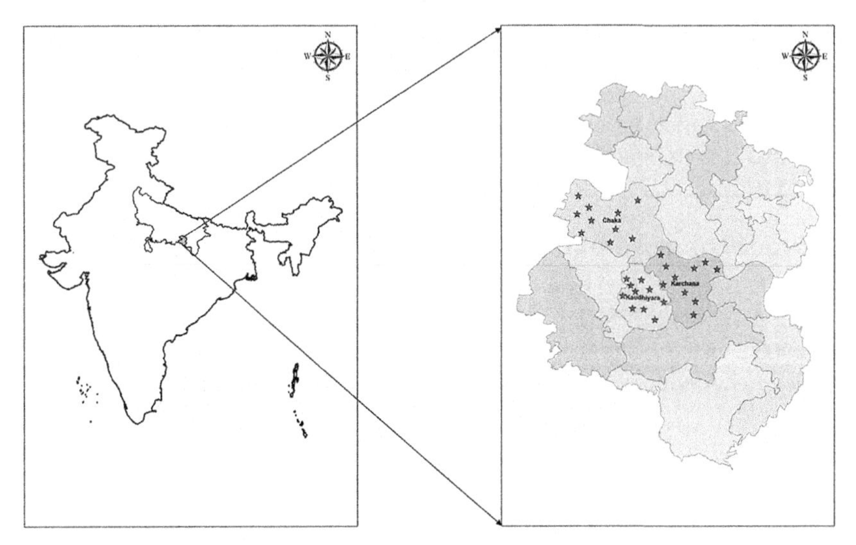

FIGURE 7.1 Study area map. The sampling locations are showed by (*).

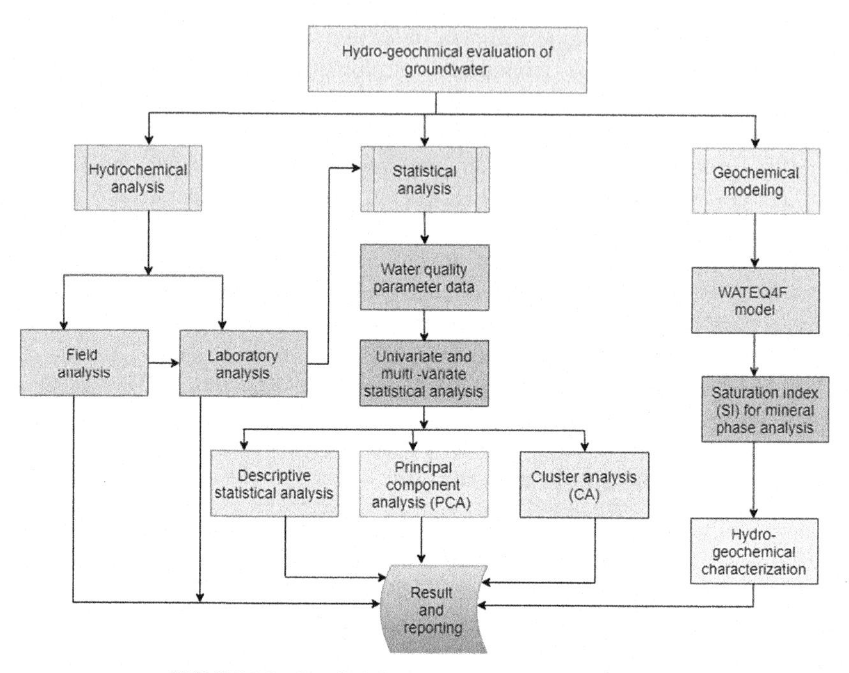

FIGURE 7.2 Detailed flow chart of adopted methodology.

2.3 Irrigation indices

2.3.1 Sodicity

Sodicity is the effect of irrigation water and can alter the chemical and physical properties of the soil due to an accumulation of Na^+. Excess of Na^+ can affect plants in three ways: (i) by degrading soil structure after each rainfall and irrigation due to crust formation which reduces water movement (permeability) and aeration in the soil, (ii) toxic effects when absorbed by leaves/roots (iii) K^+ and Ca^{2+} deficiencies may arise if the soil or irrigation water has a high concentration of Na^+. Therefore, evaluation of the sodicity hazard of irrigation water is important.

2.3.2 Sodium adsorption ratio (SAR)

The SAR is the relative ratio of Na^+ ions to Ca^{2+} and Mg^{2+} ions in the water sample (Rawat et al., 2018c). The SAR is used to estimate the potential of Na^+ to accumulate in the soil primarily (water movement) at the expense of Ca^{2+}, Mg^{2+} and K^+ as a result of regular use of sodic water. It is formulated as Eq. (7.1):

$$SAR = \frac{Na^{+}}{\sqrt{\frac{(Ca^{2+} + Mg^{2+})}{2}}} \tag{7.1}$$

where Na^{+}, Ca^{2+} and Mg^{2+} are in meq/L

On the basis of SAR ranges, irrigation water can be classified into four classes as SAR<10 (Ideal or Excellent), 10–18 (Good), 18–26 (Doubtful), >26 (Unsuitable) (Singh et al., 2015).

SAR also influences percolation time of water in the soil. Therefore the low value of SAR of irrigation water is desirable.

2.3.3 Residual sodium carbonate (RSC)/Residual alkalinity (RA)

RSC represents the amount of sodium carbonate ($NaCO_3$) and sodium bicarbonate ($NaHCO_3$) in the irrigation water, if the concentration of carbonate (CO_3^{2-}) and bicarbonate (HCO_3^{-}) ions exceeds the concentrations of Ca^{2+} and Mg^{2+} ions then alkalinity of water increases (Raghunath, 1987). If the carbonates are less than alkaline earths (Ca^{2+}+Mg^{2+}), it outlined the residual $NaCO_3$ is absent. Generally, RSC is expressed as milliequivalents per litre (meq/L) of $NaCO_3$. An excess of CO_3^{2-} and HCO_3^{-} causes precipitation of soil Ca^{2+} and Mg^{2+}, impairing the soil structure as well as potentially activating soil sodium. On the basis of RSC range sodium hazard has been classified into three classes as follows: RSC<1.25 (low), 1.25–2.5 (medium) and >2.5 (high) (Singh et al., 2015). RSC is expressed as Eq. (7.2):

$$RSC = (HCO_3^{-} + CO_3^{2-}) - (Ca^{2+} + Mg^{2+}) \tag{7.2}$$

A high range of RSC in irrigation water means an increase in the adsorption of sodium in the soil. Water having RSC>5 has not been recommended for irrigation because of damaging effects on plant growth. Generally any source of water whose RSC is higher than 2.5 is not considered suitable for agriculture purpose, and water <1.25 is recommended as safe for irrigation purpose. A negative value of RSC reveal that concentration of Ca^{2+} and Mg^{2+} are in excess. A positive RSC denotes that Na^{+} existence in the soil is possible. RSC calculation is also important in context to calculate the required amount of gypsum or sulphuric acid per acre-foot in irrigation water to neutralise residual carbonates effect.

2.3.4 Percent sodium (%Na) or sodium hazard

The %Na is also used in classifying water for irrigation purpose. Na^{+} concentration is important in categorising any source of water for irrigation because Na^{+} makes chemical bonding with soil to reduce water movement capacity of the soil (Ayers and Westcot, 1985). Percent Na^{+} concentration is a factor to assess its suitability for irrigation purposes (Wilcox, 1948); Na^{+} reacts with CO_3^{2-} and forms alkaline soils, while Na^{+} reacts with chloride ions and forms saline soils. Sodium affected soil (alkaline/saline) retards crop growth (Todd, 1980). When Na^{+} concentration in irrigation water is high, the ions tend towards the clay particles, removing Ca^{2+} and Mg^{2+} ions by a base-exchange reaction. In this condition, air and water cannot move freely or are restricted during wet conditions, and such soils become hard when dry (Collins and Jenkins, 1996; Saleh et al., 1999). The %Na can be calculated using Eq. (7.3):

$$\%Na = \frac{Na^{+}}{Ca^{2+} + Mg^{2+} + Na^{+} + K^{+}} \times 100 \quad (7.3)$$

The classification of water based on %Na is as follows: Excellent (<20%), Good (20%–40%), Permissible (40%–60%), Doubtful (60%–80%) and Unsuitable (>80%) (Khodapanah et al., 2009).

2.3.5 Kelly's ratio (KR) or Kelly's index (KI)

Kelly (1940) has introduced another factor to assess quality and classification of water for irrigation purpose based on the concentration of Na^{+} against Ca^{2+} and Mg^{2+}. It can be calculated using Eq. (7.4)

$$KR = \frac{Na^{+}}{Ca^{2+} + Mg^{2+}} \quad (7.4)$$

KR/KI > 1 indicate an excess level of Na^{+} in water. Therefore, water with a KI ≤ 1 has been recommended for irrigation, while those with >1 are not recommended for irrigation due to alkali hazards (Karanth, 1987).

2.3.6 Permeability index (PI)

The permeability index (PI) is an indicator to study the suitability of water for irrigation purpose. Water movement capability in soil (permeability) is influenced by the long-term use of irrigation water (with a high concentration of salt) as it is affected by Na^{+}, Ca^{2+}, Mg^{2+} and HCO_3^{-} ions of the soil. PI formula has been developed by Doneen (1964a,b), to assess water movement capability in the soil as the suitability of any kind of source of water for irrigation and it is formulated as Eq. (7.5):

$$PI = \frac{Na^{+} + \sqrt{HCO_3^{-}}}{Ca^{2+} + Mg^{2+} + Na^{+}} \times 100 \quad (7.5)$$

According to Doneen (1964a,b), PI can be categorised into three classes, class I (>75%, Suitable), class II (25%–75%, Good) and class III (<25%, Unsuitable).

2.3.7 Magnesium adsorption ratio (MAR)

Usually, alkaline earths (Ca^{2+} and Mg^{2+}) are in an equilibrium state in groundwater. Both Ca^{2+} and Mg^{2+} ions are linked with soil friability and aggregation, but both are also essential nutrients for the crop. The high value of Ca^{2+} and Mg^{2+} in water can increase soil pH (therefore leading to saline nature of the soil; Joshi et al., 2009), resulting in decrease in the availability of phosphorous (Al-Shammiri et al., 2005). Magnesium in groundwater affects the soil quality by converting it into alkaline, and decreases the crop yield (Singh et al., 2013; Gautam et al., 2015). According to agriculturists, excess amount of Mg^{2+} ions in water damages the soil quality, which causes low crop production (Ramesh and Elango, 2011; Narsimha et al., 2013). Szabolcs and Darab (1964) projected MH value for irrigation water and it is calculated using Eq. (7.6):

$$MH = \frac{Mg^{2+}}{Ca^{2+} + Mg^{2+}} \times 100 \quad (7.6)$$

Where, agnesium adsorption ratio or magnesium hazard (MH) is magnesium hazard, $MH > 50$ is not recommended for irrigation purposes (Khodapanah et al., 2009).

2.3.8 Potential salinity (PS)

PS is another water quality parameter based index (Doneen, 1964a,b) for categorisation of water for agriculture use. $PS < 3$ meq/L is an indication of the suitability of water for irrigation. The PS can be calculated using following Eq. (7.7):

$$PS = Cl^{-} + 0.5 \times SO_4^{2-} \quad (7.7)$$

2.3.9 Chloroalkaline indices (CAI1 and CAI2)

Information about changes in chemical composition of the groundwater during underground travel is also vital (Sastri, 1994). The chemical reaction in which ion exchange between the groundwater and the aquifer occurs during the movement and rest condition of water can be analysed through the chloroalkaline indices. The CAI1 and CAI2 are evaluated as expressed by Eqs (7.8) and (7.9):

$$CAI1 = \frac{Cl^{-} - (Na^{+} + K^{+})}{Cl^{-}} \quad (7.8)$$

$$CAI2 = \frac{Cl^- - (Na^+ + K^+)}{(SO_4^{2-} + CO_3^{2-} + HCO_3^- + NO_3^-)} \tag{7.9}$$

The CAI1 and CAI2 indices may be negative or positive depending on the exchange process of Na^+ and K^+ from the rock with Mg^{2+} and Ca^{2+} in water and vice versa. If a direct exchange process (DEP) happens between Na^+ and K^+ in water with Mg^{2+} and Ca^{2+} in rocks, CAI ratio will be positive. If a reverse exchange process (REP) occurs (Na^+ and K^+ in water with Mg^{2+} and Ca^{2+} in rocks), then CAI ratio will be negative.

2.3.10 Corrosivity ratio (CR)

The corrosivity ratio gives information about water supply. Any source of water with CR<1 is recommended to be transported through any kind of pipes, whereas CR>1 shows corrosive nature of water and therefore such water is not to be transported through metal pipes (Balasubramanian, 1986; Aravindan, 2004). The CR can be estimated using Eq. (7.10):

$$CR = \frac{\left(\frac{Cl^-}{35}\right) + 2\left(\frac{SO_4^{2-}}{96}\right)}{\left(\frac{CO_3^{2-} + HCO_3^-}{100}\right)} \tag{7.10}$$

The rate of corrosion depends upon some physical parameters like pressure, temperature and rate of flow of water. In addition to the higher value of Cl^- , and SO_4^{2-} also increases the corrosion rate.

2.3.11 Total dissolved salts (TDS)

In natural water dissolved solids are mixtures of CO_3^{2-}, HCO_3^-, Cl^-, SO_4^{2-} and PO_4^{3-}. Weathering or dissolution of soil and rocks generates ions in water (Singh et al., 2013). After evaporation of water, accumulation of salts at the root zone makes obstacles and plants are not capable to absorb water from soil, resulting in moisture stress (Modi, 2000). For irrigation the TDS has been classified as TDS <450 mg/L—preferred for irrigation, TDS >450–2000 mg/L—slight to moderate and TDS >2000 mg/L—unsuitable for agricultural purpose (FAO, 2006).

2.3.12 Total hardness (TH)

Water hardness is a result of existence of divalent metallic cations (Ca^{2+} and Mg^{2+}), and it can be calculated as the sum of Ca^{2+} and Mg^{2+} concentration as meq/L equivalent to $CaCO_3$ (Todd, 1980) and expressed by (Eq. 7.11):

$$TH = 2.5 \times Ca^{2+} + 4.1 \times Mg^{2+} \tag{7.11}$$

The high value of TH is due to encrustation on water supply systems (pipes) for distribution. Moderate values of TH is useful for plumbing system from corrosion. Total hardness as 100 mg/L provides good control over corrosion and is the usually acceptable limit. TH is usually classified (EPA, 1986) as soft (0–60 mg/L), moderately hard (60–120 mg/L), hard (120–180 mg/L) and very hard (>180 mg/L).

2.3.13 Classification of groundwater

Soltan (1998) has suggested the classification of different sources of water facies based on the concentration (meq/L) of a particular ion. Soltan (1999) has suggested a new classification method based on base exchange indices and it can be calculated using Eq. (7.12):

$$r_1 = \frac{Na^+ - Cl^-}{SO_4^{2-}} \tag{7.12}$$

The groundwater of study area can be categorized as $Na^+{-}HCO_3^-$ type if $r_1 > 1$ and $Na^+{-}SO_4^{2-}$ type if $r_1 < 1$.

The groundwater has also been categorised based on meteoric genesis index, which was calculated (Soltan, 1999) using Eq. (7.13):

$$r_2 = \frac{(K^+ + Na^+) - Cl^-}{SO_4^{2-}} \tag{7.13}$$

If the value of $r_2 < 1$, it means the groundwater is of deep meteoric percolation (DMP) type, and if the value of $r_2 > 1$ means water is of shallow meteoric percolation (SMP) type.

2.4 Statistical analysis

The spatial distribution of variable was calculated using univariate statistical analysis. The effect of the variables and their statistical correlation at a given time was calculated by multivariate analysis. Principal Component Analysis (PCA) was applied on data standardised through z-scale transformation to avoid misclassification due to wide differences in data-dimensionality (Singh et al., 2009). Correlation matrix was used to analyse dependency among the physicochemical parameters (Maliqi et al., 2020). Cluster analysis (CA) was also applied, which groups the parameters into different classes or clusters based on similarities within a class and dissimilarities between the classes (Singh et al., 2009). Descriptive and multivariate statistics were applied for the classification, modelling and interpretation of large datasets. Chemometric analysis was performed using Principal Component Analysis (PCA) and Cluster Analysis (CA) with Statistical Package for Social Sciences (SPSS) software package (SPSS Inc., version 18).

2.5 WATEQ4F model

WATEQ4F is a geochemical model, which computes the major and trace elements speciation and mineral saturation for natural water. It models the thermodynamic speciation of major and important minor inorganic ions and complex species in given water samples and in-situ measurements of temperature, pH and redox potential. From this model, the states of reaction of the water with solid and gaseous phases are calculated. The saturation index (SI) of a mineral is obtained from Eq. (7.14) (Appelo and Postma, 2005).

$$SI = \log\left(\frac{IAP}{K_t}\right) \tag{7.14}$$

where IAP and K_t is the ion activity product and the solubility product (tabulated in the WATEQ4F database for a wide variety of mineral phases) (Ball and Nordstrom, 1991).

When SI is below zero, the water is undersaturated with respect to the mineral in question. When SI is zero, it means water is in equilibrium with the mineral, whereas SI greater than zero means a supersaturated solution with respect to the mineral in question.

3. Result and discussions

3.1 Hydrogeochemistry of water

An overall descriptive statistics of the physicochemical properties is shown in Table 7.1. The pH of water samples of Chaka, Karchana and Kaundhiyaar blocks are presented in Table 7.2 and it show that pH of the water samples ranges from 7.89 to 8.33 in Chaka, 7.82 to 8.52 in Karchana and 7.98 to 8.43 in Kaundhiyara block. The pH of water samples of all blocks wide range. Irrigation water pH (Normal Range 6.5 − 8.4) is not an acceptable criterion as the pH of tending to be buffered by soil and crops can tolerate a wider range of pH (Zaman et al., 2018). As pH the water samples are slightly alkaline in nature. Water with a high pH level indicate that high level of alkaline minerals presence in a water. The most prevalent mineral compound causing alkalinity is calcium carbonate, which comes from rocks such as limestone or can be leached from dolomite and calcite in the soil (Webster et al., 2013).

EC of water sample range from 545 to 1289 μS/cm in Chaka block, 564 to 948 μS/cm in Karchana and 723 to 987 μS/cm in Kaundhiyara block (Table 7.2). The EC of water samples of all blocks are within permissible limit. EC values can be used to estimate the dissolved solids which may affect the taste of water and suitability for various uses. Higher conductivity indicates higher dissolved solids in water. As the concentration of dissolved salts (usually salts of sodium, calcium and magnesium, bicarbonate, chloride, and sulphate) increases in water, electrical conductivity increases (Kelin et al., 2005).

TABLE 7.1 Descriptive statistics of the physicochemical properties of the samples in the study area.

	Range	Minimum	Maximum	Mean		Std. deviation	Variance	Skewness		Kurtosis	
Variables	Statistic	Statistic	Statistic	Statistic	Std. error	Statistic	Statistic	Statistic	Std. error	Statistic	Std. error
pH	0.70	7.82	8.520	8.11	0.03	0.19	0.09	0.43	0.42	−0.91	0.83
EC	744.00	545.00	1289.00	829.23	28.60	156.68	24,550.59	0.64	0.42	1.76	0.83
TDS	517.00	235.00	752.00	370.00	19.45	106.58	11,360.13	1.88	0.42	5.07	0.83
Sodium	121.00	16.00	137.00	75.63	5.87	32.15	1034.10	0.15	0.42	−0.47	0.83
Potassium	53.00	3.000	56.00	13.70	3.311	18.13	328.97	1.58	0.42	0.69	0.83
Ca_Hardness	184.00	70.00	254.00	150.86	10.32	56.52	3195.15	−0.02	0.42	−1.42	0.83
Mg_Hardness	293.00	339.00	632.00	445.73	16.07	87.63	7680.54	0.55	0.42	−1.07	0.83
Total hardness	294.00	510.00	804.00	597.10	11.41	62.54	3911.74	1.28	0.42	2.71	0.83
Alkanity	288.00	355.00	643.00	433.26	11.94	65.40	4277.85	1.13	0.42	2.05	0.83
Chloride	250.00	35.00	285.00	105.56	9.18	50.31	2531.49	1.77	0.42	4.94	0.83
Sulphate	831.00	114.00	945.00	617.733	33.77	184.96	34,212.75	−0.58	0.42	0.57	0.83
Nitrate	4.58	6.09	10.67	8.45	0.194	1.06	1.13	−0.47	0.42	0.07	0.83
Anhydrite	1.314	−2.34	−1.02	−1.34	0.050	0.27	0.07	−1.52	0.42	4.36	0.83

Continued

TABLE 7.1 Descriptive statistics of the physicochemical properties of the samples in the study area.—cont'd

	Range	Minimum	Maximum	Mean		Std. deviation	Variance	Skewness		Kurtosis	
Variables	Statistic	Statistic	Statistic	Statistic	Std. error	Statistic	Statistic	Statistic	Std. error	Statistic	Std. error
Aragonite	0.608	0.66	1.27	0.92	0.026	0.14	0.02	0.21	0.42	0.07	0.83
Brucite	1.452	−3.41	−1.96	−2.75	0.08	0.43	0.19	0.32	0.42	−1.11	0.83
Calcite	0.607	0.810	1.41	1.07	0.026	0.14	0.02	0.20	0.42	0.07	0.83
Dolomite_d	1.082	1.95	3.04	2.43	0.05	0.28	0.08	0.08	0.42	−0.70	0.83
Dolomite_c	1.082	2.50	3.59	2.98	0.05	0.28	0.08	0.08	0.42	−0.70	0.83
Epsomite	0.863	−3.46	−2.60	−2.84	0.03	0.16	0.02	−1.79	0.42	5.20	0.83
Gypsum	1.314	−2.12	−0.80	−1.12	0.05	0.27	0.07	−1.57	0.42	4.67	0.83
Magnesite	0.735	0.98	1.72	1.33	0.04	0.22	0.05	0.11	0.42	−1.45	0.83

TABLE 7.2 Physio-chemical characteristics of water quality of (a) Chaka block (b) Karchana block and (c) Kaundhiyara block.

Site	pH	EC	TDS	Na	Ca hardness	Mg hardness	K	Total hardness	Alkalinity	Chloride	Sulphate	Nitrate
(a)												
C 1 (Chak Imam Ali)	8.13	739	370	101	254	354	4	608	380	110	501	8.2
C 2 (NH 30, Dandi)	8.05	881	324	103	182	416	5	599	355	75	469	9.63
C 3 (Chaka Rana Tiwari)	8.01	616	301	65	171	377	3	548	375	104	945	9.17
C 4 (Naini Taluka)	7.95	756	440	66	220	415	4	635	435	105	703	8.92
C 5 (Dabhaon)	7.92	545	416	16	198	365	9	563	503	105	817	6.09
C 6 (Chak Garibdas)	8.11	608	235	17	171	362	11	534	373	65	753	9.41
C 7 (Tignauta)	7.89	827	292	110	151	505	5	657	361	89	612	8.78
C 8 (Naini Dadri)	8.33	829	311	113	102	544	6	646	498	160	764	8.12
C 9 (Indalpur Road)	8.06	1289	436	119	172	632	46	804	511	105	456	8.83
C 10 (Sandwa, Naini)	8.02	1143	351	125	114	496	47	611	530	95	755	6.74
Minimum	7.89	545	235	16	102	354	3	534	355	65	456	6.09

Continued

TABLE 7.2 Physio-chemical characteristics of water quality of (a) Chaka block (b) Karchana block and (c) Kaundhiyara block.—cont'd

Site	pH	EC	TDS	Na	Ca hardness	Mg hardness	K	Total hardness	Alkalinity	Chloride	Sulphate	Nitrate
Maximum	8.33	1289.00	440.00	125.00	254.00	632.00	47.00	804.00	530.00	160.00	945.00	9.63
Average	8.05	823.30	347.60	83.50	173.50	446.60	14.00	620.50	432.10	101.30	677.50	8.39
Standard Deviation	0.13	235.91	67.81	40.65	45.32	93.62	17.30	76.46	71.22	25.32	162.90	1.15
(b)												
K 1 (Baraha)	8.52	918	235	67	156	392	56	548	355	45	383	9.9
K 2 (Hindupur, Karchana)	7.96	914	246	66	129	400	53	529	356	35	421	8.45
K 3 (Bardaha)	8.05	814	372	34	205	398	47	603	455	110	315	7.43
K 4 (Ligadahiya)	8.03	719	307	32	210	383	45	593	445	60	419	8.52
K 5 (Chanaini)	7.85	948	371	92	210	346	3	556	481	114	664	6.66
K 6 (Bhaterawa)	8.25	889	284	95	192	339	3	532	408	140	765	8.95
K 7 (Ghatwa)	7.96	886	431	49	232	357	4	589	448	115	687	8.15
K 8 (Kareha)	8.02	816	319	52	194	369	3	563	380	100	708	8.82
K 9 (Kaowa)	7.82	679	365	54	198	376	4	575	413	104	648	9.38
K 10 (Gandhiyawa)	7.85	564	332	63	203	352	3	556	420	80	601	6.49

Minimum	7.82	564	235	32	129	339	3	529	355	35	315	6.49
Maximum	8.52	948.00	431.00	95.00	232.00	400.00	56.00	603.00	481.00	140.00	765.00	9.90
Average	8.03	814.70	326.20	60.40	192.90	371.20	22.10	564.40	416.10	90.30	561.10	8.28
Standard Deviation	0.21	124.42	61.08	21.14	29.50	22.08	24.41	25.26	42.58	34.07	160.15	1.12
(c)												
KO 1 (Kaundhiyara Road)	8.41	823	310	76	85	586	3	671	446	59	556	8.5
KO 2 (Belawa)	8.37	987	397	79	96	542	5	638	461	110	713	10.67
KO 3 (kaundhiyara)	8.16	839	602	59	95	551	4	647	450	285	942	8.04
KO 4 (Belsara)	7.98	928	752	57	72	506	6	579	643	210	783	8.63
KO 5 (Baraha)	8.43	835	376	82	81	428	7	510	408	110	495	9.49
KO 6 (Khaptiha)	8.14	931	430	79	83	548	6	634	491	94	643	8.05
KO 7 (Dhari)	8.21	726	494	65	116	584	4	701	473	165	731	9.13
KO 8 (Piparawan)	8.29	723	380	62	74	547	7	621	395	109	114	7.09
KO 9 (Piparhatta)	8.43	829	313	137	90	446	3	536	373	39	696	9.29
KO 10 (Dewara)	8.36	876	308	134	70	456	5	527	376	70	473	8.12
Minimum	7.98	723	308	57	70	428	3	510	373	39	114	7.09
Maximum	8.43	987.00	752.00	137.00	116.00	586.00	7.00	701.00	643.00	285.00	942.00	10.67

Continued

TABLE 7.2 Physio-chemical characteristics of water quality of (a) Chaka block (b) Karchana block and (c) Kaundhiyara block.—cont'd

Site	pH	EC	TDS	Na	Ca hardness	Mg hardness	K	Total hardness	Alkalinity	Chloride	Sulphate	Nitrate
Average	8.28	849.70	436.20	83.00	86.20	519.40	5.00	606.40	451.60	125.10	614.60	8.70
Standard Deviation	0.15	85.06	143.94	29.09	13.85	57.39	1.49	65.02	78.71	75.18	224.75	0.99

Note: Units are in mg/l, except pH and EC (μS/cm).

If water has EC (<750 μS/cm) means no detrimental effects on crop, if EC (750−1500 μS/cm) then irrigation water may have detrimental effects on few sensitive crops, if EC (1500−3000 μS/cm) then irrigation water may have adverse effects on crops and if EC (1500−7500 μS/cm) is only suitable for salt tolerant plant (Zaman et al., 2018).

The total dissolved solids (TDS) of water samples ranges from 235 to 440 mg/L in Chaka block, 235−431 mg/L in Karchana and 308−752 mg/L in Kaundhiyara block (Table 7.2). If TDS is less than 450 mg/L (no degree of restriction on irrigation use), 450−2000 mg/L (slight to moderate degree of restriction on irrigation use) and greater 2000 mg/L (severe degree of restriction on irrigation use). The water with high TDS needs proper management strategies before use. The TDS in water comprises mainly inorganic salts and small amount of organic matter such as carbonate, bicarbonate, chloride, sulphate, nitrate, sodium, potassium, calcium and magnesium (Singh et al., 2009, 2015). The most common source of dissolved solids in water is from the weathering of sedimentary rocks and the erosion of the Earth's surface (Kaushik and Saksena, 1999). Groundwater usually has higher levels of TDS than surface water, since it has a longer contact time with the underlying rocks and sediments. The concentration of dissolved solids in natural water is usually less than 500 mg/L, while water with more than 500 mg/L is undesirable for drinking, irrigation and many industrial uses; high value of TDS influences the taste, hardness and corrosive property of the water (Singh et al., 2009, 2015; Gautam et al., 2018). In Chaka block Na range from 16−125 mg/L, whereas in Karchana 32−95 mg/L and Kaundhiyara block 57−137 mg/L, respectively. Sodium have specific ion toxicity to sensitive crops. The usual range of Na in irrigation water is 0−40 me/L.

The calcium hardness (Ca) of water sample range from 102 to 254 mg/L in Chaka block, 129−232 mg/L in Karchana and 70−116 mg/L in Kaundhiyara block (Table 7.2). The calcium hardness of water samples were within permissible limit except in few sites of Chaka C1 (Chak Imam Ali) and C4 (Naini Taluka Naini Dadari) and Karchana K3 (Bardaha), K4 (Ligadahiya), K5 (Chanaini), K7 (Ghatwa) and K10 (Gandhiyawa). Calcium is one of the elements which exist in divalent form Ca^{2+} ion in water (Goel, 2000). The presence of Ca^{2+} ions in the groundwater may reflect weathering of rocks such as limestone and aragonite, and thus may be considered as the major cation contributor to the water available in the region (Singh et al., 2009).

The magnesium hardness (Mg) of the water sample range from 354 to 632 mg/L in Chaka block, 339−400 mg/L in Karchana and 428−586 mg/L in Kaundhiyaar block (Table 7.2). The magnesium hardness (Mg) of water samples of the selected block exceeds the permissible limit. The magnesium is derived from dissolution of magnesium calcite, gypsum and dolomite from source rocks (Singh et al., 2015). If the concentration of magnesium in drinking water is more than the permissible limit, it causes unpleasant taste to the water (Sarala and Ravi Babu, 2012). The potassium range from 3−47 mg/L in Chaka, 3–56 mg/L in Karchana and 3–7 mg/L in Kaundhiyara block. The usual range of potassium in irrigation water is recommended as 0–2 mg/L.

Total hardness of the water samples ranges from 534 to 804 mg/L in Chaka block, 529–603 mg/L in Karchana and 510–701 mg/L in Kaundhiyara block (Table 7.2). The total hardness of water samples exceeds the permissible limit. Total hardness is primarily caused by the presence of cations such as calcium and magnesium and anions such as carbonate, bicarbonate, chloride and sulphate in water (Rao and Rao et al., 1991). The plants grow well when the total hardness is in the range of 100–150 mg/L. Water hardness has no known adverse effects; however, some evidence indicates its role in heart diseases and hardness of 150–300 mg/L and above may cause kidney problems and kidney stone formation. It causes unpleasant taste and reduces ability of soap to produce lather (Sarala and Ravi Babu, 2012).

Alkalinity of water samples ranges from 355 to 530 mg/L in Chaka block, 355–481 mg/L in Karchana and 373–643 mg/L in Kaundhiyara block (Table 7.2). Alkalinity of water samples of most of the blocks were within the permissible limit except the site Ko4 (643 mg/L). The high alkalinity of groundwater in certain locations in the study area may be due to the presence of bicarbonate and some salts. The alkaline water may decrease the solubility of metals (Goel, 2000). Alkalinity is an important parameter in evaluating the optimum dose of coagulant. Excess alkalinity affect crop yield and gives bitter taste to drinking water (Kataria et al., 2006). However, some alkalinity is required in drinking water to neutralise the acids such as lactic acid and citric acid produced in the body (Kaushik and Saksena, 1999).

The chloride of water samples ranges from 65 to 160 mg/L in Chaka block, 35–140 mg/L in Karchana block and 39–285 mg/L in Kaundhiyara block (Table 7.2). The chloride is present in all the natural waters, mostly at low concentrations (Patel and Sinha, 1998). The usual range of chloride in irrigation water is 0–30 me/L. Higher concentration of chloride in irrigation water may reduce crop yield. Most of the plants can tolerate chloride as 100 mg/L, however few plants are very sensitive to low value of chloride. Chloride imparts a salty taste and sometimes higher consumption causes the development of essential hypertension, risk for stroke, left ventricular hypertension, osteoporosis, renal stones and asthma in human beings (Sarala and Ravi Babu, 2012).

Sulphate of water sample range from 456 to 945 mg/L in Chaka block, 315–765 mg/L in Karchana and 114–942 mg/L in Kaundhiyara block (Table 7.2). The sulphate in all the samples is higher. The usual range of sulphate in irrigation water is 0–30 me/L. Irrigation water high in sulfate ions reduces phosphorus availability to plants. Sulphate concentration is less than 400 mg/L desired, whereas greater than 400 mg/L will acidify the soil and may have negative impacts. The standard desirable limit of sulphate for drinking water is 200 mg/L prescribed by BIS (1991) and is within the permissible limit of 400 mg/L. Sulphates are a combination of sulfur and oxygen and are a part of naturally occurring minerals in some soil and rock formations that contain groundwater. The mineral dissolves over time and is released into groundwater. Gypsum is an important salt in many aquifers having high concentrations of sulphate.

Nitrate of the water samples ranges from 6.09 to 9.63 mg/L in Chaka block, 6.49–9.9 mg/L in Karchana block and 7.09–10.67 mg/L in Kaundhiyara block (Table 7.2). The usual range of nitrate in irrigation water is 0–10 mg/L. Natural nitrate levels in groundwater are generally very low as they move relatively slow in soil and groundwater. Nitrate contamination in groundwater arises from point sources such as livestock facilities, sewage disposal systems, including septic tanks, and nonpoint sources such as fertilised cropland or naturally occurring sources of nitrogen (Nemčić-Jurec et al., 2013; Nemčić-Jurec and Jazbec, 2016). The acute health hazards associated with drinking water with nitrate creates the condition known as methemoglobinemia (blue baby syndrome) in which blood lacks the ability to carry sufficient oxygen to the individual's body cells (Hegesh and Shiloah, 1982).

3.2 Correlation matrix

Correlation matrix show relationship between the variables. The EC show positive relationship with Na ($r = 0.57$). pH show negative correlation with nitrate ($r = -0.41$). TDS shows relationship with alkalinity ($r=0.73$) and chloride ($r = 0.77$). Ca hardness shows significant positive correlation ($r = 0.76$) with the anhydrite mineral whereas Mg hardness shows positive correlation ($r = 0.76$) with total hardness. Alkalinity shows positive relationship ($r = 0.52$) with chloride whereas chloride shows significant positive correlation with TDS ($r = 0.77$). Sulphate shows positive correlation with anhydrite ($r = 0.60$), epsomite ($r = 0.9$) and with gypsum ($r = 0.59$). Mineral brucite show positive relation with pH ($r = 0.98$), Mg hardness ($r = -0.52$), magnesite ($r=0.93$), and dolomite (c & d) ($r = 0.88$), respectively. Calcite has positive relationship with dolomite (c & d) ($r = 0.63$) and dolomite (c & d) has significant positive relationship with pH ($r = 0.89$), magnesite ($r = 0.87$), brucite ($r = 0.88$) and aragonite ($r = 0.63$). The mineral epsomite shows relationship with sulphate ($r = 0.91$) whereas gypsum shows positive correlation with Ca hardness, sulphate and anhydrite ($r = 0.75$, 0.59 and 0.99), respectively. The mineral magnesite shows positive correlation with pH ($r = 0.88$), Mg hardness ($r = 0.71$), calcite ($r = 0.93$) and dolomite (c & d) ($r = 0.87$), respectively. Fig. 7.3 shows the heat map and hierarchical clustering tree of Spearman correlation analysis of sampling locations and physicochemical water-quality parameters.

3.3 Principal component analysis

PCA includes correlated variables with the purpose of reducing the numbers of variables and explaining the same amount of variance with fewer variables (principal components). The new variables created, the principal components

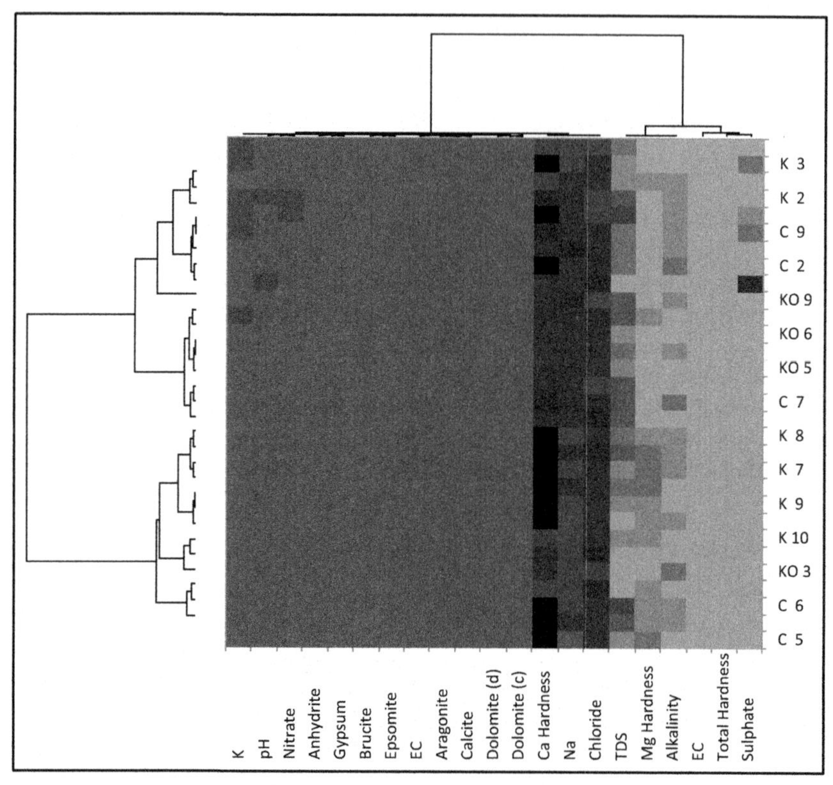

FIGURE 7.3 The heat map and hierarchical clustering tree of Spearman correlation analysis of sampling locations and physicochemical water quality parameters.

scores (PCS), are orthogonal and uncorrelated to each other, being linear combinations of the original variables. They are obtained in such a way that the first PC explains the largest fraction of the original data variability; the second PC explains a smaller fraction of the data variance than the first one and so forth (Ielpo et al., 2012; Singh et al., 2018). Our results show that PC1 explains 34.066% of the total variability and PC2 accounts only for 20.65%, PC3 accounts 14.29%, PC4 have 9.62% and PC5 accounts 7.44% with cumulative variance 86.088% (Table 7.3). Love et al. (2004) in the two case studies determined difference in groundwater quality due to all analysed variables too. The anthropogenic influences in some areas and influence on variability of groundwater quality and high mineralisation and conductivity are determined. The component matrix involving PC1 (pH, Mg hardness, brucite and dolomite c and d), PC2 (aragonite, calcite, dolomite c, d and Ca hardness), PC3 (TDS, chloride, alkalinity and total hardness), PC4 (sulphate, epsomite, gypsum) and PC5 (EC, sodium and total hardness) is shown in Table 7.4.

TABLE 7.3 The total explained variance during the PCA analysis of the samples.

Total variance explained									
	Initial eigenvalues			Extraction sums of squared loadings			Rotation sums of squared loadings		
Component	Total	% of variance	Cumulative %	Total	% of variance	Cumulative %	Total	% of variance	Cumulative %
1	7.154	34.066	34.066	7.154	34.066	34.066	6.431	30.626	30.626
2	4.338	20.657	54.723	4.338	20.657	54.723	3.420	16.285	46.911
3	3.002	14.297	69.020	3.002	14.297	69.020	3.084	14.685	61.596
4	2.020	9.620	78.640	2.020	9.620	78.640	2.911	13.863	75.459
5	1.564	7.448	86.088	1.564	7.448	86.088	2.232	10.629	86.088

Extraction method: principal component analysis.

TABLE 7.4 The PCA loadings of rotated component matrix.

Rotated component matrix[a]					
	Component				
Variables	1	2	3	4	5
pH	**0.953**	0.205	−0.191	0.011	0.011
EC	0.178	0.027	0.130	−0.058	**0.875**
TDS	0.026	−0.213	**0.870**	0.148	−0.022
Sodium	0.304	−0.196	−0.287	0.188	**0.654**
Potassium	−0.105	0.293	−0.071	−0.549	0.449
Ca_Hardness	−0.758	**0.567**	−0.129	−0.039	−0.122
Mg_Hardness	**0.570**	−0.334	0.448	−0.033	0.488
Total hardness	0.114	0.040	**0.512**	−0.079	**0.574**
Alkalinity	−0.028	−0.023	**0.856**	0.075	0.227
Chloride	0.075	−0.095	**0.779**	0.332	−0.121
Sulphate	−0.183	−0.047	0.264	**0.899**	−0.087
Nitrate	0.373	0.150	−0.347	0.361	0.131
Anhydrite	−0.696	0.423	−0.090	0.563	−0.090
Aragonite	0.190	**0.963**	−0.154	−0.050	−0.028
Brucite	**0.978**	0.110	−0.113	−0.037	0.102
Calcite	0.190	**0.963**	−0.154	−0.049	−0.027
Dolomite_d	**0.841**	**0.502**	0.069	−0.062	0.160
Dolomite_c	**0.841**	**0.503**	0.069	−0.061	0.161
Epsomite	−0.031	−0.082	0.233	**0.911**	0.174
Gypsum	−0.686	0.426	−0.097	**0.568**	−0.099
Magnesite	**0.948**	0.036	0.185	−0.038	0.226

Extraction method: Principal component analysis, Rotation method: Varimax with Kaiser normalisation.
[a]*Rotation converged in 9 iterations.*
Bold value represents the significance of the principal component at $p<0.05$.

Component matrix shows correlation between variables and the extracted components. One parameter and the corresponding principal component will have more significant correlation if the absolute value of load factor is near to 1. It presents positive correlation when the coefficient is positive and negative correlation if the coefficient is negative (Singh et al., 2009). From Table 7.4 it is seen that component 1 has positive load factors on the variables pH, Mg hardness, brucite, dolomite c and d, epsomite and magnesite and while Ca hardness, anhydrite and gypsum have negative factor loading on PC1 is the inclusive measurement for the salinisation of groundwater. High positive load factor on Cl^- might be due to the leaching of domestic and organic wastes (Singh et al., 2013, 2015). Similarly, high positive factor load of ions indicate leaching from the soil and industrial waste sites, organic and anthropogenic pollution (Singh et al., 2013, 2015). Similarly, PC2 has larger positive load factor on the variables of Ca hardness, aragonite, calcite, dolomite c and d indicating that it is the inclusive measurement of alkalisation of the groundwater (Singh et al., 2013, 2015). PC 3 has positive loading of TDS, total hardness, alkalinity, chloride and dolomite c and d. PC 4 has positive loading of alkalinity, sulphate, anhydrite, epsomite and gypsum. PC 5 has positive loading of EC, total hardness. Similarly the negative loading is also reported for different variables in different PCs (Singh et al., 2013, 2015). The scree plot shows the significant principal components (Singh et al., 2013, 2015, Fig. 7.4).

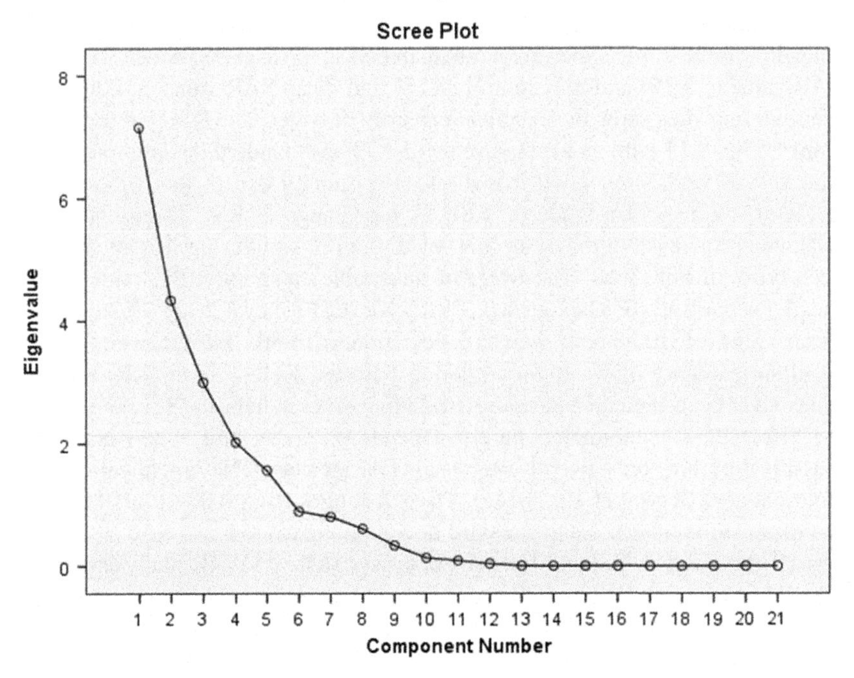

FIGURE 7.4 Shows the significant principal component.

3.4 Cluster analysis

Cluster analysis (CA) was applied to the datasets to find the existing dissimilarity groups between the sampling stations. A dendrogram was produced through hierarchical agglomerative algorithms using squared Euclidean distance which allows the backward or forward tracing to any individual case or clustering at any level. The CA helps in grouping all sampling stations into statistically meaningful clusters (Singh et al., 2013, 2015). The results showed that the CA technique showed its usefulness in the classification of water samples (Singh et al., 2013, 2015). Hence, the number of sampling sites and respective cost could be diminished in future monitoring plans. The following sites formed cluster for water samples (Fig. 7.5) on the basis of similarity/dissimilarity in pollution. Five clusters were formed on the basis of sampling sites. Cluster 1 includes sampling sites (C1, C2, K1, K2, K3, K4, Ko5, Ko8, and Ko10), Cluster 2 (C3, C5, C6, K9 and K10), Cluster 3 (Ko3, Ko4, Ko7), Cluster 4 has a independent site (C9) and sampling locations (C4, C7, K5, K6, K7, K8, Ko1, Ko2, Ko6 and Ko9) belongs to Cluster 5. Thus, the dendrogram provides representative picture of the nature of the sample sites so that future water resources analysis could be eased and sampling locations could be substantially optimised in regards to the similarity between the sampling locations.

3.5 Irrigation indices

Table 7.5 presents the overall results of irrigation indices and descriptive statistics. Table 7.6 has the index range and class of irrigation water. The min SAR value is 0.954 and maximum is 7.157 and mean SAR value is 4.062. All the sampling sites show the excellent category of water. The RSC value ranges from 1.55 to 2.11 with an average value of 1.73 and standard deviation of 0.10. Majority of wells have good irrigation water quality except few which have doubtful water quality based on RSC. A high range of RSC in groundwater indicate an increase in the adsorption of sodium in soil during irrigation from such types of bore well. The water of unsuitable category (RSC value > 2.5 meq/L) is harmful for plant growth. The %Na (ESP) shows 2.72–20.26 with a mean value of 10.88 and standard deviation of 4.30. All the wells show excellent category of water for irrigation purpose. Kelly's ratio (KR)/Kelly's index (KI) is an indicator to assess irrigation water suitability. It is free from the effect of K^+ parameter, purely depends on Ca^{2+}, Mg^{2+} and Na^+. Its classification bin (only two classes) is also easier than %Na classification bin (four classes) (Rawat et al., 2018c). The KR ranges from 0.028 to 0.256 with the mean value of 0.127 and standard deviation is also low as 0.056. It shows that all the water is suitable for irrigation uses ($KR < 1$). KR analysis shows alkali hazards; therefore, during irrigation, the use of pure gypsum is recommended to reduce effect of Na. The permeability index (PI) is also Na^+ dominate-based index, and it also represents in percent value like %Na index value. The soil permeability is affected by the extensive use of irrigation water

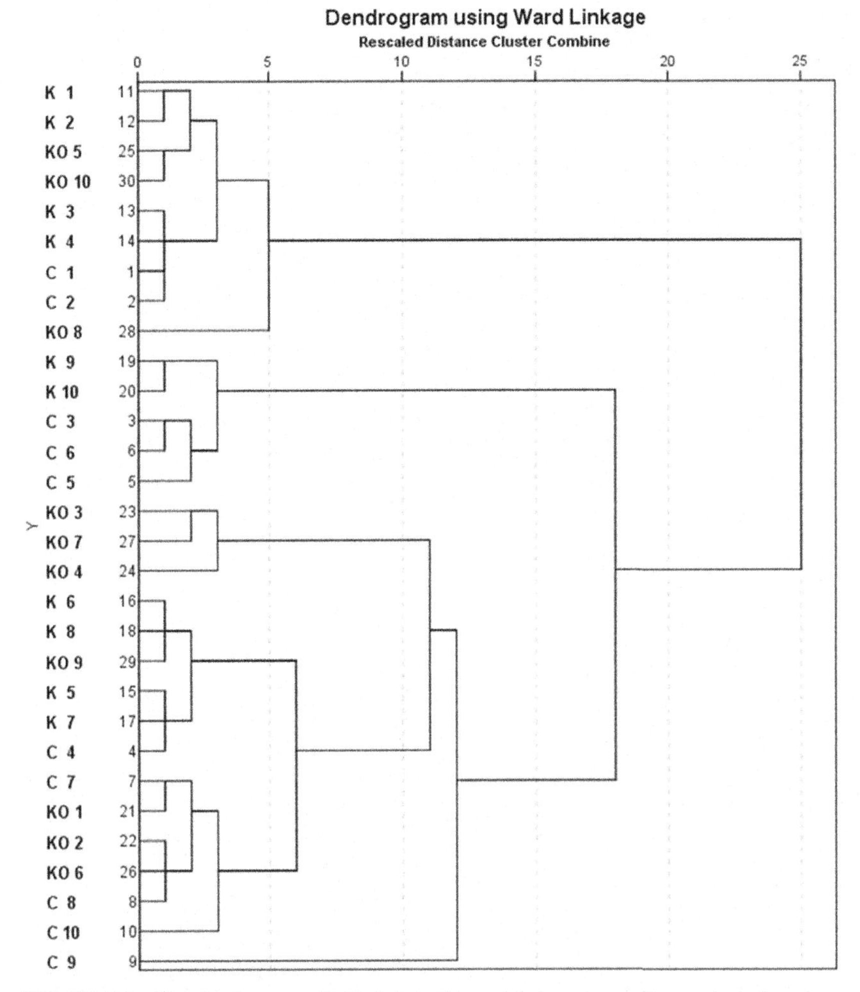

FIGURE 7.5 The dendrogram obtained from hierarchical agglomerative analysis based on sampling site forms five major clusters in study area.

as it is influenced by Na^{+}, Ca^{2+}, Mg^{2+} and HCO_3^- content of the water (Gautam et al., 2015). The PI ranges from 2.84 to 25.56 with a mean value of 12.72 and standard deviation of 5.58. It suggests that majority of wells fall in unsuitable category. MH/MAR is Mg^{2+} and Ca^{2+}-based index and it also represents in percent and contains only two classes MH<50% and MH>50%, suitable and unsuitable, respectively. The mean value of MH during the study period was 74.40 and with ranges 58.22–88.04. This fluctuation clearly indicate some degree of leaching process of Mg^{+} (from rock to groundwater) after rainfall (Singh et al., 2012). If MH < 50% then it is considered as safe, and if >50% then it is unsafe for irrigation use. All sample were found unsuitable for irrigation as per MH. The CAI1 and CAI2 indexes are indicating

TABLE 7.5 The values of irrigation indices (Wilcox, 1948; Kelly, 1940; Todd, 1980; USSL, 1954) and descriptive statistics of irrigation indices.

Sites/ Statistics	SAR	RSC	ESP	KR	PI	MH	PS	CAI1	CAI2	CR	TH	r1	r2
1	5.79	1.63	14.17	0.17	16.61	58.22	360.5	0.05	0.003	1.37	2086.4	−0.018	−0.01
2	5.96	1.60	14.59	0.17	17.22	69.57	309.5	−0.44	−0.022	1.24	2160.6	0.060	0.07
3	3.93	1.68	10.55	0.12	11.86	68.80	576.5	0.35	0.019	2.45	1973.2	−0.041	−0.04
4	3.70	1.69	9.36	0.10	10.39	65.35	456.5	0.33	0.019	1.64	2251.5	−0.055	−0.05
5	0.95	1.89	2.72	0.03	2.84	64.83	513.5	0.76	0.042	1.87	1991.5	−0.109	−0.10
6	1.04	1.70	3.03	0.03	3.19	67.92	441.5	0.57	0.022	1.93	1911.7	−0.064	−0.05
7	6.07	1.55	14.27	0.17	16.77	76.98	395	−0.29	−0.015	1.50	2448	0.034	0.04
8	6.29	1.77	14.77	0.17	17.49	84.21	542	0.26	0.020	1.79	2485.4	−0.062	−0.05
9	5.94	1.64	12.28	0.15	14.80	78.61	333	−0.57	−0.032	0.95	3021.2	0.031	0.13
10	7.16	1.87	15.98	0.20	20.49	81.31	472.5	−0.81	−0.038	1.61	2318.6	0.040	0.10
11	4.05	1.65	9.99	0.12	12.23	71.53	236.5	−1.73	−0.058	1.02	1997.2	0.057	0.20
12	4.06	1.67	10.19	0.12	12.48	75.61	245.5	−2.40	−0.061	1.10	1962.5	0.074	0.20
13	1.96	1.75	4.97	0.06	5.64	66.00	267.5	0.26	0.021	0.91	2144.3	−0.241	−0.09
14	1.86	1.75	4.78	0.05	5.40	64.59	269.5	−0.28	−0.011	1.00	2095.3	−0.067	0.04
15	5.52	1.87	14.13	0.17	16.55	62.23	446	0.17	0.011	1.64	1943.6	−0.033	−0.03

16	5.83	1.77	15.10	0.18	17.89	63.84	522.5	0.30	0.023	2.12	1869.9	−0.059	−0.05
17	2.86	1.76	7.63	0.08	8.32	60.61	458.5	0.54	0.035	1.69	2043.7	−0.096	−0.09
18	3.10	1.67	8.41	0.09	9.24	65.54	454	0.45	0.026	1.86	1997.9	−0.068	−0.06
19	3.19	1.72	8.54	0.09	9.41	65.51	428	0.44	0.027	1.66	2036.6	−0.077	−0.07
20	3.78	1.76	10.14	0.11	11.35	63.42	380.5	0.18	0.009	1.51	1950.7	−0.028	−0.02
21	4.15	1.66	10.13	0.11	11.33	87.33	337	−0.34	−0.011	1.18	2615.1	0.031	0.04
22	4.42	1.72	10.94	0.12	12.38	84.95	446.5	0.24	0.014	1.63	2462.2	−0.043	−0.04
23	3.28	1.70	8.32	0.09	9.13	85.29	776	0.78	0.106	2.53	2496.6	−0.240	−0.24
24	3.35	2.11	8.89	0.10	9.86	87.54	601.5	0.70	0.071	1.82	2254.6	−0.195	−0.19
25	5.14	1.80	13.71	0.16	16.11	84.09	357.5	0.19	0.014	1.46	1957.3	−0.057	−0.04
26	4.45	1.78	11.03	0.13	12.52	86.85	415.5	0.10	0.005	1.42	2454.3	−0.023	−0.01
27	3.47	1.68	8.45	0.09	9.29	83.43	530.5	0.58	0.049	1.69	2684.4	−0.137	−0.13
28	3.52	1.64	8.99	0.10	9.98	88.08	166	0.37	0.034	0.54	2427.7	−0.412	−0.35
29	3.59	1.70	20.27	0.26	25.56	83.21	387	−2.59	−0.058	1.71	2053.6	0.141	0.15
30	3.45	1.72	20.15	0.25	25.48	86.69	306.5	−0.99	−0.046	1.31	2044.6	0.135	0.15
Min.	0.954	1.55	2.72	0.028	2.84	58.22	166	−2.590	−0.061	0.54	1869.90	−0.412	−0.351
Max	7.157	2.11	20.26	0.256	25.56	88.08	756	0.779	0.106	2.53	3021.20	0.141	0.204
Mean	4.062	1.73	10.88	0.127	12.72	74.40	414.43	−0.095	0.007	1.54	2204.67	−0.051	−0.020
Standard deviation	1.544	0.10	4.30	0.056	5.58	10.036	124.66	0.859	0.038	0.44	278.788	0.115	0.122

TABLE 7.6 The irrigation index, their range and class (Wilcox, 1948; Kelly, 1940; Todd, 1980; USSL, 1954).

Index	Range	Class
SAR	<10	Excellent
	10–18	Good
	18–26	Doubtful
	>26	Unsuitable
RSC/RA	<1.25	Good
	1.25–2.5	Doubtful
	>2.5	Unsuitable
%Na	>20	Excellent
	20–40	Good
	40–60	Permissible
	60–80	Doubtful
	>80	Unsuitable
KR/KI	<1	Suitable
	>1	Unsuitable
PI	>75%	Suitable
	25%–75%	Good
	<25%	Unsuitable
MAR/MH	<50	Suitable
	>50	Unsuitable
PS	<3	Suitable
	>3	Unsuitable
CAI1	Negative	Reverse exchange process (REP)
	Positive	Direct exchange process (DEP)
CAI2	Negative	Reverse exchange process
	Positive	Direct exchange process
CR	<1	Suitable
	>1	Unsuitable
TDS	<450	Best
	450–2000	Moderate
	>2000	Hazard

TABLE 7.6 The irrigation index, their range and class (Wilcox, 1948; Kelly, 1940; Todd, 1980; USSL, 1954).—cont'd

Index	Range	Class
TH	<75	Soft
	75–150	Moderate
	150–300	Hard
	>300	Very hard
r1	<1	$Na^+–SO_4^{2-}$
	>1	$Na^+–HCO_3^-$
r2	<1	Deep meteoric percolation (DMP)
	>1	Shallow meteoric percolation (SMP)

the direction of reaction between groundwater and aquifer. From the average value of CAI1 (0.85) and CAI2 (0.03) from the table, it was found that most of the year's direct exchange process (DEP) happens between groundwater and aquifer in study area over study period. The majority of wells show CAI1 positive value, which means direct exchange process (DEP) and few show negative, which means reverse exchange process (REP). Similarly the CAI2 also reported that few wells show direct exchange process and few show reverse exchange process. CR is Cl dominate index like PS, CAI1 and CAI2. As per the PS all the samples shows poor quality irrigation water. Similarly majority of wells show CR greater than 1, which means that the water is not suitable for irrigation. The water of high TDS value is not recommended for irrigation. Water hardness is represented by TH and it is Ca^{2+} and Mg^{2+} dominate index. TH index shows that all wells fall under the category of 'very hard'. Hence the well water needs special treatment. Soltan's (1998) index suggests that if r1 is < 1 means water type ($Na^+–SO_4^{2-}$). From Table (7.5) (r1), water type is $Na^+–SO_4^{2-}$ because the average value of r1 belongs to less than 1 (−0.05). Soltan (1998) also suggested another index (r2) for classification of water system based on the yield of K^+, Na^+, Cl^- and SO_4^{2-} ions in water. On the basis of r2 all the water is of deep meteoric percolation type (DMP).

3.6 Saturation index

Change in the saturation index help to distinguish between various stages of hydrogeochemical evolution (Chidambaram et al., 2011). In addition, it helps to determine which geochemical process is important in controlling water chemistry of that particular area (Singh et al., 2013, 2015). This information is vital for protecting and remediating the water bearing system. Generally, as groundwater moves through an aquifer, it initially dissolves the rocks and releases minerals at varying rate due to rock-water interactions in the water systems. The Saturation Index (SI) was computed for the water samples which showed that, in general, the minerals like anhydrite, brucite, epsomite and gypsum solutions showed high to very high degree of undersaturation, which indicate that they are affected largely by the dilution (Chidambaram et al., 2006; Chidambaram et al., 2007a; Chidambaram et al., 2007b, Chidambaram et al., 2009; Singh et al., 2012, 2015). Aragonite, calcite, dolomite c and d and magnesite minerals were found saturated with respect to water (Fig. 7.6; Table 7.7). For well water, dolomite c and d mineral was highly saturated. Also, the minerals aragonite, calcite and magnesite show saturation, though not highly, at most of the sampling locations. Table 7.7 show mineral saturation value of the studied area.

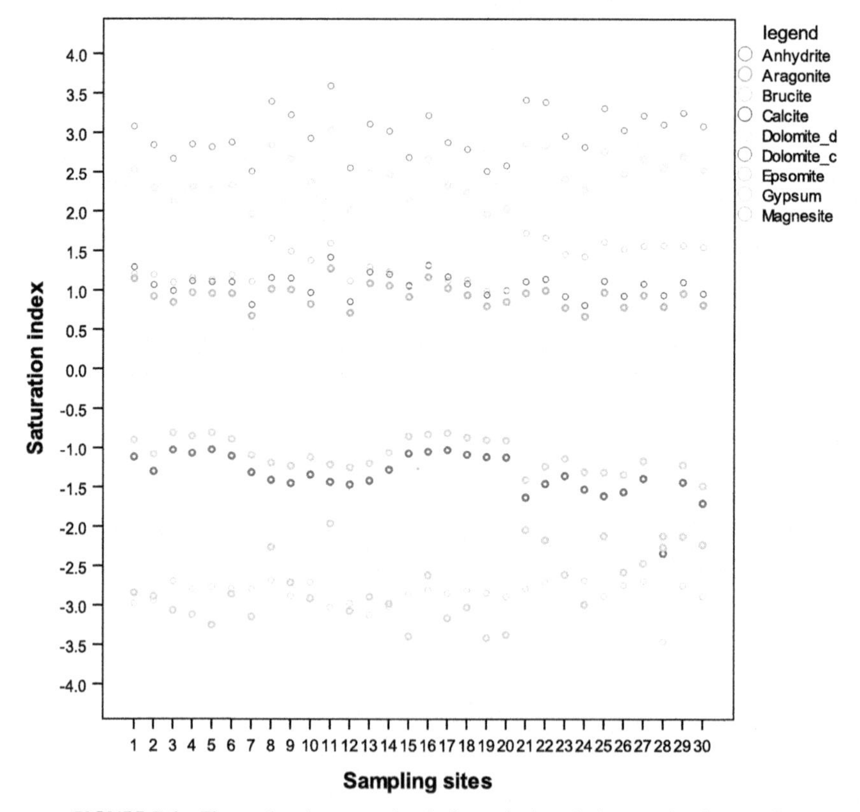

FIGURE 7.6 Figure showing saturation indices of minerals for samples from well.

TABLE 7.7 Mineral saturation index value of (a) Chaka block (b) Karchana block and (c) Kaundhiyara block.

Site	Anhydrite	Aragonite	Brucite	Calcite	Dolomite (d)	Dolomite (c)	Epsomite	Gypsum	Magnesite
(a)									
C 1	−1.121	1.147	−2.844	1.291	2.528	3.078	−2.978	−0.901	1.206
C 2	−1.303	0.922	−2.889	1.066	2.292	2.842	−2.945	−1.084	1.195
C 3	−1.029	0.847	−3.068	0.991	2.119	2.669	−2.696	−0.81	1.097
C 4	−1.071	0.97	−3.122	1.114	2.303	2.853	−2.799	−0.851	1.157
C 5	−1.027	0.959	−3.253	1.103	2.267	2.817	−2.769	−0.808	1.133
C 6	−1.107	0.959	−2.861	1.103	2.328	2.878	−2.787	−0.887	1.194
C 7	−1.314	0.673	−3.145	0.817	1.959	2.509	−2.792	−1.094	1.11
C 8	−1.411	1.014	−2.26	1.157	2.844	3.394	−2.684	−1.191	1.656
C 9	−1.452	1.007	−2.711	1.151	2.674	3.224	−2.883	−1.233	1.492
C 10	−1.343	0.826	−2.915	0.97	2.377	2.927	−2.708	−1.124	1.377
Minimum	−1.452	0.673	−3.253	0.817	1.959	2.509	−2.978	−1.233	1.097
Maximum	−1.03	1.15	−2.26	1.29	2.84	3.39	−2.68	−0.81	1.66
Average	−1.22	0.93	−2.91	1.08	2.37	2.92	−2.80	−1.00	1.26
Standard deviation	0.16	0.13	0.28	0.13	0.26	0.26	0.10	0.16	0.19

Continued

TABLE 7.7 Mineral saturation index value of (a) Chaka block (b) Karchana block and (c) Kaundhiyara block.—cont'd

Site	Anhydrite	Aragonite	Brucite	Calcite	Dolomite (d)	Dolomite (c)	Epsomite	Gypsum	Magnesite
(b)									
K 1	−1.435	1.274	−1.963	1.417	3.041	3.591	−3.031	−1.215	1.593
K 2	−1.469	0.711	−3.074	0.855	2.002	2.552	−2.979	−1.249	1.116
K 3	−1.419	1.088	−2.896	1.231	2.557	3.107	−3.129	−1.2	1.294
K 4	−1.28	1.058	−2.981	1.202	2.468	3.018	−3.019	−1.061	1.235
K 5	−1.075	0.917	−3.395	1.061	2.136	2.686	−2.863	−0.855	1.044
K 6	−1.05	1.168	−2.616	1.312	2.668	3.218	−2.808	−0.83	1.325
K 7	−1.03	1.026	−3.165	1.17	2.325	2.875	−2.848	−0.811	1.124
K 8	−1.088	0.938	−3.029	1.081	2.239	2.789	−2.814	−0.868	1.127
K 9	−1.12	0.798	−3.415	0.942	1.96	2.51	−2.846	−0.9	0.987
K 10	−1.126	0.854	−3.376	0.997	2.032	2.582	−2.892	−0.907	1.003
Minimum	−1.469	0.711	−3.415	0.855	1.96	2.51	−3.129	−1.249	0.987
Maximum	−1.03	1.27	−1.96	1.42	3.04	3.59	−2.81	−0.81	1.59
Average	−1.21	0.98	−2.99	1.13	2.34	2.89	−2.92	−0.99	1.18
Standard deviation	0.17	0.17	0.44	0.17	0.34	0.34	0.11	0.17	0.18

(c)									
KO 1	−1.632	0.964	−2.044	1.108	2.861	3.411	−2.79	−1.412	1.722
KO 2	−1.46	0.995	−2.173	1.139	2.833	3.383	−2.708	−1.241	1.663
KO 3	−1.36	0.778	−2.611	0.922	2.406	2.956	−2.601	−1.141	1.453
KO 4	−1.53	0.666	−2.996	0.81	2.266	2.816	−2.686	−1.311	1.425
KO 5	−1.614	0.972	−2.121	1.116	2.758	3.308	−2.889	−1.314	1.612
KO 6	−1.563	0.785	−2.58	0.929	2.48	3.03	−2.744	−1.344	1.52
KO 7	−1.393	0.937	−2.466	1.081	2.667	3.217	−2.692	−1.173	1.555
KO 8	−2.341	0.794	−2.263	0.938	2.556	3.103	−3.464	−2.122	1.567
KO 9	−1.441	0.958	−2.126	1.102	2.7	3.25	−2.748	−1.221	1.567
KO 10	−1.706	0.813	−2.229	0.957	2.532	3.082	−2.891	−1.487	1.544
Minimum	−2.341	0.666	−2.996	0.81	2.266	2.816	−3.464	−2.122	1.425
Maximum	−1.36	1.00	−2.04	1.14	2.86	3.41	−2.60	−1.14	1.72
Average	−1.60	0.87	−2.36	1.01	2.61	3.16	−2.82	−1.38	1.56
Standard deviation	0.28	0.11	0.30	0.11	0.19	0.19	0.24	0.28	0.09

4. Conclusion

The work outlines the inherent hydrochemical characteristics of groundwater in the research area with estimates of irrigation suitability of such source with respect to values of groundwater-quality parameters. Such an evaluation will provide the necessary guidance for water quality management and policy initiative. The study combined field and laboratory-based physicochemical analysis, statistical analysis and hydrogeological modelling (WATEQ4F model) for the computation of saturation index (SI).

Correlation matrix highlighted significant relationships of different variables. PCA analysis of these samples gave rise to five principal components with cumulative contribution rate up to 86.88%. Meanwhile, high extraction communalities were obtained for most of the variables. Variable factors obtained from PC by varimax rotation indicate that the parameters responsible for water quality variations are mainly related to trace metals (leaching from soil and industrial waste sites), dissolved salts (natural), organic pollution and nutrients (anthropogenic). PCA helped in identifying the factors and sources which are responsible for water quality variations at different sites.

Grouping of groundwater on the dendrogram elucidate common composition and origins. Such multivariate statistical techniques as hierarchical cluster analysis were applied to identify similarity and dissimilarity in monitoring sites. The hierarchical cluster analysis grouped different sampling sites into major clusters of similar water quality characteristics on the basis of physicochemical parameters. Based on the obtained information, it is possible to design optimal sampling strategy which could reduce the number of sampling stations and the cost of sampling in the future.

Based on irrigation index it is concluded that groundwater of study area is $Na^{+}-SO_4^{2-}$ and deep meteoric percolation (DMP) type. Majority of the wells fall under moderate to suitable category of water for irrigation purposes based on irrigation indices value. The saturation indices for various minerals help to identify the impacts from further mineral intrusion into the aquifer.

References

Al-Shammiri, M., Al-Saffar, A., Bohamad, S., Ahmed, M., 2005. Waste water quality and reuse in irrigation in Kuwait using microfiltration Technology in treatment. Desalination 185 (1–3), 213–225.

Amin, A., Fazal, S., Mujtaba, A., Singh, S.K., 2014. Effects of land transformation on water quality of Dal lake, Srinagar, India. J. Indian Soc. Remote Sens. 42, 119–128.

APHA, 1998. Standard Methods for the Examination of Water and Wastewater. American Public Health Association Inc., Washington, DC.

Appelo, C.A.J., Postma, D., 2005. Geochemistry, groundwater and pollution, 2nd. Balkema, Rotterdam.

Aravindan, S., 2004. Groundwater quality in the hard rock area of the Gadilem river basin, Tamilnadu. Tour Geo Soc. 63, 625–635.

Astel, M., Bizuik, A., Prazyjazny, J., Nameisink, J., 2006. Chemometrics in monitoring spatial and temporal variations in drinking water quality. Water Res. 40, 1706–1716.

Ayers, R.S., Westcot, D.W., 1985. Water Quality for Agriculture. Irrigation and Drainage Paper No. 29. Food and agriculture organization of the United Nations, Rome, pp. 1–117.

Balasubramanian, A., 1986. Hydrogeological Investigation of Tambraparani river Basin, Tamilnadu. Unpublished PhD Thesis. University of Mysore, p. 349.

Ball, J.W., Nordstrom, D.K., 1991. User's manual for WATEQ4F, with revised thermodynamic data base and test cases for calculating speciation of major, trace, and redox elements in natural waters. In: U.S. Geological Survey, Open-File Report, vols. 91–183, p. 189. Washington DC.

BIS, 1991. Specifications for drinking water, IS:10500: 1991. Bureau of Indian Standards, New Delhi, India.

Chidambaram, S., Anandhan, P., Srinivasamoorthy, K., Prasanna, M.V., Vasudeven, S., 2006. A study on the hydrogeochemical modeling of the Neyveli aquifer, Tamil Nadu, India. Indian J. Geochem. 21 (1), 229–245.

Chidambaram, S., Vijayakumar, V., Srinivasamoorthy, K., Anandhan, P., Prasanna, M.V., Vasudeven, S., 2007a. A study on variation in ionic composition of aqueous system in different lithounits around Perambalure Region, Tamil Nadu. J. Geol. Soc. India 70, 1061–1069.

Chidambaram, S., Ramanathan, A.L., Prasanna, M.V., Anandhan, P., Srinivasamoorthy, K., Vasudeven, S., 2007b. Identification of hydrogeochemically active regimes in groundwaters of Erode District, Tamil Nadu: a statistical approach. Asian J. Water Environ. Pollut. 5 (3), 93–102.

Chidambaram, S., Prasanna, M.V., Ramanathan, A.L., Vasu, K., Hameed, S., Warrier, U.K., Srinivasamoorthy, K., Manivannan, R., Tirumalesh, K., Anandhan, P., Johnsonbabu, G., 2009. A study on the factors affecting the stable isotopic composition in precipitation of Tamil Nadu, India. Hydrol. Process. 23, 1792–1800.

Chidambaram, S., Karmegam, U., Prasanna, M.V., Sasidhar, P., Vasanthavigar, M., 2011. A study on hydrochemical elucidation of coastal groundwater in and around Kalpakkam region, southern India. Environ. Earth Sci. https://doi.org/10.1007/s12665-011-0966-3.

Collins, R., Jenkins, A., 1996. The impact of agricultural land use on Stream chemistry in the middle hills of the Himalayas, Nepal. J. Hydrol. 185, 71–86.

Doneen, L.D., 1964a. Notes on Water Quality in Agriculture. Published as a Water Science and Engineering Paper 4001. Department of Water Science and Engineering, University of California.

Doneen, L.D., 1964b. Water Quality for Agriculture. Department of Irrigation, University of Calfornia, Davis, 48p.

FAO (Food and Agriculture Organization), 2006. Prospects for Food, Nutrition, Agriculture, and Major Commodity groups. World Agriculture: Toward 2030–2050, Interim Report. Global Perspective Studies Unit. FAO, Rome.

Gajbhiye, S., Singh, S.K., Sharma, S.K., 2015. Assessing the effects of different land use on water qualify using multi-temporal Landsat data. In: Siddiqui, A.R., Singh, P.K. (Eds.), Book Resource Management and Development Strategies: A Geographical Perspective. Pravalika Publication, Allahabad, Uttar Pradesh, India, pp. 337–348, 91 8-93-84292-21-8.

Gautam, S.K., Maharana, C., Sharma, D., Singh, A.K., Tripathi, J.K., Singh, S.K., 2015. Evaluation of groundwater quality in the Chotanagpur Plateau region of the Subarnarekha River basin, Jharkhand state. India Sustain. Water Qual. & Ecol. 6, 57–74. https://doi.org/10.1016/j.swaqe.2015.06.001.

Gautam, S.K., Singh, A.K., Tripathi, J.K., Singh, S.K., Srivastava, P.K., Narsimlu, B., Singh, P., 2016. Appraisal of surface and groundwater of the Subarnarekha river basin, Jharkhand, India: using remote sensing, irrigation indices and statistical techniques. In: Srivastva, P.K., Pandey, P.C., Kumar, P., Raghubanshi, A.S.H.D. (Eds.), Geospatial Technol Water Resour Appl. CRC Press, Boca Raton, FL, pp. 144–169.

Gautam, S.K., Tziritis, E., Singh, S.K., Tripathi, J.K., Singh, A.K., 2018. Environmental monitoring of water resources with the use of PoS index: a case study from Subarnarekha river basin, India. Environ. Earth Sci. 77, 70. https://doi.org/10.1007/s12665-018-7245-5.

Goel, P.K., 2000. Water Pollution-Causes, Effects and Control. New Age Int. (P) Ltd, New Delhi.

Hegesh, E., Shiloah, J., 1982. Blood nitrates and infantile methaemoglobinaemia. Clin. Chim. Acta 125, 107–115.

Ielpo, P., Cassano, D., Lopez, A., Pappagallo, P., Uricchio, V.F., De Napoli, P.A., 2012. Source apportionment of groundwater in Apulian agricultural sites using multivariate statistical analyses: case study of Foggia province. Chem. Cent. J. 6 (Suppl. 2), S/5.

Jacintha, T.G.A., Rawat, K.S., Mishra, A., Singh, S.K., 2016. Hydrogeochemical characterization of groundwater of penninsular Indian region using multivariate statistical techniques. Appl. Water Sci. 1–13. https://doi.org/10.1007/s13201-016-0400-9.

Joshi, D.M., Kumar, A., Agrawal, N., 2009. Assessment of the irrigation water quality of river Ganga in Haridwar district India. J. Chem. 2 (2), 285–292.

Karanth, K.R., 1987. Groundwater Assessment Development and Management. Tata McGraw Hill publishing company Ltd, New Delhi, p. 725.

Kataria, H.C., Gupta, M., Kumar, M., Kushwaha, S., kashyap, S., Trivedi, S., Bhadoriya, R., Bandewar, N.K., 2006. Study of physico- chemical parameters of drinking water of Bhopal city with reference to health impacts. Curr. World Environ. 6 (1), 95–99.

Kaushik, S., Saksena, D.N., 1999. Physical and Chemical Limnology of Certain Water Bodies of Central India. Daya Publication House, Delhi, pp. 1–58.

Keesari, T., Ramakumar, K.L., Chidambaram, S., Pethperumal, S., Thilagavathi, R., 2016a. Understanding the Hydrochemical Behavior of Groundwater and its Suitability for Drinking and Agricultural Purposes in Pondicherry Area, South India – A Step towards Sustainable Development.

Keesari, T., Ramakumar, K.L., Chidambaram, S., Pethperumal, S., Thilagavathi, R., 2016c. Understanding the hydrochemical behavior of groundwater and its suitability for drinking and agricultural purposes in Pondicherry area, South India – a step towards sustainable development. Groundwater Sustain. Dev. 2–3, 143–153. https://doi.org/10.1016/j.gsd.2016.08.001.

Keesari, T., Sinha, U.K., Deodhar, A., Krishna, S.H., Ansari, A., Mohokar, H., 2016b. High fluoride in groundwater of an industrialized area of Eastern India (Odisha): inferences from geochemical and isotopic investigation. Environ. Earth Sci. 75, 1090. https://doi.org/10.1007/s12665-016-5874-0.

Kelin, H., Yuang, H.F., Hong, L., Robert, W.E., 2005. Spatial variability of shallow groundwater level, electrical conductivity and nitrate concentration and risk assessment of nitrate contamination in North China plain. Environ. Int. 31, 896–903.

Kelly, W.P., 1940. Permissible composition and concentration of irrigated waters. In: Proceedings of the A.S.C.F, p. 607.

Khodapanah, L., Sulaiman, W.N.A., Khodapanah, D.N., 2009. Groundwater quality assessment for different purposes in Eshtehard district, tehran, Iran. Eur. J. Sci. Res. 36 (4), 543–553.

Love, D., Hallbayer, D., Amos, A., Haranova, R., 2004. Factor analysis as a tool in groundwater quality management: two southern African case studies. Phys. Chem. Earth 29, 1135–1143.

Maliqi, E., Jusufi, K., Singh, S.K., 2020. Assessment and spatial mapping of groundwater quality parameters using metal pollution indices, graphical methods and geoinformatics. Anal. Chem. Lett. 10 (2), 152–180.

Milovanovic, M., 2007. Water quality assessment and determination of pollution sources along the Axios/Vardar river, Southeast Europe. Desalination 213, 159–173.

Modi, P.N., 2000. Water Engineering and Water Power Engineering. New Delhi Press, pp. 339–360.

Moharir, K., Pande, C., Singh, S.K., Choudhari, P., Kishan, R., Jeyakumar, L., 2019. Spatial interpolation approach-based appraisal of groundwater quality of arid regions. J. Water Supply Res. Technol.—AQUA 68 (6), 431–447.

Narsimha, A., Sudarshan, V., Swathi, P., 2013. Groundwater and its assessment for irrigation purpose in Hanmakonda area, Warangal district, Andhra Pradesh, India. Int. J. Res. Chem. Environ. 3, 195–199.

Nemčić-Jurec, J., Jazbec, A., 2016. Point source pollution and variability of nitrate concentrations in water from shallow aquifers. Appl. Water Sci. 7 (3), 1337–1348.

Nemčić-Jurec, J., Konjačić, M., Jazbec, A., 2013. Monitoring of nitrates in drinking water from agricultural and residential areas of Podravina and Prigorje (Croatia). Environ. Monit. Assess. 185 (11), 9509–9520.

Nemčić-Jurec, J., Singh, S.K., Jazbec, A., Gautam, S.K., Kovac, I., 2017. Hydrochemical investigations of groundwater quality for drinking and irrigational purposes: two case studies of Koprivnica-Križevci County (Croatia) and district Allahabad (India). Sustain. Water Resour. Manag 5 (2), 467–490. https://doi.org/10.1007/s40899-017-0200-x.

Patel, N.K., Sinha, B.K., 1998. Study of the pollution load in the ponds of Burla area near Hirakud Dam of Orrisa. J. Environ. Pollut. 5 (2), 157–160.

Raghunath, H.M., 1987. Ground Water, second ed. Wiley Eastern Limited, New Delhi, India, p. 353.

Ramesh, K., Elango, L., 2011. Groundwater quality and its suitability for domestic and agricultural use in Tondiar river basin, Tamil Nadu, India. Environ. Monit. Assess. 184 (6), 3887–3899.

Rao, S.N., Rao, K.G., 1991. Intensity of pollution of groundwater in Visakhapatnam area, A.P. India. J. Geol. Soc. India 36, 670–673.

Rawat, K.S., Singh, S.K., 2018b. Water Quality Indices and GIS-based evaluation of a decadal groundwater quality. Geol. Ecol. & Landsc. 2 (4), 240–255. https://doi.org/10.1080/24749508.2018.1452462.

Rawat, K.S., Singh, S.K., Gautam, S.K., 2018c. Assessment of groundwater quality for irrigation use: a peninsular case study. Appl. Water Sci. 8 (8), 233.

Rawat, K.S., Singh, S.K., Jacintha, T.G.A., Nemčić-Jurec, J., Tripathi, V.K., 2018a. Appraisal of long term groundwater quality of peninsular India using water quality index and fractal dimension. J. Earth Syst. Sci. 126 (8), 122. https://doi.org/10.1007/s12040-017-0895-y.

Rawat, K.S., Tripathi, V.K., Singh, S.K., 2017. Groundwater quality evaluation using numerical indices: a case study (Delhi, India). Sustain. Water Resour. Manag 4 (4), 875–885. https://doi.org/10.1007/s40899-017-0181-9.

Rowe, D.R., Abdel-Magid, I.M., 1995. Handbook of Wastewater Reclamation and Reuse. CRC Press, Inc, 550p.

Saleh, A., Srinivasula, S.M., Acharya, S., Fishel, R., Alnemri, E.S., 1999. Cytochrome c and dATP-mediated oligomerization of apaf-1 is a prerequisite for procaspase-9 activation. J. Biol. Chem. 274, 17941–17945.

Sarala, C., Ravi Babu, P., 2012. Assessment of groundwater quality parameters in and around Jawaharnagar, Hyderabad. Int. J. Sci. & Res. Publ. 2 (10), 1–6.

Sastri, J.C.V., 1994. Groundwater Chemical Quality in River Basins, Hydrogeochemical Modeling. Lecture Notes—Refresher Course, School of Earth Sciences. Bharathidasan University, Tiruchirapalli, Tamil Nadu, India.

Singh, S.K., Laari, P.B., Mustak, S.K., Srivastava, P.K., Szabó, S., 2018. Modelling of land use land cover change using earth observation data-sets of Tons River Basin, Madhya Pradesh, India. Geocarto Int. 33 (11), 1202–1222.

Singh, S., Singh, C., Kumar, K., Gupta, R., Mukherjee, S., 2009. Spatial-temporal monitoring of groundwater using multivariate statistical techniques in Bareilly district of Uttar Pradesh, India. J. Hydrol. Hydromech. 57, 45–54.

Singh, S.K., Srivastava, P.K., Gupta, M., Mukherjee, S., 2012. Modeling mineral phase change chemistry of groundwater in a rural-urban fringe. Water Sci. Technol. 66, 1502–1510.

Singh, S.K., Srivastava, P.K., Pandey, A.C., Gautam, S.K., 2013. Integrated assessment of groundwater influenced by a confluence river system: concurrence with remote sensing and geochemical modelling. Water Resour. Manag. 27 (12), 4291–4313. https://doi.org/10.1007/s11269-013-0408-y.

Singh, S.K., Srivastava, P.K., Singh, D., Han, D., Gautam, S.K., Pandey, A.C., 2015. Modeling groundwater quality over a humid subtropical region using numerical indices, earth observation datasets, and X-ray diffraction technique: a case study of Allahabad district, India. Environ. Geochem. Health 37 (1), 157–180. https://doi.org/10.1007/s10653-014-9638-z.

Soltan, M.E., 1998. Characterisation, classification, and evaluation of some ground water samples in upper Egypt. Chemosphere 37 (4), 735–745.

Soltan, M.E., 1999. Evaluation of ground water quality in Dakhla Oasis (Egyptian Western Desert). Environ. Monit. Assess. 57, 157–168.

Srivastava, P.K., Singh, S.K., Gupta, M., Thakur, J.K., Mukherjee, S., 2013. Modeling impact of land use change trajectories on groundwater quality using remote sensing and GIS. Environ. Eng. & Manag. J. Agric. Phys. 12 (12), 2343–2355.

Szabolcs, I., Darab, C., 1964. The influence of irrigation water of high sodium carbonate content of soils. In: Proceedings of 8th International Congress of ISSS. Transaction II, pp. 803–881.

Thakur, J.K., Diwakar, J., Singh, S.K., 2015. Hydrogeochemical evaluation of groundwater of Bhaktapur Municipality, Nepal, Environ. Earth Sci. 74 (6), 4973–4988.

Todd, D.K., 1980. Groundwater Hydrology, second ed. John Wiley and Sons, Inc., New York, p. 535.

Ujević Bošnjak, M., Capak, K., Jazbec, A., Casiot, C., Sipos, L., Poljak, V., Dadić, Z., 2012. Hydrochemical characterization of arsenic contaminated alluvial aquifers in Eastern Croatia using multivariate statistical techniques and arsenic risk assessment. Sci. Total Environ. 15 (420), 100–110. https://doi.org/10.1016/j.scitotenv.2012.01.021.

United States Environmental Protection Agency (U.S. EPA), May 1, 1986. Quality Criteria for Water 1986. EPA 440/5-86-001.

USSL, 1954. Diagnosis and improvement of saline and alkali soils, 60. USDA Hand Book, p. 147.

Vasanthavigar, M., Srinivasamoorthy, K., Vijayaragavan, K., Rajiv Ganthi, R., Chidambaram, S., Sarama, V.S., Anandhan, P., Manivannan, R., Vasudevan, S., 2010. Application of water quality index for groundwater quality assessment: thirumanimuttar sub-basin, Tamilnadu, India. Environ. Monit. Assess. 171 (1-4), 595–609.

Webster, N.S., Uthicke, S., Botte, E.S., Flores, F., Negri, A.P., 2013. Ocean acidification reduces induction of coral settlement by crustose coralline algae. Glob. Change Biol. 19, 303–315.

Wilcox, L.V., 1948. The Quality of Water for Irrigation Use. US Department of Agricultural Technical Bulletin 1962, Washington.

Zaman, M., Shahid, S.A., Heng, L., 2018. Guideline for salinity assessment, mitigation and adaptation using nuclear and related techniques. Springer Nature, p. 164.

Part III

Earth observation techniques

Chapter 8

Estimation of potential evapotranspiration using INSAT-3D satellite data over an agriculture area

Prachi Singh[1], Prashant K. Srivastava[1,2], R.K. Mall[2]
[1]*Remote Sensing Laboratory, Institute of Environment and Sustainable Development, Banaras Hindu University, Varanasi, Uttar Pradesh, India;* [2]*DST-Mahamana Centre of Excellence in Climate Change Research, Institute of Environment and Sustainable Development, Banaras Hindu University, Varanasi, Uttar Pradesh, India*

1. Introduction

Evapotranspiration is considered as one of the most important components of the hydrological cycle. On the Earth's surface, evapotranspiration plays an important role in context of water-energy balance and irrigation, as well as agriculture practices. The watershed hydrology is influenced by the global climate change as a result of varying evapotranspiration (ET) processes at different scales (Rao et al., 2011; Srivastava et al., 2013). Evapotranspiration (ET) is the combined loss of water in the form of evaporation from the soil surface and transpiration from the plant through the stomata (Kar et al., 2016). There are several methods available to estimate ET, but the most robust method is the Hamon's method, which needs minimal amount of data to calculate this variable. Hydro-meteorological applications such as assessment of climate and analysis of human-induced effects on natural and agricultural ecosystem require various parameters at local as well as regional scales (Rao et al., 2011; Srivastava et al., 2016).

Evapotranspiration (ET) is one of the most useful hydrological fluxes used for maintaining water balance of the terrestrial ecosystems. Reliable and accurate quantification of changes in ET is imperative for effective irrigation management, crop yield forecast, environmental assessment, ecosystem modelling and solar energy system (Almorox and Hontoria, 2004; Khoob, 2008; Singh and Pawar, 2011; Amatya et al., 2014; Petropoulos et al., 2018). For assessment of ET some conventional methods have been used such as

Agricultural Water Management. https://doi.org/10.1016/B978-0-12-812362-1.00008-4

weighing lysimeter, Energy Balance Bowen Ratio (EBBR), eddy covariance techniques, pan-measurement, sap flow, and scintillometer. They are mainly based on a variety of complex models and are limited to local, field and landscape scales (Liou and Kar, 2014). Potential evapotranspiration (PET) is mainly used to measure the actual evapotranspiration that is difficult to assess using these conventional methods (Liou and Kar, 2014). Assessment of accurate PET is useful for many applications including irrigation scheduling, drought monitoring and understanding climate change impacts (Allen et al., 2007; Senay et al., 2007; Wagle et al., 2017; Petropoulos et al., 2015).

However assessing or modelling PET is the most difficult task especially at global scales. Reason is that the traditional approaches can accurately measure ET over homogeneous areas but cannot be directly extended to large-scale ET, due to natural heterogeneity of the land surface and complexity of hydrologic processes and because of the need for a variety of surface measurements and land surface parameters (Thakur et al., 2011; Srivastava et al., 2017). In this context, remote sensing proves to be a cost-effective approach to assess PET at both regional and global scales. Satellite-derived remote sensing images are a promising source, which provides data for mapping regional- and meso-scale patterns of PET on the Earth's surface and surface temperature helps to establish the direct link between surface radiances and energy balance components (Bartholic and Wiegand, 1970; Idso et al., 1975a; Idso et al., 1975b,c; Jackson, 1985; Caselles et al, 1992; Kustas and Norman, 1996). Information can be obtained from different regions of EMR viz. visible, near-infrared, and thermal infrared regions which can be utilized to determine the land surface temperature (LST) and atmospheric temperature. These important surface and atmospheric parameters then serve as inputs to simulate surface fluxes and PET based on the energy balance equation (Srivastava et al., 2020). Remote sensing tool extends a large and continuous spatial coverage within a short period. It is a cost-effective way of measuring ET compared to the conventional measurements, and it is the only approach for ungauged areas where manual measurements are extremely difficult to conduct (Rango, 1994; Schultz and Engman, 2012). Spatially retrieved surface temperature can provide a surface measurement from a resolution of a few cm^2 to several km^2 from certain satellites (Hatfield, 1983). Remote sensing–based ET estimation and its development have been reviewed from time to time (Moran and Jackson, 1991; Kustas and Norman, 1996' Quattrochi and Luvall, 1999).

Bastiaanssen developed SEBAL model for estimation of ET and used satellite remote sensing techniques for knowledge of spatiotemporal distribution of ET on large scale, and it can provide important information on issues related such as evaluating water distributions, water use by different land surfaces, water allocations, water rights, consumptive water use and planning, and also better management of ground and surface water resources

(Singh et al., 2008). As we know for the study of water rights management and water regulation and also for the quantification of ET irrigated projects have been playing a main role. Traditionally, ET has been calculated by using weather-based reference ET by crop coefficients (K_c) that depend upon the crop type and the crop growth stage (Allen et al., 2005). This study provided the estimation of evapotranspiration using temperature data of INSAT-3D over an agricultural area of Northern India.

2. Materials and methodology

2.1 Study area

In this study, Varanasi district of Uttar Pradesh was used as a study area, which lies geographically between 25°14′ 54.94″ N to 25.17′ 06.57″ N and 82° 58′ 30″ E to 83° 00′ 35″E and mean elevation of 80.71 m from the Mean Sea Level (MSL) (Cai et al., 2009). This study area mainly consists of agriculture landscape and is also part of Indo-Gangetic plain that supports good agricultural productivity, and the land is composed of very fertile alluvial soil deposited by the River Ganga and Varuna. Varanasi lies in the humid subtropical climatic zone that is characterized by the hot summer having a temperature between 22 to 46°C and winters with a temperature drop of up to 5°C. The mean annual rainfall is about 1056 mm (+172 SD) (Mall and Gupta, 2002). According to the study of Rao et al. (1971) Potential ET estimated approximately 1525 mm/year. The percentage distribution of annual rainfall for Varanasi district was recorded season-wise. Firstly, for monsoon season (June to September) it was about 88%; likewise, for winter season 7.7% (October to February) and lastly for summer season (March to May) 4.3%. whereas, about 90% of the total rainfall takes place in monsoon season from June to September. Lowest temperature recorded in the last week of December and first week of January is about 9.3°C, and highest temperature reaches maximum by the end of May or early June. During winter season normally wheat crops are planted and harvested in April to May month. The soil of the study area is alluvial in origin. Ustochrepts and Ustifluvents groups define the six categories of soil for Varanasi district. Those groups belong to USDA soil taxonomy (Singh et al., 1989). Soil texture found for this study area is 15%–30% clay and 30%–70% sand. The profile is 1.2 m deep (Mall and Gupta). In this study, data was collected from department of Agriculture Farm of Institute of Agricultural Sciences, B.H.U., Varanasi. Gravimetric lysimeter (mechanical weighing) instrument was used for daily ET data collection. India Meteorological Department installed this instrument which comes in fully fabricated design. Lysimetric data are available for long periods, especially for crop season. For the estimation of ET data weather data temperature were collected from the study site.

2.2 INSAT-3D

INSAT series of multipurpose spacecrafts have been providing meteorological services over Indian Ocean since early eighties through Very High Resolution Radiometer instruments. Kalpana-1 launched in 2002 was the first dedicated met satellite of ISRO. ISRO plans to deploy a new generation dedicated met satellite INSAT-3D over Indian Ocean in 2007–2008 time frame. This chapter presents an overview of the INSAT 3D mission and a detailed description of its Met Payloads, viz., Imager and Sounder (Rani and Prasad, 2013).

Indian National Satellite System (INSAT) is a set of geostationary satellites launched by ISRO in which INSAT-3D is an advanced geostationary meteorological satellite launched on 26 July 2013 from French Guiana using ARIANE rocket (Hawkins et al., 2008). The major applications of INSAT-3D are to improve imaging system for meteorological observations, land surface monitoring and vertical profile generation for weather forecasting and disaster predictions. It is one of the famous satellites developed by ISRO to enhance domestic weather forecasting and also useful for tracking cyclones and monsoons originating from Bay of Bengal and Arabian Sea. INSAT-3D is located at 82° east. It carries a multispectral imager (optical radiometer) for meteorological operations and produces images of Earth in six wavelength bands (Mishra et al., 2014).

This instrument has meteorological payloads on board the spacecraft:

- Six-channel imaging radiometer intended to estimate radiant and solar reflected energy from Earth.
- Nineteen-channel sounder for measuring vertical temperature profiles, humidity as well as for ozone distribution.
- Data relay transponder (DRT) useful to provide data collection and data dissemination using data collection platforms
- Satellite-aided search and rescue (S&SR) system (Katti et al., 2006).

Present study is based on the LST product of INSAT-3D Imager which provides data per day with a temporal resolution of $0.5° \times 0.5°$ which can be downloaded from MOSDAC website (https://mosdac.gov.in/data/).

2.2.1 Instrument detail

2.2.1.1 Imager

INSAT-3D has four meteorological payloads that are mainly designed to observe solar and radiant energy for sampled areas of the Earth's surface. Imager payloads have six imaging radiometer channels, in which one is visible, and five are infrared channels. These channels are designed to observe radiant and solar reflected energy (Pandya et al., 2011). INSAT-3D

Imager can be considered an enhanced version of Very High Resolution Radiometer (VHRR), Series of INSAT five original instruments flown on INSAT 2A through INSAT 3A satellites and on Kalpana-1.

INSAT-3D Imager spectral channels are very comparable to those of five channel of NOAA GOES Imager but diverse for additional channels in the Short Wave IR (SWIR) band. INSAT-3D generates Earth image in six spectral channels and for scanning the Earth's disk it uses scan mirror ascended on two axes Gimbal. The Visible and SWIR channels hold ground resolution at nadir 1 and 8 km in water vapour band. Sufficient radiometric resolution and dynamic range are provided for all channels to understand the application science goals. INSAT-3D is also capable of generating full Earth disk image in 26 min. A flexible scan pattern allows tradeoff between the coverage and the imaging periodicity. Visible band is useful for the monitoring of mesoscale phenomena and severe local storms. And the other two new bands, SWIR with resolution of 1 km and MWIR 4 km, will enable improved land-cloud discrimination and detection of surface features like snow. INAST-3D has TIR channel that holds 4 km resolution with two separate windows in 10.2−11.2 and 11.5−12.5 μm regions. INSAT-3D has some modules, one of which is the EO (electro-optics) module which containing the telescope, scan assembly and detectors with cooler facility. This module is mounted on the external part of spacecraft, and in the internal part of the spacecraft all electronic packages are installed. It has a total mass of ~130 kg (Misra and Kirankumar 2014).

2.2.1.2 Sounder

INSAT-3D sounder is the first geostationary INSAT series instrument being developed by ISRO. The main objective of this instrument is to measure temperature and humidity profiles that are also known as vertical distributions. Vertical profiles of temperature and humidity are mainly enabled by sounder. These vertical profiles can then be used to obtain various atmospheric stability indices and other parameters such as atmospheric water vapour content and total column ozone amount. It can obtain three-dimensional representation of the atmosphere. Sounder has 19 spectral channels which are used to measure radiation; it is useful to sense specific data parameters for atmospheric vertical temperature and moisture profiles, surface and cloud top temperature and ozone distribution. Eighteen narrow channels are distributed over three IR bands (seven long-wave (LW), five mid-wave (MW), six short-wave (SW)), while one is a broad visible channel (Katti et al., 2006). Sounder uses two axes gimbaled scan mirror for measuring radiance in 18 IR and one visible channel oven an area of 10*40 km at nadir every 100 ms. INSAT 3D Sounder is very similar to the NOAA GOES Sounder instruments. Sounder provides acceptable radiometric resolution that is useful for application of science. The operation of the Sounder is limited by ground commands using some parameters like gain, sounding area location and other such parameters, and sounding area is defined in terms of east–west and north–south 'blocks'. Like imager payload, sounder also has EO (electro-optics) module as well as collection of electronic packages with power supply modules. These modules hold telescope, scan assembly, filter wheels and coolers and work like imager instrument mounted in the internal part of spacecraft. It contains a total mass of ~145 kg (Prasad et al., 2009).

2.2.1.3 DRT (Data Relay Transponder)

Data Relay Transponder is a meteorological payload of INSAT-3D satellite; its main focus is to receive global meteorological, hydrological and oceanographic data from automatic DCPs (Data Collection Platforms). That data stores in the ground segment and relays back to downlink in extensive C-band. The satellites which have enabled DCPs can provide easily a great solution for gathering large amount of meteorological data from all over the country including inaccessible and remote areas. In the collaboration of IMD (Indian Meteorological Department) and ISRO more than 1800 DCPs have been established that will be more helpful for meteorological data collection (Bhattacharya et al., 2009).

2.2.1.4 SAS&R (Satellite Aided Search & Rescue)

SAS&R (Satellite Aided Search & Rescue) is meteorological payload that mainly focuses on relaying a distress signal/alert detection from the beacon transmitters for search and rescue purposes with global receive coverage in UHF band and it operates at 406 MHz (Rani et al., 2016). The downlink operates in extended C-band. The data are transmitted to INMCC (Indian Mission Control Center), located at ISTRAC (ISRO Telemetry, Tracking and Command Network), Bangalore (Prabhu, 2017).

2.3 Hamon's method

In this study, Hamon's method is used as a standard method for calculation of PET. As we know, it is very difficult to get large number of data like solar radiation, wind speed, rainfall. So, to overcome this problem Hamon's method was used in this study that used only minimal amount of data. Hamon's method used only temperature data for estimation of PET. Because of using minimal amount of data Hamon's method is a very simple and robust method for calculating ET (McCabe et al., 2015). PET is calculated using INSAT-3D satellite data, as well as compared with the observed PET Eq. (8.1).

Hamon's equation can be expressed as follows:

$$PET = K * 0.165 * 216.7 * n * \left(\frac{e_s}{T + 273.3}\right) \tag{8.1}$$

where *PET* is in (mm/day); *k* is the proportionality coefficient; *N* is the daytime length; e_s is the saturation vapour pressure (mb) and *T* is the average monthly temperature. Saturation vapour pressure (e_s) can be calculated using Eq. (8.2).

$$e_s = 6.108_e\left(\frac{17.27T}{T + 273.3}\right) \tag{8.2}$$

3. Performance analysis

In this study PET was assessed from land surface temperature data of INSAT-3D satellite and compared with the observed dataset. For understanding the performance analysis between datasets some statistics used correlation coefficient (r), absolute bias and RMSE (root mean square error). Correlation coefficient measured association between two variables that variables indicate the strength and suitable direction of their relationship (Matrix and Variable).

The correlation coefficient (r) is calculated using this Eq. (8.3).

$$r = \frac{n\sum xy - (\sum x)(\sum y)}{\sqrt{\left[n\sum x^2 - (\sum x)^2\right]\left[n\sum y^2 - (\sum y^2)\right]}} \tag{8.3}$$

To measure model performance in meteorology, air quality and climate research, studies root mean square error (RMSE) has been used as a standard statistical metric (Chai and Draxler, 2014).

RMSE (Root Mean Square Error) is calculated using this Eq. (8.4):

$$RMSE = \sqrt{\left(\frac{1}{n}\sum_{i=1}^{n}[y_i - x_i]^2\right)} \tag{8.4}$$

Bias used to understand under- or overestimate of the true value. Absolute bias is mainly useful to remove faulty measuring devices or procedures and also absolute bias calculates to understand the positive or negative deviation from the actual and observed value (Koch et al., 1982; Walther and Moore, 2005).

Bias is calculated using this Eq. (8.5):

$$\boldsymbol{Bias} = (\bar{\boldsymbol{y}} - \bar{\boldsymbol{x}}) \tag{8.5}$$

where *x* and y are observed and estimated values respectively. $\bar{\boldsymbol{x}}$ and $\bar{\boldsymbol{y}}$ are the mean of observed and estimated measurements, respectively and n is the total no. of observations.

4. Results and discussions

4.1 Evaluation of temperature data from INSAT-3D and ground base

In this study, land surface temperature was extracted from satellite data and observed datasets. Comparison of temperature derived from INSAT-3D and observed data is showing gradual variations that can be shown in (Fig. 8.1). In line graph, it can be clearly seen that both the temperatures are showing very close relation with each other. Performance statistics are shown in (Fig. 8.2). Correlation (r = 0.733), Bias (−2.030) and RMSE (4.993) was

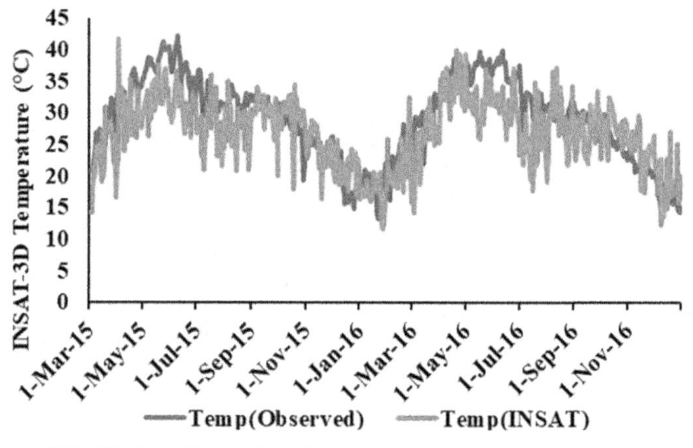

FIGURE 8.1 INSAT-3D daily temperature with observed datasets.

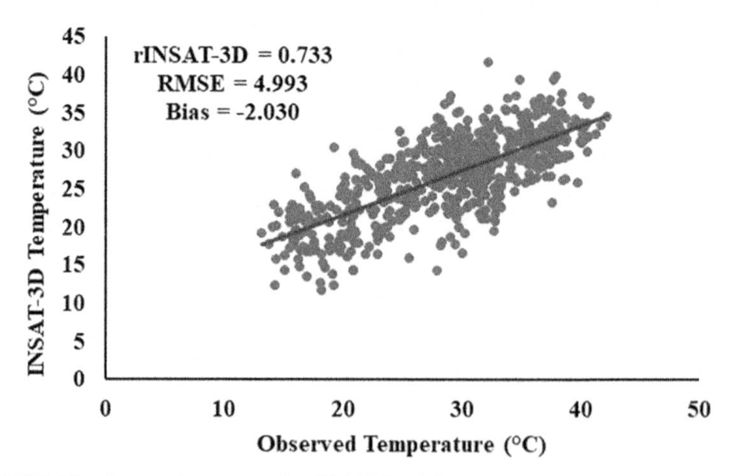

FIGURE 8.2 Scatter plot representing INSAT-3D daily temperature with observed datasets.

observed when INSAT-3D data was compared with that of observed temperature data indicating a close agreement of temperature with in-situ observation. A higher temperature was recorded in summer months (April to June), with gradual increase in both the temperature datasets. In terms of RMSE and bias, INSAT-3D showed a satisfactory performance with observed temperatures datasets (Fig. 8.2).

4.2 Comparative assessment of evapotranspiration products

PET calculated using land surface temperature of INSAT-3D satellite data over agricultural area in Northern India was compared with the observed dataset. The relative plot of INSAT-3D data with observed PET is shown in Fig. 8.3. The PET increases during the months of summer season and decreases in winter season. INSAT-3D data showed an underestimation most of the time (Fig. 8.4). Higher PET was observed from April to July months; this may be due to the very high temperatures in these months. Correlation (r) = 0.572, Bias = 0.524 and RMSE = 0.834 are found between PET of INSAT-3D and observed data. Overall, INSAT-3D data showed highest discrepancies (Fig. 8.4). PET estimation has been done on an annual basis using Hamon's method and values are compared with the annual observed PET. It is clear that PET estimated using INSAT-3D is in marginal agreement with observed PET in annual datasets. The INSAT-3D PET values were always found to be higher

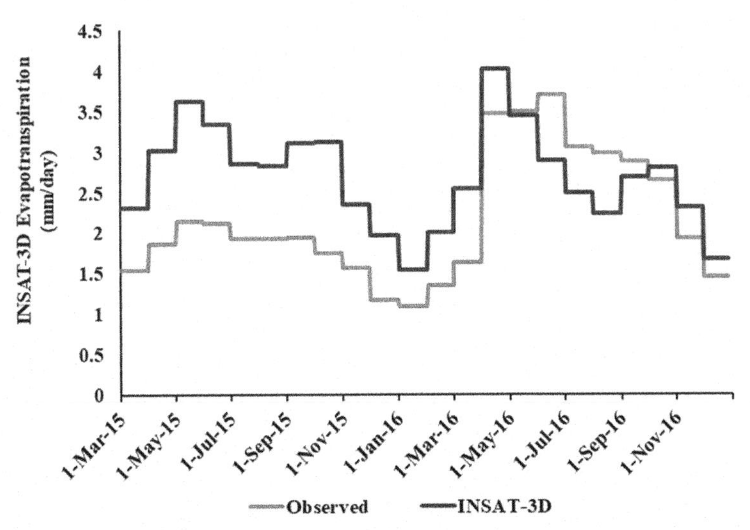

FIGURE 8.3 INSAT-3D daily PET with observed datasets.

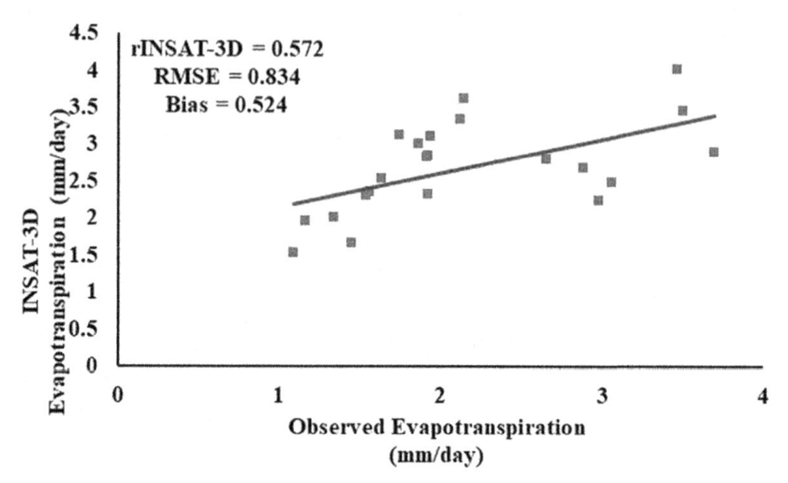

FIGURE 8.4 Scatter plot representing INSAT-3D daily PET with observed datasets.

in comparison to observed PET values. It is clearly seen in Fig. 8.4. Bias and RMSE values were higher when compared with the observed dataset. so, it indicates INSAT-3D dataset could be promising, but needs more research regarding the quality. Further, as Hamon's method has some limitations, better models like Penman Monteith may improve the estimation of PET.

5. Conclusions

In hydrological modelling, weather research and prediction of flood and drought and accurate estimation of potential evapotranspiration could be very helpful. Evapotranspiration plays a very important role in hydrological modelling. Land surface temperature data of INSAT-3D satellite is used for calculation of potential evapotranspiration (PET). This study has brought out useful information about evapotranspiration in Varanasi district. The PET values are calculated using the Hamon's method. Hamon's method does not depend upon any other factor; only climatic factors are enough for this estimation. The PET values estimated from the INSAT-3D showed some overestimation as compared to observed PET values from station datasets. Comparison of INSAT-3D evapotranspiration shows a promising match with the observed dataset in terms of trend. The comparison of PET showed quite a close correlation. This study can improve forecasting application and effectiveness of hydro-meteorological modelling. Accurate information of evapotranspiration can provide incredible support in the study of sustainable water resource management (Srivastava et al., 2016). However, the study has lot of potential and could be extended over other regions where relevant data are available.

References

Allen, R.G., Tasumi, M., Morse, A., Trezza, R., 2005. A Landsat-based energy balance and evapotranspiration model in Western US water rights regulation and planning. Irrigat. Drain. Syst. 19 (3), 251–268.

Allen, R.G., Tasumi, M., Trezza, R., 2007. Satellite-based energy balance for mapping evapotranspiration with internalized calibration (METRIC)—model. J. Irrigat. Drain. Eng. 133 (4), 380–394.

Almorox, J., Hontoria, C., 2004. Global solar radiation estimation using sunshine duration in Spain. Energy Convers. Manag. 45 (9–10), 1529–1535.

Amatya, D.M., Harrison, C.A., Trettin, C.C., 2014. Comparison of Potential Evapotranspiration (PET) Using Three Methods for a Grass Reference and a Natural Forest in Coastal Plain of South Carolina.

Bartholic, J., Wiegand, C., 1970. Remote Sensing in Evapotranspiration Research on the Great Plains.

Bhattacharya, B., Mallick, K., Padmanabhan, N., Patel, N., Parihar, J., 2009. Retrieval of land surface albedo and temperature using data from the Indian geostationary satellite: a case study for the winter months. Int. J. Remote Sens. 30 (12), 3239–3257.

Cai, X., Wang, D., Laurent, R., 2009. Impact of climate change on crop yield: a case study of rainfed corn in central illinois. J. Appl. Meteorol. & Climatol. 48 (9), 1868–1881.

Caselles, V., Sobrino, J., Coll, C., 1992. On the use of satellite thermal data for determining evapotranspiration in partially vegetated areas. Int. J. Remote Sens. 13 (14), 2669–2682.

Chai, T., Draxler, R.R., 2014. Root mean square error (RMSE) or mean absolute error (MAE)? –arguments against avoiding RMSE in the literature. Geosci. Model Dev. (GMD) 7 (3), 1247–1250.

Cheruku, D.R., 2010. Satellite Communication. IK International Pvt Ltd.

Hatfield, J., 1983. Evapotranspiration obtained from remote sensing methods. Adv. Irrig. 2, 395–416. Elsevier.

Hawkins, G., Sherwood, R., Barrett, B., Wallace, M., Orr, H., Matthews, K., Bisht, S., 2008. High-performance infrared narrow-bandpass filters for the Indian National Satellite System meteorological instrument (INSAT-3D). Appl. Optic. 47 (13), 2346–2356.

Idso, S., Schmugge, T., Jackson, R., Reginato, R., 1975a. The utility of surface temperature measurements for the remote sensing of surface soil water status. J. Geophys. Res. 80 (21), 3044–3049.

Idso, S.B., Jackson, R.D., Reginato, R.J., 1975b. Detection of soil moisture by remote surveillance: difficult problems limit immediate applications, but the potential social benefits call for serious attempts at their solution. Am. Sci. 63 (5), 549–557.

Idso, S.B., Jackson, R.D., Reginato, R.J., 1975c. Estimating evaporation: a technique adaptable to remote sensing. Science 189 (4207), 991–992.

Jackson, R.D., 1985. Evaluating evapotranspiration at local and regional scales. Proc. IEEE 73 (6), 1086–1096.

Kar, S.K., Nema, A., Singh, A., Sinha, B., Mishra, C., 2016. Comparative study of reference evapotranspiration estimation methods including Artificial Neural Network for dry sub-humid agro-ecological region. J. Soil Water Conserv. 15 (3), 233–241.

Katti, V., Pratap, V., Dave, R., Mankad, K., 2006. INSAT-3D: An Advanced Meteorological Mission over Indian Ocean. GEOSS and Next-Generation Sensors and Missions. International Society for Optics and Photonics.

Khoob, A.R., 2008. Comparative study of Hargreaves's and artificial neural network's methodologies in estimating reference evapotranspiration in a semiarid environment. Irrigat. Sci. 26 (3), 253–259.

Koch, G.G., Kotz, S., Johnson, N., Read, C., 1982. Encyclopedia of Statistical Sciences, vol. 213. J Wiley, New York, p. 7.

Kustas, W., Norman, J., 1996. Use of remote sensing for evapotranspiration monitoring over land surfaces. Hydrol. Sci. J. 41 (4), 495–516.

Liou, Y.-A., Kar, S.K., 2014. Evapotranspiration estimation with remote sensing and various surface energy balance algorithms—a review. Energies 7 (5), 2821–2849.

Mall, R., Gupta, B., n.d. Comparison of Evapotranspiration Models.

Mall, R., Gupta, B., 2002. Comparison of evapotranspiration models. Mausam 53 (2), 119–126.

Matrix, C., Variable, I. n.d.Correlation Coefficient (r).

McCabe, G.J., Hay, L.E., Bock, A., Markstrom, S.L., Atkinson, R.D., 2015. Inter-annual and spatial variability of Hamon potential evapotranspiration model coefficients. J. Hydrol. 521, 389–394.

Mishra, M.K., Rastogi, G., Chauhan, P., 2014. Operational retrieval of aerosol optical depth over Indian subcontinent and Indian ocean using INSAT-3D/imager and product validation. ISPRS–Int. Arch. Photogramm. Remote Sens. & Spat. Inf. Sci. 8, 277–282.

Misra, T., Kirankumar, A., 2014. A glimpse of ISRO's EO programme [Space Agencies]. IEEE Geosci. & Remote Sens. Mag. 2 (4), 46–53.

Moran, M.S., Jackson, R.D., 1991. Assessing the spatial distribution of evapotranspiration using remotely sensed inputs. J. Environ. Qual. 20 (4), 725–737.

Navalgund, R.R., Jayaraman, V., Roy, P., 2007. Remote sensing applications: an overview. Curr. Sci. 93 (12), 00113891.

Pandya, M., Shah, D., Trivedi, H., Panigrahy, S., 2011. Simulation of at-sensor radiance over land for proposed thermal channels of imager payload onboard INSAT-3D satellite using MODTRAN model. J. Earth Syst. Sci. 120 (1), 19–25.

Petropoulos, G.P., Ireland, G., Cass, A., Srivastava, P.K., 2015. Performance assessment of the SEVIRI evapotranspiration operational product: results over diverse mediterranean ecosystems. IEEE Sens. J. 15 (6), 3412–3423.

Petropoulos, G.P., Srivastava, P.K., Piles, M., Pearson, S., 2018. Earth observation-based operational estimation of soil moisture and evapotranspiration for agricultural crops in support of sustainable water management. Sustainability 10 (1), 181.

Prabhu, B., 2017. A Research on Sensors and Positioning of INSAT-3D.

Prasad, M., Akkimaradi, B., Selvan, T., Rastogi, S., Badrinarayana, K., Bhandari, D., Sugumar, M., Mallesh, B., Rajam, K., Rajgopalan, I., 2009. Development of Sunshield Panels for Passive Radiant Cooler on Board Meteorological Instruments of ISRO.

Quattrochi, D.A., Luvall, J.C., 1999. Thermal infrared remote sensing for analysis of landscape ecological processes: methods and applications. Landsc. Ecol. 14 (6), 577–598.

Rango, A., 1994. Application of remote sensing methods to hydrology and water resources. Hydrol. Sci. J. 39 (4), 309–320.

Rani, S.I., Prasad, V., 2013. Simulation and Validation of INSAT-3D Sounder Data at NCMRWF.

Rani, S.I., Prasad, V., Rajagopal, E., Basu, S., 2016. Height of warm core in very severe cyclonic storms Phailin: INSAT-3D perspective. In: Remote Sensing and Modeling of the Atmosphere, Oceans, and Interactions VI. International Society for Optics and Photonics.

Rao, K., George, C., Ramasastri, K., 1971. Potential evapotranspiration over India. India Met. Dept. Sci. Rep 136.

Rao, L., Sun, G., Ford, C., Vose, J., 2011. Modeling potential evapotranspiration of two forested watersheds in the southern Appalachians. Trans. ASABE 54 (6), 2067–2078.

Schultz, G.A., Engman, E.T., 2012. Remote Sensing in Hydrology and Water Management. Springer Science & Business Media.

Senay, G.B., Budde, M., Verdin, J.P., Melesse, A.M., 2007. A coupled remote sensing and simplified surface energy balance approach to estimate actual evapotranspiration from irrigated fields. Sensors 7 (6), 979–1000.

Singh, G., Agrawal, H., Singh, M., 1989. Genesis and classification of soils in an alluvial pedogenic complex. J. Indian Soc. Soil Sci. 37 (2), 343–354.

Singh, R.K., Irmak, A., Irmak, S., Martin, D.L., 2008. Application of SEBAL model for mapping evapotranspiration and estimating surface energy fluxes in south-central Nebraska. J. Irrigat. Drain. Eng. 134 (3), 273–285.

Singh, R.K., Pawar, P., 2011. Comparative study of reference crop evapotranspiration (ETo) by different energy based method with FAO 56 Penman-Monteith method at New Delhi, India. Int. J. Eng. Sci. Technol. 7861–7868.

Srivastava, P.K., Han, D., Islam, T., Petropoulos, G.P., Gupta, M., Dai, Q., 2016. Seasonal evaluation of evapotranspiration fluxes from MODIS satellite and mesoscale model downscaled global reanalysis datasets. Theor. Appl. Climatol. 124 (1–2), 461–473.

Srivastava, P.K., Han, D., Rico Ramirez, M.A., Islam, T., 2013. Comparative assessment of evapotranspiration derived from NCEP and ECMWF global datasets through Weather Research and Forecasting model. Atmos. Sci. Lett. 14 (2), 118–125.

Srivastava, P.K., Han, D., Yaduvanshi, A., Petropoulos, G.P., Singh, S.K., Mall, R.K., Prasad, R., 2017. Reference evapotranspiration retrievals from a mesoscale model based weather variables for soil moisture deficit estimation. Sustainability 9 (11), 1971.

Srivastava, P.K., Singh, P., Mall, R., Pradhan, R.K., Bray, M., Gupta, A., 2020. Performance assessment of evapotranspiration estimated from different data sources over agricultural landscape in Northern India. Theor. Appl. Climatol. 140 (1–2), 145–156.

Thakur, J.K., Srivastava, P., Pratihast, A.K., Singh, S.K., 2011. Estimation of evapotranspiration from wetlands using geospatial and hydrometeorological data. In: Geospatial Techniques for Managing Environmental Resources. Springer, Netherlands, pp. 53–67. Springer.

Wagle, P., Bhattarai, N., Gowda, P.H., Kakani, V.G., 2017. Performance of five surface energy balance models for estimating daily evapotranspiration in high biomass sorghum. ISPRS J. Photogramm. Remote Sens. 128, 192–203.

Walther, B.A., Moore, J.L., 2005. The concepts of bias, precision and accuracy, and their use in testing the performance of species richness estimators, with a literature review of estimator performance. Ecography 28 (6), 815–829.

Chapter 9

Estimation of evapotranspiration using surface energy balance system and satellite datasets

Garima Shukla[1], Prasoon Tiwari[1], Vikas Dugesar[1], Prashant K. Srivastava[2,3]

[1]*Department of Geography, Banaras Hindu University, Varanasi, Uttar Pradesh, India;* [2]*Remote Sensing Laboratory, Institute of Environment and Sustainable Development, Banaras Hindu University, Varanasi, Uttar Pradesh, India;* [3]*DST-Mahamana Centre of Excellence in Climate Change Research, Institute of Environment and Sustainable Development, Banaras Hindu University, Varanasi, India*

1. Introduction

Ever increasing demand for fresh water is making fresh water scarcer over the time. Therefore, sustainable planning for efficient use of water resources will be very much needed in the near future. Agriculture accounts for approximately 70% of the global fresh water withdrawal on average (FAO, 2017). FAO (2017) projects around 50% increase in the irrigated food production by 2050, but the amount of water withdrawn can be limited to only 10% by using improved agricultural practices. One of the major factors in irrigation water management and water resource planning for different land-use types, particularly for agricultural crops, is the knowledge of actual water use. However, actual evapotranspiration (ET) is an effective way to determine actual water use for different land uses. ET is a procedure which is a composite of two different procedures, viz., evaporation from bare land surfaces and transpiration from plants (Petropoulos et al., 2015, 2016). Solar radiation is consumed in both of these processes that convert the water present on the earth into vapour. Water, be it from precipitation, irrigation or depressions, is withdrawn and rebounded to the atmosphere.

Because of the processes involved at the soil, vegetation and atmosphere interfaces, the estimation of ET under actual field conditions is still a tough task for water managers and scholars (Jovanovic and Israel, 2004; Sett et al., 2018).

Agricultural Water Management. https://doi.org/10.1016/B978-0-12-812362-1.00009-6

It is essential to assess ET quantitatively to comprehend the hydrological cycle and for sustainable management of water resources. Conventionally ET is being computed using climate data taking alfalfa as reference crop. Then, using area specific crop coefficient (K) water requirement for different crop stages is estimated (Nistor et al., 2020). Various methods (Allen et al., 1998; Doorenbos and Pruitt, 1977) are used to estimate ET using meteorological data. Over time, due to these models the precision of ET estimation has increased (Brutsaert, 1979; Srivastava et al., 2020). But, point data-based models are often practically not feasible to record broad-scale spatial variations and not able to provide a good estimation of ET at large scale (Tuya et al., 2005; Srivastava et al., 2014, 2015). Access to in situ meteorological parameters is not possible all the time in some ungauged catchments. To overcome this constraint various researchers explored downscaled forecasted weather data using medium-scale forecasting models (Srivastava et al., 2013, 2016, 2017).

The hydrological approaches are mainly based on the water balance concept—a fixed type of system—in which outgoing water is subtracted from the incoming water and the remnant is the water storage capacity (Szolgay et al., 2003). In this system, if all the components of the outgoing water are known and just ET is unknown in the water balance, at that point ET can be evaluated. Hypothetically, this is the least complex technique for determining ET, but practically this method is imprecise because of some difficulties; for example, the inability to estimate the difference in ground water storage, and also the erroneous and inconsistent estimations of the other outgoing water components (Senay et al., 2011). Past researches have used conventional strategies in assessment of ET, based on hydrological approaches which are appropriate only for point-scale estimations (Senay et al., 2011).

Smaller-scale meteorological methods have additionally been developed and used to evaluate water loss. These incorporate the utilization of scintillometry, eddy covariance system, Bowen ratio and the surface restoration techniques (Odhiambo and Savage, 2009). Micro-meteorological system needs costly specific observation tools, which are rarely accessible in rural and remote regions of any state. Therefore, remote sensing–based techniques give a useful alternative to the spatial portrayal of ET at various spatial and temporal scales. Remote sensing–based systems provide representative estimates of different Earth surface physical attributes at various scales, from a point to a continent (Dube et al., 2014). Consequently, remote sensing techniques can precisely estimate ET over an extensive variety of conditions, at both the satellite overpass time and daily time scale, in a cost-effective manner (Dube et al., 2014; Ruhoff et al., 2012). Furthermore, technological developments and requirement for timely and precise assessments of ET have brought about tremendous enhancements to approaches that depend on remote sensing information (Srivastava et al., 2016). Remotely detected energy balance models have therefore been produced to gauge the spatial variations of ET at various spatial scales. Energy balance models assess ET, in view of the energy balance approach.

Broadly utilized models incorporate the Surface Energy Balance System (SEBS) (Su, 2002), the Surface Energy Balance Algorithm for Land (SEBAL) (Bastiaanssen et al., 1998) and the Mapping Evapotranspiration at High Resolution with Internalized Calibration (METRIC) model (Allen et al., 2007). The SEBS model has been broadly utilized in light of the fact that it is relevant at all spatial scales (i.e., local to regional); it is uninhibitedly accessible and it has been approved in a few investigations (Su et al., 2007). The SEBS models likewise comprise of an arrangement of algorithms to determine surface bio-physical properties (i.e., albedo, emissivity, temperature, vegetation cover), which are derived from satellite reflectance and radiance (Su et al., 1999). It additionally has a broadened model for determining roughness length for heat exchange; and another detailing for determining evaporative fraction based on energy balance at limiting cases (Su, 2002), when contrasted with other energy balance models. Since its formulation, the SEBS model has been broadly applied in assessing ET from local to continental scales (Glenn et al., 2007; Rwasoka et al., 2011) under different climatic conditions and land cover characteristics, for example, farming fields, natural forests, wetlands and plantations. For example, Jin et al. (2013) used the SEBS model for assessing ET in the Qaidam Basin and its eight hydrological subregions in northwest China. Then again, Chen et al., (2013) tried and assessed the robustness of SEBS in evaluating ET on typical land surfaces of the Tibetan Plateau, in view of a time-series of observations at four sites with uncovered soil, sparse canopy, dense canopy and snow surface, respectively. Nevertheless, there are extremely limited researches that have endeavoured to evaluate ET in the semi-arid, plateau regions of Bundelkhand, MP. Little is comprehended with respect to the conceivable variation of ET among the consistently changing land uses, because of atmosphere fluctuations and human interactions. Remote sensing technology can provide land surface parameters, for example, albedo, vegetation lists and surface temperature, which are imperative to remote detecting—based energy balance models for scaling-up ET and surface energy transitions to bigger spatial and longer temporal scales. It is perceived as the best way to retrieve ET at a few temporal and spatial scales (Liou and Kar, 2014; Zhang et al., 2016). To this end, there has been a noteworthy effort in the course of recent years to develop techniques that are based on remote sensing and give spatially dispersed surface fluxes maps utilizing airborne and satellite information (Khaldi et al., 2011; Petropoulos et al., 2018). ET gives the connection between the net energy flux and water budgets on the Earth surface. The SEBS model, which gauges turbulent heat fluxes of the atmosphere and evaporative fractions using radiation fluxes and land surface temperatures as determined by the satellite, is utilized to evaluate surface energy fluxes over the Chhatarpur and Panna districts of Madhya Pradesh during the end of winters (i.e., February of both years).

The primary objective of this study is to estimate actual evapotranspiration of two districts namely Chhatarpur and Panna of Madhya Pradesh, India, using the various bands of Landsat TM and OLI data with the help of SEBS method. Secondary objectives include (a) radiometric correction of the Landsat images for both the years (1995 and 2015), (b) to identify data and parameters which are required for the estimation of evapotranspiration and (c) to perform land-use classification using SVM algorithm and also compute vegetation indices to apply the SEBS approach step by step to calculate ET.

2. Materials and methods

2.1 Study area

Chhatarpur and Panna districts lie in the northernmost part of the state Madhya Pradesh of India and the study area which comprises of both the districts extends between the parallels of 24° 06′ and 25° 05′ North latitudes and between the meridians of 78° 59′ and 80° 40′. The highest peak lies at Ban Pathar (24° 37′: 79° 45′) in the district at 607 m average mean sea level. The central plateau runs to the north as an offshoot of Panna range. Ken River and its tributaries drain almost entire Panna district.

The climate in Chhatarpur and Panna is mild, and generally warm and temperate, having an average annual diurnal temperature range of 33°C as maximum and 18°C as minimum. The normal maximum temperature seen is during the month of May and is 42.30°C and minimum is during the month of January, which is 7.10°C. The normal annual rainfall of the districts is 1070 mm, and these districts receive their maximum rainfall (around 85% of the total rainfall) in the period of June to September that is during the southwest monsoon (Fig. 9.1).

Panna and Chhatarpur districts include ecologically rich forest pockets of north and south forest divisions. The area is occupied by forests, and vegetation belonging to seasonal as well as aquatic and marshy categories. The forests include mainly dry deciduous forests. However based on vegetation composition it may be further divided into six forest types, viz., southern dry deciduous teak forest, northern dry deciduous mixed forest, dry deciduous scrub forest, Salai forest, dry bamboo forest and Kardhai forest. Rainy, winter and summer are three main vegetative seasons in the region. While agriculture is the dominant occupation in the study area, land used for cultivation of crops in the region is impressively lower than other agriculture belts of the nation.

Wheat accounts for the largest area under cultivation in both the districts. In rabi season wheat is sown as the major crop while in kharif season rice is the most sown crop in both the districts. Jowar is sown in both the seasons as secondary crop in Chhatarpur while in Panna, gram and lentil are chosen to be the secondary crops in the rabi season.

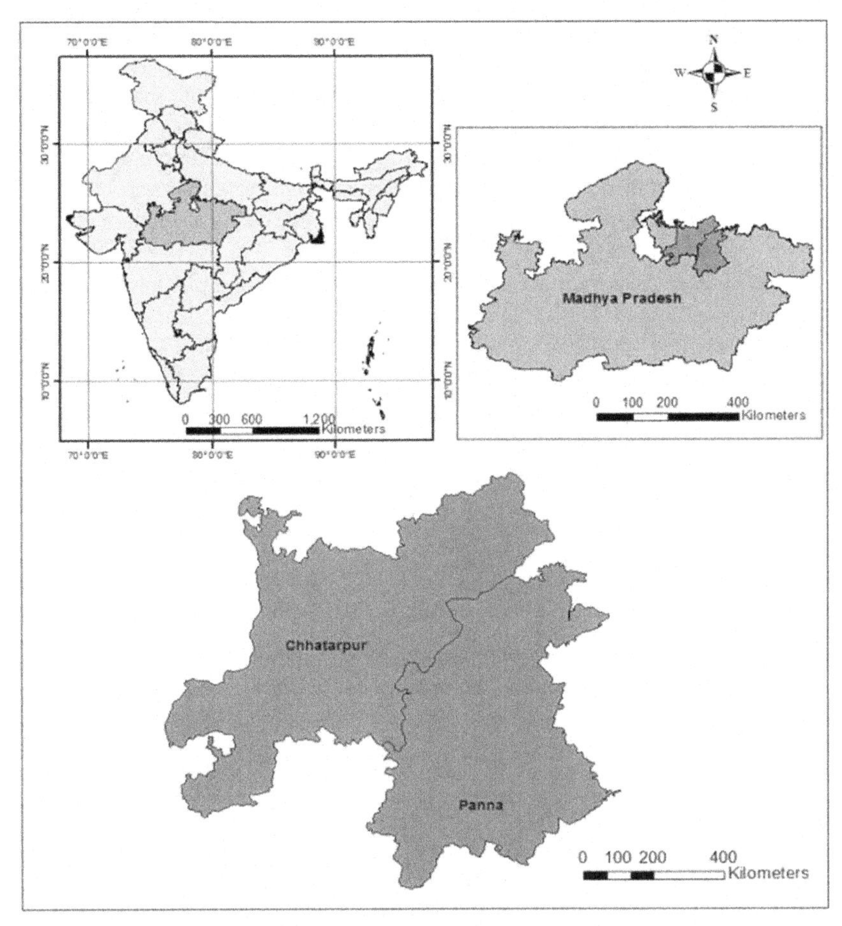

FIGURE 9.1 Area map.

3. Data used and methodology

Temporal remote sensing data of 23 February 1995 (Landsat TM 5 with 30m resolution) and 14 February 2015 (Landsat 8 OLI with 30 m resolution) were downloaded from https://earthexplorer.usgs.gov and used as primary data for the study. ASTER DEM 30 m data was used to derive topographic elevation information for the study.

It is necessary that the images used should be free from clouds. ET cannot be calculated for the cloud covered data, as the thermal band measurements drop and result in erroneous calculation of heat fluxes for even a thin film of cloud. Therefore, all satellite images were closely evaluated for the existence of cloud cover. SEBS requires a remotely sensed satellite image for the estimation of actual ET. A land-use/land-cover map for the study area is also useful in determining the variation of ET with varying land cover.

Landsat 8-OLI bands 2–7 data for the visible and near infrared spectrum were utilised. The pixel size for these bands is 30 m by 30 m. TIRS Band 10 and 11 provide data for longwave (thermal) radiation. The pixel size for this band is 100 m by 100 m for Landsat 8-OLI, which is rescaled at 60 m resolution for level 1 data. Where as in Landsat 5 TM bands 1–5 data for the visible and near infrared spectrum were used having 30 m by 30 m resolution. Thermal band 6 gives data for longwave radiation having a pixel size of 60 m by 60 m.

For this study, the Surface Energy Balance System (SEBS) model was selected to estimate the instantaneous evapotranspiration in Chhatarpur and Panna districts of Madhya Pradesh, India. SEBS model is a surface energy balance model proposed by (Su, 2002) to estimate land surface fluxes and evaporative fraction which can be driven by a combination of satellite and elevation data.

4. Theory of SEBS

Surface energy balance forms the basis of SEBS model for the calculation of ET flux from the satellite images. Through SEBS one estimates actual ET flux for the image acquisition time as the satellite images can provide information for the satellite overpass time only. The ET flux is calculated for individual pixels of the imagery as a 'remnant' of the surface energy balance equation:

$$\boldsymbol{\lambda \cdot ET = (R_n - G) - \mathbf{H}}$$

where λET is the latent heat flux (W/m^2), R_n is the net radiation flux at the surface (W/m^2), G is the soil heat flux (W/m^2) and H is the Sensible Heat Flux (W/m^2).

4.1 Surface radiation balance equation

As the first step in the SEBS procedure, the net surface radiation flux (R_n) was calculated using the surface radiation balance. The net radiation flux (R_n) symbolises the actual radiant energy available at the surface and was calculated by deducing all outgoing radiant fluxes from all incoming radiant fluxes, i.e., '**Net surface radiation = radiation gains −radiation losses'.** This is given by the surface radiation balance equation:

$$\boldsymbol{R_n = (R_{s\downarrow} - \alpha R_{s\downarrow}) + (R_{L\downarrow} - R_{L\uparrow}) - (1 - \varepsilon_o) \cdot R_{L\downarrow}} \tag{9.1}$$

where $R_{s\downarrow}$ is the incoming shortwave radiation (W/m^2), α is the surface albedo (dimensionless), $R_{L\downarrow}$ is the incoming long wave radiation (W/m^2), $R_{L\uparrow}$ is the outgoing long wave radiation (W/m^2) and ε_o is the surface thermal emissivity (dimensionless).

4.1.1 Surface albedo (α)

Surface albedo can be described as the ratio of the reflected radiations to the incident shortwave radiation. It was calculated in SEBS using the following steps:

The spectral radiance for each band (L_λ) was calculated by using band math tool. It is the outgoing radiation energy of the band at the top of the atmosphere as measured by the satellite. It was calculated using the following equation given for Landsat 8.

$$\mathbf{L_\lambda = M_L \cdot Qcal + A_L} \quad \text{(For Landsat 8 OLI)} \tag{9.2}$$

$$\mathbf{L_\lambda = \left(\frac{Lmax - Lmin}{255}\right) \cdot DN + Lmin} \quad \text{(For Landsat 5 TM)} \tag{9.3}$$

where L_λ is Top of Atmosphere (TOA) spectral radiance (W/($m^2 \cdot$ srad $\cdot$ μm)), M_L is band-specific multiplicative rescaling factor from the metadata (Radiance_Mult_Band_x, where x is the band number), A_L is band-specific additive rescaling factor from the metadata (Radiance_Add_Band_x, where x is the band number), Qcal is quantized and calibrated standard product pixel values (DN).

The reflectivity for each and every one band (ρ_λ) was calculated. The reflectivity of a surface is defined as the ratio of the reflected radiation flux to the incident radiation flux. It was calculated using the following equation given for Landsat images:

$$\rho_\lambda = M_p \cdot Qcal + A_p \quad \text{(For Landsat 8 OLI)} \tag{9.4}$$

$$\rho_\lambda = \frac{\pi \cdot L_\lambda}{E_{sun} \cdot \cos\theta \ \mathrm{d}r} \quad \text{(For Landsat 5 TM)} \tag{9.5}$$

where M_p is band-specific multiplicative rescaling factor from the metadata (Reflectance_Mult_Band_x, where x is the band number), A_p is band-specific additive rescaling factor from the metadata (Reflectance_Add_Band_x, where x is the band number), Qcal is quantized and calibrated standard product pixel values (DN), L_λ is the spectral radiance for each band computed, $ESUN_\lambda$ is the mean solar exo-atmospheric irradiance for each band (W/m^2/μm), cos θ is the cosine of the solar incidence angle (from nadir) and d_r is the inverse squared relative Earth-Sun distance.

Cosine θ was computed using the header file data on Sun elevation angle (β) where θ = (90° − β). The term d_r is defined as $1/d^2_{e\text{-}s}$ where $d_{e\text{-}s}$ is the relative distance between the Earth and the Sun in astronomical units. d_r was computed using the following equation by Duffie and Beckman (1980), also given in FAO 56 paper:

$$\mathbf{d_r = 1 + 0.033 \cdot \cos\left(DOY \frac{2\pi}{365}\right)} \tag{9.6}$$

where DOY is the sequential day of the year (given in SEBAL manual, 2000), and the angle (DOY × 2π/365) is in radians. Values for d_r range from 0.97 to 1.03 and are dimensionless.

The albedo at the top of the atmosphere (α_{toa}) was calculated and is the albedo unadjusted for atmospheric transmissivity and was computed as follows:

$$\alpha_{toa} = \Sigma(\omega_\lambda \times \rho_\lambda) \tag{9.7}$$

where ρ_λ is the reflectivity computed earlier and ω_λ the weighting coefficient for each band is computed as follows:

$$\omega_\lambda = \frac{E_{sun_\lambda}}{\sum E_{sun_\lambda}} \tag{9.8}$$

The final step is to compute the surface albedo. Surface albedo is computed by correcting the α_{toa} for atmospheric transmissivity:

$$\alpha = \frac{\alpha_{toa} - \alpha_{path\_radiance}}{\tau_{sw}^2} \tag{9.9}$$

where $\alpha_{path\_radiance}$ is the average portion of the incoming solar radiation across all bands that is back-scattered to the satellite before it reaches the Earth's surface and τ_{sw} is the atmospheric transmissivity.

Values for $\alpha_{path\_radiance}$ ranges between 0.025 and 0.04 and for SEBS we recommend a value of 0.03 based on Bastiaanssen (2002).

Atmospheric transmissivity is described as the proportion of incident radiation that is transmitted by the atmosphere, and it shows the effects of absorption and reflection by the atmosphere. This affects both the incoming radiation and outgoing radiation and was therefore squared. τ_{sw} includes transmissivity of both direct solar beam radiation and diffuse (scattered) radiation to the surface. We calculated τ_{sw} assuming cloud-free and relatively dry conditions using an elevation-based relationship from FAO-56 paper:

$$\tau_{sw} = 0.75 + 2\times10^{-5}\times z \tag{9.10}$$

where z is the elevation above sea level (m). This elevation should best represent the area of interest, for elevation DEM data has been used.

4.1.2 Incoming shortwave radiation ($R_{S\downarrow}$)

Incoming shortwave radiation is the direct and diffuse solar radiation flux that actually reaches the Earth's surface (W/m^2). It was calculated, assuming cloud-free conditions, as a constant for the image time using:

$$R_{s\downarrow} = G_{sc} \times cos\theta \times d_r \times \tau_{sw} \tag{9.11}$$

where G_{sc} is solar constant (1367 W/m^2), and τ_{sw} is the atmospheric transmissivity. This calculation is done in band math calculator in ENVI. Values for $R_{s\downarrow}$ can range from 200 to 1000 W/m^2 depending on the time and location of the image.

4.1.3 Outgoing long wave radiation ($R_{L\uparrow}$)

The outgoing long wave radiation is the thermal radiation flux emitted from the Earth's surface to the atmosphere (W/m^2). It was calculated following the given steps.

- **NDVI:** Normalized Difference Vegetation Index (NDVI) was computed using the reflectivity values in the band math calculator. The NDVI is the ratio of the difference in reflectivities for the near-infrared band and the red band to their sum:

$$NDVI = (\rho_{NIR} - \rho_R)/(\rho_{NIR} + \rho_R) \tag{9.12}$$

where ρ_{NIR} and ρ_R are reflectivities for Near Infrared and Red bands of Landsat TM and OLI.

The *NDVI* is a sensitivity indicator of the amount, condition and growth stage of green vegetation. Values for *NDVI* range between −1 and +1. Green surfaces have an *NDVI* between 0 and 1 and water and cloud are usually less than zero.

- **Surface Emissivity (ε_o):** Surface emissivity (ε_o) is defined as the effectiveness of a surface to radiate thermal energy compared to energy radiated by a blackbody at the same temperature. The emissivity representing surface behaviour for thermal emission in the relatively narrow band 10 and 11 of Landsat 8 (10.60 to 12.51 μm), expressed as ε_o and was used in calculation of surface temperature (T_s) for Landsat 5 TM data and later on to calculate total long wave radiation emission from the surface.

$$\varepsilon_o = \mathbf{0.004PV + 0.986} \tag{9.13}$$

where PV is proportional vegetation and is computed with the help of NDVI

$$PV = \left[\frac{NDVI - NDVI_{min}}{NDVI_{max} - NDVI_{min}}\right]^2 \tag{9.14}$$

where $NDVI_{min}$ and $NDVI_{max}$ are the minimum and maximum values of *NDVI*, respectively, as computed using band math.

The surface temperature (T_s) is computed by using the following modified Planck equation:

$$T = \frac{K_2}{\ln\left(\frac{K_1}{L_\lambda}\right) + 1} \text{ (For Landsat 8 TIRS data)} \tag{9.15}$$

$$T = \frac{K_2}{\ln\left(\frac{\varepsilon_o \cdot K_1}{L_\lambda}\right) + 1} \text{(For Landsat 5 TM data)} \quad (9.16)$$

where L_λ is the corrected thermal radiance from the surface using thermal bands 10 and 11 for Landsat 8 OLI and thermal band 6 of Landsat 5 TM. K_1 and K_2 are constants for Landsat images.

- **Outgoing Long wave Radiation ($R_{L\uparrow}$):** Finally, the outgoing long wave radiation ($R_{L\uparrow}$) was calculated using the Stefan-Boltzmann equation:

$$R_{L\uparrow} = \varepsilon_o \cdot \sigma \cdot T_s^4 \quad (9.17)$$

where ε_o is the surface emissivity (dimensionless), σ is the Stefan-Boltzmann constant (5.67×10^{-8} W/m^2/K^2), and T_s is the surface temperature (K).

Values for $R_{L\uparrow}$ can range from 200 to 700 W/m^2 depending on the location and time of the image.

4.1.4 Incoming long wave radiation ($R_{L\downarrow}$)

The incoming long wave radiation is the downward thermal radiation flux from the atmosphere (W/m^2). It is computed using the Stefan-Boltzmann equation:

$$R_{L\downarrow} = \varepsilon_a \cdot \sigma \cdot T_a^4 \quad (9.18)$$

where ε_a is the atmospheric emissivity (dimensionless), σ is the Stefan-Boltzmann constant (5.67×10^{-8}W/m^2/K^4) and T_a is the near surface air temperature (K). The following empirical equation for ε_a by Bastiaanssen (1995) is applied for our conditions:

$$\varepsilon_a = \mathbf{0.85} \cdot (-\ln\tau_{sw})^{\mathbf{0.09}} \quad (9.19)$$

where τ_{sw} is the atmospheric transmissivity calculated from equation for τ_{sw}.

Using above equations and T_{cold}, i.e., LST corresponding to wet conditions as calculated from the corelation established between albedo and LST yields the following equation:

$$R_{L\downarrow} = 0.85 \cdot (-ln\tau_{sw})^{0.09} \cdot \sigma \cdot T_{cold}^4 \quad (9.20)$$

This computation was done in ENVI using band math tool. Values for $R_{L\downarrow}$ can range from 200 to 500 W/m^2, depending on the location and time of image. Surface albedo (α), outgoing long wave radiation ($R_{L\uparrow}$), surface emissivity ($_o$) and atmospheric emissivity ($_a$) were the inputs along with the incoming short wave radiation ($R_{S\downarrow}$) and the incoming long wave radiation ($R_{L\downarrow}$) and band math was performed.

4.2 Soil heat flux (G)

Soil heat flux is defined as the rate of storage of heat into the soil and vegetation by the process of conduction. Through SEBS, at first, the ratio G/R_n was calculated using the following empirical equation developed by (Bastiaanssen, 2000) representing values near mid-day:

$$\frac{G}{R_n} = \frac{Ts}{\alpha} \cdot (0.0038 \cdot \alpha + 0.0074 \cdot \alpha^2) \cdot (1 - 0.98 \cdot NDVI^4) \tag{9.22}$$

where T_s is the surface temperature (°C), α is the surface albedo and *NDVI* is the Normalized Difference Vegetation Index. G is then readily calculated by multiplying G/R_n by the value for R_n calculated.

Soil heat flux is a difficult term to calculate and therefore was calculated carefully. Land classification and soil type affects the value of G and a land-use map is thus important for identifying the various surface types. The equation was utilised for the estimation of G for irrigated crops near Kimberly, Idaho, and was found that it predicted quite accurately (Allen et al., 2002).

4.3 Sensible heat flux (H)

Sensible heat flux is defined as the change in heat lost to the air by convection and conduction with time, as a result of difference in temperature. It can be calculated by using the following equation for heat transport:

$$H = (\rho \cdot cp \cdot dT)/r_{ah}$$

where ρ is air density (kg/m^3), cp is air specific heat (1004 J/kg/K), dT (K) is the temperature difference ($T_1 - T_2$) between two heights (z_1 and z_2) and r_{ah} is the aerodynamic resistance to heat transport (s/m).

The sensible heat flux (H) is a function of the temperature gradient, surface roughness and wind speed. It is difficult to calculate the sensible heat flux as there are two unknowns, r_{ah} and dT. Among all the parameters of surface energy balance equation, the estimation of H is the most complex component. So another effective parameter, i.e., evaporative fraction ($\wedge$) was constrained by the hot and cold regions and formulated by interpolating the albedo dependent surface temperature between albedo-dependent maximum and minimum surface temperatures (Danodia et al., 2017).

$$\wedge = \frac{T_{hot} - T_s}{T_{hot} - T_{cold}} \tag{9.23}$$

where T_{hot} is the land surface temperature corresponding to dry condition and represents the minimum latent heat flux and maximum sensible heat flux. T_{cold} is land surface temperature analogous to wet condition and depicts the maximum latent heat flux and minimum sensible heat flux over the region (Fig. 9.2).

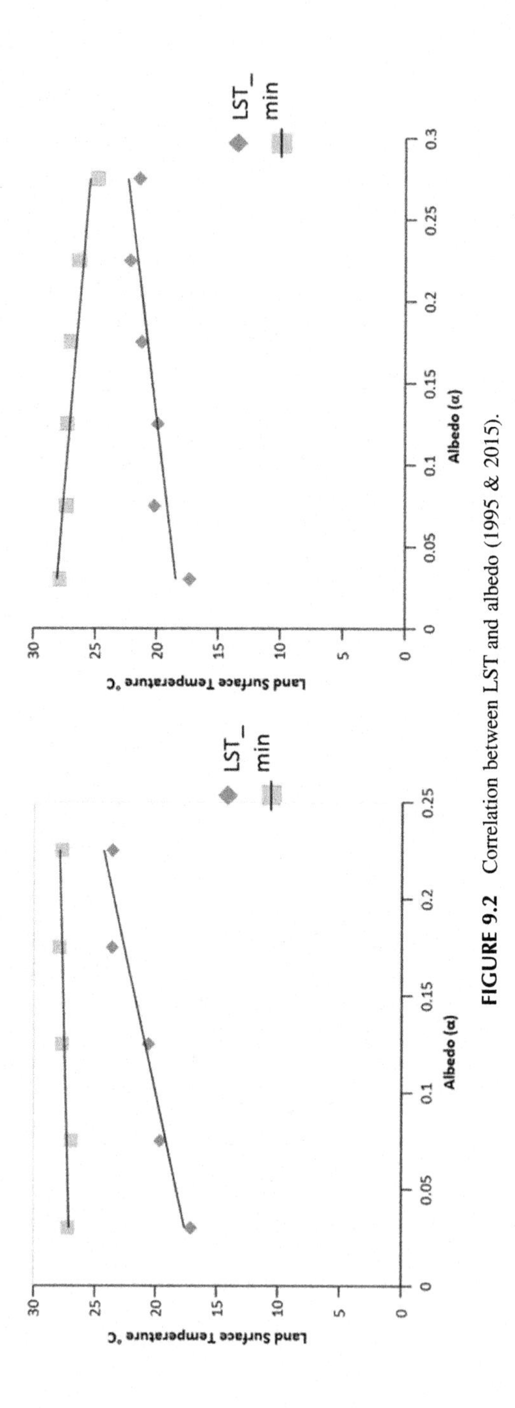

FIGURE 9.2 Correlation between LST and albedo (1995 & 2015).

Using the following regression equation, T_{hot} and T_{cold} can be respectively calculated:

$$T_{hot} = X_{max} + Y_{max} \cdot \alpha \tag{9.24}$$

$$T_{cold} = X_{min} + Y_{min} \cdot \alpha \tag{9.25}$$

where the empirical coefficients X_{max}, Y_{max}, X_{min} and Y_{min} are estimated from the scatter plot of LST and albedo over the study area.

4.4 Latent heat flux ($\lambda \cdot ET$), instantaneous ET (ET_{inst})

The soil heat flux (G) is deducted from the net radiation flux at the surface (R_n) and the resultant is multiplied with evaporative fraction ($\wedge$) to calculate the 'remnant' energy available for evapotranspiration ($\lambda \cdot ET$).

$$\wedge = \frac{\lambda \cdot ET}{R_n - G} \tag{9.26}$$

Finally, sensible heat flux (H) was calculated from evaporative fraction ($\wedge$) given below:

$$\wedge = \frac{\lambda ET}{\lambda ET + H} \tag{9.27}$$

Once the latent heat flux ($\lambda \cdot ET$) was calculated for each pixel, an equal amount of instantaneous ET (mm/h) was readily calculated.

$$ET_{inst} = 3600 \frac{\lambda \cdot ET}{\lambda} \tag{9.28}$$

where ET_{inst} is the instantaneous ET (mm/h), 3600 is the time conversion from seconds to hours and λ is the latent heat of vaporization or the heat absorbed when a kilogram of water evaporates (J/kg).

5. Results and discussion

5.1 Land-use/Land-cover classification (LULC)

A land-use map is not a requirement for SEBS but is highly recommended, since it helps us in deriving relation between the land use/land cover and the various flux parameters (LST, albedo, soil heat flux, etc.). The land-use map delineates different classes of land use such as agriculture-cover, built-up, water bodies, barren and forest to outline the pattern and behaviour of land surface interaction with solar radiation elements. Support Vector Machine (SVM) classifier has been used for classification of the study area in ENVI 5.3. The SVM classifier acts as a powerful, modern, supervised classification tool, which is capable of handling a segmented raster data or a standard image. In comparison to other classification techniques the SVM classifier is rather a new classification tool that is broadly utilised (Srivastava et al., 2012;

Singh et al., 2014; Banerjee and Srivastava, 2013; Nandi et al., 2017). For standard image inputs, the tool accepts multiple-band imagery with any bit depth, and performs the SVM classification on a pixel basis, based on the input training feature file (Wang et al., 2017).

A hierarchical classification system was used to classify the data in land-use, land-cover classes. The LULC classification of the study area clearly shows the temporal changes which occurred over the 20 years (1995–2015). In 1995 there was lesser interaction of humans in the region thus most of the areas were either forest (43.73%) or barren surface (25.64%), whereas the agricultural land use was about 13.67%. In 2015 changes are clearly visible showing the effects of human interaction on land-use practices increasing the agricultural land to 45.28% from 13.67%, whereas forest showed deduction as a result of deforestation from 43.73% to 24.09% of the total land cover (Fig. 9.3).

5.2 Albedo

Albedo is the measurement of how well a surface reflects or absorbs the incoming radiations. The results show that the albedo for sand bar and rock out crop is the highest as they reflect larger amount of incident radiation than the other land surfaces open to radiation, whereas water bodies absorb most of the incoming radiations followed by forest cover showing lower albedo values (Table 9.1). Albedo was found to have greater range values in 1995 than in 2015 as the radiations were found to interact with much larger bare land surfaces in the former year than in 2015 (Fig. 9.4).

5.3 Normalized difference vegetation index (NDVI)

NDVI generally differentiates the vegetation from the nonvegetative areas. As the agricultural areas in the region of study have grown extensively over 20 years, major changes in the NDVI values can be seen in the agriculture occupied areas. Main forests being tropical dry teak forest, tropical dry deciduous mixed forest, dry deciduous scrub forest and bamboo forest, the NDVI values of the forest region of the study area are lower than the NDVI values of well-irrigated agricultural land use. The values of NDVI for both 1995 and 2015 years were found to be greater in agricultural region than in forest areas while the water bodies show negative values of NDVI (Fig. 9.5).

5.4 Land surface temperature (LST)

Land surface temperature is the response of mixture of vegetation and bare soil temperature to incoming solar radiation as both have their own contribution. The warmth rising of the region's landscape influences (and is influenced by) the weather conditions as well as land-use change. The variation in the land

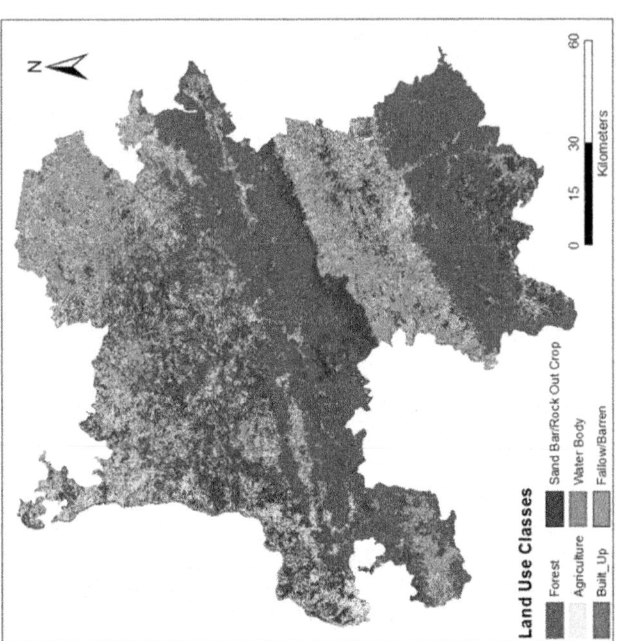

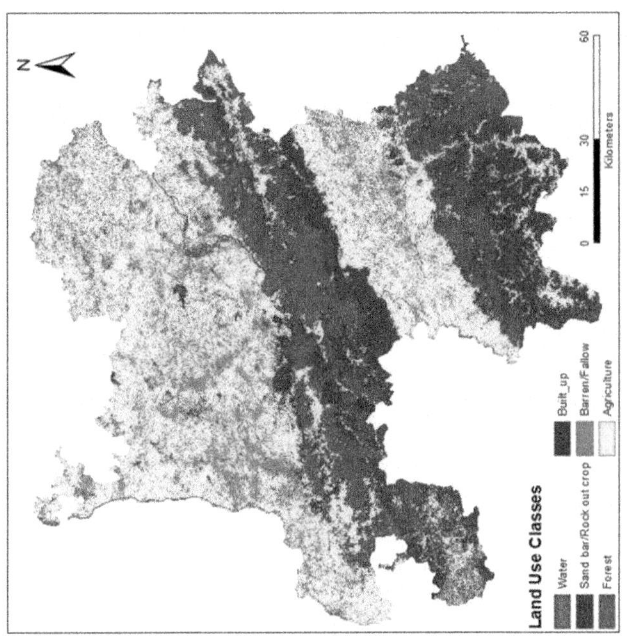

FIGURE 9.3 Land-use/Land-cover maps of the study area (1995 & 2015).

TABLE 9.1 Comparative values of different parameters.

Parameters		1995	2015
LULC	Forest	43.73%	24.09%
	Agriculture	13.67%	45.28%
	Barren	25.64%	10.46%
Albedo		0.0075–0.3004	0.0462–0.4316
NDVI		−0.5695–0.7897	−0.3778–0.7912
LST (°C)		14.8024–32.7184	15.3984–35.3984
Net radiation (W/m^2)		330.888–580.481	385.047–723.909
Soil heat flux (W/m^2)		25.9576–61.6445	37.4177–87.6418
Sensible heat flux (W/m^2)		0–467.338	0–581.287
Instantaneous ET (mm/h)		0–0.871	0–1.065

use and land cover caused the LST of the region chosen for study to register an increment of 3°C (approx.) over the 20 years from 1995 to 2015. The percentage of forest canopy covers a larger area in 1995 to check the interaction of solar radiation with bare surface. Thus forest canopy acts as a sink for the radiations resulting in a lower land surface temperature. Compared to 1995, LST was found to have greater values as a result of deforestation and exposed bare surfaces to solar radiation (Fig. 9.6). Temporally the surface temperature showed an increase of 3°C (approx.) from 1995 to 2015 which was an impact of forest degradation and deforestation.

5.5 Net radiation flux (R_n)

Earth's net radiation, also known as net radiation flux, is the difference between the total incoming energy in form of radiation and total outgoing radiation at the top of the atmosphere. It is the aggregate energy that is accessible to impact the climate of a region. Energy comes in to the system when daylight enters the highest point of the air. Energy goes out in two different ways: reflection by clouds, particulate matters like aerosol, or the Earth's surface; and warm radiation, heat emitted by the surface and the air, including clouds. The extremes of net radiation flux are found to be higher in 2015 than in 1995 as the percentage of bare surface in 2015 was greater than in 1995, which reflect back larger percentage of incoming radiation back to the atmosphere (Table 9.1). Increase in the agriculture cover over the area of Chhatarpur and Panna region results in absorption of more incoming radiations and lesser reflection of radiation back to atmosphere (Fig. 9.7).

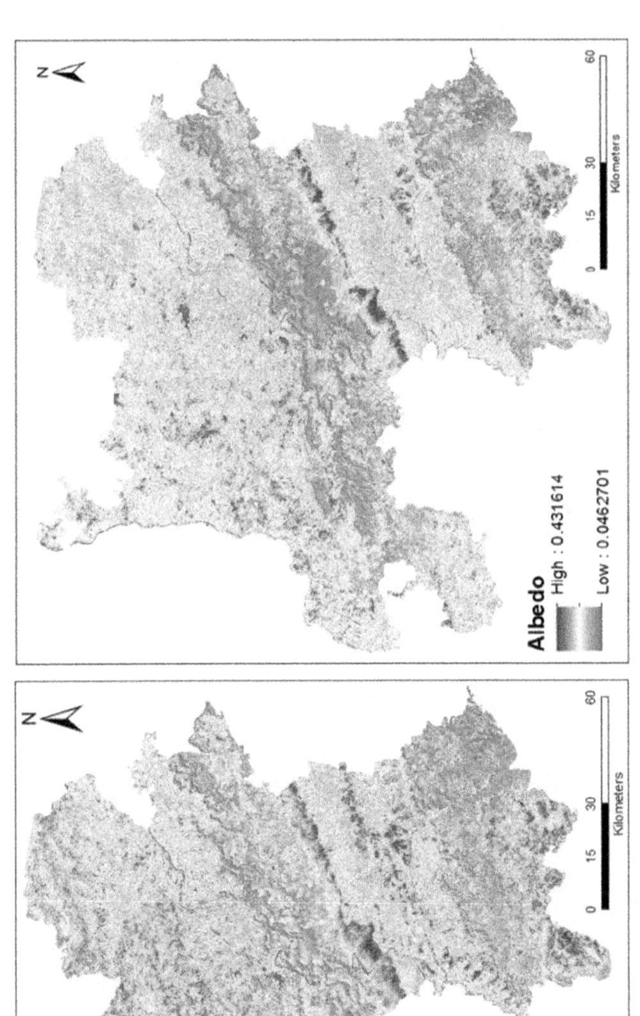

FIGURE 9.4 Albedo maps of the study area (1995 & 2015).

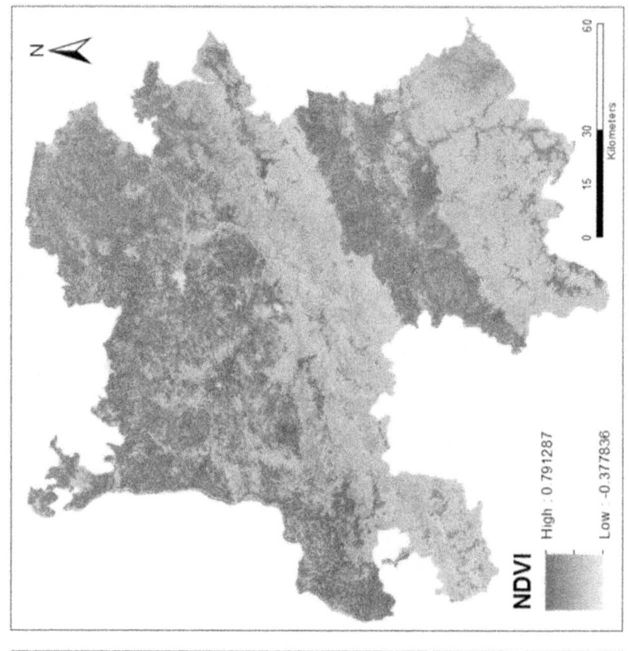

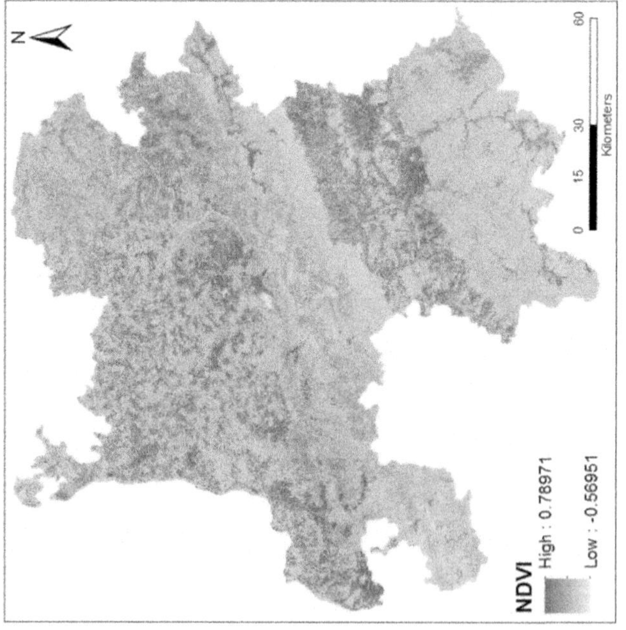

FIGURE 9.5 NDVI maps of the study area (1995 & 2015).

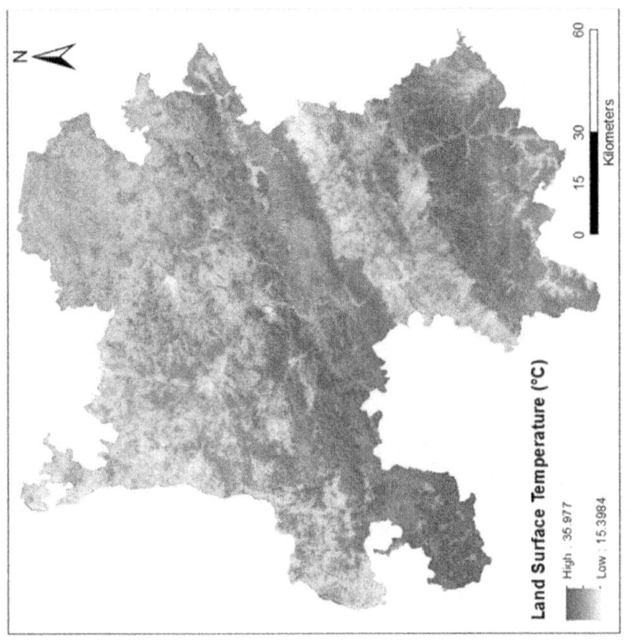

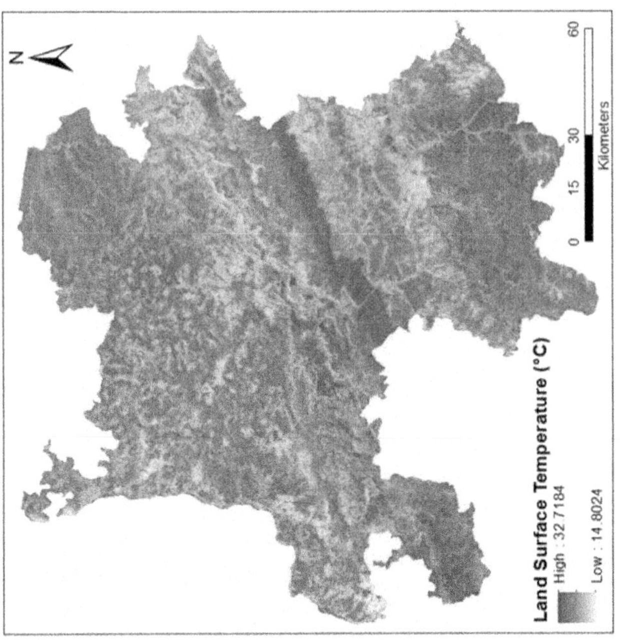

FIGURE 9.6 Land surface temperature maps of the study area (1995 & 2015).

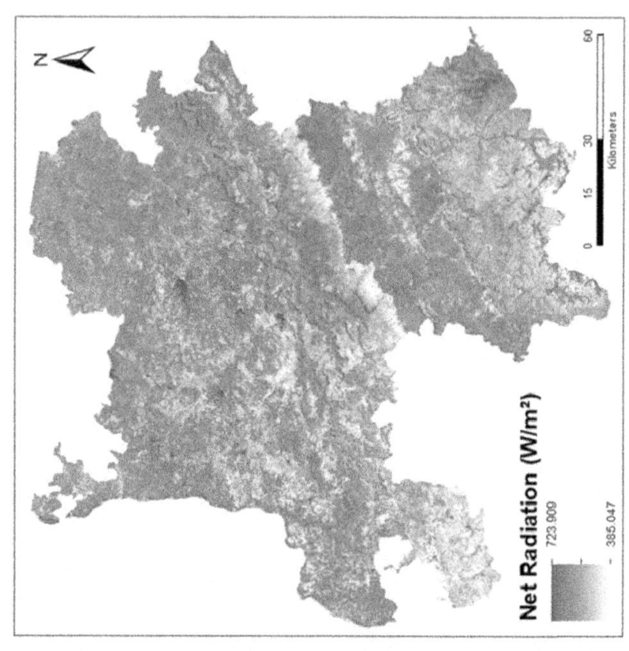

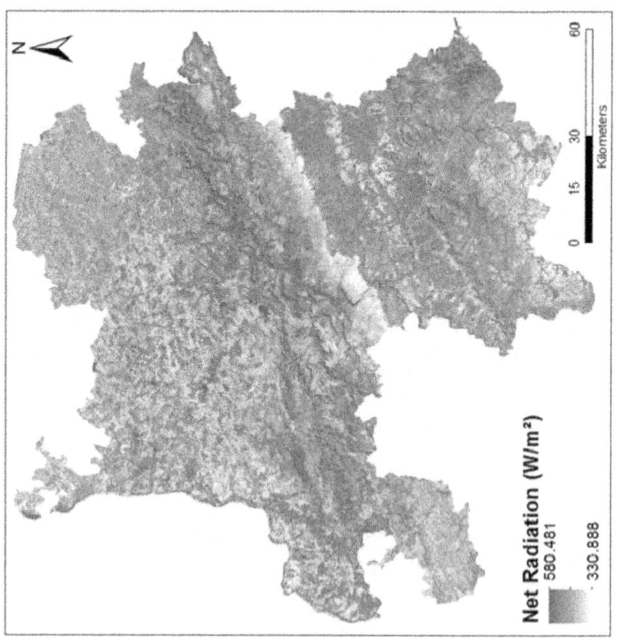

FIGURE 9.7 Net radiation maps of the study area.

5.6 Soil heat flux (G)

As a component of the surface energy balance, the magnitude of soil heat flux (G) varies with surface cover, soil moisture content and solar irradiance (Khaldi et al., 2011). Forest degradation or deforestation has exposed the soil to the direct solar irradiance increasing the flux values from 61 to 87 W/m^2 for corresponding soil surfaces, while the increase in agricultural land use in 2015 as compared to 1995 has resulted in lower soil heat flux values for barren surfaces which were exposed to direct solar irradiance in 1995 (Fig. 9.8).

5.7 Instantaneous evapotranspiration (ET)

The instantaneous or actual evapotranspiration in vegetative area depends on the climatic potential as well as various limiting factors such as moisture availability, types of vegetation and their physiological characteristics among others. The temporal variation in ET and other parameters over a period of 20 years has been given in Table 9.1.

Instantaneous ET estimated based on SEBS method and Landsat-8 imagery showed a distinct pixel-wise variation along different land use and land covers. The main land features controlling the ET characteristics are agriculture and forest cover, which have an area extent of 45.28% and 24.09%, respectively, in 2015, whereas in 1995 agriculture cover was just 13.67% and the forest cover was shading 43.73% of the total study area. ET as a function of land use and land cover was found maximum at the places with good agricultural land use with proper irrigation and minimum at the bare surfaces, i.e., barren land and rock outcrop. Since the land use under agricultural class has a major increment from 13.67% in 1995 to 45.28% in 2015, the ET has also increased and now the water from a larger area is going to the atmosphere. The variation of ET for different land-cover classes show the hourly ET range from 0 at bare land and rock outcrops to 0.533 mm/h at forest area and a maximum of 0.871 mm/h over few cropland areas having dense vegetation with proper water availability while in 2015 the hourly ET range increased to a maximum of 1.065 mm/h. An important factor which also affects the ET of a region is the direct solar radiation. The Tropic of Cancer passes through the study area, i.e., the study area falls in the region of high solar radiation during the peak summers. The forest covers have the faunal diversity all belonging to tropical dry region, viz., tropical dry teak forest, tropical dry deciduous mixed forest, dry deciduous scrub forest and bamboo forest, thus having narrow leaf trees and therefore not resulting in higher transpiration as general tropical mixed or tropical rain forest do (Fig. 9.9).

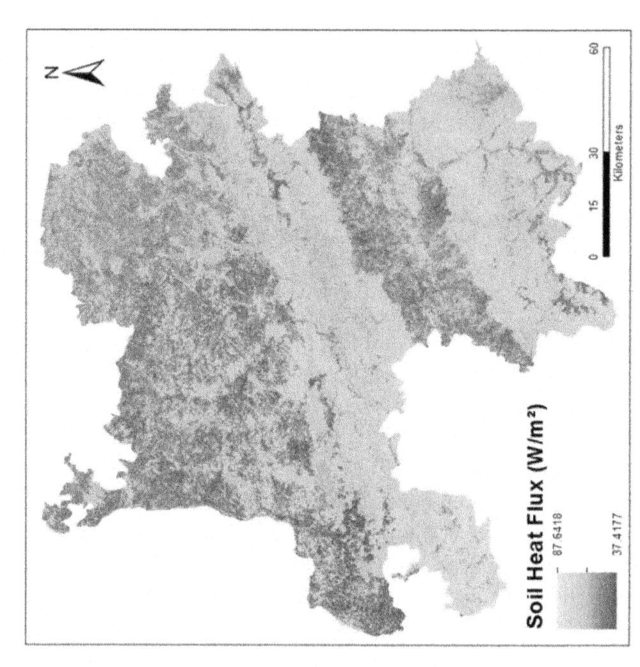

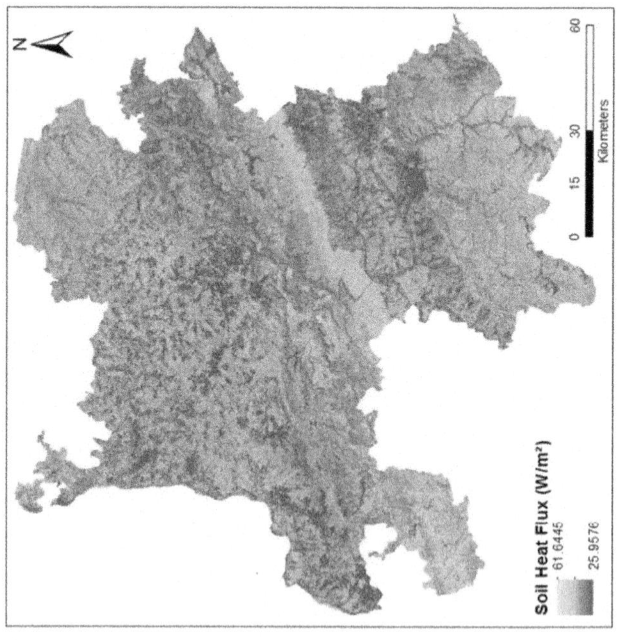

FIGURE 9.8 Soil heat flux maps of the study area (1995 & 2015).

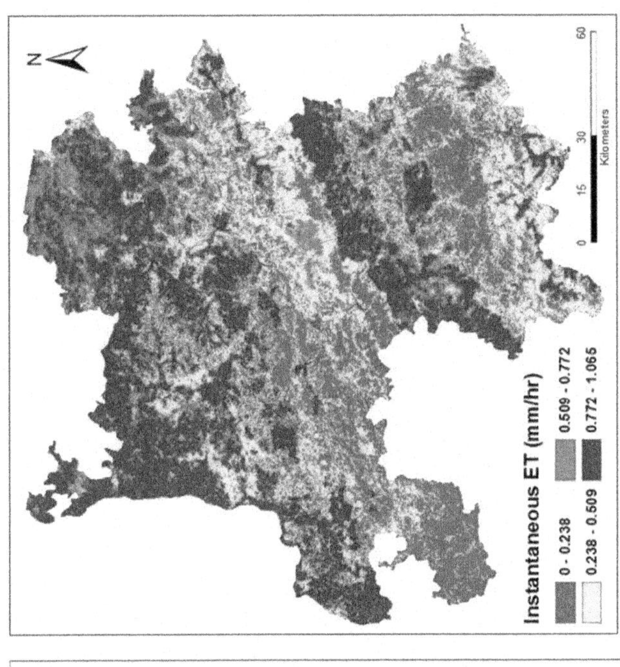

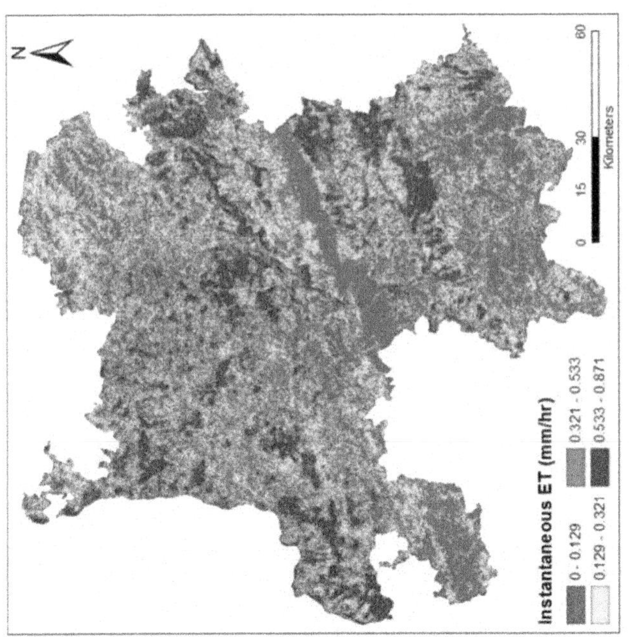

FIGURE 9.9 Actual evapotranspiration maps of the study area (1995 & 2015).

6. Conclusion

The study of evapotranspiration being a key element of the hydrological cycle, it is necessary to estimate actual evapotranspiration for the management of the water resources and their sustainable use. Minimal difference in hourly evapotranspiration showed up in the bare land region and rock outcrops among the land-use classes, which are characterized by the absence of vegetation cover and have shown increased in areal extent over the years. The results define clear relation among the heterogeneity of land-use/land-cover classes and solar radiation following the impact of change in vegetative cover over the 20 years on the values of evapotranspiration for each pixel. Estimated results of instantaneous evapotranspiration based on SEBS method and using Landsat-8 imagery showed a distinct pixel-wise variations along different land use and land covers. The land surface temperature showed a variation spatially as well as temporally following heterogeneity of surface features. The spatial variation is caused due to variable land cover over the study area with water bodies showing lowest surface temperature, whereas the bare land surface and the rock outcrop tends to show maximum surface temperature. Forest covers work as a sink for solar radiation and check the incident sun rays reaching the bare surface, but due to anthropogenic activities, it is reduced now. It is well known that energy and availability of water are the most dominating factors in the process of evaporation and transpiration. Therefore, surface water and solar radiation both act as major factors controlling these processes in the region. This concludes that changing patterns of land use have a direct impact on the climatic conditions and thus impact the evapotranspiration of the region. The change in land cover over the period of years has exposed the land surface to the direct incident solar radiations of the Sun affecting the evapotranspiration rate of the region.

References

Allen, R.G., Morse, A., Tasumi, M., Trezza, R., Bastiaanssen, W., Wright, J.L., Kramber, W., 2002. Evapotranspiration from a satellite-based surface energy balance for the snake plain aquifer in Idaho. Proc. USCID Conf. 167–178.

Allen, R.G., Pereira, L.S., Raes, D., Smith, M., Ab, W., 1998. Fao,1998. Irrigation and Drainage Paper No. 56. FAO. https://doi.org/10.1016/j.eja.2010.12.001.

Allen, R.G., Tasumi, M., Morse, A., Trezza, R., Wright, J.L., Bastiaanssen, W., Kramber, W., Lorite, I., Robison, C.W., 2007. Satellite-based energy balance for mapping evapotranspiration with internalized calibration (METRIC)—applications. J. Irrigat. Drain. Eng. 133 (4), 395–406. https://doi.org/10.1061/(asce)0733-9437.

Banerjee, R., Srivastava, P.K., 2013. Reconstruction of contested landscape: detecting land cover transformation hosting cultural heritage sites from Central India using remote sensing. Land Use Policy **34**, 193–203.

Bastiaanssen, W.G.M., 1995. Regionalization of surface flux densities and moisture indicators in composite terrain: a remote sensing approach under clear skies in Mediterranean climates. SC-DLO.

Bastiaanssen, W.G.M., 2000. SEBAL-based sensible and latent heat fluxes in the irrigated Gediz Basin, Turkey. J. Hydrol. 229, 87–100.

Bastiaanssen, W.G.M., Ahmad, M.U., Chemin, Y., 2002. Satellite surveillance of evaporative depletion across the Indus Basin. Water Resour. Res. 38 (12), 1–9. https://doi.org/10.1029/2001WR000386.

Bastiaanssen, W.G.M., Menenti, M., Feddes, R.A., Holtslag, A.A.M., 1998. A remote sensing surface energy balance algorithm for land (SEBAL): 1. Formulation. J. Hydrol. 198–212. https://doi.org/10.1016/S0022-1694(98)00253-4.

Brutsaert, W., 1979. Heat and mass transfer to and from surfaces with dense vegetation or similar permeable roughness. Boundary-Layer Meteorol. 365–388. https://doi.org/10.1007/BF03335377.

Chen, X., Su, Z., Ma, Y., Yang, K., Wen, J., Zhang, Y., 2013. An improvement of roughness height parameterization of the surface energy balance system (SEBS) over the Tibetan plateau. J. Appl. Meteorol. Climatol. 607–622. https://doi.org/10.1175/JAMC-D-12-056.1.

Danodia, A., Patel, N.R., Chol, C.W., Nikam, B.R., Sehgal, V.K., Taylor, P., September, 2017. Application of S-SEBI model for crop evapotranspiration using Landsat-8 data over parts of North India. Geocarto Int. 34, 114–131. https://doi.org/10.1080/10106049.2017.1374473.

Doorenbos, J., Pruitt, W.O., 1977. Guidelines for predicting crop water requirements. In: FAO Irrigation and Drainage Paper.

Dube, T., Mutanga, O., Elhadi, A., Ismail, R., 2014. Intra-and-inter species biomass prediction in a plantation forest: testing the utility of high spatial resolution spaceborne multispectral rapideye sensor and advanced machine learning algorithms. Sensors 14 (8), 15348–15370. https://doi.org/10.3390/s140815348.

Duffie, J.A., Beckman, W.A., 1980. Solar Engineering of Thermal Processes. Wiley, New York, p. 16951.

FAO, 2017. Water for sustainable food and agriculture water for sustainable food and agriculture. In: A Report Produced for the G20 Presidency of Germany.

Glenn, E.P., Huete, A.R., Nagler, P.L., Hirschboeck, K.K., Brown, P., 2007. Integrating remote sensing and ground methods to estimate evapotranspiration. In: Critical Reviews in Plant Sciences. https://doi.org/10.1080/07352680701402503.

Jin, X., Guo, R., Xia, W., 2013. Distribution of actual evapotranspiration over Qaidam basin, an Arid area in China. Rem. Sens. 6976–6996. https://doi.org/10.3390/rs5126976.

Jovanovic, N., Israel, S., 2004. Critical Review of Methods for the Estimation of Actual Evapotranspiration in Hydrological Models.

Khaldi, A., Hamimed, A., Mederbal, K., Seddini, A., 2011. Obtaining evapotranspiration and surface energy fluxes with remotely sensed data to improve agricultural water management. Afr. J. Food Nutr. Sci. 11 (1) https://doi.org/10.4314/ajfand.v11i1.65881.

Liou, Y.A., Kar, S.K., 2014. Evapotranspiration estimation with remote sensing and various surface energy balance algorithms-a review. Energies 7 (5), 2821–2849. https://doi.org/10.3390/en7052821.

Nandi, I., Srivastava, P.K., Shah, K., 2017. Floodplain mapping through support vector machine and optical/infrared images from Landsat 8 OLI/TIRS sensors: case study from Varanasi. Water Resour. Manag. **31** (4), 1157–1171.

Nistor, M.M., Rai, P.K., Dugesar, V., Mishra, V.N., Singh, P., Arora, A., Kumra, V.K., Carebia, I.A., 2020. Climate change effect on water resources in Varanasi district, India. Meteorol. Appl. 27 (1) https://doi.org/10.1002/met.1863.

Odhiambo, G.O., Savage, M.J., 2009. Surface layer scintillometry for estimating the sensible heat flux component of the surface energy balance. South Afr. J. Sci. 208–216. https://doi.org/10.4102/sajs.v105i5/6.92.

Petropoulos, G.P., Srivastava, P.K., Piles, M., Pearson, S., 2018. Earth observation-based operational estimation of soil moisture and evapotranspiration for agricultural crops in support of sustainable water management. In: Sustainability (Switzerland). https://doi.org/10.3390/su10010181.

Petropoulos, G.P., Ireland, G., Cass, A., Srivastava, P.K., 2015. Performance assessment of the SEVIRI evapotranspiration operational product: results over diverse mediterranean ecosystems. IEEE Sens. J. 15 (6), 3412–3423.

Petropoulos, G.P., Ireland, G., Lamine, S., Griffiths, H.M., Ghilain, N., Anagnostopoulos, V., North, M.R., Srivastava, P.K., Georgopoulou, H., 2016. Operational evapotranspiration estimates from SEVIRI in support of sustainable water management. Int. J. Appl. Earth OBS 49, 175–187.

Ruhoff, A.L., Paz, A.R., Collischonn, W., Aragao, L.E.O.C., Rocha, H.R., Malhi, Y.S., 2012. A MODIS-based energy balance to estimate evapotranspiration for clear-sky days in Brazilian tropical savannas. Rem. Sens. 703–725. https://doi.org/10.3390/rs4030703.

Rwasoka, D.T., Gumindoga, W., Gwenzi, J., 2011. Estimation of actual evapotranspiration using the surface energy balance system (SEBS) algorithm in the upper manyame catchment in Zimbabwe. Phys. Chem. Earth 736–746. https://doi.org/10.1016/j.pce.2011.07.035.

Senay, G.B., Budde, M.E., Verdin, J.P., 2011. Enhancing the simplified surface energy balance (SSEB) approach for estimating landscape ET: validation with the METRIC model. Agric. Water Manag. 606–618. https://doi.org/10.1016/j.agwat.2010.10.014.

Sett, T., Nikam, B.R., Nandy, S., Danodia, A., Bhattacharjee, R., Dugesar, V., 2018. Estimation of instantaneous evapotranspiration using remote sensing based energy balance technique over parts of North India. Int. Arch. Photogramm. Remote Sens. Spatial Inf. Sci. 42 (5), 345–352. https://doi.org/10.5194/isprs-archives-XLII-5-345-2018.

Srivastava, P.K., Han, D., Rico-Ramirez, M.A., Bray, M., Islam, T., 2012. Selection of classification techniques for land use/land cover change investigation. Adv. Space Res. **50** (9), 1250–1265.

Singh, S.K., Srivastava, P.K., Gupta, M., Thakur, J.K., Mukherjee, S., 2014. Appraisal of land use/land cover of mangrove forest ecosystem using support vector machine. Environ. Earth Sci. 71 (5), 2245–2255.

Srivastava, P.K., Han, D., Islam, T., Petropoulos, G.P., Gupta, M., Dai, Q., 2016. Seasonal evaluation of evapotranspiration fluxes from MODIS satellite and mesoscale model downscaled global reanalysis datasets. Theor. Appl. Climatol. 124 (1–2), 461–473. https://doi.org/10.1007/s00704-015-1430-1.

Srivastava, P.K., Han, D., Rico Ramirez, M.A., Islam, T., 2013. Comparative assessment of evapotranspiration derived from NCEP and ECMWF global datasets through weather research and forecasting model. Atmos. Sci. Lett. 14 (2), 118–125. https://doi.org/10.1002/asl2.427.

Srivastava, P.K., Han, D., Rico-Ramirez, M.A., Islam, T., 2014. Sensitivity and uncertainty analysis of mesoscale model downscaled hydro-meteorological variables for discharge prediction. Hydrol. Process. 28 (15), 4419–4432.

Srivastava, P.K., Islam, T., Gupta, M., Petropoulos, G., Dai, Q., 2015. WRF dynamical downscaling and bias correction schemes for NCEP estimated hydro-meteorological variables. Water Resour. Manag. 29 (7), 2267–2284.

Srivastava, P.K., Han, D., Yaduvanshi, A., Petropoulos, G.P., Singh, S.K., Mall, R.K., Prasad, R., 2017. Reference evapotranspiration retrievals from a mesoscale model based weather variables for soil moisture deficit estimation. Sustainability 9 (11). https://doi.org/10.3390/su9111971.

Srivastava, P.K., Singh, P., Mall, R.K., Pradhan, R.K., Bray, M., Gupta, A., 2020. Performance assessment of evapotranspiration estimated from different data sources over agricultural landscape in Northern India. Theor. Appl. Climatol. 1–12.

Su, H., Wood, E.F., Mccabe, M.F., 2007. Evaluation of remotely sensed evapotranspiration over the CEOP EOP-1 reference sites. J. Meteorol. Soc. Jap. 85.

Su, Z., 2002. The surface energy balance system (SEBS) for estimation of turbulent heat fluxes. Hydrol. Earth Syst. Sci. 85–99. https://doi.org/10.5194/hess-6-85-2002.

Su, Z., Pelgrum, H., Menenti, M., 1999. Aggregation effects of surface heterogeneity in land surface processes. Hydrol. Earth Syst. Sci. 3 (4), 549–563. https://doi.org/10.5194/hess-3-549-1999.

Szolgay, J., Hlavčová, K., S. K.-J, of H., 2003. Regional estimation of parameters of a monthly water balance model. Dlib.Lib.Cas.Cz 256–273. Retrieved from. http://dlib.lib.cas.cz:8080/1899/.

Tuya, S., Batjargal, Z., Kajiwara, K., Honda, Y., 2005. Satellite-derived estimates of evapotranspiration in the arid and semi -arid region of Mongolia. Int. J. Environ. Stud. 517–526. https://doi.org/10.1080/00207230500289347.

Wang, X., Wu, J., Lu, F., Jiao, H., Zhang, H., 2017. Optimal selection of algorisms for denoising ICESat-GLAS waveform data and development of a forest crown height estimation model. Linye Kexue/Scientia Silvae Sinicae 62–72. https://doi.org/10.11707/j.1001-7488.20171207.

Zhang, Y., Peña-arancibia, J.L., Mcvicar, T.R., Chiew, F.H.S., Vaze, J., Liu, C., Lu, X., Zheng, H., Wang, Y., Liu, Y.Y., Miralles, D.G., Pan, M., 2016. Multi-decadal Trends in Global Terrestrial Evapotranspiration and its Components. Nature Publishing Group, pp. 1–12. https://doi.org/10.1038/srep19124. August, 2015.

Chapter 10

Large-scale soil moisture mapping using Earth observation data and its validation at selected agricultural sites over Indian region

Dharmendra Kumar Pandey, Deepak Putrevu, Arundhati Misra

Microwave Techniques Development Division (MTDD), Advanced Microwave and Hyperspectral Techniques Development Group (AMHTDG), Earth, Ocean, Atmosphere and Planetary Science Applications Area (EPSA), Space Applications Centre (SAC), Indian Space Research Organization (ISRO), Ahmedabad, Gujarat, India

1. Introduction

Soil moisture is one of the important parameters in the hydrological cycle to drive weather conditions, plant growth, groundwater storage, etc.; thus, it has a role in global climate (Vereecken et al., 2008). Soil moisture consists of only 0.05% of the total water in the global hydrological cycle (Robinson et al., 2008) and 0.001% of the total available freshwater (Drinkwater et al., 2009), but it has been declared as one of the Essential Climate Variables (Dorigo et al., 2011) due to its important role in the hydrological cycle.

Accurate information about soil moisture temporal and spatial variation is therefore important for further application (Vereecken et al., 2016), e.g., for flood and drought forecasts (Wanders et al., 2014; Sheffield et al., 2014), as well as for climate impact studies (Senevirantne et al., 2010). Coarse scale, but moderate temporal resolution global surface soil moisture can be obtained by satellite remote sensing, mostly by microwave sensors (Wagner et al., 2007a,b; Mohanty et al., 2017). Currently several satellite missions provide global surface soil moisture products, such as: The Soil Moisture Active Passive (SMAP) (Enthekhabi et al., 2010a,b), the Soil Moisture and Ocean Salinity (SMOS) (Kerr et al., 2010), the METOP-A/B Advanced Scatterometer

Agricultural Water Management. https://doi.org/10.1016/B978-0-12-812362-1.00010-2

(ASCAT) (Bartails et al., 2007; Wagner et al., 2013), and the Advanced Microwave Scanning Radiometer 2 (AMSR2) (Parinussa et al., 2015). Recently, Space Applications Centre (SAC), Indian Space Research Organization (ISRO), also adopted a modified version of ASCAT operational algorithm (Bartails et al., 2007; Wagner et al., 2013) based on change detection (time-series methodology) and developed daily operational soil moisture products using SMAP L-band brightness temperature data at 12.5 km grid resolution over India, which is available at Visualization of Earth Observation Data and Archival System (VEDAS) and Meteorological & Oceanographic Satellite Data Archival Centre (MOSDAC) web portal of SAC (ISRO) (Dharmendra et al., 2016). These Soil Wetness Index (SWI) and Soil Moisture (SM) data products have wide applications in agriculture productivity assessment, crop-water stress assessment, flood and drought monitoring and meteorological applications. Before their applications to solving various scientific or societal problems, satellite-derived soil moisture data products have to be evaluated, and their validity and accuracy have to be assessed by using in-situ or reference data.

According to Crow et al. (2012), Gruber et al. (2013) and Miralles et al. (2010), the validation of satellite-based soil moisture products typically faces the problem of scale mismatch, and recommendations for the measurement density requirements for ground-based soil moisture networks are given. As the recommended measurements density cannot always be met due to financial constraints, satellite soil moisture products are also validated against hydrological model simulations (Montzka et al., 2013; Juglea et al., 2010). Reviews about coarse-scale soil moisture remote sensing accuracies can be found in Wagner et al. (2013) and Kerr et al. (2016).

The objective of this study is development of large-scale Surface Soil Moisture (SSM) maps and the use of long time-series in-situ data to validate the soil moisture (SAC-ISRO) data products at different agricultural fields of validation sites over India for a period of almost 2 years.

2. Study area

The study area is Berambadi watershed which includes four validation sites (1) Beechanhalli (2) Beemanbeedu (3) Maddur and (4) Madyanhundi, located in the Kabini River basin of South India. The Kabini River basin lies in the south of Karnataka state, India. The location of the Berambadi watershed within the Kabini River basin (AMBHAS Manual Sujala, 2015). The Berambadi watershed has an approximate area of 84 km^2. The Berambadi watershed has a semi-arid climate with a mean annual dominant southwest monsoon rainfall of 800 mm (coefficient of variation is 0.28) and an annual PET of 1100 mm (coefficient of variation is 0.05). There are mainly two types of soils in the watershed comprising of black (Calcic Vertisols) and red (Ferralsols and Calcic Luvisols), underlain by granitic/gneissic rocks, as identified by the geophysical studies. The soils in the Berambadi watershed are representative of the soil types of the granitic/gnessic lithology found in Gundlupet taluk of the Kabini River basin. The Berambadi watershed comprises of land use such as dense/closed forest, scrub forest, land with scrub, Kharif crop, double crop and plantation classes. During Kharif (monsoon rainy season, summer) and Rabi (dry season, winter), crops such as maize, marigold, sunflower, finger millet, garlic, sugarcane and horse gram are grown in various plots. The main cropping period is from May to August (Kharif season) and the second cropping period is from September to December (Rabi season). In the watershed, the surface soil moisture, profile soil moisture, ground water levels and canopy variables were measured. The Installation of hydra probe stations was done at agricultural fields at Beechanhalli, Beemanbeedu, Maddur and Madyanhundi. These stations provide soil moisture, soil temperature and electrical conductivity at two soil depth, i.e., 5 and 50 cm depth.

Table 10.1 shows validation sites with their locations.

3. Data and methodology

The data and methodology used to validate Soil Moisture (SAC-ISRO) data products derived from SMAP radiometer data are described in this section. Time-series plots and statistical analysis were performed on the in-situ soil moisture data with quality checking. Finally, the accuracy of the Soil Moisture (SAC-ISRO) data products was assessed against soil moisture measured in the field.

3.1 SMAP-derived soil moisture data products (SAC-ISRO)

Soil Wetness Index (SWI) and Soil Moisture (SM) data products were developed using SMAP L-band radiometer data. The algorithm for SWI data products as a primary product was adopted from ASCAT soil moisture product (Wolfgang et al., 2013) based on change detection approach (time-series methodology). The TU Wien change detection algorithm is from a mathematical point of view less complex than semi-empirical modelling approaches

TABLE 10.1 List of validation sites over berambadi watershed (Karnataka, India).

Sr. No.	Validation sites	Time period (year)
1.	Beechanhalli (berambadi watershed)	2016–18
2.	Beemanbeedu (berambadi watershed)	2016–18
3.	Maddur (berambadi watershed)	2016–18
4.	Madyanhundi (berambadi watershed)	2016–18

built upon the Cloud Model (Wagner et al., 2013). It can be inverted analytically and therefore soil moisture can be estimated directly from the Scatterometer measurements without the need for iterative adjustment processes. Because of this, it is also quite straightforward to perform an error propagation to estimate the retrieval error for each land surface pixel (Naeimi et al., 2009). With above advantage of simplistic approach, less requirement of ancillary data and analytical solution to soil moisture inversion from time-series data, same algorithm was adopted and implemented using time-series SMAP L-band Tb data to derive large-scale soil moisture as a SWI the value of which ranges from 0 to 1, showing driest to saturation condition (Dharmendra et al., 2016). An absolute soil moisture W(t) at time can also be derived from SWI at time (t) if the $W_{\min}$ and $W_{\max}$ corresponding to minimum and maximum soil moisture values (gravimetric or volumetric) are available (Thapliyal et al., 2005 and Sasmita et al., 2012),

$$\boldsymbol{W(t) = W_{min} + SWI(t) * (W_{max} - W_{min})}$$

The $W_{\min}$ and $W_{\max}$ represent the Permanent Wilting Point (PWP), the field capacity (FC) of soil which can derived from soil texture information. Using above approach, soil moisture as a secondary product was derived using SWI data product and soil texture information from Harmonized World Soil Database (HWSD), the value of which covers the top 5 cm of the soil column, ranges from 0 to 0.55 m^3/m^3.

Currently, Soil Wetness Index (SWI) and Soil Moisture (SM) act as primary and secondary products that are generated as daily operational products which are available as Visualization of Earth Observation Data and Archival System (VEDAS) and Meteorological & Oceanographic Satellite Data Archival Centre (MOSDAC) web portal of SAC (ISRO) as givenin Fig. 10.1. The soil moisture data products are available since April 2015 (Dharmendra et al., 2016) with spatial resolution of 0.125° (12.5 km).

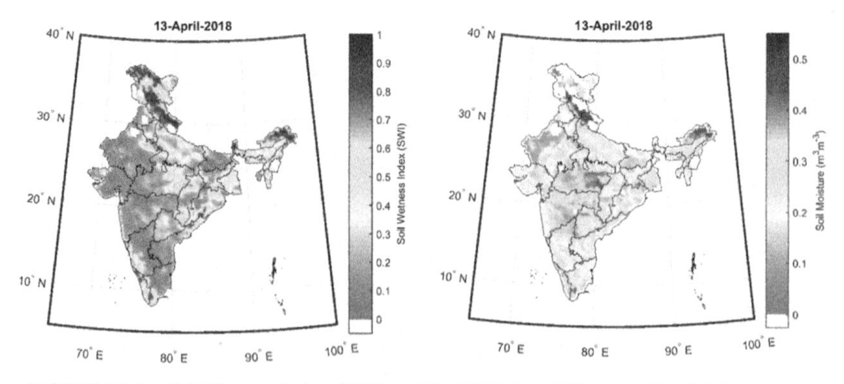

FIGURE 10.1 Soil Wetness Index (SWI) and Soil Moisture daily operational data products.

3.2 In-situ soil moisture data

In-situ soil moisture data were acquired from four soil moisture stations, spatially distributed at different locations over Berambadi Watershed (Karnataka, India) as given in Table 10.1. All hydro probe stations were installed with two probes at 5 cm soil depth at two different points to get representative averaged soil moisture and one probe at 50 cm soil depth. Along with soil moisture at three different soil depths, soil temperature and electrical conductivity are also being measured on an hourly and daily time scale basis. In-situ field data including soil moisture, soil temperature and electrical conductivity at soil depth is directly disseminated to SAC (ISRO) email server through GSM technology on hourly and daily basis. Time-series of daily average soil moisture and soil temperature at three different soil depths for each station are shown for respective period from Figs. 10.2–10.5. The soil moisture observation sites acquired data are available every day and every month, respectively; however, some stations have no reported values on some days and months.

Soil moisture data which are temporally discontinuous were excluded from the analysis. To obtain enough samples for analysis, stations without continuous daily values were retained.

4. Validation of soil moisture (SAC-ISRO) products using in-situ data

SMAP mission has ~ 3-day revisit time globally with sample diurnal cycle at consistent time of day (6 a.m./6 p.m. equator crossing) (Enthekhabi et al., 2010a). In this context, daily soil moisture (SAC-ISRO) data products are primarily generated using past 3 days SMAP L1C brightness temperature data in descending orbits (6.00 a.m. overpass). However, in case of SMAP data gap over particular region in descending orbit, ascending orbit (6.00 p.m.) is also selected to fill the gap areas for generating 3-day composite all India soil moisture products. Therefore, in order to validate daily (3-day composite) soil moisture all India products, the soil moisture values extracted from the soil moisture products were compared with daily mean soil moisture values measured by stations located over different validation sites within each cell. Prior to comparison, the quality of the in-situ stations data and soil moisture (SAC-ISRO) data products were checked by removing any outliers and invalid values presented in data. After the quality of the in-situ and soil moisture data products were checked, accuracy assessments including the computation of bias, root mean square error (RMSE), unbiased root mean square error (ubRMSE) and coefficient of determination (R^2) were performed at each validation sites.

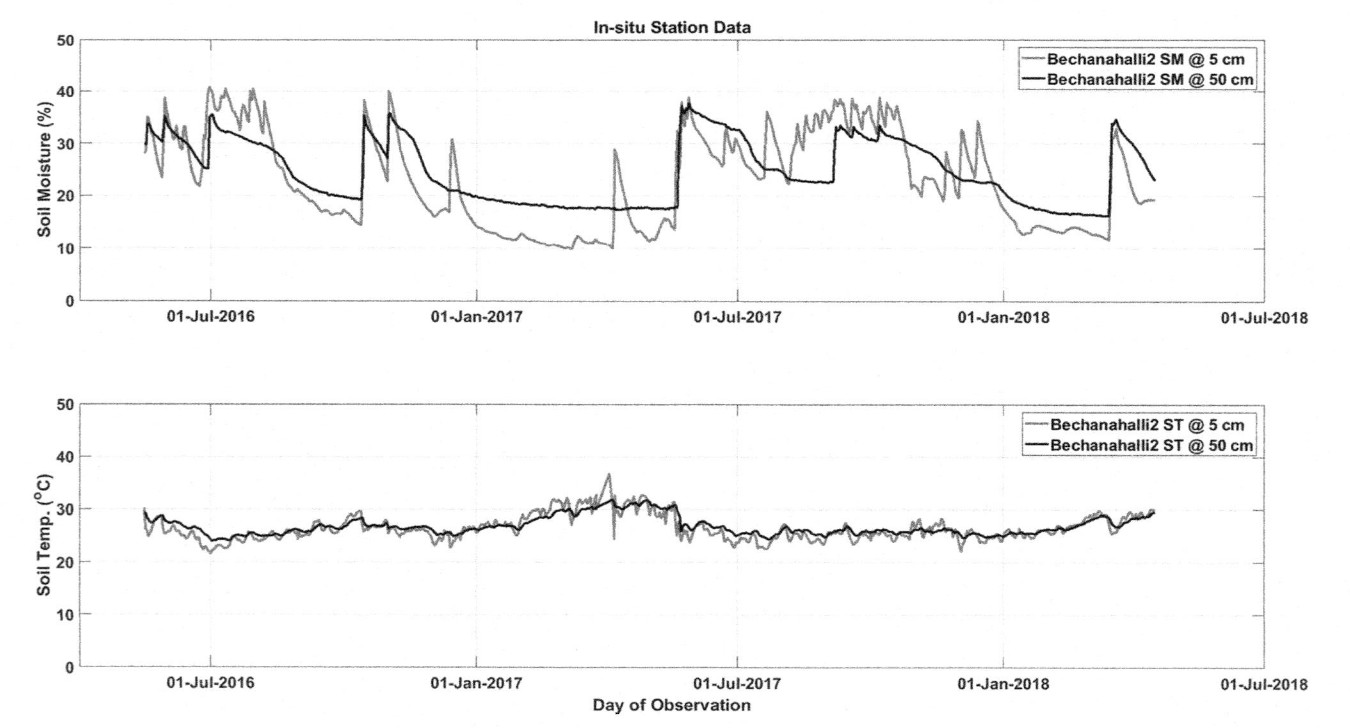

FIGURE 10.2 Time-series of daily average station data (A) *(Top)* Soil Moisture (B) *(Down)* Soil Temperature at 5 and 50 cm soil depth at Beechanhalli (Karnataka).

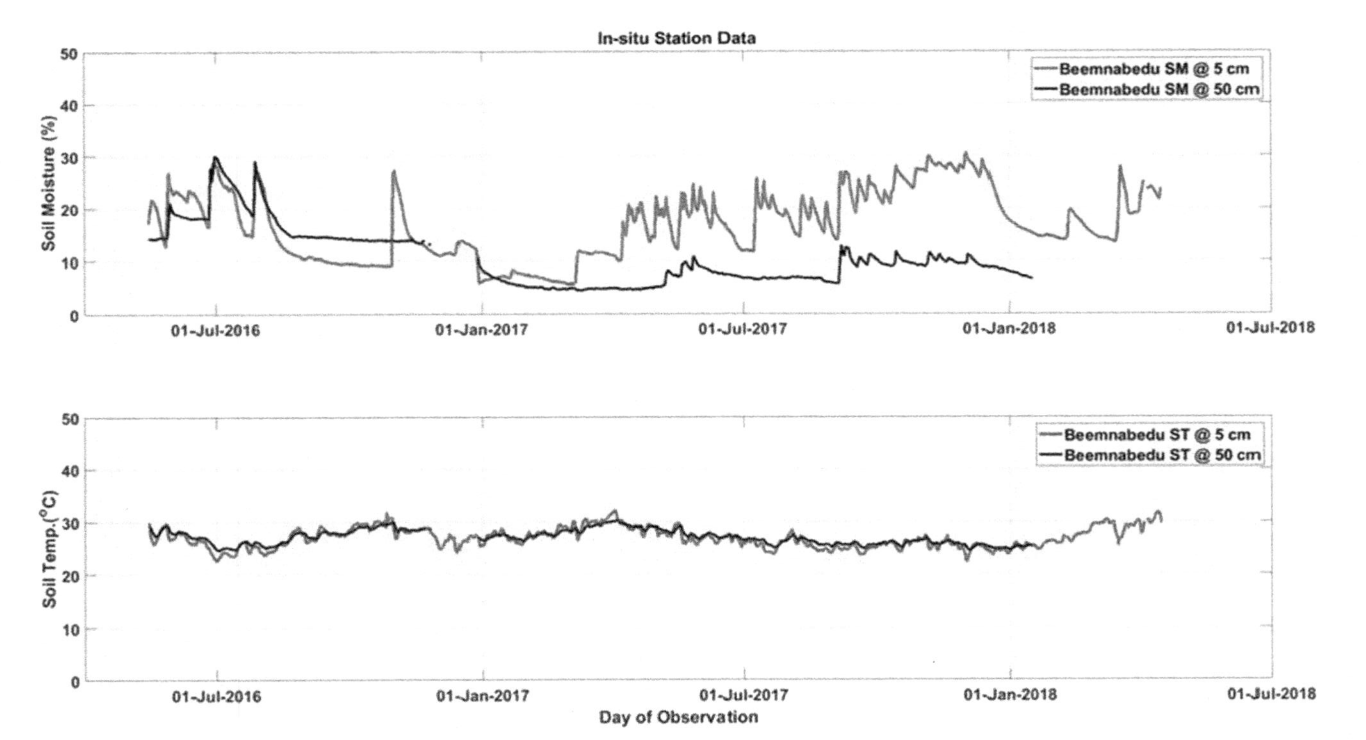

FIGURE 10.3 Time-series of daily average station data *(Top)* Soil Moisture *(Down)* Soil Temperature at 5 and 50 cm soil depth at Beemanbeedu (Karnataka).

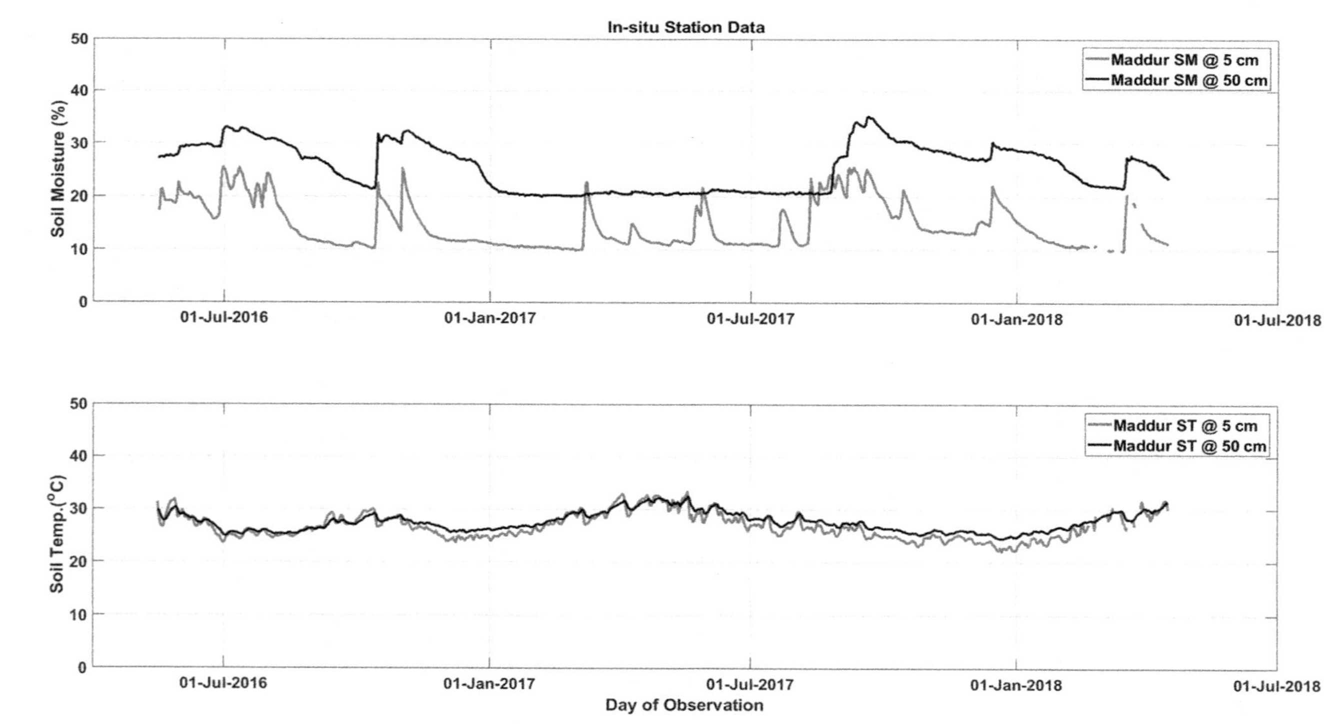

FIGURE 10.4 Time-series of daily average station data (A) (*Top*) Soil Moisture (B) (Down) Soil Temperature at 5 and 50 cm soil depth at Maddur (Karnataka).

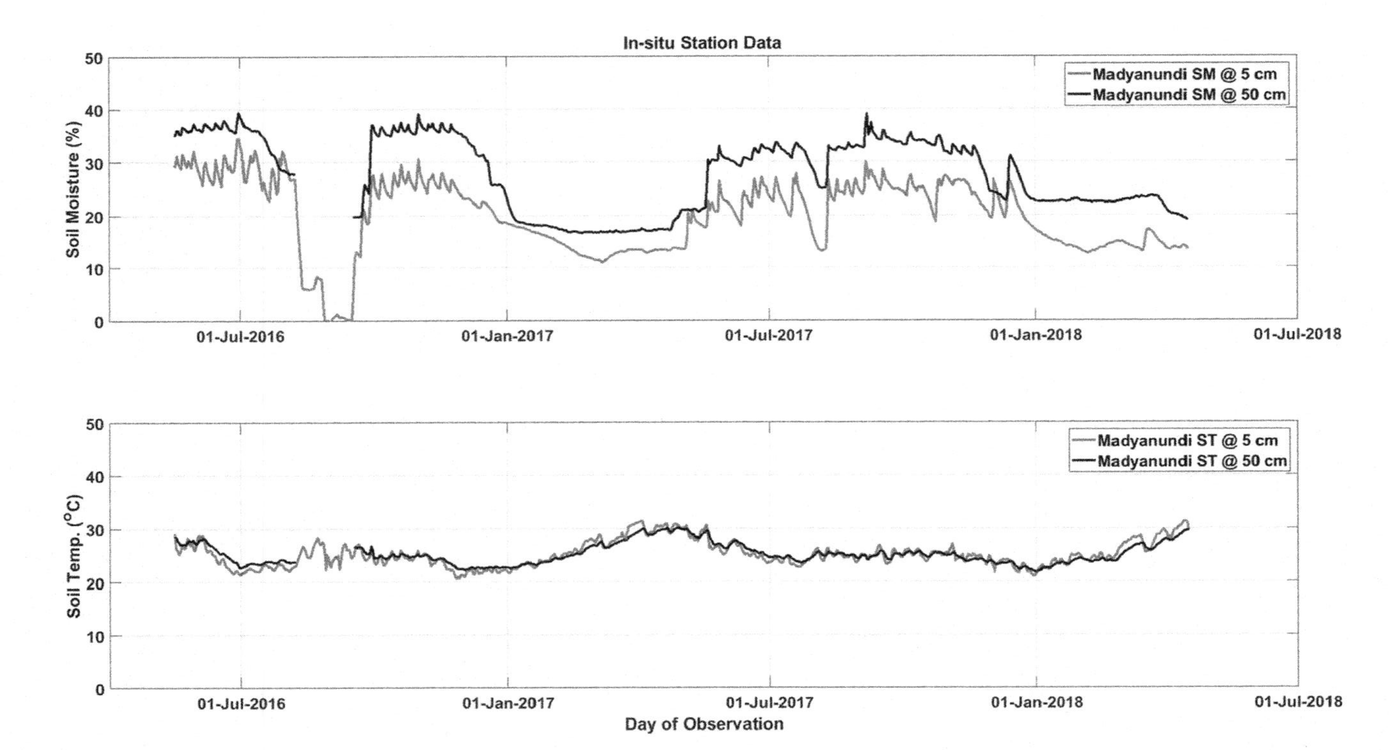

FIGURE 10.5 Time-series of daily average station data (A) (*Top*) Soil Moisture (B) (*Down*) Soil Temperature at 5 and 50 cm soil depth at Madyanhundi (Karnataka).

4.1 Accuracy assessment based on metrics computation

To assess and evaluate the accuracy of the soil moisture (SAC-ISRO) data products using in-situ field data, four metrics are computed over each alidation sites: ubRMSE, bias, RMSE and Pearson correlation coefficient (Entekhabi et al., 2010b). If the true surface volumetric soil moisture (at a given scale) is defined as θ_{true} and the corresponding estimated retrieval is θ_{est} then the RMS error metric is simply expressed as:

$$\mathrm{RMSE} = \sqrt{E\left[(\theta_{\mathrm{est}} - \theta_{\mathrm{true}})^2\right]}$$

where $E[]$ is the expectation operator. This metric quadratically penalises deviations of the estimate with respect to the true soil moisture (in units of volumetric soil moisture) and is a compact and easily understood measure of estimation accuracy. This metric, however, is severely compromised if there are biases in either the mean or the amplitude of fluctuations in the retrieval. If it can be estimated reliably, the mean bias $b = E[\theta_{est}] - E[\theta_{true}]$ can easily be removed by defining the unbiased RMSE:

$$\mathrm{ubRMSE} = \sqrt{E\{[(\theta_{\mathrm{est}} - \mathrm{E}[\theta_{\mathrm{est}}]) - (\theta_{\mathrm{true}} - \mathrm{E}[\theta_{\mathrm{true}}])2\}}$$

The RMSE and the unbiased RMSE are related through:

$$\mathrm{RMSE}^2 = \mathrm{ubRMSE}^2 + b^2$$

which implies $RMSE \geq |b|$ and underscores the shortcomings of the RMSE metric in the presence of mean bias. As the bias in soil moisture may vary with season, it is straightforward to generalise the relationships to account for such slowly varying bias. Therefore, it is assumed that ubRMSE reflects the RMSE of soil moisture anomalies that are computed by removing the mean seasonal cycle (Entekhabi et al., 2010b).

Given the percentile-based transformation that is typically applied when assimilating retrievals into a model, an additional metric is needed that is insensitive to any retrieval mean bias and bias in amplitude of fluctuations (as expressed in the statistical variance). The metric Pearson correlation coefficient is the sample time-series correlation, expressed as

$$r = \frac{E[(\theta_{\mathrm{est}} - E[\theta_{\mathrm{est}}])(\theta_{\mathrm{true}} - E[\theta_{\mathrm{true}}])]}{\sigma_{\mathrm{est}}\sigma_{\mathrm{true}}}$$

where σ_{est}^2 and σ_{true}^2 are the time variances of the estimated (retrieval) and true soil moisture for the remote sensing pixel, respectively. Finally, Coefficient of Determination (R^2) is estimated which is square of Pearson correlation coefficient (r). The correlation metric captures the correspondence in phase between retrieval estimates and truth, and therefore the coherent phasing information even if there is mean bias and/or differences in variance. In this sense, it provides a different perspective on retrieval performance than RSME (Entekhabi et al., 2010b).

It is preferable to report the data product uncertainty foremost using ubRMSE, while the rest of the metrics are used to gain additional insight into the performance of the data product and help in further adjustments of the algorithms. The summary of all metrics is presented as the average over the metric value for each validation site and also as the RMSE over the biases determined for each validation site.

4.2 Validation approach for coarse-scale satellite-derived data products based on scaling method

Soil moisture observations from satellite sensors are becoming widely available at the global scale with a good accuracy and a spatial-temporal resolution suitable for many applications (Wagner et al., 2007; Seneviratne et al., 2010). However, the climatology of satellite-derived and modelled/in-situ soil moisture observations can be very different because of uncertainties affecting both data sources (Enthekhabi et al., 2010b). Moreover, the difference between the spatial extent of in situ ($\sim 0.1\ m^2$) and coarse resolution satellite ($\sim 600\ km^2$) observations is very large. Because of the high soil moisture spatial variability (Brocca et al., 2010), point measurements are expected to be not representative of the actual absolute value (i.e., in m^3/m^3) of average soil moisture at the satellite pixel scale. Therefore, the spatial characteristics of these in-situ stations/networks are not ideal for the evaluation of coarse-scale satellite soil moisture products. However, the well-known scaling properties of soil moisture (Vachaud et al., 1985; Brocca et al., 2011) have showed that in-situ measurements can capture the large-scale temporal dynamics. In order to validate coarse-scale satellite-derived data products, the most widely used scaling technique, Cumulative Distribution Function (CDF) matching approach (e.g., Drusch et al., 2005), was adopted that allows matching the complete CDF of satellite and in-situ observations by applying a nonlinear operator. Systematic differences between remote sensing–derived and site-specific data of soil moisture prevent an absolute agreement between the two-time series. Consequently, comparison of the remotely sensed and site-specific time series is often aided by normalizing the remotely sensed data to match the distribution of ground data.

4.2.1 Cumulative Density Function (CDF) matching approach

The CDF matching approach (Reichle and Koster, 2004; Drusch et al., 2005) can be considered as an enhanced nonlinear technique for removing systematic differences between two datasets. Through this method, the satellite time-series is rescaled in such a way that its CDF matches the CDF of in-situ measurements. There are several ways to apply this method. Among them, one is based on, first, the computation of the differences in soil moisture values between the corresponding elements of each ranked dataset. Then, these differences are plotted against the satellite data and a fitting function f is used to calculate the CDF-corrected soil moisture datasets, SATCDF (Brocca et al., 2011):

$$SAT_{CDF} = SAT + f(SAT)$$

A polynomial function is used for generating *f*, suitable to fit the relationship between the ranked differences and the satellite data. On the one hand, it is also evident that at least 1 year of data is needed to obtain robust results with the CDF matching. If a lower number of months is available, the linear approaches provide better findings.

5. Results and discussion

The time-series of soil moisture (SAC-ISRO) data products from June 2016 to April 2018 was extracted over different validation sites given as per Table 10.1 and converted from m^3m^{-3} to % v/v volumetric percentage. For quality evaluation of SMAP-derived soil moisture products in long time-series, in-situ stations were selected over validation sites where daily data is available for the period of last two years (from May 2016 to April 2018). As soil moisture products having grid resolution of 12.5 km, all four stations of point observations i.e., Bechanhalli, Beemanbeedu, Maddur and Madyanhundi lies in same satellite pixel. Therefore, it was preferred to compute mean of all four stations over satellite pixel scale and then compare for evaluation so that it would be more representative of satellite pixel. Individual station observations over Berambadi watershed were also compared with soil moisture data products, along with mean in-situ soil moisture observation over Berambadi watershed.

Validation exercises were divided into two phases for quality evaluation of Soil Moisture (SAC-ISRO) data products. In the first phase of validation exercise, In-situ soil moisture observations were used without any scaling methods (cdf) and four metrics were computed over each validation sites: ubRMSE, bias, RMSE and coefficient of determination (R^2) (Enthekhabi et al., 2010b) to evaluate the performance of original soil moisture data products.

In the second phase of validation exercise, CDF-based approaches were implemented for scaling of SMAP-derived soil moisture observations to match with in-situ (reference data) soil moisture observations over selected validation sites to correct the satellite-derived soil moisture products. Over validation sites, in-situ data from Jan 2017 to Dec 2017 period (dry and wet period) was selected as reference data for CDF matching and generation of polynomial coefficient on monthly basis from January to December. Multiple order polynomial from first to seventh order was selected for CDF matching and evaluated on basis of RMSE, and it was observed that first order polynomial was the best fit with lowest RMSE which minimized the difference between soil moisture data products and in-situ reference data as shown in Fig. 10.6 and produced the corrected soil moisture data products. In Fig. 10.6, soil moisture data products represent as blue line (black line in printed version), in-situ data (References data) as grey line and corrected soil moisture data as red line (black dotted line in printed version). After generating monthly coefficients, testing of CDF matching was done over 2016 and 2018 data of soil moisture data products with corresponding in-situ data.

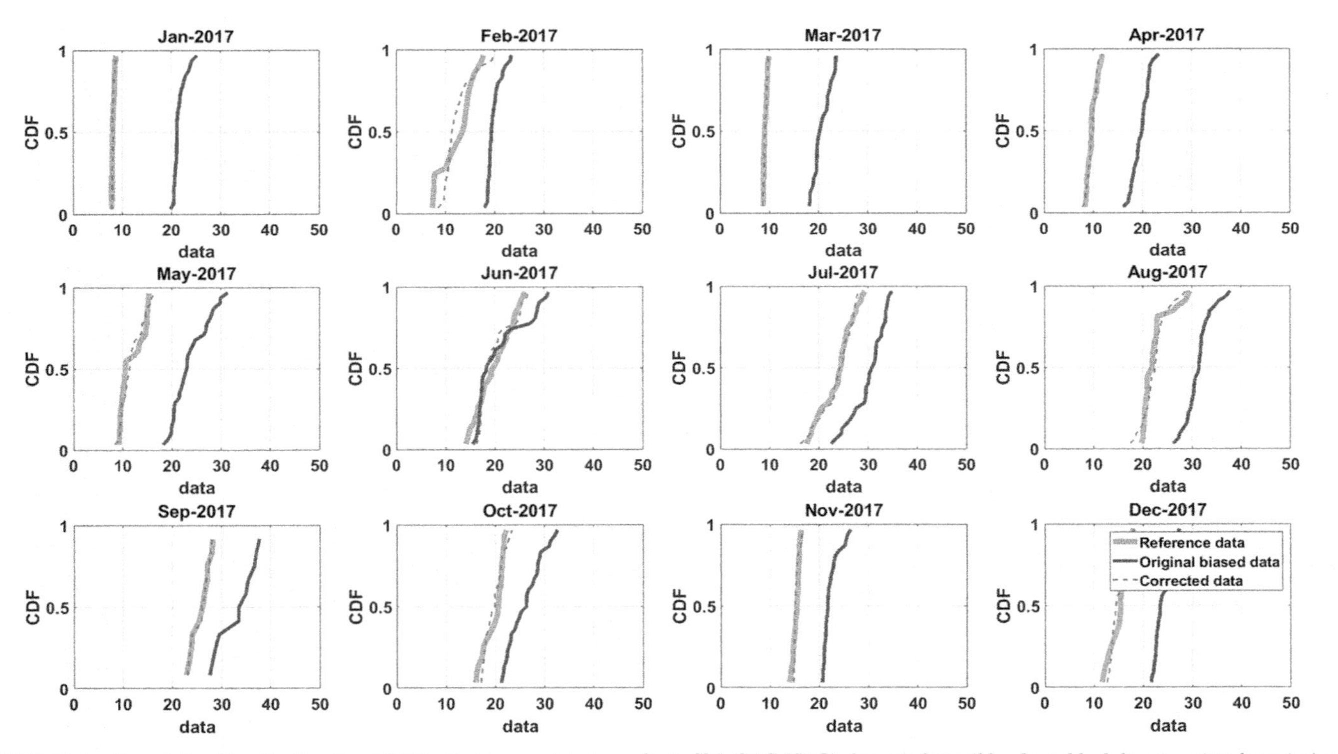

FIGURE 10.6 Cumulative Density Function (CDF) of reference data (*grey line*), SM (SAC-ISRO) data products (*blue line: black line in printed version*) and corrected data (*red dotted line: black dotted line in printed version*) for year 2017.

Fig. 10.7 shows the comparison between the in-situ daily averaged soil moisture data over various validation sites as listed in Table 10.2 and retrieved daily soil moisture data products corresponding to (a) Berambadi watershed (Avg. of four stations) (c) Beechanhalli, (d) Beemanbeedu, (e) Maddur and (f) Madyanhundi. The time-series plots in Fig. 10.7 include, in addition to the in-situ soil moisture at hydra probe station (black line) and SMAP retrieved soil moisture (SAC-ISRO) as blue line (gray line in printed version) and corrected SMAP retrieved soil moisture (SAC-ISRO) after CDF based matching approach as red line (dark gray line in printed version). Generally, an over-estimation (shown by the positive bias values as shown in Table 10.2) of Soil Moisture (SAC-ISRO) products was found for all validation sites. In general, the overestimation of SMAP derived soil moisture values are more apparent between October and June (dry spell and mostly non-monsoon period) for 2016, 2017 and 2018 as compared to monsoon/wet period (July to September) which reflected in time-series plot of Fig. 10.7 for all hydra probe stations at different locations within Berambadi watershed (Karnataka, India).

Table 10.2 summarize the performance metrics that characterize the retrieval performance of original soil moisture (SAC-ISRO) data products at different validation sites without any scaling/bias correction approach.

Stations over Berambadi watershed show similar range of performance accuracy (ubRMSE), which ranges from 5.17% to 6.91%. Out of all four stations in Berambadi watershed, Maddur sites provided the best accuracy (ubRMSE ~ 5.17%) with high R^2 of 0.37 as Table 10.2. Mean of all four stations over Berambadi watershed has good agreement with mean in-situ of all four stations. It supports that average in-situ soil moisture represents more spatial heterogeneity and compares well with satellite-scale data as having lowest ubRMSE of 5.01% and highest coefficient of determinant (R^2) of 0.38. Fig. 10.8 shows the scatter plots between in-situ soil moisture and satellite-derived soil moisture with correction (*blue: black in printed version*) and without correction (*red: gray in printed version*) for same sites.

Overall, the performance accuracy of all validation sites in terms of ubRMSE varies from 5.01% (best) to 6.91% (worst) which is still very good agreement with in-situ field data without implying any scaling (upscaling/downscaling) approaches. Point-wise in-situ observations have to be upscaled to properly estimate the true mean of the soil moisture fields at the coarse-scale (several kilometers) of the satellite products.

In order to overcome the problem of scale difference between satellite and in-situ point observations, CDF matching techniques were implemented to minimize the systematic difference between satellite-derived soil moisture and in-situ soil moisture data as reference data and to properly validate the soil moisture products. In other words, soil moisture (SAC-ISRO) data products were corrected with respect to in-situ point observations.

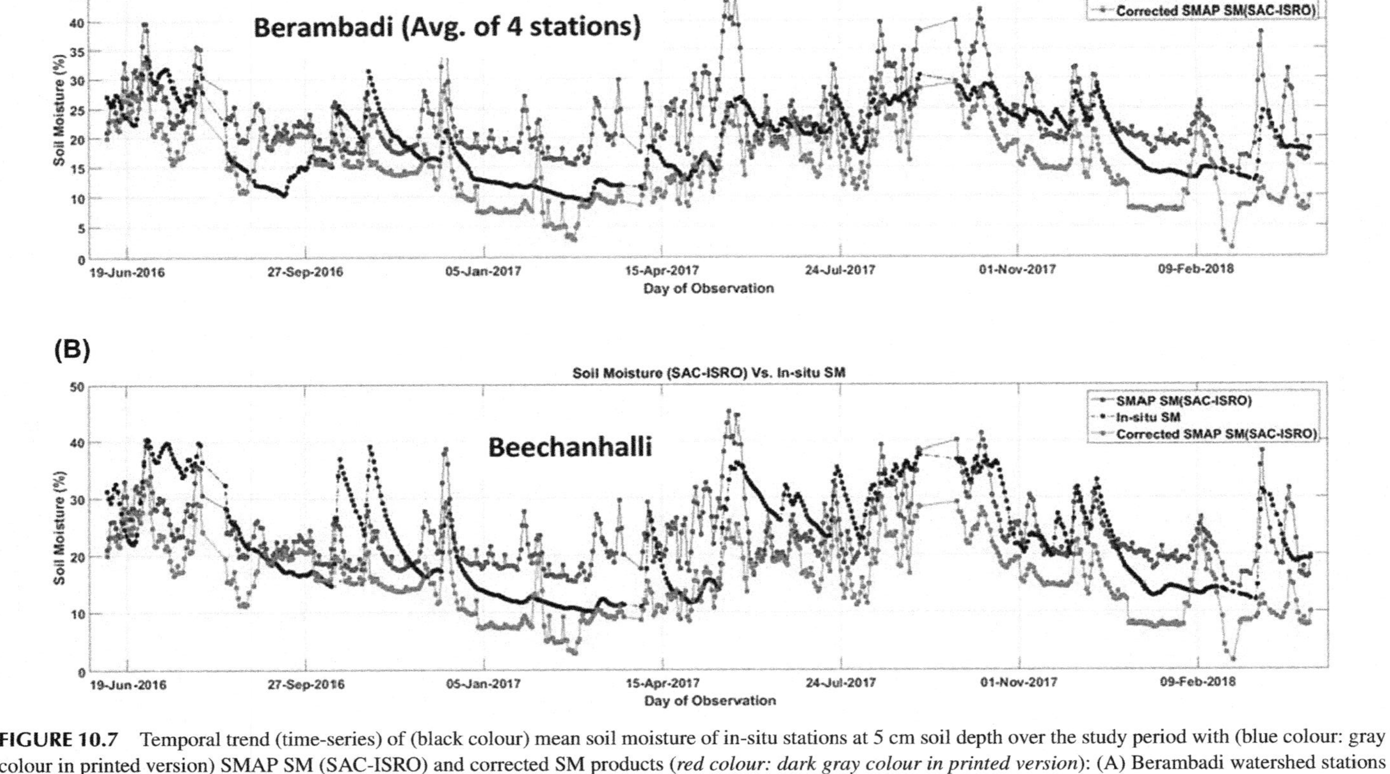

FIGURE 10.7 Temporal trend (time-series) of (black colour) mean soil moisture of in-situ stations at 5 cm soil depth over the study period with (blue colour: gray colour in printed version) SMAP SM (SAC-ISRO) and corrected SM products (*red colour: dark gray colour in printed version*): (A) Berambadi watershed stations (Avg. of 4 Stations), (B) Beechanhalli, (C) Beemanbeedu, (D) Maddur and (E) Madyanhundi.

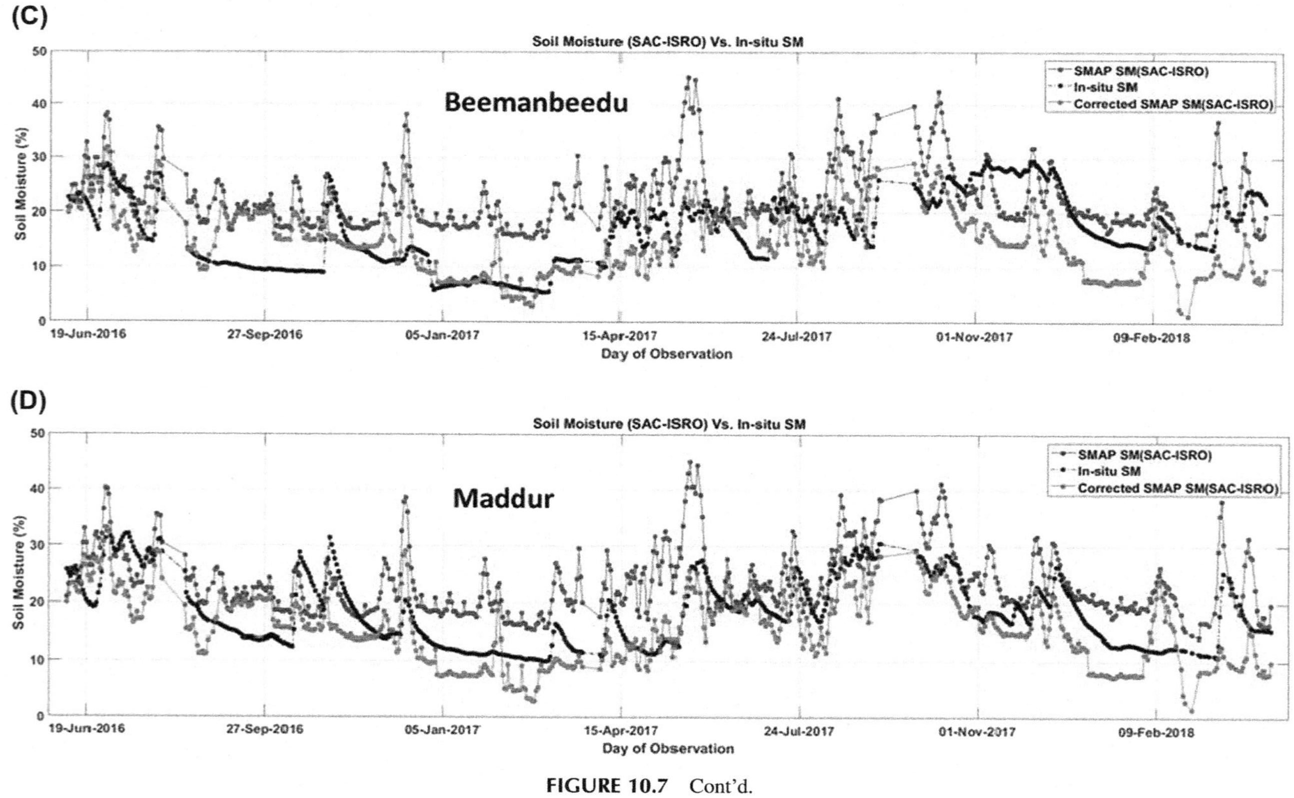

FIGURE 10.7 Cont'd.

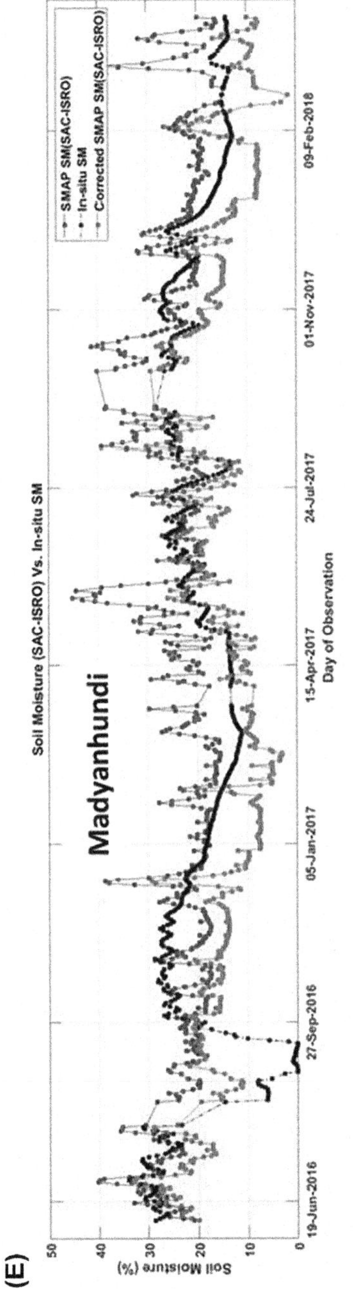

FIGURE 10.7 Cont'd.

TABLE 10.2 Comparison between daily soil moisture (SAC-ISRO) data products and daily mean in-situ soil moisture observations (without scaling/preorrection).

Sr. No.	Validation sites	ubRMSE (% v/v)	Bias (% v/v)	RMSE (% v/v)	R^2
1.	Beechanhalli (berambadi watershed)	6.91	1.10	6.99	0.36
2.	Beemanbeedu (berambadi watershed)	6.37	6.24	8.92	0.21
3.	Maddur (berambadi watershed)	5.17	5.34	7.43	0.37
4.	Madyanhundi (berambadi watershed)	6.89	4.18	8.06	0.15
5.	Berambadi (avg. of 4 stations)	5.01	4.22	6.55	0.38

Table 10.3 summarizes the performance metrics of validation sites after scaling (CDF matching), which demonstrate significant improvement in coefficient of determination (R^2) to explain temporal variability of soil moisture over all validation sites. Performance accuracy (ubRMSE) varies from 4.10% to 6.10% and coefficient of determination (R^2) from 0.50 to 0.60. The best improvement was observed in Madyanhundi station in Berambadi watershed of which ubRMSE improved from ubRMSE of 6.89%–4.22% and R^2 from 0.15 to 0.60. Overall, very significant improvement was observed in soil moisture (SAC-ISRO) data products over all validation sites by applying scaling which are very close to SMAP mission requirement (ubRMSE $\leq$ 4%) and shown a very good agreement with in-situ observations as shown in Fig. 10.8.

6. Conclusion

SAC has developed an daily operational soil moisture (SAC-ISRO) data product by adopting and modifying the algorithm from ASCAT soil moisture product, which is based on change detection approach (time-series methodology) using SMAP L-band brightness temperature. This daily soil moisture product derived from SMAP radiometer brightness temperature data is available on MOSDAC and VEDAS web portal of SAC (ISRO) on 12.5 km grid resolution since April 2015. Validation of satellite-derived geophysical data products is a critical task in supporting the success of Earth science missions and their utilisation of different applications. However, this task is often hindered by two important issues: (1) the difference in scales between coarse resolution (~ km scale) mission products and localized point-scale in-situ sensor measurements and (2) the inability of in-situ networks to fully capture the spatial heterogeneity of geophysical variables.

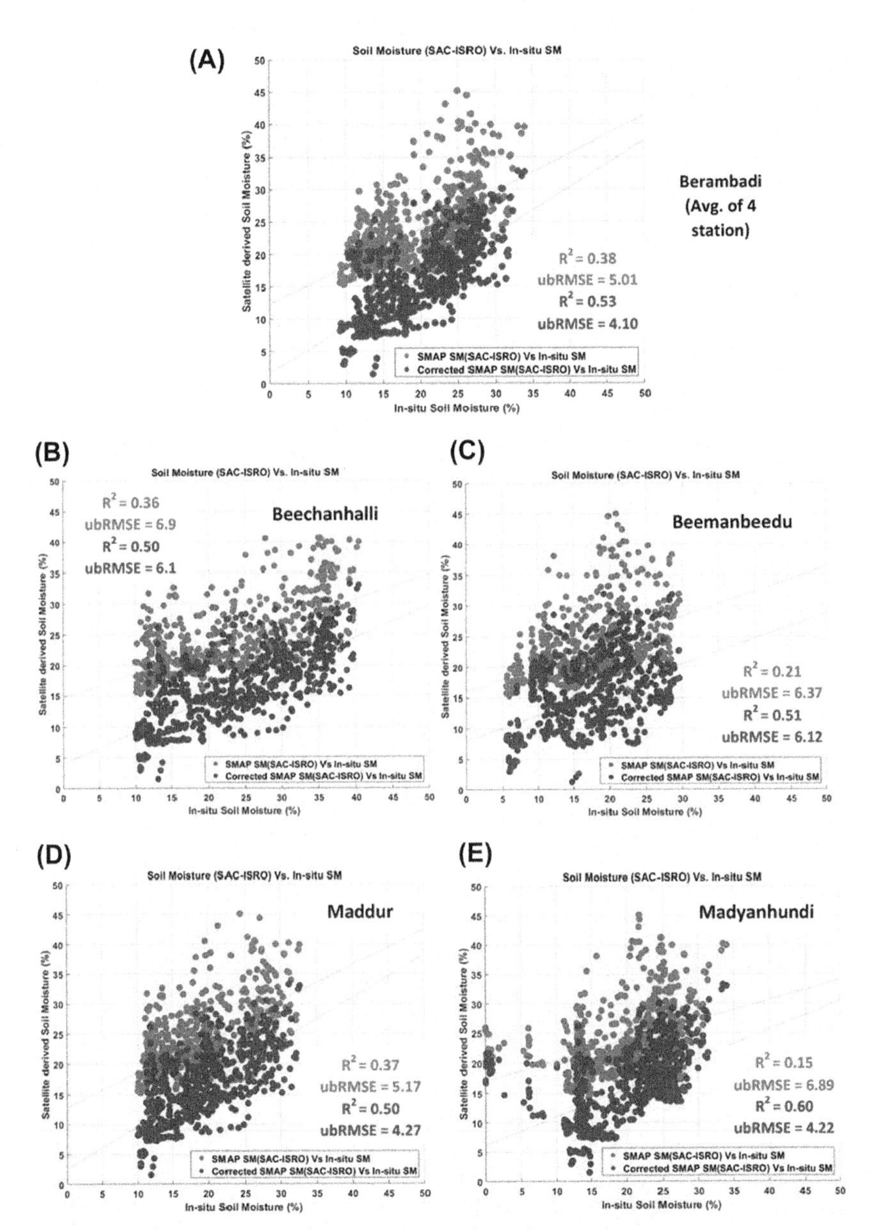

FIGURE 10.8 Scatter plot between the in-situ soil moisture (% v/v) and satellite-derived SM (% v/v): (red colour: gray colour in printed version) uncorrected and (*blue colour: black colour in printed version*) corrected SM data products: (A) Berambadi watershed (avg. of 4 stations), (B) Beechanhalli, (C) Beemanbeedu, (D) Maddur and (E) Madyanhundi.

TABLE 10.3 Comparison results after CDF-based scaling approach (postcorrection).

Sr. No.	Validation sites	ubRMSE (% v/v)	R^2
1.	Beechanhalli (berambadi watershed)	6.10	0.50
2.	Beemanbeedu (berambadi watershed)	6.12	0.51
3.	Maddur (berambadi watershed)	4.27	0.50
4.	Madyanhundi (berambadi watershed)	4.22	0.60
5.	Berambadi (avg. Of 4 stations)	4.10	0.53

In this work, validation of soil moisture (SAC-ISRO) data product was assessed at different agricultural field sites over Berambadi watershed (Karnataka, India). The validation plan was designed by collecting in-situ soil moisture data over selected agricultural fields of validation sites, which provides the continuous field soil moisture data at different depths with near real time. The data from the selected sites are quality controlled before application of site-specific spatial scaling function and comparison with the soil moisture (SAC-ISRO) data products. The time-series soil moisture data collected from four in-situ stations over Berambadi watershed (Karnataka, India) were analysed statistically and according to their temporal patterns. The validation exercise was focused on daily mean soil moisture data product comparison with daily mean in-situ soil moisture observation over selected validation sites from May 2016 to April 2018 (~period of 2 years). An overestimation of the soil moisture products was observed with bias (including dry and wet) ranging from 1.10% to 6.24%, unbiased RMSE from 4.10% to 6.12%, RMSE from 6.55% to 8.92% and R^2 from 0.50 to 0.60 over validation sites. Overall, very significant improvement was observed in soil moisture (SAC-ISRO) data products over all validation sites by applying scaling approaches which were very close to SMAP mission requirement (ubRMSE $\leq$ 4%) and with a very good agreement with in-situ observations as shown in Table 10.3.

The validation results presented in this work demonstrate that the performance of soil moisture (SAC-ISRO) data products derived from SMAP L-band radiometer data are closely matching to SMAP science mission requirements. As the retrieval algorithm is continuously under development and improvement, the accuracy of soil moisture (SAC-ISRO) data product is intended to be improved from time to time. The augmentation and collection of new in-situ soil moisture station data can also fill the paucity of soil moisture information over different regions and aid the improvement of soil moisture estimation over the Indian region. The new versions of the soil moisture retrieval algorithms will be implemented based on this validation exercise, which will ensure better results as expected in the near future. The use of localized data such as soil moisture, soil temperature, weather information and soil and land-use information to improve and develop more robust soil moisture retrieval is the key to improving soil moisture retrieval at local scale, and also to provide improved soil moisture information over different regions over India. A validation and inter-comparison of other satellite-derived soil moisture data products from SMAP, SMOS, ASCAT and AMSR-2 over Indian regions along with soil moisture (SAC-ISRO) data products is also planned and will be executed in near future.

Acknowledgements

The authors are grateful to Shri D. K. Das, Director (SAC) and Dr Raj Kumar, Deputy Director, Earth, Ocean, Atmosphere, Planetary Sciences and Applications Area (EPSA) of Space Applications Centre for their full support, guidance and keen interest in this work. Authors also highly acknowledge Indian Institute of Science (Bengaluru) for their full support to establish soil moisture station over different validation sites Berambadi watershed (Karnataka, India). Authors are also thankful Dr. Bimal Kr. Bhattacharya for fruitful discussion about validation process and overall guidance.

References

Agro-Hydrological Monitoring, 2015. A Manual for Agro-Hydrological Monitoring in Pilot Experimental Watersheds, August 2015. Indian Institute of Science Bengaluru, AMBHAS manual, Sujala.

Bartalis, Z., Wagner, W., Naeimi, V., Hasenauer, S., Scipal, K., Bonekamp, H., Figa, J., Anderson, C., 2007. Initial soil moisture retrievals from the METOP-A Advanced Scatterometer (ASCAT). Geophys. Res. Lett. 34.

Brocca, L., Melone, F., Moramarco, T., Morbidelli, R., 2010. Spatial-temporal variability of soil moisture and its estimation across scales. Water Resour. Res. 46, W02516.

Brocca, L., Hasenauer, S., Lacava, T., Melone, F., Moramarco, T., Wagner, W., Dorigo, W., et al., 2011. Soil moisture estimation through ASCAT and AMSR-E sensors: an intercomparison and validation study across Europe. Rem. Sens. Environ. 115 (12), 3390–3408.

Crow, W.T., Berg, A.A., Cosh, M.H., Loew, A., Mohanty, B.P., Panciera, R., de Rosnay, P., Ryu, D., Walker, J.P., 2012. Upscaling sparse ground-based soil moisture observations for the validation of coarse-resolution 671 satellite soil moisture products. Rev. Geophys 50. RG2002, 670.

Dharmendra, P., et al., 2016. Development of a Time Series–Based Methodology for Estimation of Soil Wetness Using SMAP Radiometer Data: Preliminary Results. SAC/EPSA/GHCAG/MHTD/TR/05/2016 (Available on request).

Dorigo, W.A., Wagner, W., Hohensinn, R., Hahn, S., Paulik, C., Xaver, A., Gruber, A., Drusch, M., Mecklenburg, S., van Oevelen, P., 2011. The international soil moisture network: a data hosting facility for global in situ soil moisture measurements. Hydrol. Earth Syst. Sci. 15 (5), 1675–1698. https://doi.org/10.5194/hess-15-1675-2011.

Drinkwater, M., Kerr, Y., Font, J., Berger, M., 2009. Exploring the water cycle of the blue planet. In: The Soil Moisture and Ocean Salinity Mission, vol. 137. ESA Bulletin, Noordwijk, pp. 7–15.

Drusch, M., Wood, E.F., Gao, H., 2005. Observation operators for the direct assimilation of TRMM microwave imager retrieves soil moisture. Geophys. Res. Lett. 32, L15403.

Entekhabi, D., Njoku, E.G., O'Neill, P.E., Kellogg, K.H., Crow, W.T., Edelstein, W.N., Entin, J.K., Goodman, S.D., Jackson, T.J., Johnson, J., et al., 2010a. The soil moisture active passive (SMAP) mission. Proc. IEEE 98, 704–716.

Entekhabi, D., Reichle, R.H., Koster, R.D., Crow, W.T., 2010b. Performance metrics for soil moisture retrievals and application requirements. J. Hydrometeorol. 11, 832–840.

Gruber, A., Dorigo, W.A., Zwieback, S., Xaver, A., Wagner, W., 2013. Characterizing coarse-scale representativeness of in situ soil moisture measurements from the international soil moisture network. Vadose Zone J. 12.

Juglea, S., Kerr, Y., Mialon, A., Lopez-Baeza, E., Braithwaite, D., Hsu, K., 2010. Soil moisture modelling of a SMOS pixel: interest of using the persiann database over the Valencia anchor station. Hydrol. Earth Syst. Sci. 14, 1509–1525.

Kerr, Y.H., Waldteufel, P., Wigneron, J.P., Delwart, S., Cabot, F., Boutin, J., Escorihuela, M.J., Font, J., Reul, N., Gruhier, C., et al., 2010. The smos mission: new tool for monitoring key elements of the global water cycle. Proc. IEEE 98, 666–687.

Kerr, Y.H., Al-Yaari, A., Rodriguez-Fernandez, N., Parrens, M., Molero, B., Leroux, D., Bircher, S., Mahmoodi, A., Mialon, A., Richaume, P., et al., 2016. Overview of SMOS performance in terms of global soil moisture monitoring after six years in operation. Remote Sens. Environ. 180, 40–63.

Miralles, D.G., Crow, W.T., Cosh, M.H., 2010. Estimating spatial sampling errors in coarse-scale soil moisture estimates derived from point-scale observations. J. Hydrometeorol. 11, 1423–1429.

Mohanty, B.P., Cosh, M., Lakshmi, V., Montzka, C., 2017. Soil Moisture Remote Sensing—State-Of-The-Science. Vadose Zone J.

Montzka, C., Bogena, H.R., Weihermüller, L., Jonard, F., Bouzinac, C., Kainulainen, J., Balling, J.E., Loew, A., Dall'Amico, J.T., Rouhe, E., et al., 2013. Brightness temperature and soil moisture validation at different scales during the smos validation campaign in the rur and erft catchments, Germany. IEEE Trans. Geosci. Rem. Sens. 51, 1728–1743.

Naeimi, V., Scipal, K., Bartalis, Z., Hasenauer, S., Wagner, W., 2009. An improved soil moisture retrieval algorithm for ERS and METOP Scatterometer observations. IEEE Trans. Geosci. Rem. Sens. 47, 1999–2013.

Parinussa, R.M., Holmes, T.R.H., Wanders, N., Dorigo, W.A., de Jeu, R.A.M., 2015. A preliminary study toward consistent soil moisture from amsr2. J. Hydrometeorol. 16, 932–947.

Reichle, R.H., Koster, R.D., 2004. Bias reduction in short records of satellite soil moisture. Geophys. Res. Lett. 31, L19501.

Robinson, D.A., Campbell, C.S., Hopmans, J.W., Hornbuckle, B.K., Jones, S.B., Knight, R., Ogden, F., Selker, J., Wendroth, O., 2008. Soil moisture measurement for ecological and hydrological watershed-scale observatories: a review. Vadose Zone J. 7 (1), 358–389. https://doi.org/10.2136/Vzj2007.0143.

Sasmita, C., Thapliyal, P.K., Pal, P.K., 2012. Application of a time series-based methodology for soil moisture estimation from AMSR-E observations over India. Geosci. Rem. Sens. Lett. IEEE 9 (5), 814–821.

Seneviratne, S.I., Corti, T., Davin, E.L., Hirschi, M., Jaeger, E.B., Lehner, I., Orlowsky, B., Teuling, A.J., 2010. Investigating soil moisture-climate interactions in a changing climate: a review. Earth Sci. Rev. 99, 125–161.

Sheffield, J., Wood, E.F., Chaney, N., Guan, K.Y., Sadri, S., Yuan, X., Olang, L., Amani, A., Ali, A., Demuth, S., et al., 2014. A drought monitoring and forecasting system for Sub-Sahara African water resources and food security. Bull. Am. Meteorol. Soc. 95, 861–882.

Thapliyal, P.K., Pal, P.K., Narayanan, M.S., Jan. 2005. Development of a time series based methodology for estimation of large area soil wetness over India using IRS-P4 microwave radiometer data. J. Appl. Meteorol. 44 (1), 127–143.

Vachaud, G., Passerat de Silans, A., Balabanis, P., Vauclin, M., 1985. Temporal stability of spatially measured soil water probability density function. SSSA (Soil Sci. Soc. Am.) J. 49, 822–828.

Vereecken, H., Huisman, J.A., Bogena, H., Vanderborght, J., Vrugt, J.A., Hopmans, J.W., 2008. On the value of soil moisture measurements in vadose zone hydrology: a review. Water Resour. Res. 44 https://doi.org/10.1029/2008wr006829.

Vereecken, H., Schnepf, A., Hopmans, J.W., Javaux, M., Or, D., Roose, T., Vanderborght, J., Young, M.H., Amelung, W., Aitkenhead, M., et al., 2016. Modeling soil processes: review, key challenges, and new perspectives. Vadose Zone J. 15, 57.

Wagner, W., Naeimi, V., Scipal, K., de Jeu, R., Martinez-Fernandez, J., 2007a. Soil moisture from operational meteorological satellite. Hydrogeol. J. 15 (1), 121–113.

Wagner, W., Bloschl, G., Pampaloni, P., Calvet, J.C., Bizzarri, B., Wigneron, J.P., Kerr, Y., 2007b. Operational readiness of microwave remote sensing of soil moisture for hydrologic applications. Nord. Hydrol 38, 1–20.

Wagner, W., Hahn, S., Kidd, R., Melzer, T., Bartalis, Z., Hasenauer, S., Figa-Saldana, J., de Rosnay, P., Jann, A., Schneider, S., et al., 2013. The ASCAT soil moisture product: a review of its specifications, validation results, and emerging applications. Meteorol. Z. 22, 5–33.

Wanders, N., Karssenberg, D., de Roo, A., de Jong, S.M., Bierkens, M.F.P., 2014. The suitability of remotely sensed soil moisture for improving operational flood forecasting. Hydrol. Earth Syst. Sci. 18, 2343–2357.

Chapter 11

A preliminary evaluation of the 'simplified triangle' with Sentinel-3 images for mapping surface soil moisture and evaporative fluxes: results obtained in a Spanish savannah environment

George P. Petropoulos[1], Ionuţ Şandric[2], Andrew Pavlides[1], Dionissios T. Hristopulos[1]

[1]*Geostatistics Laboratory, School of Mineral Resources Engineering, Technical University of Crete, Chania, Greece;* [2]*Faculty of Geography, University of Bucharest, Bucharest, Romania*

1. Introduction

The land surface and atmosphere interact over a wide range of space and time scales; such interactions involve numerous complex natural processes which influence the global climate system (Stoyanova and Georgiev, 2013; Petropoulos et al., 2016). Globally, climate change is facilitating large-scale changes within the atmosphere, biosphere, geosphere and hydrosphere (Steinhauser et al., 2012). The quantification and management of such change have become urgent and important research directions within numerous scientific disciplines (Coudert et al., 2008), in addition to providing essential information for politicians, policymakers and the wider global community (IPCC, 2009). In this context, the accurate monitoring of parameters such as evaporative fraction (defined as the ratio of instantaneous latent heat flux (LE) to net radiation (Rn)) and surface soil moisture (SSM) has a high priority within current EU frameworks, particularly communities in water-limited environments or areas which rely on rain-fed agriculture, such as the Mediterranean (Amri et al., 2014; European Commission, 2009).

Agricultural Water Management. https://doi.org/10.1016/B978-0-12-812362-1.00011-4

To date Earth Observation (EO) allows obtaining temporally consistently coverage of both EF and SSM at different spatial scales (Piles et al., 2016; Srivastava et al., 2019). Several EO-based approaches have been proposed for this purpose utilising spectral information acquired in different regions of the electromagnetic spectrum (see reviews by Petropoulos et al., 2015, 2018). Methods based in particular on the physical relationships between a satellite-derived surface temperature (Ts) and vegetation index (VI) have been very promising in that respect. Assuming conditions of full variability in fractional vegetation cover within the sensor's field of view, when plotting Ts and VI in a scatterplot a triangular (or trapezoidal) shape emerges, due to Ts being less sensitive to water content at the surface in vegetated areas than in areas of exposed soil. Such a scatterplot encapsulates several key biophysical variables (e.g., see Gillies et al., 1997; Petropoulos et al., 2009; Maltese et al., 2015).

It has been demonstrated that the derivation of spatially distributed EF and/or SSM using the Ts/VI domain is feasible using a variety of approaches (see review by Petropoulos et al., 2009). Recently, Carlson & Petropoulos (2019) proposed a new technique for estimating both SSM and EF, which they named 'simplified triangle'. Silva-Fuzzo et al. (2020) demonstrated its application coupled with a crop prediction and a climatological water balance model for predicting soybean yield using MODIS data. To our knowledge, implementation and verification of this new technique using ESA's Sentinel-3 has not been conducted in detail as yet. This despite its promising potential of this new 'simplified triangle' technique. Such an investigation would be undoubtedly of key importance, since it would inform on the potential usefulness of this technique when combined with one of the most sophisticated EO satellites currently in orbit.

In purview of the above, the present study aims to explore, to our knowledge for the first time, the ability of the 'simplified triangle' method used synergistically with Sentinel-3 data for predicting the spatiotemporal variability of both EF and SSM, at one experimental site located in Spain, belonging to the FLUXNET global in-situ monitoring network.

2. Materials

2.1 Study sites and in-situ data

Our experimental site consisted of the Albuera ('ES-Abr') experimental site located in Spain (38.702 Lat & -6.786 Lon, see Fig. 11.1). The site is representative of a typical Mediterranean savannah ecosystem type and is a relatively flat area (279 m asl). In the site is installed a dense ground monitoring instrumentation network for the long-term measurement of several parameters characterising land surface interactions. ES-Alb is part of the CarboEurope monitoring network, which is part of FLUXNET, the largest global observational network today acquiring micrometeorological fluxes and several ancillary parameters (Baldocchi et al., 1996). In FLUXNET, all ground measurements are conducted using standardised instrumentation across the network sites.

FIGURE 11.1 Study sites' geographical location in Italy (*left*) and Spain (*right*). *Background image source: ArcGIS Online.*

In this study, ground measurements of the required parameters (i.e., LE, Rn, SSM at surface layer) were data collected from the ICOS (Integrated Carbon Observation System) database (http://www.europe-fluxdata.eu/icos/home) at Level 2 processing, to allow consistency and interoperability. Following the data acquisition, preprocessing that was applied to the data included the extraction of the specific days for which were available Sentinel-3 images at the experimental sites for that year, the computation of EF (as defined previously, i.e., LE/Rn). The final dataset of the in-situ measurements consisted of a total of 11 calendar days spanning the period from June to September 2018.

2.2 Sentinel data: acquisition and pre-processing

Sentinel-3 is an EO satellite constellation developed by the European Space Agency (ESA) as part of the Copernicus Programme. It consists of two satellites, Sentinel-3A and Sentinel-3B. The Sentinel-3 satellite constellation allows a short revisit time of fewer than 2 days for the OLCI instrument and less than 1 day for SLSTR at the equator. The Sentinel-3 product used in this study included the Level 2 product named 'SL_2_LST' (Birks, 2011). This product is the Land Surface Temperature (LST or Ts as defined previously) parameters provided to the users. The SL_2_LST product contains 10 annotation files. The Fr is included among them. This Fr product was used in our study together with the LST/Ts (SLSTR ATBD Land surface Temperature, 2012). For the current study, this product was obtained for a total of 11 days of the year 2018, spanning from the start of the summertime period to the early autumn. The specific dates of the images used are the following: 23/06, 06/07, 09/07, 17/08, 25/08, 13/09, 20/09, 21/09, 24/09, 5/09 and 28/09. To implement the method, first the Sentinel-3 images for the dates mentioned above were downloaded from CREODIAS (https://creodias.eu/). For each image, a spatial subset was first implemented covering the wider area of Spain only. Then, in each image product were retained only the layers of LST, Fr and normalized difference vegetation index (NDVI). Then, each of those bands was masked for clouds and inland water using the masks already provided in each Sentinel-3 product. An example of this is illustrated in Fig. 11.2 for a selected day.

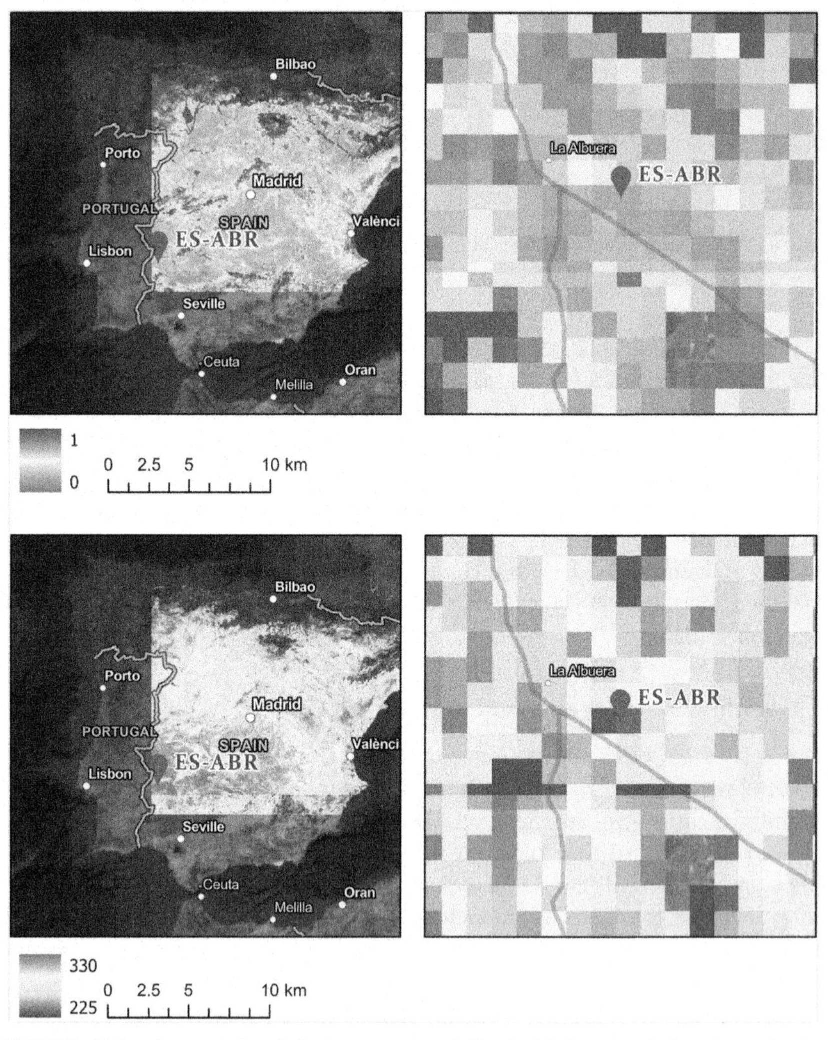

FIGURE 11.2 An example of final pre-processed Sentinel-3 images used as input in the 'simplified triangle' implementation. The Fr map is shown on the top and the LST map on the bottom in each case. Sentinel-3 image acquisition date is 25/08/2018. The geographical location of the study site is also indicated within the image.

3. Methods

Briefly, the method allows the estimation of two parameters, i.e., the surface wetness (Mo), and (EF) which is defined as the ratio of evapotranspiration to net radiation (Rn). Mo applies only to the top few millimetres of the bare soil surface. Briefly, the method is based on constructing a scatterplot of radiometric surface temperature (Ts, or equally LST) versus the Fr. The basis of the method operation is illustrated in Fig. 11.3.

The method requires an estimate of the Fr, which Carlson and Petropoulos (2019) propose can be derived from scaling the normalized difference vegetation index (NDVI). This required defining the NDVIo, NDVIs, which represent the NDVI values for bare soil and full vegetation cover, respectively (see Fig. 11.3) and then scaling NDVI using the formulae below (Gilles et al., 1997; Carlson, 2007):

$$Fr = \left\{ \frac{NDVI - NDVI_0}{NDVIs - NDVI_0} \right\}^2 \quad (11.1)$$

However, in Fr estimation, any other method can be equally used. Since the Fr layer was already provided in the Sentinel product, no further computation of Fr was performed.

NDVIs and Tmin, represent dense vegetation, define the lower left (wet or cold) vertex and the so-called 'wet edge' (or 'cold edge') of the triangle (refer to Fig. 11.3). Similarly, $NDVI_0$ and Ts[max] define the lower right vertex of the triangle. The wet edge represents the limit of soil wetness and corresponds to the values of Mo and EF equal to 1.0. Another highly important feature, the 'dry edge' or 'warm edge' (also shown in Fig. 11.3), represents the limit of soil dryness where Mo = 0 and extends from Ts [max] and NDVIo to NDVIs,

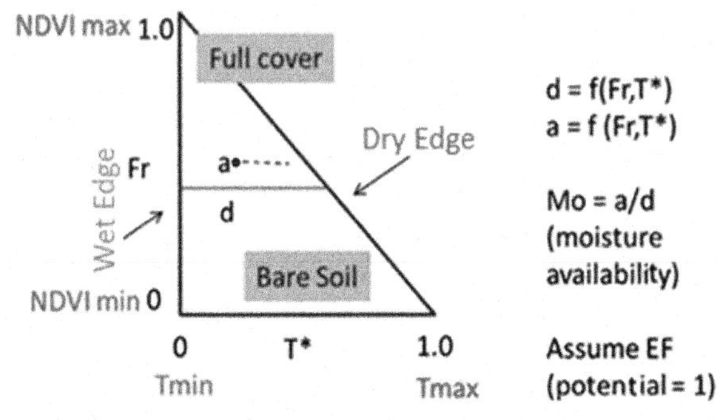

FIGURE 11.3 Simple geometry of the triangle. NDVI varies between its minimum and maximum values, respectively, $NDVI_0$ and NDVIs, where NDVI is here scaled in Fr. *After Carlson, T.N., Petropoulos, G.P., 2019. A new method for estimating of evapotranspiration and surface soil moisture from optical and thermal infrared measurements: the simplified triangle. Int. J. Rem. Sens. 40 (20), 7716–7729. https://doi.org/10.1080/01431161.2019.1601288.*

which, for a triangle with a well-defined upper vertex, occurs at Ts[min]. Note that while Mo equals zero along the dry edge EF itself is nonzero along the dry edge except at the lower right vertex.

The next step in the method implementation involves the scaling of Ts to a variable named T*. To do this, the Ts for dry/bare soil needs to be determined, which is representative of the highest values of Ts for pixels found over dry/bare soil (Ts [max]) and the value of the minimum Ts representative of cool, wet pixels (Ts[min]) such as found over dense vegetation. Ts varies between its limits of Ts[min] and Ts[max]. The variable T* is scaled between 0 and 1 as defined below.

$$T^* = \{Ts - Ts(min)\}/\{(Ts(max) - Ts(min)\} \tag{11.2}$$

In our study, T* was derived from the Fr/Ts scatterplot and by scaling the LST layer of each Sentinel-3 image (using Eq. (11.1) above).

In the next step, Mo and EF are derived directly from Fr and T*. To do this, two important assumptions are made by the authors. The first is that transpiration (evaporation from the leaves) that always equals potential, at least when the vegetation is not at the wilting point. The second assumption is that the relation between EF and Mo varies linearly across the triangle domain. At bare soil fraction (equal to Mo), Mo is the availability of moisture on the surface, is the ratio between the lengths of a/d, both of these lengths being functions of the scaled radiometric surface temperature (T*) and Fr. Thus, Mo and EF are estimated as follows:

$$Mo = 1 - T(pixel)/T(warm\ edge) \tag{11.3}$$

$$EF = EF_{soil}(1 - Fr) + Fr\ EF_{veg} = Mo(1 - Fr) + Fr \tag{11.4}$$

where EF_{soil} refers to the ratio of soil evaporation to net radiation.

The above mathematical expressions are valid on the assumption made by Carlson and Petropoulos (2019) that both Mo and EF vary linearly within the triangle between 0 and 1.0, such as (for Mo) between the cold and warm edges of the triangle. In addition, for each value of Fr and EF from the combined vegetation and bare soil, the canopy EF is assumed to be the weighted value of EF for the vegetation fraction of the pixel ($EF_{veg} = 1.0$, by definition). In our study, the steps described above concerning the 'simplified triangle' implementation were applied on each Sentinel-3 image, which resulted into obtaining two final image products for each image that was processed, namely the EF and Mo map. Fig. 11.4 shows an example of a scatterplots set that was created for one of the exprerimental days on which Sentinle −3 data had been acquired.

3.1 Statistical analysis

Point-by-point comparisons formed the main validation approach. In order to perform the SSM comparisons, in particular, Mo was converted to SSM using

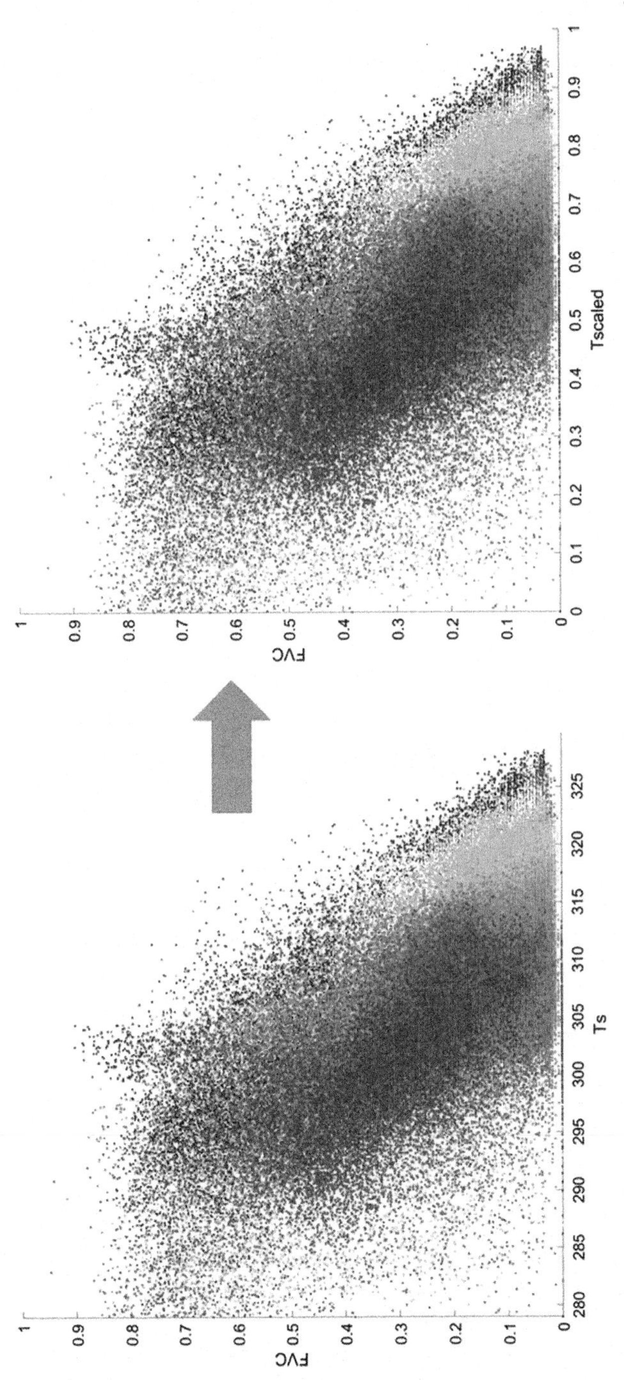

FIGURE 11.4 An example of the derived scatterplots during the technique implementation, shown here for the case of Sentinel-3 image with acquisition date of 25/08/2018. The use of colour in the scatterplots is to support visualisation only.

TABLE 11.1 Statistical measures used to assess the agreement between the predictions from the 'simplified triangle' and the in-situ observations.

Name	Description	Mathematical definition
Bias/MBE	Bias (accuracy) or mean bias error	$bias = MBE = \frac{1}{N}\sum_{i=1}^{N}(P_i - O_i)$
Scatter/SD	Scatter (precision) or standard deviation	$scatter = \frac{1}{(N-1)}\sum_{i=1}^{N}\sqrt{\left(P_i - O_i - \overline{(P_i - O_i)}\right)^2}$
MAE	Mean absolute error	$MAE = N^{-1}\sum_{i=1}^{N}\lvert P_i - O_i\rvert$
RMSD	Root mean square difference	$RMSD = \sqrt{bias^2 + scatter^2}$
R	Pearson's correlation coefficient	$R = \frac{E[(\theta_{sat} - E[\theta_{sat}])(\theta_{in-situ} - E[\theta_{in-situ}])]}{\sigma_{sat}\sigma_{in-situ}}$

Subscripts i = 1 ... N denote the individual observations, P denotes the predicted values and O denotes the 'observed' values. The horizontal bar denotes the mean value.

the soils' field capacity (an average value of which was used for each site). Similarly, in the in-situ data, the acquired volumetric moisture content (VMC, expressed as %) was converted to SSM. Also from the ground measurements, the EF was computed from the instantaneous latent heat fluxes (LE) and Rn. The statistical measures employed to quantify the agreement are summarised in Table 11.1 below:

4. Preliminary results

An example of the EF and Mo maps obtained from the 'simplified triangle' technique implementation for the Sentinel 3 image acquired on 25/08/2018 is illustrated in Fig. 11.5. As can be observed, predicted EF and Mo exhibited mostly a spatially reasonable range and also a realistic variability spatially across the area covered in the satellite field of view. This spatial variability seems to also be in agreement to land use/cover of the area, the Fr and Ts maps (see, for example, Fig. 11.5 in combination with Fig. 11.2) and the area topographical characteristics (i.e., slope, elevation). This observation, although it does not provide direct quantitative evidence of the EF product accuracy, suggests that the examined method is able to provide reasonably the spatial variability in the EF and Mo.

The main results from the quantitative comparisons obtained are summarised in Table 11.2 and also in the scatterplot shown in Fig. 11.6, which illustrated better the agreement found for the individual calendar days included in this study. As can be observed, EF has been predicted reasonably well in

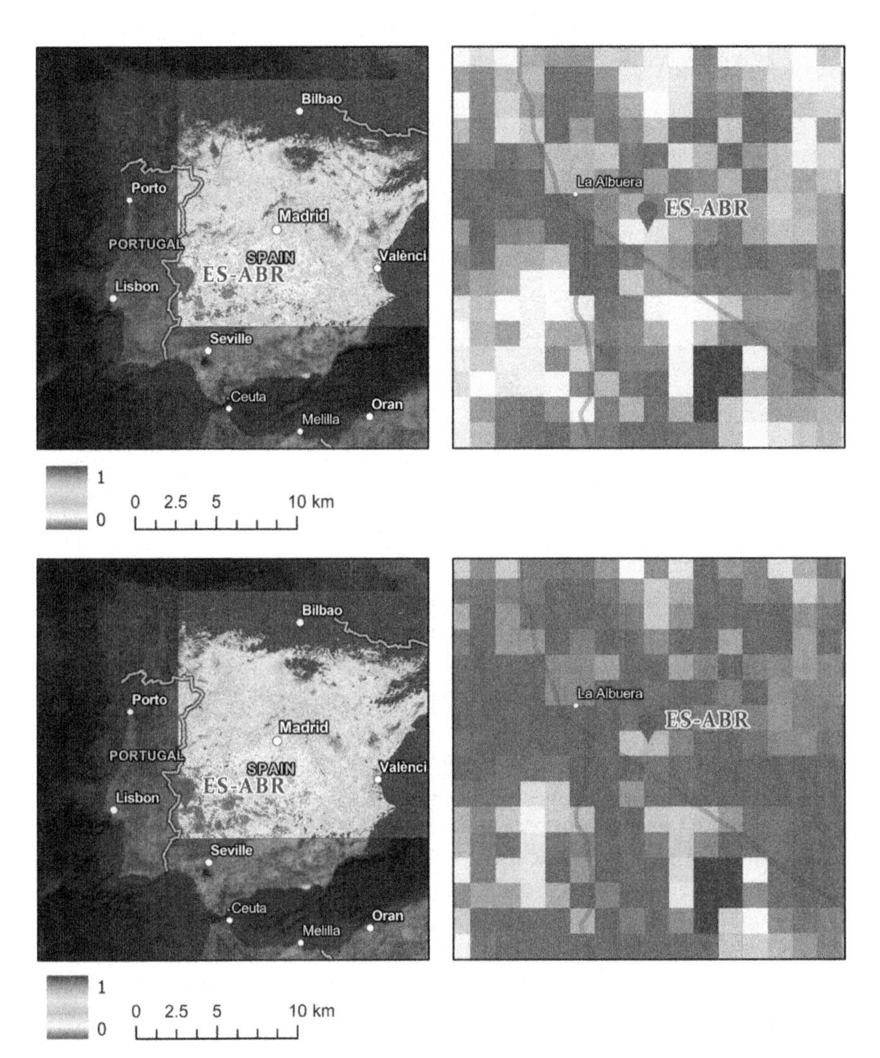

FIGURE 11.5 An example of a map of EF (top) and SSM (bottom) derived from the 'simplified triangle' implementation using the Sentinel-3 data. Image acquisition date is 25/08/2018.

TABLE 11.2 Summary of the statistical agreement for both EF and SSM for all days.

	Bias/MBE	Scatter/SD	MAE	RMSD	R
EF	0.028	0.057	0.055	0.063	0.777
SSM	−0.040	0.027	0.04	0.048	0.439

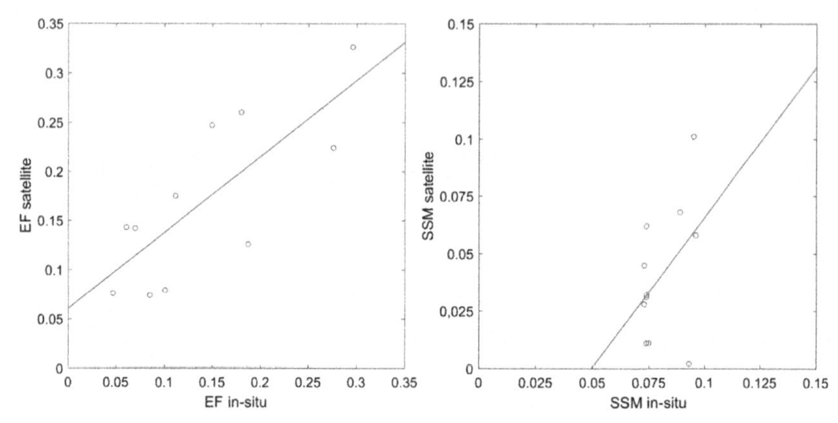

FIGURE 11.6 Comparisons for all days on which the technique was implemented between the in-situ measurements and satellite product values, for both EF (*left*) and SSM (*right*). *Red lines* (gray lines in printed version) represent the trend lines.

comparison to the reference data (i.e., ground observations), with a good R of 0.777, an RMSD of 0.063, an MBE of 0.028 and an SD of 0.057. As for the SSM comparisons, R was 0.439 (lower in comparison to the EF comparisons), whereas the RMSD was 0.048 vol vol^{-1}, well below the 0.1 vol vol^{-1} operational limit. MBE and SD were −0.04 and 0.027 vol vol^{-1}, respectively. All in all, quantitative comparisons showed that the 'simplified triangle' was able to provide predictions of both EF and SSM that were in good agreement to the collocated ground observations, which consisted the reference data. In terms of RMSD, prediction accuracy was better for EF in comparison to SSM, with the predicted EF slightly overestimated, whereas the predicted SSM was slightly underestimated. Because of the small number of tested days (11 in total), we cannot confirm that the MBE presented has statistical significance. It can be observed that the correlation between predictions and observations was significantly better for EF than the SSM comparisons.

As can be seen in Fig. 11.7, the trend line for the EF scatterplot shows a good fit, exhibiting *P*-value of 0.005. However for the SSM scatterplot trend line has a problematic *P*-value of 0.177. This was caused by the small range of SSM in-situ values that range from 0.073 to 0.096 vol vol^{-1} while the SSM predicted by Sentinel-3 presents a higher range, with values between 0.002 and 0.101 vol vol^{-1}. This disparity in the range also affects the correlation coefficient and is reflected in the relatively high bias (−0.04 vol vol^{-1} while the RMSD is 0.048).

5. Discussion

The investigation performed in this work, the first application of the 'simplified triangle technique' on Sentinel-3 EO data, resulted in promising results in

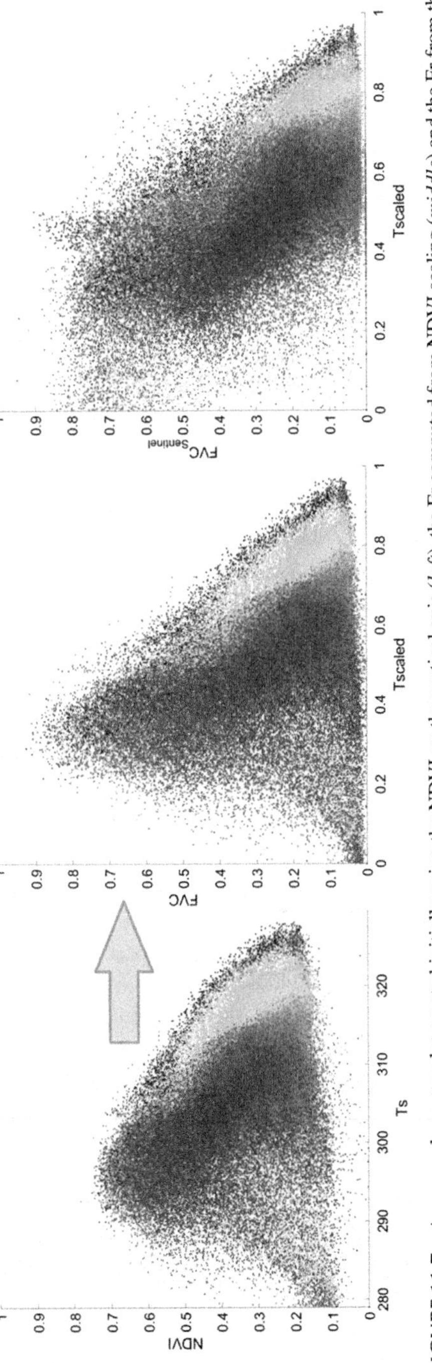

FIGURE 11.7 An example scatterplot created initially using the NDVI on the vertical axis (*left*), the Fr computed from NDVI scaling (*middle*) and the Fr from the Sentinel product. All image layers refer to the Sentinel-3 image and the acquisition date is 25/08/2018. The use of colour in the scatterplots is to support visualisation only.

deriving spatiotemporal estimates of both SSM and EF. The results support that the simplified triangle technique can provide reasonably accurate predictions of both parameters. Several studies (i.e., Chan et al., 2016; Bindlish et al., 2015) have already made a strong case for the utility of the in-situ SSM estimates in order to correctly assess the accuracy of satellite SSM products.

In regards to EF, it is noted that higher estimation accuracy was achieved for the LE/Rn and H/Rn fluxes compared with other methods that use a slightly different estimation method for EF (Peng and Loew, 2014; Lu et al., 2015). However, direct comparisons of results obtained herein against results where EF has been estimated using different approaches is not a feasible practice and could lead to erroneous conclusions. Prediction accuracy of SSM with the 'simplified triangle' was close or even improved compared to studies that use different Ts/VI methods (e.g., Carlson et al., 1995; Gillies et al., 1997).

There are various explanations for the imperfect agreement in the case of the EF comparisons, despite the estimated EF having strong correlation and low RMSD with the measured EF. Cloud cover has been identified as a critical factor influencing the stability of EF predictions during daytime (Hall et al., 1992). Despite the images being collected from May to September, cloudiness could have affected the radiation received by the validation site. Instrumentation accuracy could have also negatively impacted the agreement between the estimated and measured EF. Generally, the instrumental uncertainty regarding measurement of Rn is of the order of $\pm 10\%$, although another $\pm 10\%$ can be added due to constraints on view angle/measurement volume (particularly in cases of sloped terrain). Also typical uncertainty in the estimation of Tair is $\sim 2°C$. Uncertainty in the estimation of the turbulent fluxes by the eddy covariance system is typically in the order of 10%–15% (e.g., Petropoulos et al., 2015), and according to some researchers potentially more when the eddy covariance system is installed in nonflat terrain (e.g., Schmid and Lloyd, 1999). As of the SSM comparisons, the mismatch of the horizontal and vertical coordinates between the location of the station and the satellite pixel was shown to negatively affect correlation. Furthermore, the prediction from the satellite is responding to the soil water content in the top few millimeters of the soil, a much shallower layer than the ground measurements (Petropoulos et al., 2018; Deng et al., 2019).

Another potential factor in both the EF and SSM retrievals could be related to the method that the Fr was computed. The present study utilized the Fr that was provided directly from Sentinel-3. This Fr is computed using a different approach than the NDVI scaling suggested by Carlson and Petropoulos (2019). To our knowledge, the sensitivity of the 'simplified triangle' to the Fr and Ts computation method has not yet been sufficiently investigated to allow a quantification of the influence of the specific Fr method selected in this study. However, results of a preliminary investigation (results not shown in this study) indicate that the estimation method of Fr is affecting the predictions of EF and SSM significantly. For illustration purposes only, Fig. 11.6 presents the difference in the scatterplots between the two Fr estimation methods. On the right, there is the scatterplot using the Sentinel-derived Fr. On the middle is the Fr derived from NDVI scaling technique (Carlson and Petropoulos, 2019). The significant differences of the derived Fr consequently may have an important bearing to the 'simplified triangle' technique implementation. This is an area requiring further investigation.

Finally, the spatial resolution differences between the CarboEurope point measurements (5 m × 5 m) and Sentinel-3 pixel resolution (1 km × 1 km) increase the degree of uncertainty of the validation for both EF and SSM (Stisen et al., 2008). In-situ measurements cannot represent SSM or EF at the same spatial scale within the large footprint of the Sentinel-3 product. Thus, the averaged value of SSM is often represented as reference value. Several studies (Wagner et al., 2013; Petropoulos et al., 2015b) have shown that point-based measurements cannot sufficiently represent the absolute value of SSM for large pixels. Such representation can be achieved by upscaling the point estimates using techniques like those proposed by Srivastava (2017). Dense in-situ networks are very useful in this regard if present at the location of interest.

6. Conclusions

In this study, the ability of the so-called 'simplified triangle' technique was evaluated when used with Sentinel-3 EO data. The ability of the method to predict EF and SSM was evaluated for 11 days of the year 2018 at an experimental site in Spain belonging to the CarboEurope operational network. To our knowledge, this study represents the first attempt to examine this specific technique's accuracy using Sentinel-3 data in a typical Mediterranean savannah ecosystem.

A satisfactory agreement for both EF and EF an RMSD was reported, with Root Mean Square Error (RMSE) of 0.063 and 0.048 vol vol^{-1} and a correlation coefficient (R) of 0.777 and 0.439 for EF and SSM, respectively. This prediction accuracy is comparable to that reported in other similar studies where the same technique has been implemented with dissimilar EO data. The RMSD for the SSM was below the 0.1 vol vol^{-1} limit. Evidently, the 'simplified method' allows one to make estimates of EF and SSM over an area using just a few simple calculations in conjunction with satellite or aircraft images made at optical wavelengths and thermal infrared. The technique seems to have a significant advantage over other methods belonging in the same group of models in that it requires no land surface model or an ancillary surface or atmospheric data for its execution and is easy to apply. Yet, more work is required to evaluate its predictions over a wide range of ecosystems and environmental conditions globally, and the sensitivity of the Fr and Ts to the EF and Mo predicted by the technique. All in all, the results of this study are of considerable scientific and practical value in regards to the evaluation of the potential of the examined technique for deriving key biophysical parameters of the Earth's system.

Acknowledgements

In the present work, the participation of Dr. Petropoulos has received funding from the European Union's Horizon 2020 research and innovation programme ENViSIoN-EO under the Marie Skłodowska-Curie grant agreement No. 752094. Authors are also grateful to the anonymous reviewers for their comments that helped to improve the manuscript.

References

Amri, R., Zribi, M., Lili-Chabaane, Z., Szczypta, C., Calvet, J.C., Boulet, G., 2014. FAO-56 dual model combined with multi-sensor remote sensing for regional evapotranspiration estimations. Rem. Sens. 6, 5387–5406.

Baldocchi, D., Valentini, R., Oechel, W., Dahlman, R., 1996. Strategies for measuring and modelling carbon dioxide and water vapor fluxes over terrestrial ecosystems. Global Change Biol. 2, 159–168.

Bindlish, R., Jackson, T., Cosh, M., Zhao, T., O'Neill, P., 2015. Global soil moisture from the aquarius/SAC-D satellite: description and initial assessment. Geosci. Rem. Sens. Lett. IEEE 12, 923–927.

Birks, A., Cox, C., January 14, 2011. SLSTR: Algorithm Theoretical Basis Definition Document for Level 1 Observables. Science & Technology Facilities Council Rutherford Appleton Laboratory, p. 173.

Carlson, T.N., 2007. An overview of the "triangle method" for estimating surface evapotranspiration and soil moisture from satellite imagery. Sensors 7, 1612–1629.

Carlson, T.N., Petropoulos, G.P., 2019. A new method for estimating of evapotranspiration and surface soil moisture from optical and thermal infrared measurements: the simplified triangle. Int. J. Rem. Sens. 40 (20), 7716–7729. https://doi.org/10.1080/01431161.2019.1601288.

Carlson, T.N., Capehart, W.J., Gilies, R.R., 1995. A new look at the simplified method for remote sensing of daily evapotranspiration. Rem. Sens. Environ. 54, 161–167.

Chan, S.K., Bindlish, R., O'Neill, P.E., Njoku, E., Jackson, T., Colliander, A., et al., 2016. Assessment of the SMAP passive soil moisture product. IEEE Trans. Geosci. Rem. Sens. 54, 4994–5007.

Coudert, B., Ottlé, C., Briottet, X., 2008. Monitoring land surface processes with thermal infrared data: calibration of SVAT parameters based on the optimisation of diurnal surface temperature cycling features. Rem. Sens. Environ. 112 (3), 872–887.

Deng, K.A.K., Lamine, S., Pavlides, A., Petropoulos, G.P., Srivastava, P.K., Bao, Y., Hristopulos, D., Anagnostopoulos, V., 2019. Operational soil moisture from ASCAT in support of water resources management. Rem. Sens. 11, 579.

European Commission, 2009. White Paper, Adapting to climate change: towards a European framework for action. COM (4), 147.

Gillies, R.R., Carlson, T.N., Cui, J., Kustas, W.P., Humes, K.S., 1997. Verification of the "triangle" method for obtaining surface soil water content and energy fluxes from remote measurements of the normalized difference vegetation index NDVI and surface radiant temperature. Int. J. Rem. Sens. 18, 3145–3166.

Hall, F.G., Huemmrich, K.F., Goetz, S.J., Sellers, P.J., Nickeson, J.E., 1992. Satellite remote sensing of the surface energy balance: success, failures and unresolved issues in FIFE. J. Geophys. Res. 97 (D17), 19061–19089.

IPCC, 2009. Summary Report of the IPCC Expert Meeting on the Science of Alternative Metrics 18–20 March 2009, Oslo, Norway. IPCC-XXX/Doc.13 (31.III.2009). Available from: www.ipcc.ch/meetings/session30/doc13.pdf.

Lu, J., Tang, R., Shao, K., Li, Z.L., Zhou, G., 2015. Assessment of two temporal-information-based methods for estimating evaporative fraction over the Southern Great Plains. Int. J. Rem. Sens. 1–17.

Maltese, A., Capodici, F., Ciraolo, G., Loggia, G.L., 2015. Soil water content assessment: critical issues concerning the operational application of the triangle method. Sensors 15 (3), 6699–6718.

Peng, J., Loew, A., 2014. Evaluation of daytime evaporative fraction from MODIS TOA radiances using FLUXNET observations. Rem. Sens. 6 (7), 5959–5975.

Petropoulos, G.P., Carlson, T.N., Wooster, M.J., Islam, S., 2009. A review of ts/VI remote sensing based methods for the retrieval of land surface fluxes and soil surface moisture content. Adv. Phys. Geogr. 33 (2), 1–27.

Petropoulos, G.P., Ireland, G., Srivastava, P.K., 2015. Evaluation of the soil moisture operational estimates from SMOS in Europe: results over diverse ecosystems. IEEE Sensor. J. 15, 5243–5251.

Petropoulos, G.P., Ireland, G., Barrett, B., 2015. Surface soil moisture retrievals from remote sensing: evolution, current status, products & future trends. Phys. Chem. Earth. https://doi.org/10.1016/j.pce.2015.02.009.

Petropoulos, G.P., Ireland, G., Lamine, S., Ghilain, N., Anagnostopoulos, V., North, M.R., Srivastava, P.K., Georgopoulou, H., 2016. Evapotranspiration estimates from SEVIRI to support sustainable water management. J. Appl. Earth Obs. Geoinf. 49, 175–187. https://doi.org/10.1016/j.jag.2016.02.006.

Petropoulos, G.P., Srivastava, P.K., Feredinos, K.P., Hristopoulos, D., 2018. Evaluating the capabilities of optical/TIR imagine sensing systems for quantifying soil water content. Geocarto Int. https://doi.org/10.1080/10106049.2018.1520926.

Piles, M., Petropoulos, G.P., Sanchez, N., González-Zamora, A., Ireland, G., 2016. Towards improved spatio-temporal resolution soil moisture retrievals from the synergy of SMOS & MSG SEVIRI spaceborne observations. Rem. Sens. Environ. 180, 403–471. https://doi.org/10.1016/j.rse.2016.02.048.

Schmid, H.P., Loyd, C.R., 1999. Spatial representativness and the location bias of flux footprints over inhomogeneous areas. Agric. For. Meteorol. 93, 195–209.

Silva-Fuzzo, D., Carlson, T.N., Kourgialas, N., Petropoulos, G.P., 2020. Coupling remote sensing with a water balance model for soybean yield predictions over large areas. Earth Sci. India (in press).

SLSTR ATBD Land Surface temperature, 2012. Sentinel-3 Optical Products & Algorithm Definition. Version 2.3. Available from: https://sentinel.esa.int/web/sentinel/technical-guides/sentinel-3-slstr/level-2/calculate-vegetation-fraction.

Srivastava, P.K., 2017. Satellite soil moisture: review of theory and applications in water resources. Water Resour. Manag. 31, 3161–3176.

Srivastava, P.K., Pandey, P.C., Petropoulos, G.P., Kourgialas, N.K., Pandley, S., Singh, U., 2019. GIS and remote sensing aided information for soil moisture estimation: a comparative study of interpolation technique. Resour. MDPI 8 (2), 70.

Steinhaeuser, K., Ganguly, A.R., Chawla, N.V., 2012. Multivariate and multiscale dependence in the global climate system revealed through complex networks. Clim. Dynam. 39, 889-8950.

Stisen, S., Sandholt, I., Nørgaard, I., Fensholt, R., Jensen, K.H., 2008. Combining the triangle method with thermal inertia to estimate regional evapotranspiration—applied to MSG-SEVIRI data in the Senegal River basin. Rem. Sens. Environ. 112 (3), 1242–1255.

Stoyanova, J.S., Georgiev, C.G., 2013. SVAT modelling in support to flood risk assessment in Bulgaria. Atmos. Res. 123, 384–399.

Wagner, W., Hahn, S., Kidd, R., Melzer, T., Bartalis, Z., Hasenauer, S., et al., 2013. The ASCAT soil moisture product: a review of its specifications, validation results, and emerging applications. Meteorol. Z. 22, 5–33.

Part IV

Computational intelligence techniques

Chapter 12

Artificial neural network for the estimation of soil moisture using earth observation datasets

Sumit Kumar Chaudhary[1], Jyoti Sharma[2], Dileep Kumar Gupta[1], Prashant K. Srivastava[1,3], Rajendra Prasad[2], Dharmendra Kumar Pandey[4]

[1]*Remote Sensing Laboratory, Institute of Environment and Sustainable Development, Banaras Hindu University, Varanasi, Uttar Pradesh, India;* [2]*Remote Sensing Laboratory, Department of Physics, Indian Institute of Technology (BHU), Varanasi, Uttar Pradesh, India;* [3]*DST-Mahamana Centre of Excellence in Climate Change Research, Institute of Environment and Sustainable Development, Banaras Hindu University, Varanasi, India;* [4]*Microwave Techniques Development Division (MTDD), Advanced Microwave and Hyperspectral Techniques Development Group (AMHTDG), Earth, Ocean, Atmosphere and Planetary Science Applications Area (EPSA), Space Applications Centre (SAC), Indian Space Research Organization (ISRO), Ahmedabad, Gujarat, India*

1. Introduction

Surface soil moisture (SSM) information has great importance in Earth monitoring as a major concern of environmental conditions, irrigation management in agriculture, global climate change research and many more (Clark and Arritt, 1995; Fennessy and Shukla, 1999; Srivastava, 2017). Estimating of SSM is a highly essential variable for knowing the hydrological processes (Srivastava et al., 2013, 2015; Zhuo et al., 2015). The retrieval of soil moisture by remote sensing is very challenging and needs extra attention over the development of algorithms (Wang and Qu, 2009; Srivastava et al., 2014a,b).

Remote sensor boards on airborne/spaceborne platforms are very common over wide area for the estimation of SSM using optical/infrared and microwave data. In recent decades, microwave remote sensing becomes very famous for SSM estimation (Bindlish and Barros, 2001; Data et al., 2017; He et al., 2016; Jagdhuber et al., 2014; Walker et al., 2004). Soil moisture retrieval with active microwave data is highly influenced by soil surface roughness (Lakhankar et al., 2009) and vegetation cover heterogeneity (Rahimzadeh-Bajgiran et al.,

Agricultural Water Management. https://doi.org/10.1016/B978-0-12-812362-1.00012-6

2013). However, the passive microwave sensors can overcome the influence of dense vegetation. The combined use of active and passive microwave sensor may have great ability for the retrieval of SSM. In spite of that, NASA has launched a mission in 2015, namely soil moisture active passive (SMAP), for the retrieval of SSM using combined microwave active and passive data. Optical remote sensing has the advantage of exception from complex polarisation information and soil surface roughness, which makes the soil moisture retrieval studies very frequent using optical remote sensing (Fabre et al., 2015; Hassan-Esfahani et al., 2015a,b; Rahimzadeh-Bajgiran et al., 2013).

The land surface temperature (LST) and vegetation cover have a dependency on SSM. The SSM regulates the temperature of land surface and provides a healthy environment for vegetation growth. The land surface temperature may not be increased more than 35°C for healthy vegetation. The areas with hottest land surface temperatures are found to be vegetation-free like deserts. The SSM retrieval using optical/thermal infrared sensing is based on LST/vegetation concept started by Nemani et al. (1993). They found a strong negative relationship between LST and Normalized Difference Vegetation Index (NDVI) for various biome types with a distinct change in the slope between dry and wet days. Carlson et al. (1995a) further improved the existing idea and later presented the universal triangle method to study the relationship between SSM, LST and NDVI. Several studies have reported the relationship between SSM, LST and vegetation indices (VI) (Carlson et al., 1995b; Gillies et al., 1997; Nishida et al., 2003; Price, 1990; Song et al., 2014; Venturini et al., 2004). Li et al. (2009) and Petropoulos et al. (2009) have reported a good review paper on estimation of SSM using remotely sensed LST/vegetation indices.

Two types of LST—VI feature spaces, namely triangular and trapezoidal space, have been reported in previous studies (Carlson, 2007; Moran et al., 1994) for the retrieval of SSM. The primary step for determining the dry and wet limiting edges affects the SSM estimation accuracy using the LST-VI feature space approach. A 'wet edge' signifies the edge in LST-VI space at which the surface ET is equal to surface available energy and the adequate soil water content (Li et al., 2009). A 'dry edge' is considered at which the vegetation is subjected to water stress, evapotranspiration is minimised and the SSM is 0 (Sandholt et al., 2002). There are two common approaches that have been used to determine these limiting edges.

The first approach is termed as 'observed limiting edges', the least square method has been used to fit the extreme points to obtain these edges. This method is simple and more accurate and hence a large number of applications are performed using this approach (Nemani et al., 1993; Tang et al., 2010). The main limitation of the approach is that the edges can vary with size of the study area, and the extreme dry and wet points must be all present in the feature space (Long and Singh, 2012; Tang et al., 2010). The second approach is based on surface energy balance principle, the theoretical limiting edges have been obtained from ground meteorological data (such as air temperature, vapor pressure, wind speed and radiation) (Long et al., 2012; Moran et al., 1994). Robustness and accurate determination of feature space in true land surface conditions are the main advantages of this approach. However, the requirements of more ground-based parameters and varying the influence of water stress on vegetation with varying time are the major drawbacks of this approach. The approach is also restricted to study regions of small size.

After determination of dry and wet edges, an interpolation technique can be used to estimate the soil moisture. However, there is disagreement in use of interpolation in LST−VI/FVC feature space (Carlson, 2007). In soil moisture estimation studies, most authors assume that soil moisture changes linearly with LST (Chen et al., 2011; Gao et al., 2011; Han et al., 2010). However, Carlson et al. (1995b) proposed that the soil moisture exhibits a nonlinear relationship with LST and vegetation index using a soil vegetation atmosphere transfer (SVAT) model simulation.

Artificial Neural Networks (ANN) have the ability to learn multivariate nonlinear relations among the input variables and are also very robust for noise present in the data (Twarakavi et al., 2006). The ANNs have been widely used over a range of applications in various disciplines. In water resources and hydrology their performances are substantially recognised (Ahmad and Simonovic, 2005; Hsu et al., 1995; Maier and Dandy, 1996; Zealand et al., 1999). ANNs become a very important technique to learn the complex data among the scientist community during the past two decades. Chai et al. (2009) used the ANN for the retrieval of soil moisture using microwave data acquired from polarimetric L-band multi-beam radiometer. They concluded with remarkable improvement in the retrieval of soil moisture using the ANN model. A number of similar studies for soil moisture retrieval using microwave data were reported by various authors (Alexakis et al., 2017; Gupta et al., 2017; Kumar et al., 2019; Paloscia et al., 2013). ANN model have also been applied to retrieve soil moisture from optical data such as LST and NDVI (Ahmad et al., 2010; Kumar et al., 2019). For optical data, very few studies observed how ANN could be used for estimation of soil moisture (Chai et al., 2009). Esfahani et al. (2015) evaluated an ANN model to quantify the effectiveness to estimate SSM using spectral images. The model achieved acceptable SSM retrieval results with coefficient of correlation ($r = 0.88$) by combining field measurements.

The present study exploits an artificial neural network to learn nonlinear relationship of soil moisture with different combinations of LST and NDVI feature space to retrieve the SSM. In this study, the two ANN models namely ANN-I and ANN-II are developed using different combinations of input-output variables. The ANN-I has single input variable (LST) and single output variable (in-situ SSM). The ANN-II has single input variable (NDVI) and single output variable (in-situ SSM). The ANN-III has developed with combined LST-NDVI feature space. The ANN-III has two input variables (LST and NDVI) and single output variable (in-situ SSM). The results of each ANN model have been evaluated in terms of coefficients of correlation, bias and RMSE.

2. Materials and methods

2.1 Data used

LST and NDVI from MODIS collection 6 data are downloaded from Earth data (https://earthdata.nasa.gov). The in-situ SSM data is obtained from Stevens Hydra-probe installed at BHU campus. The Hydra-probe logged the data at every 15 min. The detailed description of datasets is given in Table 12.1. The in-situ measured SSM at 6:00 p.m. of every day is used in the study. The point measured in-situ SSM is considered for the representation of 1×1 km SSM value. The homogeneous area is available at the study site at least approximately 3×3 km. The daily and 1×1 km MODIS LST and in-situ measured SSM data are available. The MODIS NDVI data is available for every 16 days with spatial resolution of 1×1 km. To resize the temporal scale of each dataset, MODIS NDVI values are considered as same for every 16 days. The 1 year long (January to December) in-situ measured SSM, MODIS LST and MODIS NDVI datasets are considered to expect the monsoon season for the study.

2.2 Performance indices

The ANN model performance was accessed using three metrics, which are as follows:

Pearson's correlation coefficients (R): The Pearson's correlation coefficient (r) signifies the agreement or linear association between two variables. The value of R ranges from 1 to -1 with 1 as strong positive linear correlation and -1 as strong negative correlation, whereas 0 has no correlation. Mathematically,

$$r = \frac{\sum_{i=1}^{n}(O_i - \overline{O})(S_i - \overline{S})}{\sqrt{\sum_{i=1}^{n}(O_i - \overline{O})^2} \cdot \sqrt{\sum_{i=1}^{n}(S_i - \overline{S})^2}} \tag{12.1}$$

Root mean square error (RMSE): RMSE is used for measuring the average magnitude of estimated error between observed and estimated variables. The value of RMSE can vary between 0 to ∞ with zero as an ideal value. Mathematically,

$$RMSE = \sqrt{\frac{1}{n}\sum (S_i - O_i)^2} \tag{12.2}$$

TABLE 12.1 Summary of the MODIS data products used in this study associated with spatial and temporal resolution.

Product name	Platform	Variable	Period	Pixel resolution	Temporal resolution
MOD11A1 version 6	Terra	LST	Jan to Dec 2018 (except monsoon season)	1 × 1 km	Daily
MOD13A2 version 6	Terra	NDVI	Jan to Dec 2018 (except monsoon season)	1 × 1 km	16 Days
In-situ	Stevens hydra-probe	SSM	Jan to Dec 2018 (except monsoon season)	Point	Daily at 6:00 p.m.

Bias: Bias depicts the deviation between the observed and estimated values. The positive bias shows the overestimation whereas the negative value of bias shows the underestimation. Mathematically,

$$RB = \frac{\sum_{i=1}^{n} (S_i - O_i)}{\sum_{i=1}^{n} (O_i)} \tag{12.3}$$

2.3 Artificial neural network architecture and modelling

An approach based on ANN was used for soil moisture estimation with input parameters; MODIS LST, and MODIS NDVI with spatial resolution of 1 km and in-situ soil moisture. This study was carried out for the year of 2018 excluding the months of June, July and August due to the limitations of the soil sensor in the rainy season. Three different ANN architectures have been developed with the different combinations of input and output variables namely ANN-I, ANN-II and ANN-III. The two ANN models namely ANN-I and ANN-II are developed using different combinations of input (LST or NDVI)-output (in situ SSM) variables. The ANN-I has input variable (LST) and output variable (in-situ SSM). The ANN-II has input variable (NDVI) and output variable (in-situ SSM). The ANN-III has developed with input variables (LST and NDVI) and output variables (in-situ SSM).

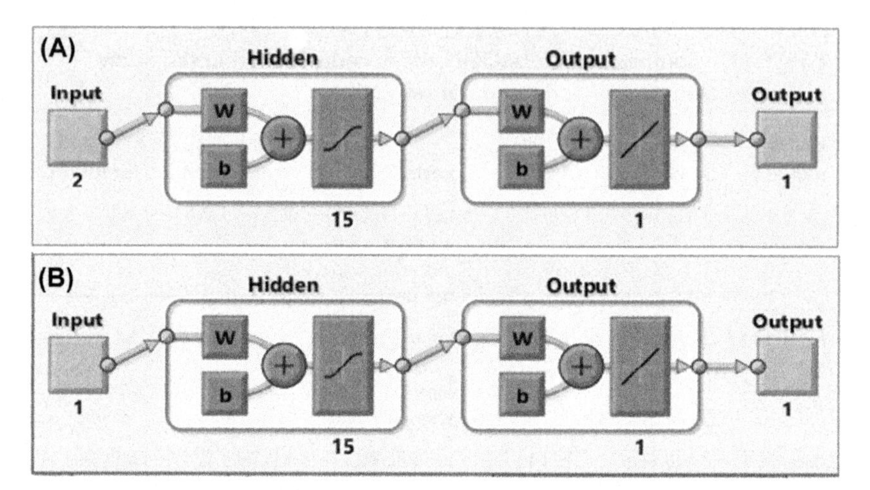

FIGURE 12.1 ANN architecture for (A) ANN-III with two input variables (B) ANN-I and ANN-II with single input variables.

Total 252 datasets are carried out for the training, testing and validation of the three different ANN models (ANN-I, ANN-II and ANN-III). The 164 (65%), 50 (20%) and 38 (15%) out of 252 have been taken for training, validation and testing datasets. ANN regression is performed three times with different inputs to examine the sensitivity of soil moisture estimation with LST and NDVI using training, validation and testing datasets. Each ANN model has a single hidden layer with 15 neurons. The Levenberg-Marquardt algorithm is used as the training algorithm. Fig. 12.1 depicts the architecture of different ANN models.

3. Results and discussions

3.1 Optimization of ANN models

The ANN models have been trained using 164 datasets, which consists in situ SSM, MODIS LST and MODIS NDVI. Generally, ANN models have three basic components named as input, hidden, and output layer. Here, LST and NDVI used as input and in situ SSM used as the output layer for the training of the models. The training and performance evaluation of this ANN models are done by using Levenberg–Marquardt optimisation algorithm. The loss function of algorithms can be expressed as the sum of the squares of the error, and it also uses the gradient vector as well as Jacobian matrix. The loss function can be defined as:

$$f = \sum_{i=1}^{m} e_i^2$$

where m is the number of datasets, and Mean Square Error (MSE), $e_i = (y_i - x_i)$; here y and x denote the desired outputs and process inputs, respectively.

During the training of algorithm, the weight and biases are varied to minimise the MSE and increase the performance of the network. The Jacobian matrix of the loss function can be expressed as the derivative of the error with respect to the weight parameters

$$J_{i,k} = \frac{\partial e_i}{\partial w_j}$$

where $i = 1, \ldots, m$ and $j = 1, \ldots, n$, n is the number of parameters in the neural network.

The gradient vector of loss function can be defined as,

$$\nabla f = 2J^T.e$$

now, the Hessian matrix and the parameter of the Levenberg–Marquardt algorithm can be estimated by using these expressions

$$Hf = 2J^T.J + \lambda I$$

and

$$w^{i+1} = w^i - \left(J^{(i)T} . J^i + \lambda I\right)^{-1} . \left(2J^{(i)T} . e^i\right)$$

where λ is a damping factor and shows the positivity of the Hessian and I is the identity matrix, whereas, i is number of iterations. By using the above expression, a sequence of parameters will be generated to reduce the loss factor in each iteration of the algorithm. The difference between the loss in two steps is known as the loss decrement. This training process will stop when certain condition or stopping criterion is satisfied.

3.2 Performance analysis of ANN models

Three different ANN models (ANN-I, ANN-II and ANN-III) are used to estimate the SSM. The performance of each ANN model is also evaluated in terms of coefficient of correlation (r), bias and RMSE during training, validation and testing of models. Fig. 12.2 shows the 1:1-line scatter plot between observed in-situ SSM and estimated SSM by ANN-I model. Fig. 12.3 shows the 1:1-line scatterplot between observed in-situ SSM and estimated SSM by ANN-II model. Fig. 12.4 shows the 1:1-line scatterplot between observed in-situ SSM and estimated SSM by ANN-III model. The values of performance indices have found during training (r 0.613, RMSE 0.0572, bias 0.132), testing (r 0.512, RMSE 0.051, bias 0.169) and validation (r 0.589, RMSE 0.049, bias 0.096) of ANN-I model. The values of performance indices have been found during training (r 0.972, RMSE 0.026, bias 0.010), testing (r 0.969, RMSE 0.026, bias –0.005) and validation (r 0.953, RMSE 0.060, bias

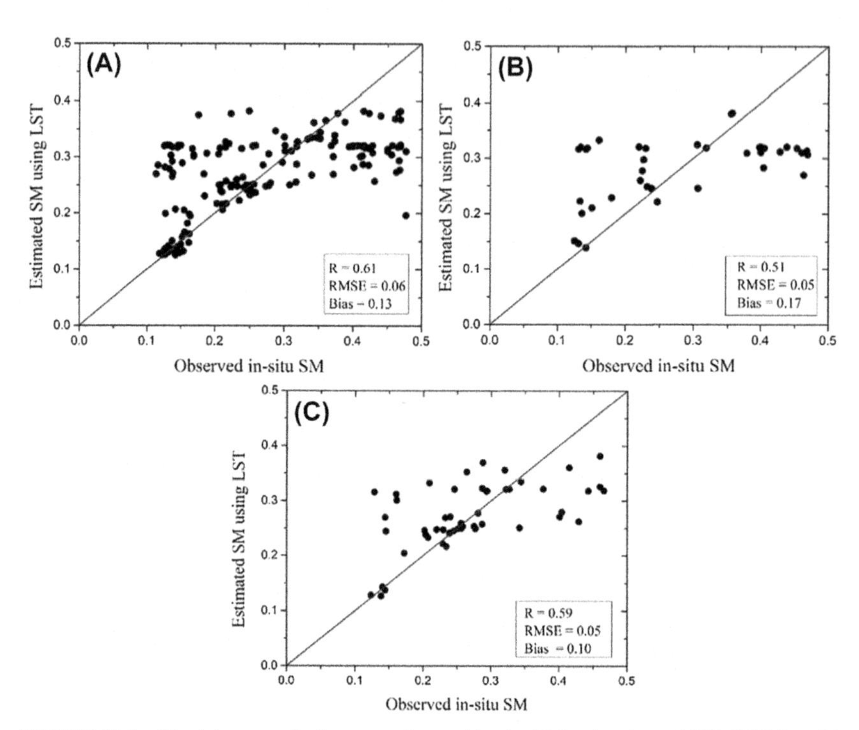

FIGURE 12.2 The 1:1 scatterplot between observed in-situ SM and estimated SM ANN-I model for (A) training data, (B) testing and (C) validation data.

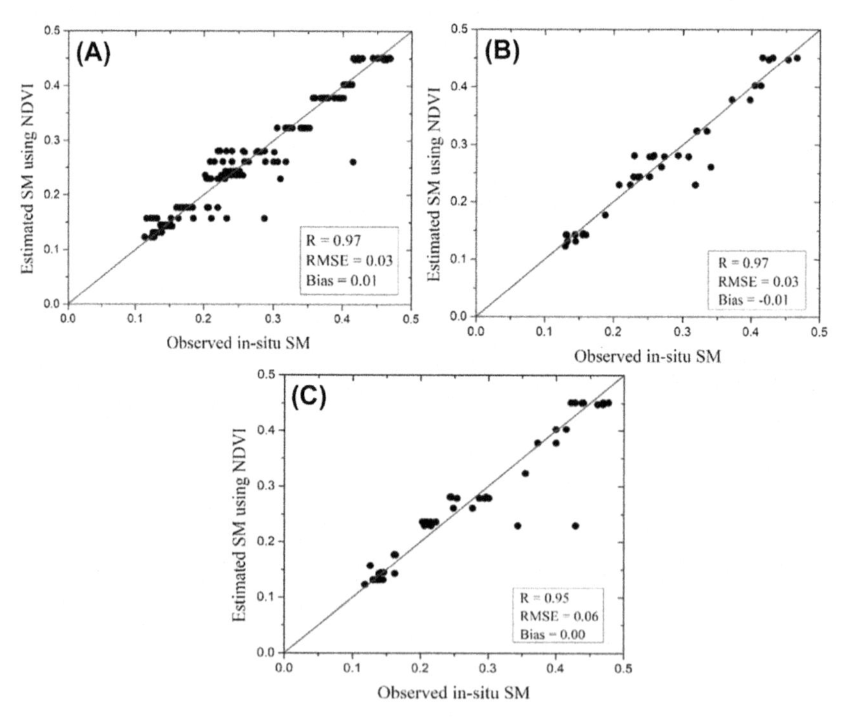

FIGURE 12.3 The 1:1 scatterplot between observed in-situ SM and estimated SM by ANN-II model for (A) training data, (B) testing and (C) validation data.

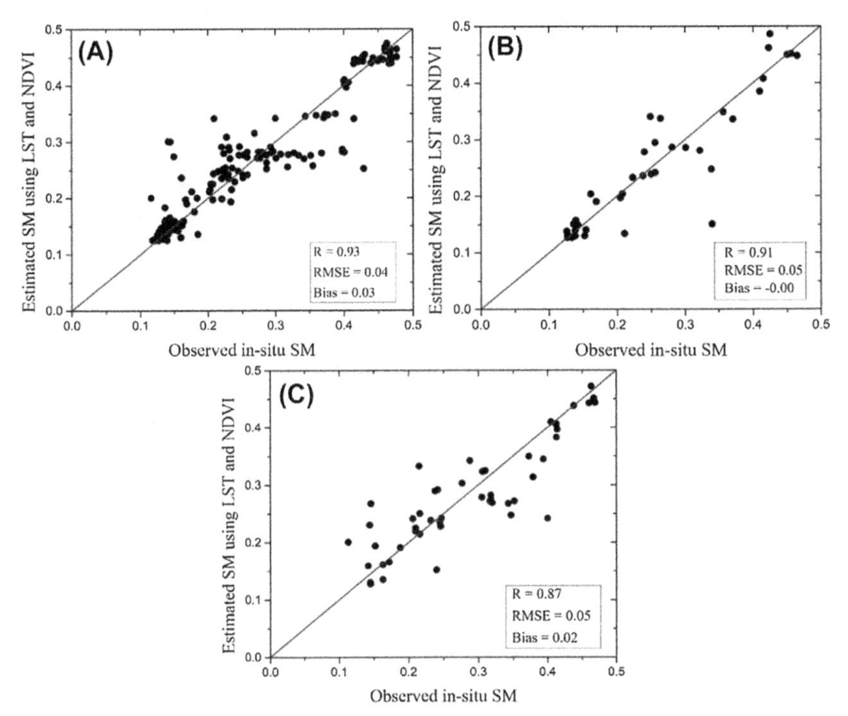

FIGURE 12.4 The 1:1 scatterplot between observed in-situ SM and estimated SM ANN-III model for (A) training data, (B) testing and (C) validation data.

0.004) of ANN-II model. The values of performance indices have been found during training (r0.935, RMSE 0.039, bias 0.032), testing (r 0.914, RMSE 0.047, bias −0.001) and validation (r 0.869, RMSE 0.046, bias 0.023) of ANN-III model. The results make it evident that the estimation of SSM by NDVI (ANN-II) is found to be better than the estimation of SSM by LST (ANN-I). The performance of ANN-II (with input NDVI) is also found to be better than the ANN-III (input combination of LST and NDVI). In comparison among three ANN models, the higher performance of ANN-II model, moderate performance of ANN-III model and lower performance of ANN-I model are found. The inclusion of LST in NDVI feature space deteriorates all the metrics of ANN models during training and testing.

4. Conclusion

ANN-based soil moisture estimations from MODIS LST/NDVI were carried out. The estimation of SSM using three different ANN models (with the different mapping of input-output variables among MODIS LST, MODIS NDVI and in-situ SSM) is performed. The results suggest that the estimation of SSM using LST is less significant than NDVI. In case of LST feature space, the lower correlation, larger RMSE and bias are observed for each training, validation and testing stage of ANN-I model. In case of NDVI feature space, the significant improvement is observed in terms of *r*, RMSE, and bias during the estimation of SSM for training, validation and testing stage of ANN-II model. The inclusion of LST in NDVI feature space deteriorates all the metrics of ANN models during training and testing. However, when both the LST and NDVI were used, the ANN-III model performance deteriorates from individual NDVI performance (ANN-II model).

Acknowledgement

Authors are thankful to Indian Space Research Organization RESPOND programme for funding this research.

References

Ahmad, S., Simonovic, S.P., 2005. An artificial neural network model for generating hydrograph from hydro-meteorological parameters. J. Hydrol. 315 (1–4), 236–251. https://doi.org/10.1016/j.jhydrol.2005.03.032.

Ahmad, S., Kalra, A., Stephen, H., 2010. Estimating soil moisture using remote sensing data: a machine learning approach. Adv. Water Resour. 33 (1), 69–80. https://doi.org/10.1016/j.advwatres.2009.10.008.

Alexakis, D.D., Mexis, F.-D.K., Vozinaki, A.-E.K., Daliakopoulos, I.N., Tsanis, I.K., 2017. Soil moisture content estimation based on sentinel-1 and auxiliary earth observation products. A Hydrological Approach. Sensors 17 (6), 1455. https://doi.org/10.3390/s17061455.

Bindlish, R., Barros, A.P., 2001. Parameterization of vegetation backscatter in radar-based, soil moisture estimation. Remote Sens. Environ. 76 (1), 130–137. https://doi.org/10.1016/S0034-4257(00)00200-5.

Carlson, T., 2007. An overview of the "triangle method" for estimating surface evapotranspiration and soil moisture from satellite imagery. Sensors 7 (8), 1612–1629. https://doi.org/10.3390/s7081612.

Carlson, T.N., Capehart, W.J., Gillies, R.R., 1995a. A new look at the simplified method for remote sensing of daily evapotranspiration. Remote Sens. Environ. 54 (2), 161–167. https://doi.org/10.1016/0034-4257(95)00139-R.

Carlson, T.N., Gillies, R.R., Schmugge, T.J., 1995b. An interpretation of methodologies for indirect measurement of soil water content. Agric. For. Meteorol. 77 (3–4), 191–205. https://doi.org/10.1016/0168-1923(95)02261-U.

Chai, S.-S., Walker, J., Makarynskyy, O., Kuhn, M., Veenendaal, B., West, G., 2009. Use of soil moisture variability in artificial neural network retrieval of soil moisture. Rem. Sens. 2 (1), 166–190. https://doi.org/10.3390/rs2010166.

Chen, J., Wang, C., Jiang, H., Mao, L., Yu, Z., 2011. Estimating soil moisture using temperature-vegetation dryness index (TVDI) in the Huang-huai-hai (HHH) plain. Int. J. Rem. Sens. 32 (4), 1165–1177. https://doi.org/10.1080/01431160903527421.

Clark, C.A., Arritt, R.W., 1995. Numerical simulations of the effect of soil moisture and vegetation cover on the development of deep convection. J. Appl. Meteorol. 34 (9), 2029–2045. https://doi.org/10.1175/1520-0450(1995)034<2029:NSOTEO>2.0.CO;2.

Data, F.F.R., Xie, Q., Meng, Q., Zhang, L., Wang, C., Sun, Y., 2017. A Soil Moisture Retrieval Method Based on Typical Polarization Decomposition Techniques for a Maize 9 (2), 168. https://doi.org/10.3390/rs9020168.

Fabre, S., Briottet, X., Lesaignoux, A., 2015. Estimation of soil moisture content from the spectral reflectance of bare soils in the 0.4–2.5 μm domain. Sensors 15 (2), 3262–3281. https://doi.org/10.3390/s150203262.

Fennessy, M.J., Shukla, J., 1999. Impact of initial soil wetness on seasonal atmospheric prediction. J. Clim. 12 (11), 3167–3180. https://doi.org/10.1175/1520-0442(1999)012<3167:IOISWO>2.0.CO;2.

Gao, Z., Gao, W., Chang, N.B., 2011. Integrating temperature vegetation dryness index (TVDI) and regional water stress index (RWSI) for drought assessment with the aid of LANDSAT TM/ETM+images. Int. J. Appl. Earth Obs. Geoinf. 13 (3), 495–503. https://doi.org/10.1016/j.jag.2010.10.005.

Gillies, R.R., Carlson, T.N., Cui, J., Kustas, W.P., Humes, K.S., 1997. A verification of the "triangle" method for obtaining surface soil water content and energy fluxes from remote measurements of the normalized difference vegetation index (ndvi) and surface e. Int. J. Rem. Sens. 18 (15), 3145–3166. https://doi.org/10.1080/014311697217026.

Gupta, D.K., Prasad, R., Kumar, P., Vishwakarma, A.K., 2017. Soil moisture retrieval using ground based bistatic scatterometer data at X-band. Adv. Space Res. 59 (4), 996–1007. https://doi.org/10.1016/j.asr.2016.11.032.

Han, Y., Wang, Y., Zhao, Y., 2010. Estimating soil moisture conditions of the greater changbai mountains by land surface temperature and ndvi. IEEE Trans. Geosci. Rem. Sens. 48 (6), 2509–2515. https://doi.org/10.1109/TGRS.2010.2040830.

Hassan-Esfahani, L., Torres-Rua, A., Jensen, A., McKee, M., 2015a. Assessment of surface soil moisture using high-resolution multi-spectral imagery and artificial neural networks. Rem. Sens. 7 (3), 2627–2646. https://doi.org/10.3390/rs70302627.

Hassan-Esfahani, L., Torres-Rua, A., Jensen, A., McKee, M., 2015b. Assessment of surface soil moisture using high-resolution multi-spectral imagery and artificial neural networks. Rem. Sens. 7 (3), 2627–2646. https://doi.org/10.3390/rs70302627.

He, L., Panciera, R., Tanase, M.A., Walker, J.P., Qin, Q., 2016. Soil moisture retrieval in agricultural fields using adaptive model-based polarimetric decomposition of SAR data. IEEE Trans. Geosci. Rem. Sens. 54 (8), 4445–4460. https://doi.org/10.1109/TGRS.2016.2542214.

Hsu, K. -l, Gupta, H.V., Sorooshian, S., 1995. Artificial neural network modeling of the rainfall-runoff process. Water Resour. Res. 31 (10), 2517–2530. https://doi.org/10.1029/95WR01955.

Jagdhuber, T., Hajnsek, I., Papathanassiou, K.P., 2014. Polarimetric Soil Moisture Retrieval Using an Iterative I, pp. 53–56.

Kumar, P., Prasad, R., Choudhary, A., Gupta, D.K., Mishra, V.N., Vishwakarma, A.K., Singh, A.K., Srivastava, P.K., 2019. Comprehensive evaluation of soil moisture retrieval models under different crop cover types using C-band synthetic aperture radar data. Geocarto Int. 34 (9), 1022–1041. https://doi.org/10.1080/10106049.2018.1464601.

Lakhankar, T., Ghedira, H., Temimi, M., Azar, A.E., Khanbilvardi, R., 2009. Effect of land cover heterogeneity on soil moisture retrieval using active microwave remote sensing data. Rem. Sens. 1 (2), 80–91. https://doi.org/10.3390/rs1020080.

Li, Z.L., Tang, R., Wan, Z., Bi, Y., Zhou, C., Tang, B., Yan, G., Zhang, X., 2009. A review of current methodologies for regional Evapotranspiration estimation from remotely sensed data. Sensors 9 (5), 3801–3853. https://doi.org/10.3390/s90503801.

Long, D., Singh, V.P., 2012. A two-source trapezoid model for evapotranspiration (TTME) from satellite imagery. Remote Sens. Environ. 121, 370–388. https://doi.org/10.1016/j.rse.2012.02.015.

Long, D., Singh, V.P., Scanlon, B.R., 2012. Deriving theoretical boundaries to address scale dependencies of triangle models for evapotranspiration estimation. J. Geophys. Res. Atmos. 117 (D5). https://doi.org/10.1029/2011JD017079.

Maier, H.R., Dandy, G.C., 1996. The use of artificial neural networks for the prediction of water quality parameters. Water Resour. Res. 32 (4), 1013–1022. https://doi.org/10.1029/96WR03529.

Moran, M.S., Clarke, T.R., Inoue, Y., Vidal, A., 1994. Estimating crop water deficit using the relation between surface-air temperature and spectral vegetation index. Remote Sens. Environ. 49, 246–263. https://doi.org/10.1016/0034-4257(94)90020-5.

Nemani, R., Pierce, L., Running, S., Goward, S., 1993. Developing satellite-derived estimates of surface moisture status. J. Appl. Meteorol. 32 (3), 548–557. https://doi.org/10.1175/1520-0450(1993)032<0548:DSDEOS>2.0.CO;2.

Nishida, K., Nemani, R.R., Running, S.W., Glassy, J.M., 2003. An operational remote sensing algorithm of land surface evaporation. J. Geophys. Res. Atmos. 108 (D9). https://doi.org/10.1029/2002jd002062.

Paloscia, S., Pettinato, S., Santi, E., Notarnicola, C., Pasolli, L., Reppucci, A., 2013. Soil moisture mapping using Sentinel-1 images: algorithm and preliminary validation. Remote Sens. Environ. 134, 234–248. https://doi.org/10.1016/j.rse.2013.02.027.

Petropoulos, G., Carlson, T.N., Wooster, M.J., Islam, S., 2009. A review of Ts/VI remote sensing based methods for the retrieval of land surface energy fluxes and soil surface moisture. Prog. Phys. Geogr. 33 (2), 224–250. https://doi.org/10.1177/0309133309338997.

Price, J.C., 1990. U sing spatial context in satellite data to infer regional scale evapotranspiration. IEEE Trans. Geosci. Rem. Sens. 28 (5), 940–948. https://doi.org/10.1109/36.58983.

Rahimzadeh-Bajgiran, P., Berg, A.A., Champagne, C., Omasa, K., 2013. Estimation of soil moisture using optical/thermal infrared remote sensing in the Canadian Prairies. ISPRS J. Photogrammetry Remote Sens. 83, 94–103. https://doi.org/10.1016/j.isprsjprs.2013.06.004.

Sandholt, I., Rasmussen, K., Andersen, J., 2002. A simple interpretation of the surface temperature/vegetation index space for assessment of surface moisture status. Remote Sens. Environ. 79 (2–3), 213–224. https://doi.org/10.1016/S0034-4257(01)00274-7.

Song, C., Jia, L., Menenti, M., 2014. Retrieving high-resolution surface soil moisture by downscaling AMSR-E brightness temperature using MODIS LST and NDVI data. IEEE J. Sel. Top. Appl. Earth Obs. Remote Sens. 7 (3), 935–942. https://doi.org/10.1109/JSTARS.2013.2272053.

Srivastava, P.K., 2017. Satellite soil moisture: review of theory and applications in water resources. Water Resour. Manag. 31 (10), 3161–3176.

Srivastava, P.K., Han, D., Ramirez, M.R., Islam, T., 2013. Machine learning techniques for downscaling SMOS satellite soil moisture using MODIS land surface temperature for hydrological application. Water Resour. Manag. 27 (8), 3127–3144.

Srivastava, P.K., Han, D., Rico-Ramirez, M.A., O'Neill, P., Islam, T., Gupta, M., Dai, Q., 2015. Performance evaluation of WRF-Noah Land surface model estimated soil moisture for hydrological application: synergistic evaluation using SMOS retrieved soil moisture. J. hydrol. 529, 200–212.

Srivastava, P.K., Han, D., Rico-Ramirez, M.A., O'Neill, P., Islam, T., Gupta, M., 2014a. Assessment of SMOS soil moisture retrieval parameters using tau–omega algorithms for soil moisture deficit estimation. J. Hydrol. 519, 574–587.

Srivastava, P.K., O'Neill, P., Cosh, M., Kurum, M., Lang, R., Joseph, A., 2014b. Evaluation of dielectric mixing models for passive microwave soil moisture retrieval using data from ComRAD ground-based SMAP simulator. IEEE J. Sel. Top. Appl. Earth Obs. Remote Sens. 8 (9), 4345–4354.

Tang, R., Li, Z.L., Tang, B., 2010. An application of the Ts-VI triangle method with enhanced edges determination for evapotranspiration estimation from MODIS data in arid and semi-arid regions: implementation and validation. Remote Sens. Environ. 114 (3), 540–551. https://doi.org/10.1016/j.rse.2009.10.012.

Twarakavi, N.K.C., Misra, D., Bandopadhyay, S., 2006. Prediction of arsenic in bedrock derived stream sediments at a gold mine site under conditions of sparse data. Nat. Resour. Res. 15 (1), 15–26. https://doi.org/10.1007/s11053-006-9013-6.

Venturini, V., Bisht, G., Islam, S., Jiang, L., 2004. Comparison of evaporative fractions estimated from AVHRR and MODIS sensors over South Florida. Remote Sens. Environ. 93 (1–2), 77–86. https://doi.org/10.1016/j.rse.2004.06.020.

Walker, J.P., Houser, P.R., Willgoose, G.R., 2004. Active microwave remote sensing for soil moisture measurement: a field evaluation using ERS-2. Hydrol. Process 18 (11), 1975–1997. https://doi.org/10.1002/hyp.1343.

Wang, L., Qu, J.J., 2009. Satellite remote sensing applications for surface soil moisture monitoring: a review. Front. Earth Sci. China 3 (2), 237–247. https://doi.org/10.1007/s11707-009-0023-7.

Zealand, C.M., Burn, D.H., Simonovic, S.P., 1999. Short term streamflow forecasting using artificial neural networks. J. Hydrol. 214 (1–4), 32–48. https://doi.org/10.1016/S0022-1694(98)00242-X.

Zhuo, L., Han, D., Dai, Q., Islam, T., Srivastava, P.K., 2015. Appraisal of NLDAS-2 multi-model simulated soil moistures for hydrological modelling. Water Resour. Manag. 29 (10), 3503–3517.

Chapter 13

Soil moisture retrieval from the AMSR-E

Dleen Al-Sharafany
Geomatics Department, College of Engineering, University of Salahaddin, Erbil, Kurdistan Region, Iraq

1. Introduction

Soil moisture measurement plays an important role in understanding the hydrologic cycle and its effect on weather and climate. It is considered as a good response of the land surface to atmospheric forces through the partitioning of rainfall into runoff and infiltration (Entekhabi et al., 1994; Lakshimi et al., 1997). Furthermore, accurate near-surface soil moisture estimations lead to better understanding and estimation of the surface energy budget which is an active area of research in hydrological and atmospheric processes (Yang et al., 2007). It is comprehensively explained in the research literatures that soil moisture is a highly variable parameter both spatially and temporally, which makes it difficult to collect accurate measurements of this important parameter. Remote sensing satellite has proven to be one of the powerful methods which are recently popular in the hydrological communities for retrieving soil moisture. It was found that satellite sensors can only measure soil moisture content in the top few centimetres (1−5 cm) of the soil surface layer. However, the role of soil moisture in the top layer of the Earth's surface is widely recognised as a key parameter in numerous environmental studies, including hydrology, meteorology, agriculture and climate change. Although this thin layer of soil water may seem insignificant when compared to the total amount of water on the global scale, it regulates the partitioning of precipitation into runoff and ground water storage, and controls the success of agriculture.

Hence, this chapter focuses on retrieving soil moisture for the Brue catchment using the Advanced Microwave Scanning Radiometer AMSR-E. A passive microwave theory is the basis behind retrieving soil moisture by AMSR-E and a physically based model known as Land Parameter Retrieval Model (LPRM) is used in the retrieving process. A new methodology is

Agricultural Water Management. https://doi.org/10.1016/B978-0-12-812362-1.00013-8

proposed to calibrate the LPRM's parameters. The assessment and validation of satellite soil moisture are challenging problems due to the lack of 'ground-truth' soil moisture measurements. On the other hand, soil moisture determination is associated with the rainfall and evapotranspiration rates. Adopting a hydrological model as an alternative tool to assess and validate satellite soil moisture is a limited research area as most and the majority of research has been focused on how the satellite soil moisture can improve the hydrological modelling. However, such an attempt is presented in this chapter based on the idea of getting benefit from abundances of hydrological data in comparison to soil moisture ground measurements, as was published in Al-Shrafany et al. (2011), in which the catchment soil storage is predicted from a calibrated observation-based hydrological model and used then to assess satellite soil moisture. The work presented in this chapter could be a step forward to a further development along this path.

2. Microwave theory

This section gives a brief overview of the passive microwave theory relevant to the present study, as it has been described in a considerable detail in some other papers (e.g., Njoku and Kong, 1977; Ulaby et al., 1982, 1986). Thermal radiation measurement from the land surface in the centimetre wave band is the operational concept of the passive microwave remote sensing. Microwave technology is a remote sensing method that permits quantitative estimation of soil moisture using physically based models such as radiative transfer models. The measured radiation is expressed as brightness temperature and is mainly determined by two components: physical temperature and the emissivity of the radiating surface (Owe et al., 2001; Njoku et al., 2003; Wang et al., 2009). The measured brightness temperature is expressed by:

$$T_b = e_0 \cdot T \tag{13.1}$$

where T_b is the observed microwave brightness temperature in Kelvin; T is the physical temperature of the emitting surface; and e_0 is smooth-surface emissivity (Njoku and Li, 1999). The complex permittivity ε of the soil depends primarily on the soil moisture content, and it is considered as the basis of the theory behind microwave remote sensing of soil moisture due to the significant variation in the dielectric properties of liquid water ($\varepsilon \approx 80$) and dry soil ($\varepsilon < 4$) (Ulaby et al., 1986; Woodhouse, 2006). The dielectric constant is a difficult quantity to measure in the field. Moreover, reproducing precise field conditions in laboratory soil samples makes laboratory analysis of the dielectric constant not entirely straightforward. Consequently, the validation of theoretical calculations is often somewhat difficult. Hence, the dielectric constant is calculated via a model developed by Hallikainen (1984) and used in the current study to calculate the dielectric constant of the soil medium.

At low soil moisture content, strong bonds are developed between the surfaces of the soil particles and the thin films of water which surround them. Therefore, in a relatively dry soil, the water is tightly bound and contributes little to the dielectric constant of the soil-water mixture. As more water is added, the molecules are able to rotate more freely. This is referred to as the free water phase. The subsequent influence of the free water on the soil dielectric constant therefore also increases. Smaller particles such as irregular fine sands silts, and clays have a higher surface area-to-volume ratio and therefore are able to hold more water molecules at higher potentials. Two soils with different textural compositions may exhibit markedly different relationships between moisture content and their respective soil dielectric constants. Soils with high clay content will generally have a lower dielectric constant than coarse sandy soils at the same moisture content, since more water is being held in the bound-water phase.

3. Surface roughness and vegetation effects

Despite the dependency of the soil reflectivity on the soil dielectric constant, surface roughness also has a significant effect on the soil reflectivity due to the scattering. It increases due to the apparent emissivity of natural surfaces, which is caused by increased scattering due to the increase in surface area of the emitting surfaces (Schmugge, 1985). Roughness also reduces the sensitivity of emissivity to soil moisture variations, and thus reduces the range in measurable emissivity from dry to wet soil conditions (Wang, 1983). Furthermore, the path through the atmosphere between the surface and the sensor depends on the elevation and also slope of the emitting surface. However, this effect is only significant at frequencies which are affected by atmospheric attenuation (i.e., $>$ 10 GHz; Mätzler and Standley, 2000). The development of the theoretical approaches that have been proposed to estimate the surface roughness parameters, where a simpler, semi-empirical expression for the rough surface reflectivity was proposed by Wang and Choudhury (1981). The statistical parameters which characterise the scale of roughness of a randomly rough surface are known as the h and Q parameters. It has been assumed from previous studies that at low frequencies when surface roughness condition is unknown, a value of zero is often assigned to the Q parameter, whereas a low value can be assigned to the h parameter (in the range 0–0.3). However, in this study, h and Q parameters are empirically calibrated.

The vegetation effect on the microwave emission as measured from above the canopy is twofold: (1) the vegetation will absorb or scatter the radiation emanating from the soil, and (2) it will also emit its own radiation. In areas of sufficiently dense canopy, the emitted soil radiation will become masked, and the observed emissivity will be due largely to the vegetation. The magnitude of the absorption by the canopy depends upon the wavelength and the vegetation water content. The most frequently used wavelengths for soil moisture sensing are in the L- and C-bandwidths ($\lambda \approx 21$ and 5 cm, respectively), although only L-band sensors are able to penetrate vegetation of any significant density. While observations at all frequencies are subject to scattering and absorption and require some correction if the data are to be used for soil moisture retrieval, shorter wave bands are especially susceptible to vegetation influences. Contributions from the soil, vegetation and atmosphere are included into the upwelling radiation from the land surface as observed from the above canopy, and are given as a radiative transfer equation which explains the relationship between the surface parameters such as surface soil moisture, vegetation water content, surface temperature and microwave brightness temperature (Jackson et al., 1982; Njoku et al., 2003). The brightness temperature at H and V polarisations are given by:

$$T_{bH} = T_s\{e_{sH}\Gamma_H + (1-\omega)(1-\Gamma_H)[1+(1-e_{sH})\Gamma_H]\} \quad (13.2)$$

$$T_{bV} = T_s\{e_{sV}\Gamma_V + (1-\omega)(1-\Gamma_V)[1+(1-e_{sV})\Gamma_V]\} \quad (13.3)$$

where the subscripts H and V refer to the horizontal and vertical polarisations, respectively; T_s is the single surface temperature; e_{sH} and e_{sV} are the soil emissivity at H and V polarisation, respectively; ω is the vegetation single scattering albedo and Γ is the transmissivity. The emission radiation from the soil as attenuated by vegetation layer is defined by the first term of the above equation. The second term accounts the direct upward radiation from the vegetation canopy, while the third term defines the downward radiation from the vegetation which is then reflected by the soil and again attenuated by the canopy.

Transmissivity of the vegetation can be defined as a function of the vegetation opacity along the observation path and it is given by:

$$\Gamma_{H,V} = \exp\left(-\frac{\tau}{\cos \Phi}\right) \quad (13.4)$$

where Φ is the incidence angle, τ is the vegetation opacity and has a linear relationship to the vegetation water content given by:

$$\tau = b_P \cdot vwc \quad (13.5)$$

where vwc is the vegetation water content in kg m^{-2}, b_p is a relationship parameter, which has a value of 0.15 at lower frequencies for most agriculture crops (Jackson and Schmugge, 1991).

Theoretical calculations show that the sensitivity of above-canopy brightness temperature measurements to variations in soil emissivity decreases with increasing optical depth or canopy thickness (Ulaby et al., 1986). This is because the soil emission is attenuated by the canopy and the emission from the vegetation canopy tends to saturate the signal with increasing optical depth. This subsequently results in decreased sensor sensitivity to soil moisture variations.

The single scattering albedo describes the scattering of the emitted radiation by the vegetation. The scattering albedo is a function of plant geometry, and consequently varies according to plant species and associations. Experimental data for this parameter are limited, and values for selected crops have been found to vary from 0.04 to about 0.12 (Brunfeldt and Ulaby, 1984; Jackson and O'Neill, 1990b; Mo et al., 1982). Values for natural vegetation are even scarcer, although Becker and Choudhury (1988) estimated a value of 0.05 for a semi-arid region in Africa. It is assumed that the single scattering albedo (ω) is negligible at lower frequencies (Jackson et al., 1982).

4. Atmospheric effects

Electromagnetic radiation emitted from the ground surface may interact with the atmosphere in two ways, as it propagates to a satellite radiometer. These are interactions between the electromagnetic radiation and (1) atmospheric gases (primarily oxygen and water vapour) and (2) water droplets existing in clouds and rain. The primary interaction mechanism is that of absorption of energy by the atmosphere. However, for frequencies below 15 GHz the effects are quite small, and for frequencies below 10 GHz the effects are negligible. The effect of water droplets in clouds and rain may be somewhat more significant, and depends largely on two factors: (1) the phase state of the particles (i.e., ice or liquid) and (2) the size of the particle relative to the wavelength (Chahine, 1983; Ulaby et al., 1982).

In addition to the atmospheric effects on the emitted surface radiation, there is also a sky background radiation component, which is reflected back to the observing instrument, and also a direct atmospheric component. Each of these components is further affected (attenuated) by the atmospheric transmissivity.

5. Soil moisture retrieval for the Brue catchment using AMSR-E

5.1 Modelling approach

The techniques adopted in this study for soil moisture retrieval provide spatially averaged soil moisture data, which is ideal for environmental and hydrological modelling. Such spatially averaged area sets are logistically and economically difficult to obtain through traditional in-situ measurement techniques. The technique only uses the horizontal and vertical polarisation brightness temperatures T_b at one frequency (6.9 GHz) observed by the Advanced Microwave Scanning Radiometer AMSR-E in a descending mode. The approach is based on simple radiative transfer equations (see Eqs 13.2 and 13.3), and the presented methodology solves for the soil moisture and vegetation optical depth simultaneously.

The AMSR-E satellite has a footprint size of 25 km at which all retrieval calculations are based on. With respect to the average soil and vegetation biophysical characteristics, a uniform footprint is assumed. Therefore, surface soil moisture and vegetation optical depth are subsequently extracted as average footprint values. The soil and vegetation temperatures are assumed to be approximately equal in the use of the AMSR-E descending measurements as the temperature and soil profiles are reasonably uniform. Moreover, the effects of the atmospheric moisture and the multiple scattering in the vegetation layer are negligible due to the AMSR-E low frequency measurements (up to X-band, i.e., ~ 10 GHz).

Generally speaking, the integral contribution of the surface roughness and vegetation canopy is more difficult to separate unless one of them is known a priori. An analytical approach developed by Meesters et al. (2005) is considered in this study for calculating vegetation optical depth from the Microwave Polarisation Difference Index (MPDI) and the dielectric constant of the soil. The MPDI effectively normalises out the effects of the surface temperature, resulting in a quantity that is highly dependent on the soil moisture and vegetation. This approach produces a significant improvement in accuracy and overall efficiency of the soil moisture estimates (Njoku et al., 2003; Wang et al., 2009).

The MPDI is defined as

$$\mathrm{MPDI} = \frac{T_{\mathrm{bV}} - T_{\mathrm{bH}}}{T_{\mathrm{bV}} + T_{\mathrm{bH}}} \tag{13.6}$$

and the vegetation optical depth is calculated from:

$$\tau = \cos \Phi \ln\left(ad + \sqrt{(ad)^2 + a + 1} \right) \tag{13.7}$$

where

$$a = 0.5 \left(\frac{e_{\mathrm{sV}} - e_{\mathrm{sH}}}{\mathrm{MPDI}} - e_{\mathrm{sV}} - e_{sH} \right) \tag{13.8}$$

$$d = 0.5 \left(\frac{\omega}{1 - \omega} \right) \tag{13.9}$$

Then by substituting Eq. (13.7) in (13.4),

$$\Gamma_{H,V} = \frac{1}{ad + \sqrt{(ad)^2 + a + 1}} \tag{13.10}$$

Hence, in this study, the brightness temperatures are converted to volumetric soil moisture values with the Land Parameter Retrieval Model LPRM (Owe et al., 2008; Wang et al., 2009).

$$MPDI = \frac{e_{sV} - e_{sH}}{e_{sV} + e_{sH}}(1-2Q)\left\{1 + \frac{2}{es_V + es_H}\left[exp(2\tau + h) - 1\right]\right\}^{-1} \tag{13.11}$$

The two surface roughness parameters (h and Q) and the vegetation parameter (ω) are the unknowns in Eq. (13.11), and the soil moisture that affects the soil emissivity is ultimately retrieved from Eq. (13.11). The Brue catchment study area is one of the rural areas in the UK, and it is mainly pasture land which is characterised as a noncomplex topography area. It has been assumed from previous studies that at low frequencies when surface roughness condition is unknown, a value of zero is often assigned to the Q parameter and a low value can be assigned to the h parameter (in the range 0–0.3), whereas, values of ω parameter were found to vary from 0.04 to 0.13.

In this study the h and Q parameters are empirically calibrated, since the lowest frequency of the AMSR-E instrument is at 6.9 GHz, and its footprint scale is large which results in no data available to quantify the regional variability of the those parameters. Therefore, in order to estimate the optimal (h and Q) values for a particular catchment area, a new approach is proposed in this study which basically uses a hydrological model in the context of catchment storage calculation in order to achieve the best correlation between the calculated storage and the retrieved soil moisture from the AMSR-E satellite. For this purpose, event-based water balance equation will firstly be adopted as it is a general formulation of the hydrological cycle and would provide a preliminary result and in turn an indication whether there is a good correlation between the satellite retrieval soil moisture and the catchment storage. Secondly, a particular and more complex physically based rainfall-runoff model will be adopted to obtain more accurate estimation and representation of the catchment storage over a continuously long time-series. The PDM model is the hydrological model that will be used for the purpose of this study.

5.2 LPRM calibration based on event-based water balance equation

As it was mentioned above in the previous section, the event-based water balance equation is used to calibrated the LPRM surface roughness parameters (h and Q) through achieving the best correlation between the changing in the catchment storage Δs calculated from water balance calculations and the changing in the volumetric soil moisture $\Delta\theta$ retrieved from AMSR-E satellite. To carry out this calibration, a total of 47 flow events over 3 years (2004–06) were selected and synchronised with the satellite measurements. From this, a total of 33 events out of the 47 were selected (years 2004–05) for calibration purposes, while the remaining 14 events (year 2006) were used for validation purposes.

For each selected event, water runoff volume is calculated. The Penman-Monteith equation is used for evapotranspiration calculation and finally the changing in the catchment storage Δs is calculated separately for each flow event using the basic water balance equation:

$$\Delta S / \Delta t = P - ET - Q \tag{13.12}$$

Simultaneously, for a given rainfall runoff event, the *vsm* is retrieved using the LPRM (Eq. 13.11), assuming given values for the parameters h and Q. Therefore, the changing in the volumetric soil moisture $\Delta\theta$ can be computed from:

$$\Delta\theta = \theta_2 - \theta_1 \tag{13.13}$$

where $\Delta\theta$ is the change in the *vsm*, θ_2 is the *vsm* after the event and θ_1 is the *vsm* before the event. Assuming a sufficient number of rainfall-runoff events, the correlation between Δs and $\Delta\theta$ can be calculated for the given parameters h and Q. A different value of correlation can be obtained by assuming a new set of values for the parameters h and Q. By randomly generating the values for the parameters h and Q within valid ranges, we can calculate the correlation curve. The value of maximum correlation indicates the corresponding optimal h and Q roughness parameters for the study catchment area (Brue).

The simulated MPDI is iteratively computed through the radiative transfer model for the feasible *vsm* range and the retrievals can be generated at any time step. The satellite brightness temperatures dataset measured at 6.9 GHz were used as inputs. In the early steps of the iteration process, a conversion of the input *vsm* into the dielectric constant is calculated by applying the dielectric mixing model developed by Hallikainen et al. (1985), which expresses the soil moisture m_v as a function of the dielectric constant ε:

$$\varepsilon = (a_0 + a_1 S + a_2 C) + (b_0 + b_1 S + b_2 C)m_v + (c_0 + c_1 S + c_2 C)m_v^2 \tag{13.14}$$

where a_i, b_i, c_i are the dielectric coefficients (see Hallikainen et al., 1985, for details) and S and C represent the percentage of soil and clay within the catchment (for the Brue catchment S = 20% and C = 50%). Note that $m_v \equiv$ *vsm*. At each iteration step, the difference between the simulated and the observed MPDI is computed and the minimum difference will indicate the optimal value of soil moisture. The methodology implemented to estimate surface soil moisture from the AMSR-E satellite is described in Fig. 13.1.

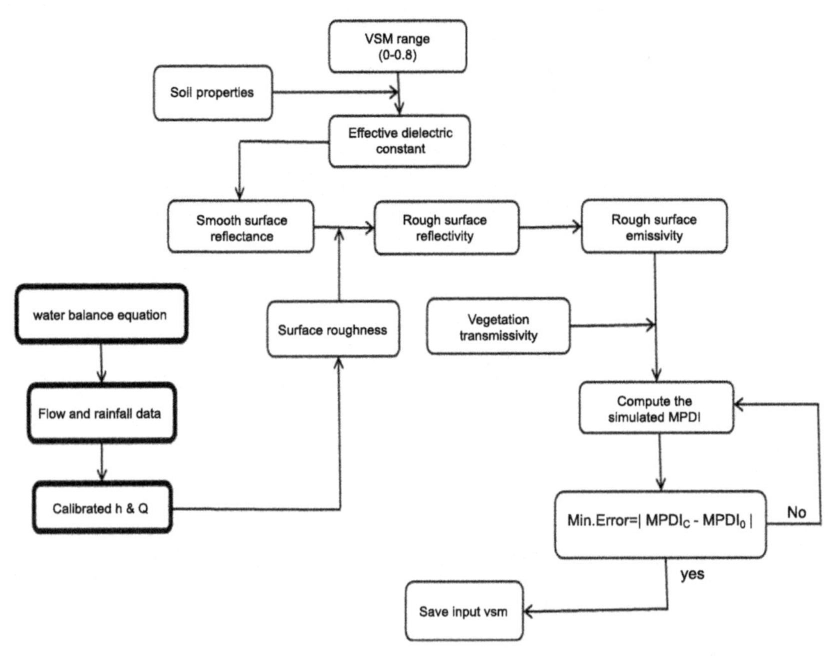

FIGURE 13.1 Volumetric soil moisture scheme from the LPRM (the new approach is highlighted with the *solid bold blocks*).

Two years of data (2004–05) from the Brue catchment were used in this study for the calibration of the surface roughness parameters. Rainfall-runoff events were synchronised with the AMSR-E measurements for the entire study period. With the selected rainfall-runoff events, the difference in the water storage Δs is calculated separately, using the water balance equation. The simulated MPDI is iteratively computed through the radiative transfer model for the *vsm* ranging from 0.0 to 0.8. Night-time AMSR-E brightness temperature measurements at 6.9 GHz for the same years were also used in the analysis. The two surface roughness parameters h and Q are calibrated using the water storage change. It was assumed that the h parameter ranges from 0 to 0.33, while the parameter Q has a range from 0 to 0.2. It is important to clarify that the values of the parameters h and Q were considered as fixed values over the whole catchment due to its noncomplex topography and its relatively small drainage area. The optimal h and Q parameters were found by maximizing the correlation between Δs *and* $\Delta\theta$ (see Fig. 13.2). It is important to mention that the correlation was used as performance indicator given the fact that the units of Δs and $\Delta\theta$ are in mm and m^3/m^3, respectively, and the use of other

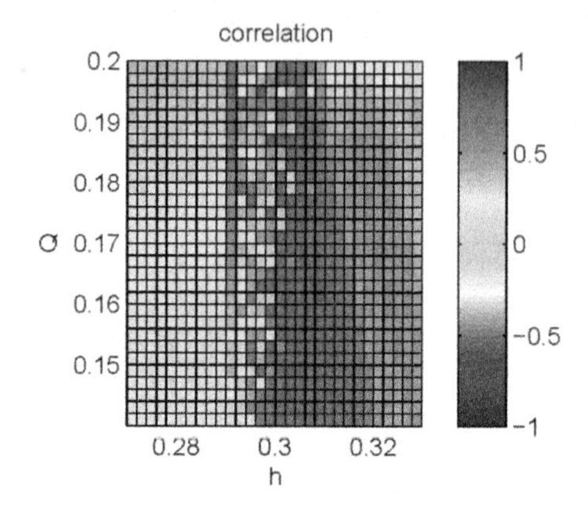

FIGURE 13.2 Correlation between Δs and $\Delta\theta$ for different values of the surface roughness parameters h and Q.

performance measures such as the root mean square error would not be appropriate. Although, the difference in units could be overcome by proper conversion, the absolute values of the change in storage and those in surface soil moisture are inherently different, even if they are both in the same units. Therefore, no attempts have been made to convert them into the same units. As shown in Fig. 13.2, the maximum values of correlations are shown in dark red (dark gray in printed version) colour. The maximum values of correlations correspond to values of the h parameter in the range 0.3–0.32. Fig. 13.2 clearly shows that even a very small increase or decrease in the value of the parameter h will produce a significant change in the correlation between Δs and $\Delta\theta$.

The results indicate that the changing in the volumetric soil moisture is very sensitive to the selection of the h parameter, but less sensitive to the selection of the Q parameter. The maximum correlation of 0.74 was obtained for $h_{\text{optimal}} = 0.304$ and $Q_{\text{optimal}} = 0.16$. A sensitivity analysis is presented here to identify how the changing in the two calibrated roughness parameters h and Q influence the changing in the satellite-based soil moisture estimation and hence the achieved correlation between Δs and $\Delta\theta$. For the purpose of this analysis, the optimal volumetric soil moisture (vsm_{optimal}) was calculated using the optimal roughness parameters (h_{optimal} and Q_{optimal}) obtained from the correlation between Δs and $\Delta\theta$. The vsm_{optimal} was used as a benchmark to compare with the $vsm_{h,Q}$ computed for given values of h and Q. Then the Normalised Error (NE) and the Normalised Bias (NB) was computed between vsm_{optimal} and $vsm_{h,Q}$ from the following equations:

$$NE = \frac{\sum_{h,Q=1}^{M} \left| vsm_{h,Q} - vsm_{opt} \right|}{\sum_{h,Q=1}^{M} vsm_{opt}} \tag{13.15}$$

$$NB = \frac{\sum_{h,Q=1}^{M} \left(vsm_{h,Q} - vsm_{opt}\right)}{\sum_{h,Q=1}^{M} vsm_{opt}} \tag{13.16}$$

where vsm_{opt} is the optimum value of the volumetric soil moisture, $vsm_{h,Q}$ is the value of the volumetric soil moisture for a particular set of h and Q parameters, M is the number of a set of (h and Q) parameters within their proposed values range.

The results are shown in Fig. 13.3. As shown, the retrieval of satellite soil moisture is very sensitive to the selection of the parameter h, which is in agreement with the results shown in Fig. 13.2. The NE is larger than 60% for values of h larger than 0.32 and the NE is larger than 40% for values of h smaller than 0.29. The NB results show that any value of h larger than $h_{optimal}$ will produce overestimation of the *vsm*, whereas values of $h<h_{optimal}$ will produce underestimation of the *vsm*. The results also indicate that the retrieval of soil moisture is less sensitive to the selection of the parameter Q as shown in Fig. 13.3.

The calibration results show a good agreement between Δs and $\Delta\theta$ (see Fig. 13.4A). It is interesting to note that the correlation between Δs and $\Delta\theta$, demonstrate the fact that the total water storage within the soil column (Δs) is highly related to the surface soil moisture retrieved from satellite measurements ($\Delta\theta$).

In order to validate the proposed approach, the year of 2006 was used for validation. The optimal parameters $h_{optimal}$ and $Q_{optimal}$ obtained in the calibration phase have been used to estimate the changing in the volumetric soil moisture $\Delta\theta$ using the validation dataset. The results are shown in Fig. 13.4B indicating that there is a good correlation between the changing in the storage and the changing in the AMSR-E soil moisture. However, it is worth to note

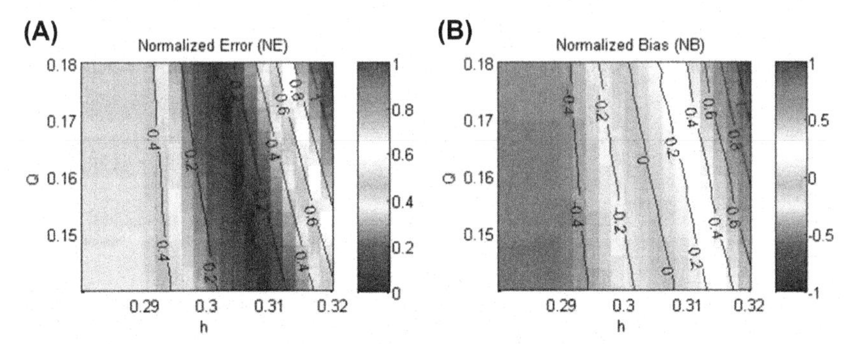

FIGURE 13.3 Sensitivity of the retrieval soil moisture from AMSR-E assuming different surface roughness parameters h and Q.

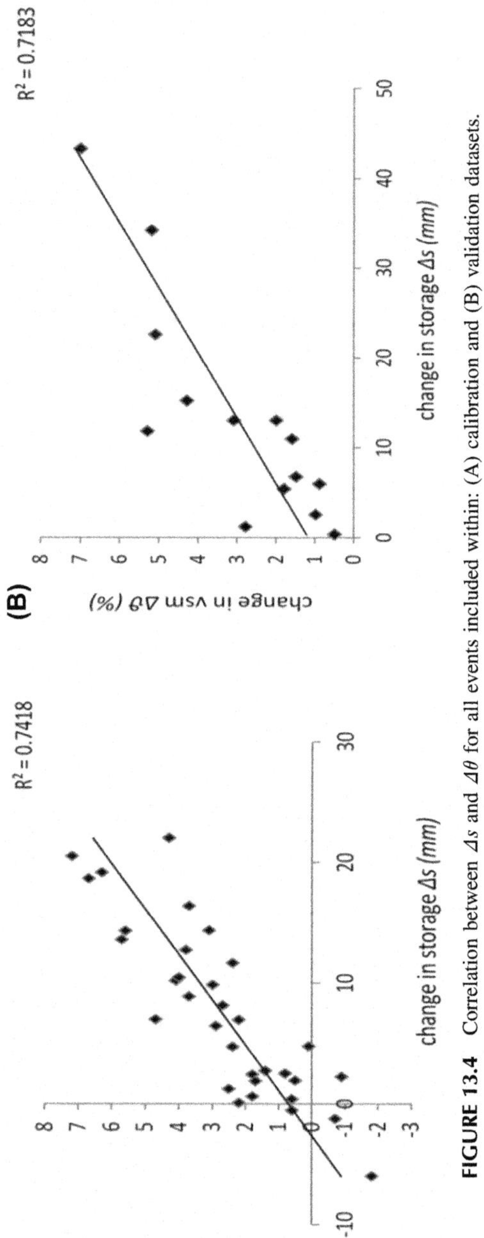

FIGURE 13.4 Correlation between Δs and $\Delta\theta$ for all events included within: (A) calibration and (B) validation datasets.

that the validation data points that represent the selected events are insufficient in comparison to the calibration case as only 1-year hydrological data was used for validation purpose. Therefore, the significance of this linear correlation is checked by conducting a statistical t-test in which 0.05 is adopted as a significance level and the estimated P-value is equal to 0.009, which is below the significance level and in turn proves that there is a good significance of the linear correlation between Δs and $\Delta\theta$ and statistically significant. Fig. 13.5 shows the time-series of precipitation, flow and temperature with the corresponding time-series of volumetric soil moisture retrieved from satellite. As shown in this figure, large values of *vsm* are associated to high flow and low temperature periods (e.g., January and February 2004 and December 2005) whereas low values of *vsm* are associated with low flow and high temperature periods. During the summer period, the flow decreases due to the gradually increasing of temperature (and therefore evapotranspiration) and the relatively low precipitation causing an overall decrease in the estimated soil moisture. It is important to mention that the direct linkage of the soil moisture to the

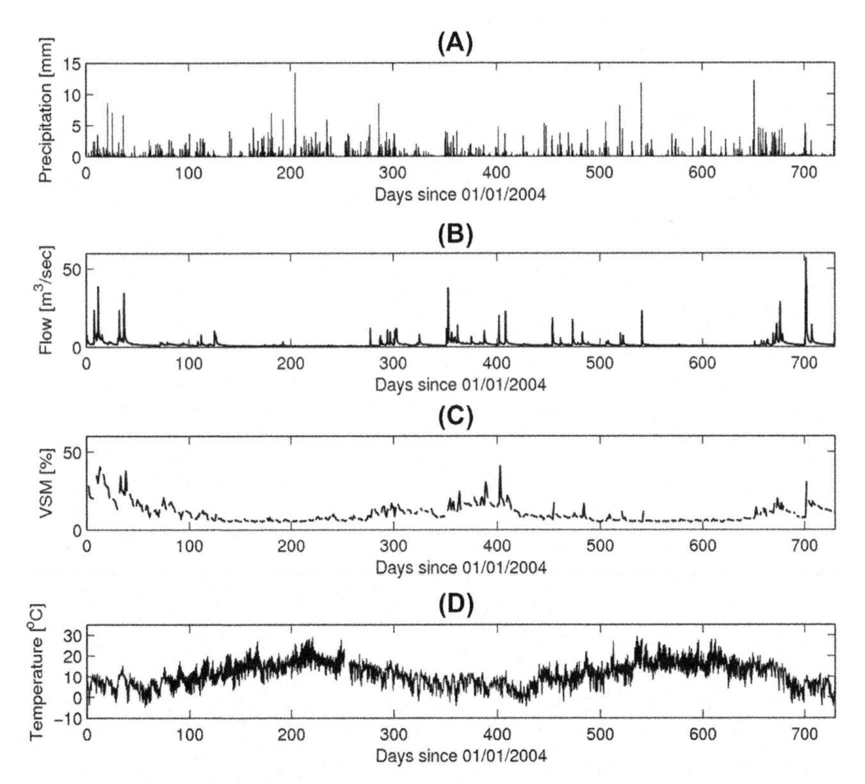

FIGURE 13.5 Brue catchment (2004–05): (A) averaged-measured precipitation; (B) measured flow; (C) AMSR-E retrieved volumetric soil moisture and (D) temperature measurement.

precipitation and runoff through the water balance equation clearly explains the miscellaneous measurements with high soil moisture values that are observed over the catchment during the summer time as those values can be explained by the corresponding high precipitation (see Fig. 13.5A). The annual retrieved soil moisture from satellite agrees well with the daily flow measurements (see Fig. 13.5B and C). On average, the mean annual volumetric soil moisture is around 10.5%. The lowest values were found during the summer season (May, June and July), which were around 5% whereas larger values were found during the winter season (December, January and February), which were around 47%. The variation in temperature is shown in 13.5D.

The summary of calibration and validation results is presented in Tables 13.1 and 13.2. Δs increases with $\Delta\theta$ because a high rainfall rate produces more free water content which in turn increases the moisture content in the soil. However, the temperature plays an important role in controlling the amount of evapotranspiration and in turn the mean flow, where relatively low temperatures produce a relatively low evapotranspiration. Table 13.1 shows that the highest values for Δs and $\Delta\theta$ were found during the winter and spring seasons due to the high rainfall rate and low evapotranspiration rate, which in turn lead to an increase in soil water storage (see events March 2005 and December 2004 and 2005). However, some particular events that occurred during the summer season such as in June 2004 and July 2004 had significantly large values of Δs, even though the evapotranspiration was high. The reason for this is because the high rainfall exceeds the evapotranspiration rate and

TABLE 13.1 Calibration results of the application of the water balance equation (2004 and 2005).

Event	Total rain (mm)	Runoff volume (mm)	Total ET (mm)	Δs (mm)	Δθ %
2004					
Jan_1	27.78	16.65	4.18	6.95	2.2
Jan_2	28.4	16.09	4.19	8.12	2.7
Feb_1	14.74	11.18	3.53	0.03	2.2
Feb._2	31.9	20.54	4.39	6.97	4.7
Mar._1	22.21	4.46	4.21	13.54	5.7
Mar._2	14.6	4.86	7.35	2.39	1.8
April	13.81	5.16	7.45	1.2	2.5

TABLE 13.1 Calibration results of the application of the water balance equation (2004 and 2005).—cont'd

Event	Total rain (mm)	Runoff volume (mm)	Total ET (mm)	Δs (mm)	Δθ %
May	34.78	11.77	10.31	12.7	3.8
June	33.42	1.39	15.66	16.37	3.7
July	37.38	3.67	19.38	14.33	3.1
Aug.	25.55	0.89	14.84	9.82	3
Sep.	4.8	0.75	10.04	−5.99	−1.8
Oct._1	31.2	4.44	8.11	18.65	6.7
Oct._2	11.51	4.92	4.74	1.85	1.7
Nov.	12.52	7.17	3.46	1.89	0.5
Dec.	36.04	18.9	2.84	14.3	5.6
2005					
Jan._1	11	7.16	3.5	0.34	0.6
Jan._2	10	6.72	2.7	0.58	1.8
Feb._1	22.77	8.88	3.72	10.17	4.1
Feb._2	24.35	10.74	4.77	8.84	3.7
March	33.53	8.32	4.72	20.49	7.2
April_1	23.39	6.27	6.69	10.43	4
April_2	14.51	7.35	7.73	−0.57	0.6
May	36.54	4.85	9.69	22	4.3
June_1	14.05	4.25	7.62	2.18	−0.9
June_2	19.32	5.5	11.35	2.47	0.8
July	19.2	0.87	11.88	6.45	2.9
Aug.	15.81	0.59	10.48	4.74	0.1
Sep.	8.78	0.44	9.59	−1.25	−0.7
Oct._1	19.02	1.63	5.74	11.65	2.4
Oct._2	15.59	4.98	7.92	2.69	1.4
Nov.	24.93	15.61	4.63	4.69	2.4
Dec.	65.59	41.98	4.48	19.13	6.3

TABLE 13.2 Validation results of the application of the water balance equation (2006).

Event	Total rain (mm)	Runoff volume (mm)	Total ET (mm)	Δs (mm)	Δθ %
2006					
Jan.	33.6	16.81	3.78	13.01	3.1
Feb.	26.2	11.84	3.43	10.93	1.6
Mar._1	30.8	14.4	4.63	11.77	5.3
Mar._2	8.4	4.09	4.01	0.3	0.5
April	21.4	7.76	6.89	6.75	1.5
May	65.8	21.38	10.24	34.18	5.2
June	19.2	1.68	12.14	5.38	1.8
July	39	1.84	14.58	22.58	5.1
Aug.	20.6	1.32	13.32	5.96	0.9
Sep.	27.6	2.48	9.91	15.21	4.3
Oct.	77.8	28.14	6.41	43.25	7.0
Nov.	51	33.77	4.24	12.99	2.0
Dec._1	15.21	9.35	3.36	2.5	1.0
Dec._2	18.35	14.31	2.87	1.17	2.8

consequently increasing the value of Δs over the catchment. This kind of weather condition was slightly noticeable in the computed *vsm* from satellite. Similarly, the validation results shown in Table 13.2 corroborate the results explained above.

Hence, it is obvious that the seasonal changes have significant effects and play a vital role in the retrieval *vsm* from the AMSR-E sensor. Therefore, an investigation of these effects in the context of the Antecedent Precipitation Index API will be adopted as the API estimation procedure and results for the Brue catchment were presented in Chapter 4.

5.3 LPRM calibration based on continuous PDM storage predictions

This section focuses on calibrating the LPRM parameters using a conceptual hydrological model (PDM). Surface soil moisture retrieval from the AMSR-E follows the same proposed procedure discussed in the previous section and illustrated in Fig. 13.1. This approach has the same common basis of the previously discussed results by achieving the best correlation between the retrieved surface soil moisture from the AMSR-E and the predicted storage from PDM. However, the difference is represented by: (1) obtaining a proper representation of the surface and vegetation parameters over a continuous time period, as PDM provide daily/hourly based catchment storage, while water balance approach is an event-based approach that could miss lots of information associated to the parameters over the catchment area; (2) PDM is a moderate complex physically based hydrological model with spatial representation of the catchment storage. Two surface roughness parameters h and Q are not the only parameters that will be calibrated in this analysis, vegetation scattering albedo ω is the third parameter that would be calibrated as well, as it was assumed to have a zero value when the water balance equation was applied. The reason behind this is due to the complexity of the PDM and its calibration procedure that takes into account all the relevant information in order to provide a high model performance and in turn an accurate prediction of catchment water storage.

Hence, the LPRM parameters should be properly calibrated in order to provide the best model performance when comparing the retrieved surface soil moisture from satellite and the predicted storage from the PDM. Two-year (2004 and 2005) hourly basis data of observed rainfall, flow and evapotranspiration were used for calibration and 1-year (2006) hourly-basis data set was used for validation. The PDM performance was examined by comparing the simulated flow against the observed flow and the model performance was mathematically assessed by adopting three statistical indicators, NSE, RMSE and NAE. The results of the PDM performance over the Brue catchment for the calibration and validation phases with a good model performance. Hence, 3 years (2004–06) of catchment water storage was predicted (see Fig. 4.7A and B) to be used in this chapter as a basis of the proposed approach for calibration of the satellite LPRM parameters.

Similar to the water balance approach, 2 years (2004 and 2005) of daily basis of PDM storage is used to calibrate three LPRM parameters (h, Q and ω), and the same iteration procedure is applied as mentioned earlier. It was assumed that the h parameter ranges from 0 to 0.33; the parameter Q has a range from 0 to 0.2 and the vegetation parameter ω has a range from 0 to 0.15. These parameters are optimised by maximising the NSE value between the retrieved surface soil moisture from the AMSR-E and the predicted storage from the PDM as shown in Fig. 13.6. In this figure, it can clearly be seen that the maximum correlations correspond to the values of the h parameter in the range 0.3–0.31. It is obvious that the retrieved volumetric soil moisture vsm from satellite is very sensitive to the change in h parameter values, which is

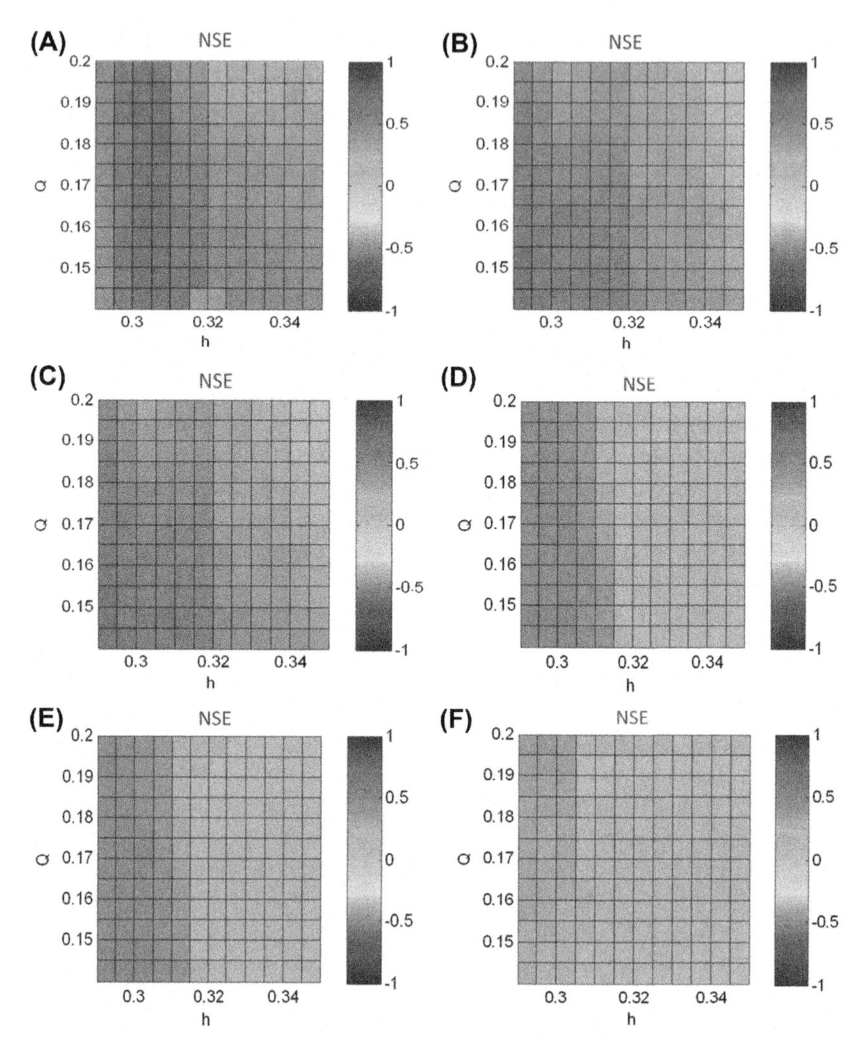

FIGURE 13.6 Correlation (NSE) between Δs and Δθ for different values of h, Q and ω as: [(A) $\omega = 0$; (B) $\omega = 0.03$; (C) $\omega = 0.06$; (D) $\omega = 0.09$; (E) $\omega = 0.12$; (F) $\omega = 0.15$].

consistent with the results shown previously. Even a small change in the value of the parameter h produces a significant change in the correlation between the PDM storage and the surface soil moisture. The figure also shows that the *vsm* is less sensitive to the selection of the Q parameter.

With regard to the vegetation parameter ω, Fig. 13.6 shows that the increase in ω values causes a decrease in the achieved NSE value as the best correlation (i.e., maximum NSE value 0.72) between the retrieved *vsm* and PDM storage is achieved when ω has a value of zero as shown in Fig. 13.6A where the maximum values of correlation are shown in a dark red (dark gray in printed version) colour. Hence, the maximum correlation was obtained for $h_{optimal} = 0.30$, $Q_{optimal} = 0.145$ and $\omega_{optimal} = 0$. The retrieved surface *vsm* from AMSR-E at the optimal case is compared against the observation-based PDM storage in Fig. 13.7 where it can been seen that there is a good agreement in both time-series, despite the fact that both quantities have different units (i.e., the retrieved *vsm* is a volumetric measurement and presented as a percentage value, while the PDM storage is a depth measurement in millimetres).

It is noticeable from Fig. 13.7 that the increase in the PDM storage corresponds to an increase in the retrieved *vsm* during the winter season (January, February and December) mainly due to the low temperature records and consequently low rates of the evapotranspiration as the storage at its maximum value (96.4 mm) in December 2005 and the *vsm* at its maximum value (40.8%) in January 2005. On the other hand, during the summer or growing season (April to October), generally speaking the decrease in the catchment storage corresponds to a decrease in the retrieved *vsm* as the minimum values of the storage and *vsm* (0.3 mm, 5%), respectively, were recorded in June and July of the years 2004 and 2005. However, it is obvious from Fig. 13.7 that during the growing season, particularly during the summer months such as June and July 2004, the AMSR-E satellite retrieved *vsm* shows very low values (5%–8%) even though the corresponding predictions of the PDM storage are relatively high values. The reason behind this is firstly due to the fact that the catchment storage predicted from the PDM is produced over the soil profile (i.e., accounting soil depth), so the high rainfall rates during the growing season lead to the increase in the catchment storage over the soil profile. Secondly, the limitation of the AMSR-E satellite to retrieve soil moisture only from the surface layer of the catchment soil. Therefore, the water in the surface layer usually evaporates rapidly when the temperature increases during the summer months.

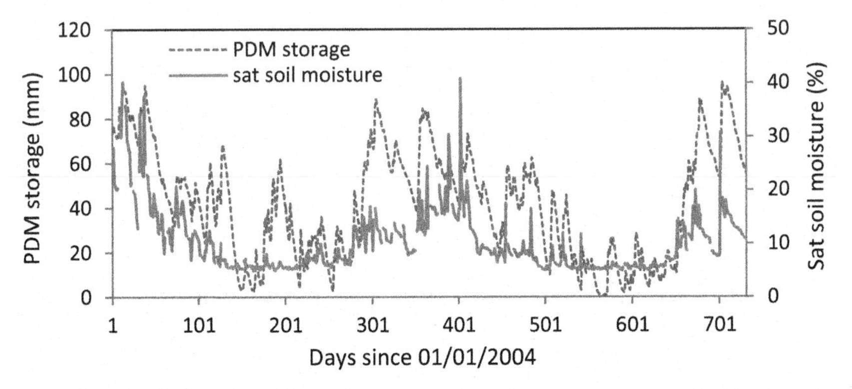

FIGURE 13.7 Calibrated satellite retrieved soil moisture time-series against the PDM storage for the Brue catchment.

For the purpose of this study, the correlation relationship between the retrieved soil moisture *vsm* from the AMSR-E satellite and the observation-based PDM storage is analysed. However, it is important to mention that some seasonal effects on the retrieved soil moisture from satellite have been revealed when using the event-based water balance approach in the previous section and also when using the PDM approach. It is proposed that investigating the mathematical relationship between the *vsm* and the observed PDM storage will be conducted following three proposed schemes. The first scheme (scheme 1) will contain one uniform model to predict catchment storage from satellite *vsm* at any time of the year (i.e., general prediction). The second scheme (scheme 2) will split the data into '*growing*' and '*nongrowing*' seasons. The growing period from April to October will be considered as the *growing* season while the months of January, February, November and December will be considered as the *nongrowing* season. In this case, scheme 2 will contain two prediction models (i.e., catchment storage will have one value during the *growing* season and another one for the *nongrowing* season). The rainfall rate is the main driver that controls the growing season and consequently the storage prediction. Equally, the antecedent catchment conditions are very important in both storage prediction and soil moisture retrieval. On the other hand, temperature is the dominant factor during the *nongrowing* season especially at freezing degrees. Hence, the third scheme (scheme 3) is proposed which will contain four models as the *growing* season data will be split into 'high and low API' daily data and the *nongrowing* season data will be split into 'frozen and nonfrozen daily data'. The analysis of the model performance is presented next in order to reveal which scheme is the best and more robust that could be used to predict the catchment storage from the satellite retrieved soil moisture. The three previously described statistical measures NSE, RMSE and NAE were used as performance indicators for all the proposed schemes. It should be mentioned that units of the RMSE and NAE in all schemes are millimetres as the target value represents the observation-based PDM storage and the estimate value represents the satellite-based storage.

For scheme 1, the correlation between the retrieved *vsm* and the observation-based PDM storage is plotted in Fig. 13.8A and B for the calibration and validation datasets, respectively. Fig. 13.8A shows that the data points are densely plotted around the fitted curve indicating a good agreement between the *vsm* and the PDM storage. This is corroborated with the performance indicators (NSE, RMSE and NAE) shown in the same Fig. 13.8.

From this figure, it was found that the catchment water storage is related to the satellite *vsm* in a logarithmic way that can be expressed with the following mathematical formula (using the calibration dataset only):

$$St = 43.05\ \ln(vsm_{sat}) - 54.09 \tag{13.17}$$

where St is the observation-based PDM storage for the Brue catchment in [mm], and vsm_{sat} is the retrieved volumetric soil moisture from the AMSR-E

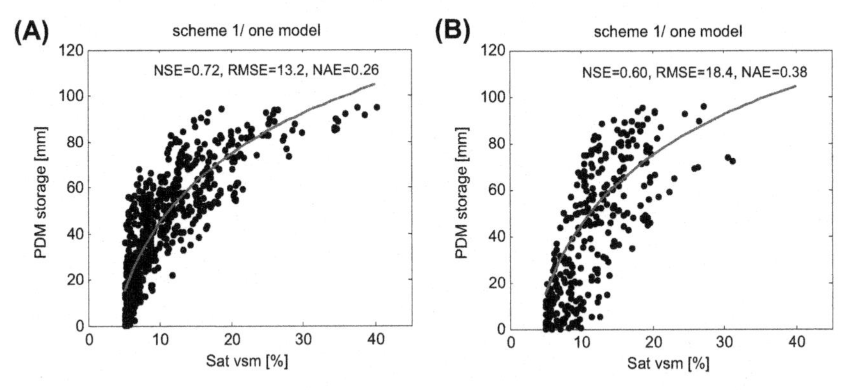

FIGURE 13.8 Scheme 1, scatterplot between the retrieved soil moisture from AMSR-E satellite *vsm* and the PDM storage. (A) calibration dataset; (B) validation dataset.

satellite in percentage. The model performance for scheme 1 is shown in Table 13.3 where it can be seen that the model shows a good performance during the calibration and validation phases.

The derived relationship was verified using the validation dataset (2006) as shown in Fig. 13.9. In this figure, the PDM storage simulated using observed

TABLE 13.3 Statistical results of the scheme 1 model performance.

	Calibration			Validation		
Scheme 1 model	NSE	RMSE	NAE	NSE	RMSE	NAE
All data	0.72	13.2	0.26	0.60	18.4	0.38

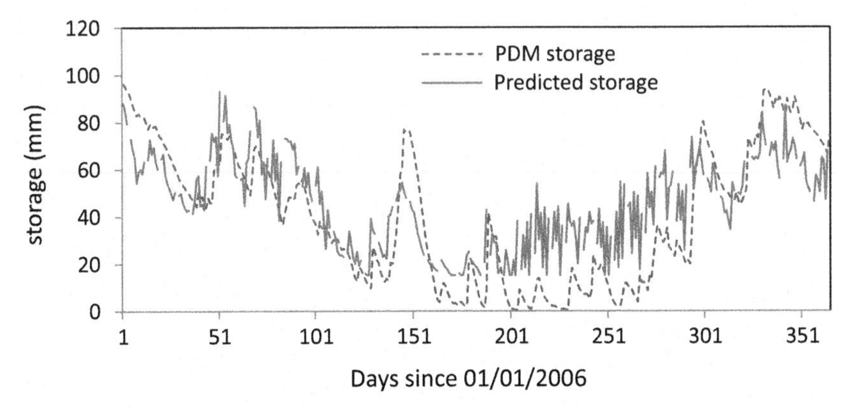

FIGURE 13.9 Validation time-series of the predicted storage from satellite *vsm*.

measurements (rainfall, flow and evapotranspiration) is compared against the predicted storage from satellite and using the model given by Eq. (13.17). Fig. 13.9 shows a good agreement between the observed and predicted storages although it is obvious that there is some variation in the values of the predicted storages, and in particular there are relatively large values during June and July. This is due to the fact that the minimum *vsm* value is 5%, which occurred more frequently during summer time. If this value is used in Eq. 13.17, that will always produce a predicted storage of around 15.2 mm which is higher than the observed PDM storage (see also Fig. 13.8).

For scheme 2, the behaviour of the satellite *vsm* during the *growing* and *nongrowing* seasons over the Brue catchment is revealed, respectively, when compared against the observed PDM storage in the context of time-series as shown in Fig. 13.10A and B. In this figure, we can see that the PDM storage and satellite *vsm* showed a good agreement following the same trend during the *nongrowing* season and slightly better than that during the *growing* season although the units are different (i.e., *vsm* in % and PDM storage in mm). It is interesting to see that during the *nongrowing* season (see Fig. 13.10B) the

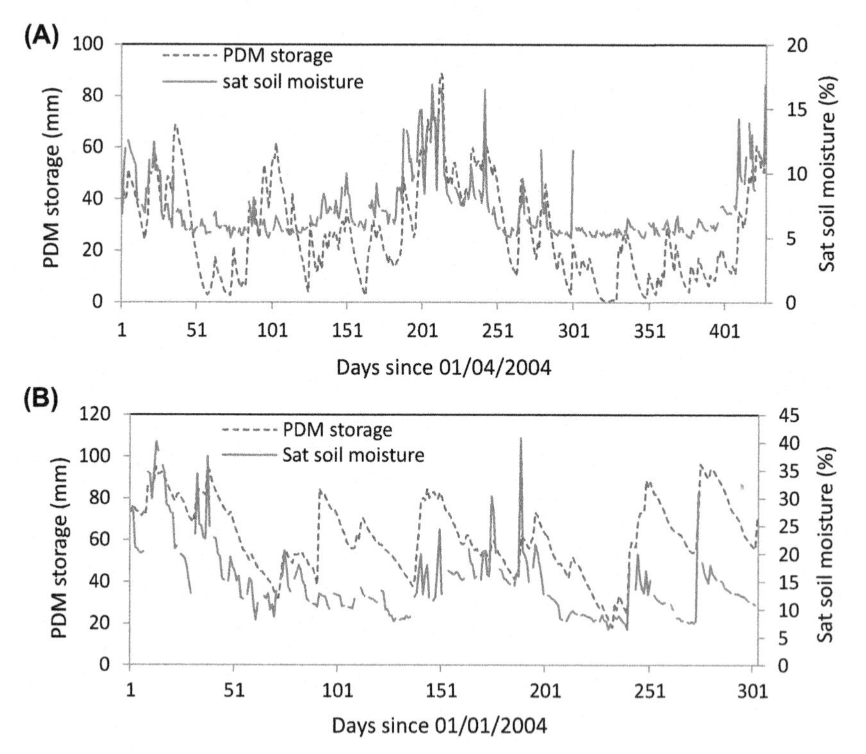

FIGURE 13.10 AMSR-E soil moisture and PDM water storage time-series in (A) *growing* and (B) *nongrowing* season/calibration phase.

increase in the retrieved *vsm* from satellite corresponds to an increase in the PDM storage, although the PDM storage range is generally higher than the satellite *vsm*. This is due to the fact that the PDM is a hydrological model predicting the storage over the entire soil profile (i.e., accounting the entire soil depth) while the AMSR-E satellite is retrieving soil moisture from the surface layer of the soil only.

The performance of all the schemes is presented in Table 13.3. As shown in Table 13.3, the *nongrowing* season model outperforms the *growing* season model in predicting catchment storage from satellite retrieved *vsm*. This means that the satellite performance is slightly better during the winter than during spring and summer, and this is mainly due to the vegetation effects, which significantly increase during the spring and summer seasons, as the vegetation represents an obstacle for the natural radiation path emitted from the soil surface and recorded by the satellite sensors.

In addition, Table 13.4 also shows that the performance of the combination of the two models is better than the performance of a single model when compared to scheme 1 (see Table 13.3). As the statistical results clearly showed that the RMSE and NAE values decreased in scheme 2 about 1.2 and 0.2 m^3s^{-1} respectively with the regard to the corresponding values produced in scheme 1. The interesting issue in Table 13.4 could be seen in the validation results in that the RMSE and NAE values are significantly decreased to the corresponding values in scheme 1 by about 3.5 and 0.11 m^3s^{-1}, respectively. The correlation between the *vsm* and the PDM storage during the *growing* and *nongrowing* seasons in calibration and validation modes is shown in Fig. 13.11A and B). It can be seen from Fig. 13.11A that the *nongrowing* season data points are almost densely plotted and smoothly follow the fitted curve, which in turn reflects the goodness of the model behaviour. Validation result in Fig. 13.11B has proven the above explanation although only 1-year (2006) data has been used for validation purposes in this study.

TABLE 13.4 Statistical results of scheme 2 models' performance.

	Calibration			Validation		
Scheme 2 models	**NSE**	**RMSE**	**NAE**	**NSE**	**RMSE**	**NAE**
Growing season	0.48	13.5	0.4	0.35	15.5	0.61
Nongrowing season	0.55	12.5	0.26	0.49	15.1	0.29
Two models' performance	0.73	12.0	0.24	0.71	14.9	0.27

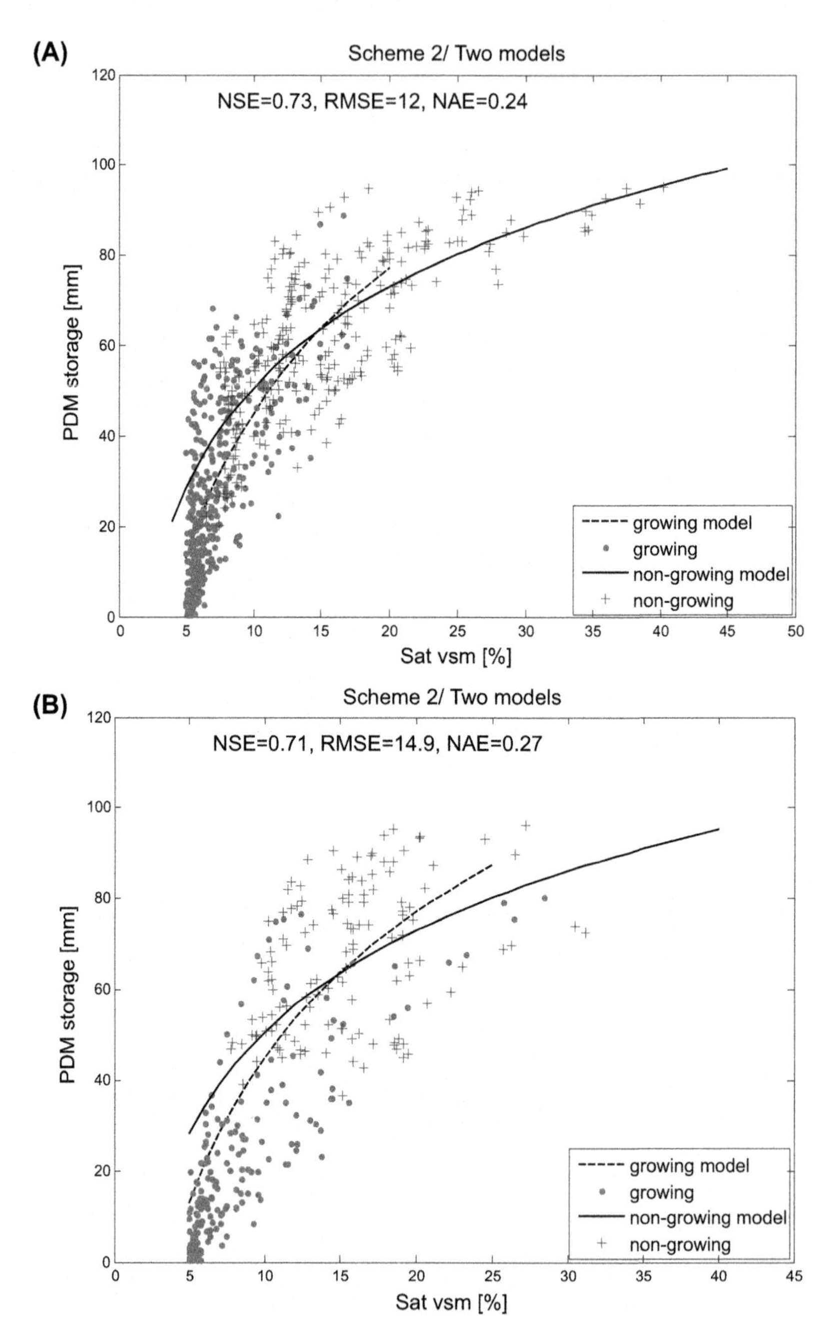

FIGURE 13.11 Scheme 2 (growing and nongrowing season): Correlation of (A) calibration and (B) validation between the retrieved *vsm* from satellite and PDM storage.

Furthermore, Fig. 13.11A shows that during the growing and nongrowing seasons, the satellite *vsm* correlates with a linear logarithmic relationship with the observed PDM storage. That relationship can be expressed mathematically by the following two equations:

$$St_{growing} = 46.33\ \ln(vsm_{sat}) - 61.66 \tag{13.18}$$

$$St_{non-growing} = 32.29\ \ln(vsm_{sat}) - 23.70 \tag{13.19}$$

where St_{growing} and $St_{\text{nongrowing}}$ represent the Brue catchment water storage during *growing* and *nongrowing* seasons. These equations have been verified using 1-year (2006) of satellite *vsm* and PDM storage data. The performance of these proposed models with the validation dataset is shown in Fig. 13.12A and B, where there is an overall agreement between the satellite-based predicted storage and the observed PDM storage. However, during the *growing season*, Fig. 13.12A shows that the predicted storage has highly variable values particularly during the summer and late summer months and this is due to the

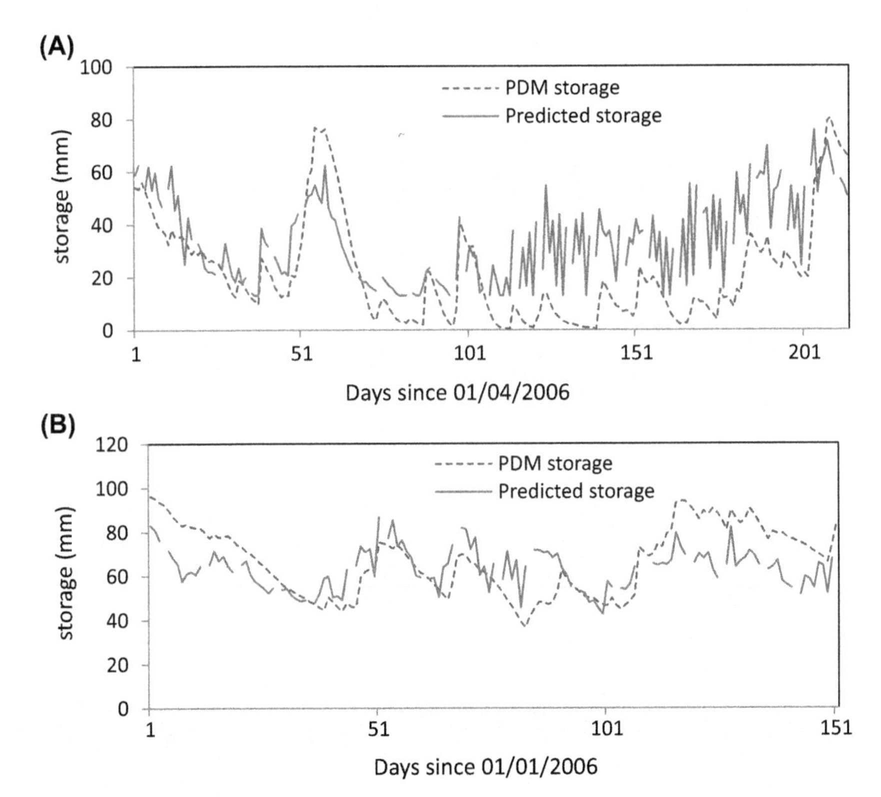

FIGURE 13.12 Validation time-series of the predicted storage from satellite *vsm* for (A) growing and (B) nongrowing season.

high variation of the satellite *vsm* values during those months in particular and also due to the vegetation canopy. Both are the controlling factors that greatly impact the retrieval of surface soil moisture values from satellite-based soil moisture retrievals. It is interesting to see that this variation is not seen during the *non-growing* season, and therefore satellite soil moisture retrievals produce better results (see Fig. 13.12B).

Moreover, the predicted storage in Fig. 13.12A records higher values than the corresponding PDM storage, particularly during the summer months (June, July and August). The reason behind this is because the satellite *vsm* minimum value is 5% for the Brue catchment during the summer time, which will produce a predicted storage higher than the minimum value of the PDM storage.

For scheme 3, it was mentioned earlier in this section that this scheme will contain four models as the antecedent precipitation index API has been calculated for the Brue catchment. This was done using 2-year daily-basis data of the *growing season* split into high and low API, while the 2-year daily-basis data of the *nongrowing* season was split into frozen and nonfrozen. Hence, this scheme will investigate the correlation between the satellite *vsm* and the observed PDM storage over those particular cases along 2-year (2004–05) time period. To start with this analysis, Fig. 13.13 shows the behaviour of the satellite *vsm* during the *growing* season when compared against the calculated daily-basis API over the Brue catchment.

Although the API and satellite *vsm* are two different variables that have different units (i.e., the API represent depth measurements of water that fall over the catchment and measured in millimetres, whereas satellite *vsm* is a volumetric measurement measured in percentage), it can be seen from Fig. 13.13 that there is a general agreement between the two time-series. Usually, it is expected that when the API increases, the *vsm* should show a corresponding increase as a normal response of the soil surface to the high

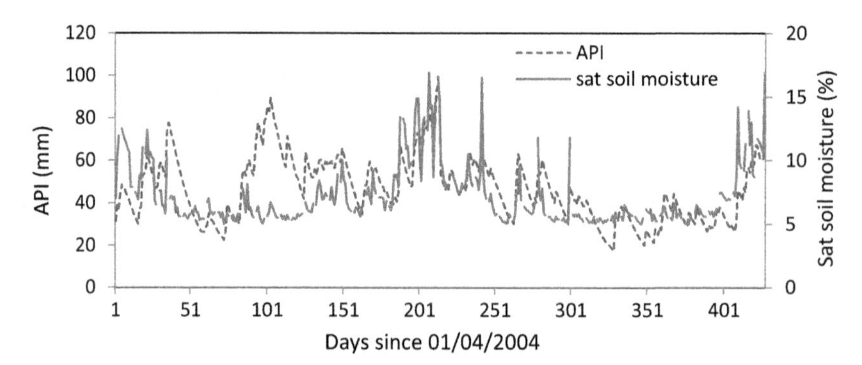

FIGURE 13.13 API and satellite soil moisture *vsm* time-series for the Brue catchment during the *growing* season.

rainfall rates. However, it is obvious that during the summer months of the year 2004 (i.e., days range 80–150) the satellite *vsm* showed low values even though the corresponding API showed relatively high values. This is again due to the fact that the AMSR-E satellite retrieves soil moisture measurements from the surface thin layer of the soil in which water content is rapidly evaporated during the summer time due to the high evapotranspiration rates.

According to Fig. 13.13, the minimum, maximum and average values of daily API measurements were found to be 17.4, 96.0 and 46.8 mm, respectively. It is obvious that the maximum API value had increased about 50 mm than the average value. Therefore, a 50% of the API average value will be added to the average value to produce the API boundary value BV between high and low API that will be adopted in this study in order to split the daily data during the growing season into low and high API. Hence, 70 mm is the API value that was considered in this study to be the BV for the Brue catchment.

Accordingly, any value of API $\leq$ BV will be considered as low API data point, whereas API values $>$ BV will be considered as high API data points. On the other hand, during the *nongrowing* season, the issue is simpler as a temperature boundary of zero degrees is adopted in order to split the daily data into frozen and nonfrozen datasets.

The behaviour of the satellite *vsm* during the growing season in the two particular cases with low and high API is shown in Fig. 13.14A and B. In this figure the retrieved satellite *vsm* is compared against the observation-based PDM storage in the context of time series. It is obvious that Fig. 13.14A is quite similar to Fig. 13.10A as the majority of the data points during the growing season are categorised as low API (i.e., API $\leq$ BV); therefore, the same explanation can be subscribed for this particular case of Fig. 13.14A.

With regard to the other case (high API) revealed from Fig. 13.14B, only 34 days during 2-year daily data of growing season have been categorised as high API (i.e., API $>$ BV). Those days of high API actually occurred during the summer months particularly in May and July. Despite the good match between the satellite *vsm* and the PDM storage in Fig. 13.14B, it should be clarified that the units are in mm and % for the PDM storage and satellite soil moisture, respectively. The satellite *vsm* values range between 5% and 9%, while the PDM storage values range between 40 and 68.4 mm. This might produce a source of uncertainty and leads to poor results.

On the other hand, the behaviour of the satellite *vsm* during the *nongrowing* season for the particular cases that are nonfrozen and frozen is shown in Fig. 13.15A and B using 2-year daily data from the calibration dataset. The figure shows the retrieved satellite *vsm* compared against the PDM storage as time-series.

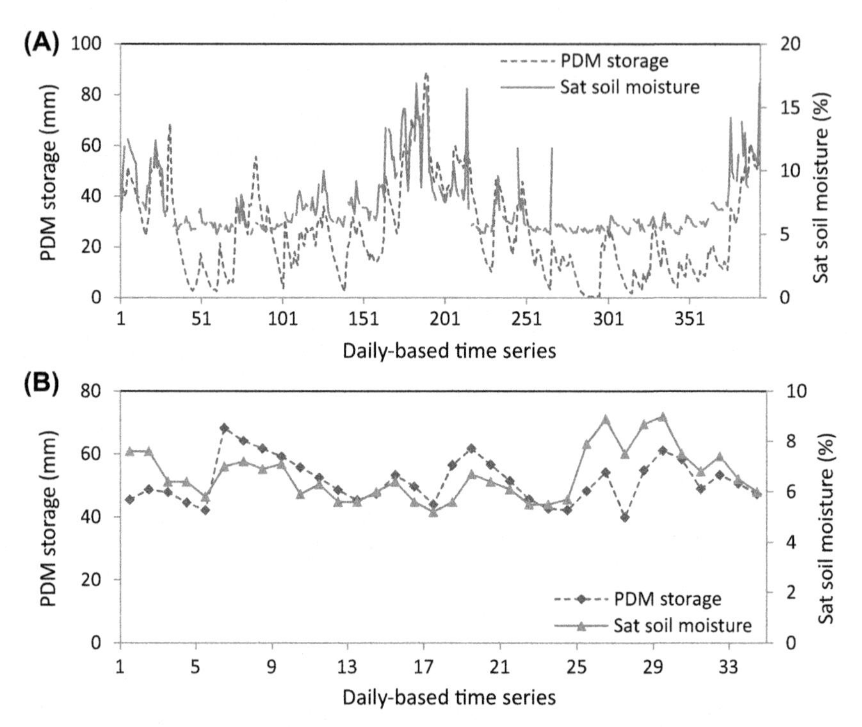

FIGURE 13.14 AMSR-E soil moisture and PDM water storage time-series for (A) low API and (B) high API cases/calibration phase.

It is clearly noticeable that Fig. 13.15A is quite similar to the one generated for the nongrowing season in scheme 2 (see Fig. 13.10B) as in scheme 3, almost the majority of the daily basis data of the nongrowing/nonfrozen case are categorised as nonfrozen days (i.e., temperature is larger than zero). The remaining dataset is categorised as frozen days (i.e., temperature is $\leq$ zero). Hence, the nonfrozen case (Fig. 13.15A) can be explained in the same way as it was previously explained in Fig. 13.10B.

Frozen/nongrowing case results are shown in Fig. 13.15B where it can be seen that only 33 daily basis data points (i.e., days) were categorised as frozen days and most of them occurred in November and December, and it can be seen that there is a good agreement in trends between the two time-series although both have different units. The difference in the range of values of the satellite *vsm* and PDM storage could be seen in Fig. 13.15B as the minimum and maximum values of them, respectively, are 7.6%−15.7% for satellite *vsm* and 43.5−90.7 mm for the PDM storage.

The overall performance results for all the proposed models in scheme 3 are presented in Table 13.5, where NSE, RMSE and NAE were used as statistical indicators.

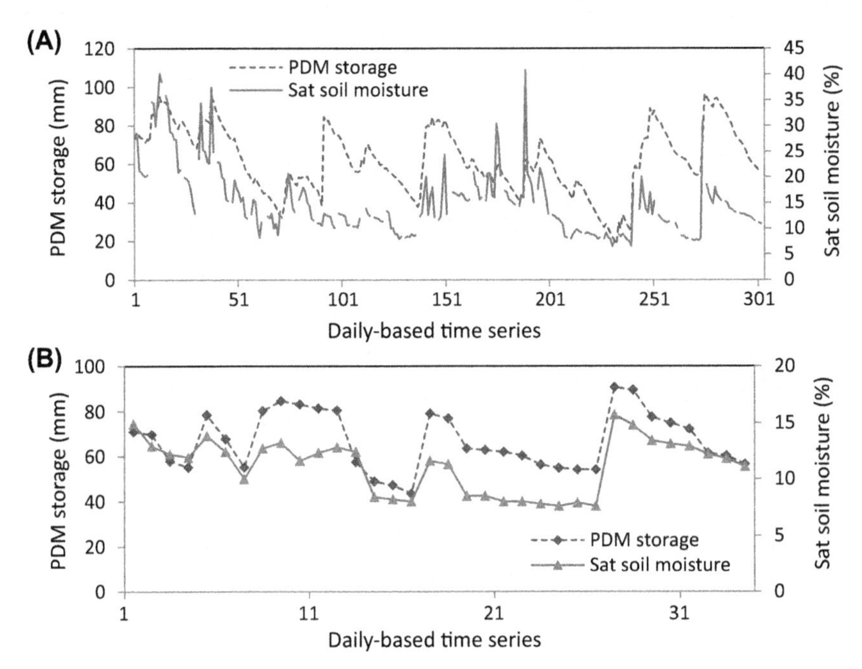

FIGURE 13.15 AMSR-E soil moisture and PDM water storage time-series for (A) nonfrozen and (B) frozen case/calibration phase.

TABLE 13.5 Statistical results of scheme 3 models' performance.

	Calibration			Validation		
Scheme 3 models	**NSE**	**RMSE**	**NAE**	**NSE**	**RMSE**	**NAE**
Low API	0.68	10.3	0.33	0.55	12.5	0.49
High API	0.56	6.7	0.16	0.47	8.7	0.18
Nonfrozen	0.65	10.5	0.14	0.53	13.7	0.17
Frozen	0.60	10.6	0.16	0.51	13.8	0.19
Four models' performance	0.83	10.0	0.17	0.78	13.2	0.26

It is obvious from Table 13.5 that during the growing season when the API is high, the AMSR-E satellite outperforms that when the API is low in terms of RMSE and NAE. It is important to mention that only 34 days over the growing season period have a high API (i.e., API > BV) and the majority of the data showed a low API (i.e., API $\leq$ BV). Interestingly, the AMSR-E satellite also shows a good performance when the temperatures is higher than zero (i.e., nonfrozen) over the nongrowing season than the frozen days. In other words, freezing temperatures affect the retrieval of soil moisture by the satellite sensors by affecting the observed emissivity of the natural radiation emitted from the soil surface. The validation of the statistical results also justified this statement. Table 13.5 shows that the combination of the four models (four model performances) produced good results as shown by the correlation between the catchment-based storage and the satellite-based storage. As a result, different models (e.g. 4 in this scheme) that each one of them reflects a particular case of many variables (such as time, rainfall and temperatures) to examine the satellite *vsm* behaviour and predicting the catchment water storage showed a better performance than treating all the variation during the whole year with only one predicting model as shown in scheme 1. Hence, Table 13.5 clearly shows that scheme 3 outperforms schemes 1 and 2, as it is obvious that the RMSE and NAE values in Table 13.5 are greatly decreased to around 2 and 0.07 m^3s^{-1}, respectively, from those corresponding in Table 13.4. Interestingly, the validation results of scheme 3 also showed a good performance when compared to those results presented for schemes 1 and 2.

The above statistical results are the reflection of the correlation relationship between the satellite *vsm* and the catchment water storage for the four cases that belong to scheme 3 and shown in Figs 13.16A and B, which represent the calibration and validation phases, respectively.

Overall, Fig. 13.16A shows that scheme 3 (four-model scheme) has a good performance as most of the data points are close to the fitted curves for all models except the growing/low API model (red [dark gray in printed version] coloured data points). In this later model, the upper part of the curve shows some data points scattered away from the fitted curve, and this was reflected in the calibration and validation results of this model as shown in Table 13.5. Besides it is obvious that for all models, there are no *vsm* values less than 5%. This is because the minimum value of the AMSR-E soil moisture retrieved over the Brue catchment soil surface is 5% and all of these values occurred during the growing season (April—October), particularly during the summer months. However, the four model correlation results showed a good improvement in the AMSR-E performance than the results of schemes 1 and 2 during both the calibration and the validation phases.

As mentioned before, for the purpose of this study, the regression-based mathematical models are derived from the correlation between the satellite *vsm* and the PDM storage in order to provide an appropriate equation to predict the Brue catchment water storage from the calibrated satellite *vsm* during *growing* and *nongrowing* seasons. Hence, four formulas have been

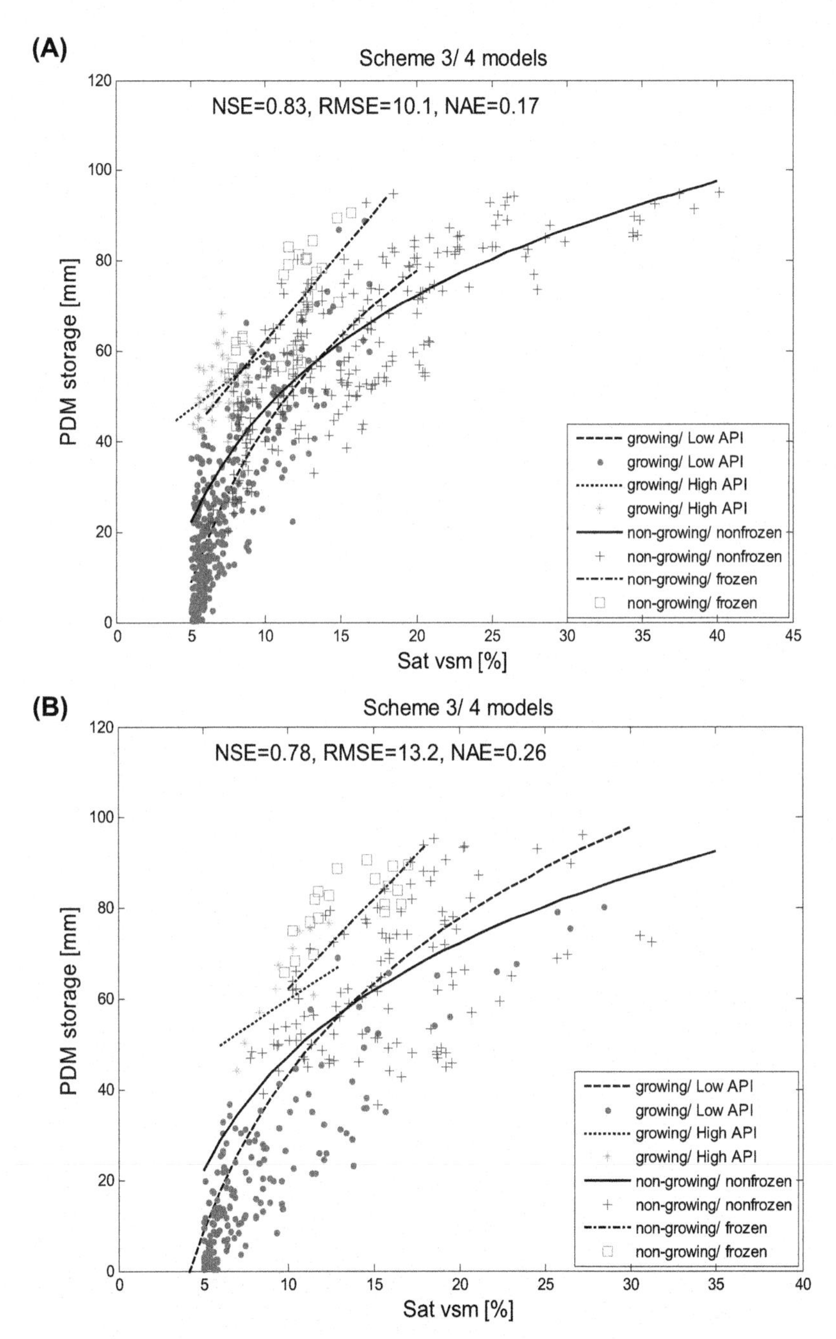

FIGURE 13.16 Scheme 3 (high and low API; nonfrozen and frozen days): Correlation of (A) calibration and (B) validation between the retrieved *vsm* from satellite and PDM water storage.

derived to represent the mathematical models that correlate satellite *vsm* to observed PDM storage during *growing* (low and high API) and *nongrowing* (nonfrozen and frozen) seasons as shown below:

$$St_{lowAPI} = 49.84\ \ln(vsm_{sat}) - 71.54 \quad (13.20)$$

$$St_{highAPI} = 2.56(vsm_{sat}) + 34.77 \quad (13.21)$$

$$St_{non-frozen} = 36.19\ \ln(vsm_{sat}) - 36.06 \quad (13.22)$$

$$St_{frozen} = 3.97(vsm_{sat}) + 22.36 \quad (13.23)$$

where St_{lowAPI}, $St_{highAPI}$, $St_{nonfrozen}$ and St_{frozen} are the predicted storage for the Brue catchment at low API, high API, nonfrozen and frozen cases, respectively. In growing/low API and nongrowing/nonfrozen cases, the observation-based PDM storage correlates with a simple logarithmic relationship to the retrieved soil moisture from the AMSR-E satellite. On the other hand, for growing/high API and nongrowing/frozen cases, a linear relationship is the best fit to predict the catchment storage from the retrieved satellite *vsm*. The four derived equations that belong to scheme 3 are verified using 1-year (2006) data of satellite *vsm*. By implementing those equations, a satellite-based predicted storage of the Brue catchment will be computed individually for each case, and the results will then be compared in the context of time-series to the observation-based PDM storage obtained by running the calibrated PDM model. Hence, the time-series of the satellite-based catchment storage against the observation-based PDM storage during the growing season (low and high API) cases are shown in Fig. 13.17A and B, respectively, while the time-series of the nongrowing season (nonfrozen and frozen) cases, respectively, are presented in Fig. 13.18A and B.

Overall, there is a good agreement between the satellite-based predicted storage and the observation-based PDM storage in the growing (low and high API) season as shown in Fig. 13.17A and B. However, it is clear from Fig. 13.17A that there is a significant variation of the satellite-based predicted storage from day 100 to the end of the growing season. In fact, that period represents the summer and late summer months (June, July, August and September). That variation originates due to the large variation that was originally produced in the satellite retrieved *vsm* over the Brue catchment during the growing season as the vegetation canopy is the dominant factor that affects the emissivity of the natural radiation from the soil surface by blocking the radiation path to be sensed by the satellite sensor. Fig. 13.17B also shows a good agreement between the predicted and the observation-based storage over the Brue catchment during the growing/high API case, as this is reflected in the validation statistical results presented in Table 13.4.

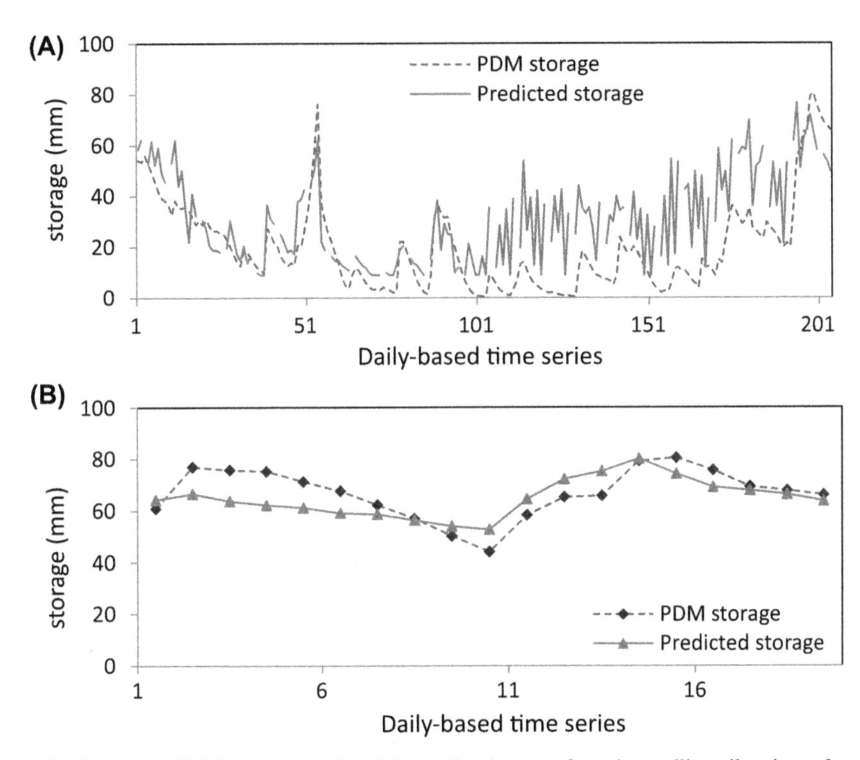

FIGURE 13.17 Validation time-series of the predicted storage from the satellite soil moisture for (A) *growing* low API and (B) *growing* high API.

On the other hand, regarding the *nongrowing* (nonfrozen and frozen) season, the validation results are shown in Fig. 13.18A and B. In this figure it can be seen that there is a general good agreement between the satellite-based predicted storage and the observation-based PDM storage although in Fig. 13.18A there is a little noticeable variation in the predicted satellite-based storage. This is due to the original variation of the retrieved soil moisture by the AMSR-E during nonfrozen days with the instantaneous observation of the AMSR-E which in turn records the rapid response of soil moisture to the rainfall. One common feature could be noted from both Fig. 13.18A and B) as both time-series showed almost the same range of the storage values around 35–100 mm for the nonfrozen case (see Fig. 13.18A) and approximately 60–100 mm for the frozen case (see Fig. 13.18B). The range limits represent the minimum and maximum values, respectively.

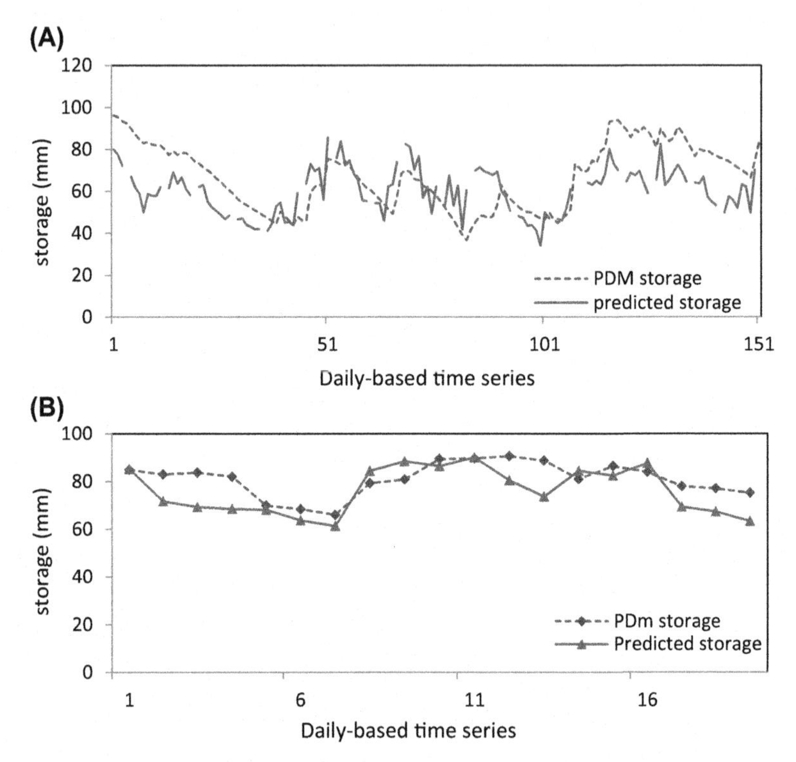

FIGURE 13.18 Validation time-series of the predicted storage from the satellite soil moisture for (A) *nongrowing* nonfrozen and (B) *nongrowing* frozen days.

6. Conclusions

A 3-year (2004–06) daily surface soil moisture was retrieved over the Brue catchment using the night-time AMSR-E brightness temperature measurements at 6.9 GHz. The retrieval procedure was based on the simulation of the MPDI, which iteratively computes the soil moisture through the land parameter retrieval model LPRM which is a physically based radiative transfer model. A new approach is proposed in this study to calibrate the vegetation (ω) and roughness parameters (h and Q) of the LPRM model based on the use of the precipitation, river flow and evapotranspiration data (i.e., hydrological model) to estimate the total water storage within the catchment. Event-based water balance was used to examine the applicability of the proposed approach by looking at the correlation between the change in the catchment water storage (Δs) over the selected rainfall-flow events and the corresponding change in the satellite soil moisture ($\Delta \theta$). As a result, the optimisation of the LPRM parameters is based on the best correlation that could be achieved between Δs and $\Delta \theta$. The second part of this chapter used the PDM model as a more complex conceptual hydrological model for the purpose of this study, as a continuous 3-year observation-based catchment storage was predicted after properly calibrating all the PDM model parameters. Three statistical indicators (NSE, RMSE and NAE) were used to examine the PDM performance during calibration and validation phases. Similar to the water balance approach, the PDM storage was used to calibrate the satellite LPRM parameters. The feedback to rainfall and soil moisture is reflected by the river runoff. Since the whole catchment is used to calibrate the satellite-sensed data, the two data types agree with each other due to their spatial averaging characteristics. In addition, since satellite measurements are calibrated using hydrological data, the resultant soil moisture data are of more practical value than those from other calibration approaches.

For the purpose of this study, the correlation between the satellite *vsm* and the observation-based PDM storage was investigated and verified, and a regression-based mathematical model was derived to predict the catchment water storage from the satellite-retrieved soil moisture. Three schemes have been proposed for the Brue catchment in order to investigate the behaviour of the satellite LPRM model during different time periods over the whole year and different weather conditions. It was concluded that scheme 3, which includes four different models, provides extremely good results for the Brue catchment during the calibration and validation phases and provides greatly improved results than the other two schemes in the context of model performance indicators (NSE, RMSE and NAE).

Although it was clear that there is a strong correlation between the retrieved soil moisture using the AMSR-E satellite sensor and the catchment water storage, there is still a significant source of uncertainty that cannot be neglected (the satellite sensor can only measure the surface soil moisture within a few centimetres while the catchment storage is extended to deeper soil layers). Therefore, it would be useful to quantify the uncertainties of the soil moisture results. It has been recognised that data uncertainty is an important factor to consider when applying these results to the real-world problems. It would be worthwhile to combine numerical weather prediction NWP modelling techniques with satellite soil moisture sensing. Many land surface modules in modern numerical weather models are able to simulate the interaction between various meteorological and hydrological variables (temperature, wind, solar radiation, evapotranspiration, vegetation, etc.). Therefore, the strong synergy between satellite soil moisture sensing and numerical weather modelling would enable us to make full use of the available information and produce soil moisture with further improved accuracy. This would be the basis of the next two chapters (6 and 7) as Chapter 6 will investigate the methodologies of estimating soil moisture over the Brue catchment at different soil depths using a land surface model (LSM) combined with an NWP. Finally, Chapter 7 will look at the synergy issue between satellite and NWP/LSM through the fusion application.

References

Al-Shrafany, D.M., Rico-Ramirez, M.A., Han, D., 2011. Calibration of roughness parameters using rainfall runoff water balance for satellite soil moisture retrieval. ASCE J. Hydrol. Eng. https://doi.org/10.1061/(ASCE)HE.1943-5584.0000508.

Becker, F., Choudhury, B.J., 1988. Relative sensitivity of normalized difference vegetation index (NDVI) and microwave polarization difference index (MPDI) for vegetation and desertification monitoring. Rem. Sens. Environ. 24, 297–311.

Brunfeldt, D.R., Ulaby, F.T., 1984. Measured microwave emission and scattering in vegetation canopies. IEEE Trans. Geosci. Rem. Sens. 22, 520–124.

Chahine, M.T., 1983. Interaction mechanisms within the atmosphere. In: Colwell, R.N. (Ed.), Manual of Remote Sensing. John Wiley, New York, USA.

Entekhabi, D., Nakamura, H., Njoku, E., 1994. Solving the inverse problem for soil moisture and temperature profiles by sequential assimilation of multifrequency remotely sensed observations. IEEE Trans. Geosci. Rem. Sens. 32, 438–448.

Hallikainen, M.T., Ulaby, F.T., Dobson, M.C., El-Rayes, M.A., 1984. Dielectric measurements of soils in the 3- to 37- GHz band between −50°C and 23°C. In: Proceeding of International Geosciences and Remote Sensing Symposium, IGARSS'84, Strasbourg, France, pp. 163–168.

Hallikainen, M.T., Ulaby, F.T., Dobson, M.C., El-Rayes, M.A., 1985. Microwave dielectric behaviour of wet soil-part 1: empirical models and experimental observations. IEEE Trans. Geosci. Rem. Sens. 23, 25–34.

Jackson, T.J., Schmugge, T.J., Wang, J.R., 1982. Passive microwave sensing of soil moisture under vegetation canopies. Water Resour. Res. 18 (4), 1137–1142.

Jackson, T.J., O'Neill, P.E., 1990. Attenuation of soil microwave emission by corn and soybeans at 1.4 and 5 GHz. IEEE Trans. Geosci. Rem. Sens. 28, 978–980.

Jackson, T.J., Schmugge, T.J., 1991. Vegetation effects on the microwave emission of soils. Rem. Sens. Environ. 36, 203–212.

Lakshimi, V., Wood, E., Choudhury, B., 1997. Evaluation of special sensor microwave/imager satellite data for regional soil moisture estimation over the Red River Basin. J. Appl. Meteorol. 36, 1309–1328.

Mätzler, C., Standley, A., 2000. Relief effects for passive microwave remote sensing. In: Mätzler, C. (Ed.), Radiative Transfer Models for Microwave Radiometry. COST Action 712: Application of Microwave Radiometry to Atmospheric Research and Monitoring. Project 1: Development of Radiative Transfer Models. European Communities, Brussels, Belgium.

Meesters, A., De Jeu, R., Owe, M., 2005. Analytical derivation of the vegetation optical depth from the microwave polarization difference index. IEEE Trans. Geosci. Rem. Sens. Lett. 2, 121–123.

Mo, T., Choudhury, B.J., Schmugge, T.J., Jackson, T.J., 1982. A model for microwave emission from vegetation-covered fields. J. Hydrol. 184, 101–129.

Njoku, E., Kong, J., 1977. Theory of passive microwave remote sensing of near-surface soil moisture. J. Geophys. Res. 82, 3108–3118.

Njoku, E., Li, L., 1999. Retrieval of land surface parameters using passive microwave measurements at 6-18 GHz. IEEE Trans. Geosci. Rem. Sens. 37, 79–93.

Njoku, E., Jackson, T.J., Lakshmi, V., Chan, T.K., Nghiem, S.V., 2003. Soil moisture retrieval from AMSR-E. IEEE Trans. Geosci. Rem. Sens. 41, 215–229.

Owe, M., De Jeu, R., Walker, J., Branch, H.S., Greenbelt, M.D., 2001. A methodology for surface soil moisture and vegetation optical depth retrieval using the microwave polarization difference index. IEEE Trans. Geosci. Rem. Sens. 39, 1643–1654.

Owe, M., De Jeu, R., Walker, J., Branch, H.S., Greenbelt, M.D., 2008. A methodology for surface soil moisture and vegetation optical depth retrieval using the microwave polarization difference index. IEEE Trans. Geosci. Rem. Sens. 39, 1643–1654.

Schmugge, T., 1985. Remote sensing of soil moisture. In: Anderson, M.G., Burt, T.P. (Eds.), Hydrological Forecasting. John Wiley and Sons, New York, USA.

Ulaby, F.T., Moore, R.K., Fung, A.K., 1982. Microwave Remote Sensing - Active and Passive, Vol. I: Microwave Remote Sensing Fundamentals and Radiometry. Artech House, Boston, USA.

Ulaby, F.T., Moore, R.K., Fung, A.K., 1986. Microwave remote sensing - active and passive, Vol. III: From Theory to Applications. Artech House, Boston, USA.

Wang, L., Wen, J., Zhang, T., Zhao, Y., Tian, H., Shi, X., Wang, X., Liu, R., Zhang, J., Lu, S., 2009. Surface soil moisture estimation from AMSR-E observations over an arid area, Northwest China. Hydrol. Earth Syst. Sci. Discuss. 6, 1055–1087.

Wang, J.R., Choudhury, B.J., 1981. Remote sensing of soil moisture content over bare field at 1.4 GHz frequency. J. Geophys. Res. 86, 5277–5282.

Wang, J.R., 1983. Passive microwave sensing of soil moisture content: the effects of soil bulk density and surface roughness. Rem. Sens. Environ. 13, 329–344.

Woodhouse, I.H., 2006. Introduction to Microwave Remote Sensing. Taylor and Francis, USA.

Yang, K., Watanabe, T., Koike, T., Li, X., Fujii, H., Tamagawa, K., Ma, Y., Ishikawa, H., 2007. Auto-calibration system developed to assimilate AMSR-E data into a land surface model for estimating soil moisture and the surface energy budget. J. Meteorol. Soc. Jpn. 85, 229–242.

Chapter 14

Bistatic scatterometer for the retrieval of soil moisture

Dileep Kumar Gupta[1], **Rajendra Prasad**[2], **Prashant K. Srivastava**[1,3]
[1]*Remote Sensing Laboratory, Institute of Environment and Sustainable Development, Banaras Hindu University, Varanasi, Uttar Pradesh, India;* [2]*Remote Sensing Laboratory, Department of Physics, Indian Institute of Technology (BHU), Varanasi, Uttar Pradesh, India;* [3]*DST-Mahamana Centre of Excellence in Climate Change Research, Institute of Environment and Sustainable Development, Banaras Hindu University (BHU), Varanasi, Uttar Pradesh, India*

1. Introduction

Soil is basically a layer of unconsolidated material found at the Earth's surface that has been influenced by the soil forming factors. The gaps between the soil particles are known as pore spaces or voids, which consist of variable amount of air and water. The amount of void space within a soil depends on the distribution of particle sizes, and is quantified by soil porosity. Besides solid particles, the soil also contains air, the amount of which may vary depending on the soil type. "Soil saturation" state is reached, when the volume of air in the soil is higher and the density of soil is lower. Soil moisture can be expressed either in gravimetric which is the mass of water/mass of solid material or in volumetric which is defined as the volume of soil/total porosity (Srivastava, 2017a).

The estimation of the soil surface parameters through microwave remote sensing is very important for prediction of climate and irrigation scheduling (Chaube et al., 2019; Srivastava et al., 2013a,b,c). The irrigation scheduling is very important for the huge production of agricultural products, which is helpful for the economic growth of any country (Srivastava et al., 2017b; Gupta et al., 2017). For this purpose, it requires a good tool which provides the temporal and spatial distribution of soil surface parameters. The microwave remote sensing may be a good tool to retrieve the soil surface parameters with higher temporal and spatial resolution (Suman et al., 2020; Srivastava et al., 2014a,b, 2015). The microwave sensors have capability to work in any weather conditions and all the time (day/night). The knowledge of spatial and temporal distribution of soil surface parameters is important at regional/global scale for drought monitoring, crop yield estimation and nature investigation and

Agricultural Water Management. https://doi.org/10.1016/B978-0-12-812362-1.00014-X

eco-environment (Deng et al., 2019; Zhuo et al., 2015). Researchers demonstrated that the retrieval of soil moisture using microwave remote sensing technique is possible due to soil moisture available at the surface layer of the soil (about 0–5 cm). In addition to that the remotely sensed surface soil moisture may be used to estimate evapotranspiration rates and root-zone soil moisture (Petropoulos et al., 2018; Schmugge, 1983).

The large dissimilarity in dielectric constant of dry soil (~4) and pure water (~80) at microwave frequencies is the main advantage to establish the relation for discrimination and estimation of water content available in the soil. The dielectric constant of soil-water mixture increases with the water content and it produces a large range of dielectric constants as the function of amount of water. Several studies have already been published for the dielectric properties of soil-water mixture at microwave frequencies (Wang and Schmugge, 1980; Dobson et al., 1985; Hallikainen et al., 1985).

Several researchers have developed empirical, semi-empirical and theoretical models for the estimation of soil surface parameters by monostatic active radar at microwave frequencies (Wang et al., 1983; Oh et al., 1992; Engman and Chauhan, 1995; Chauhan, 1997; Njoku and Li, 1999; Du et al., 2000; Baghdadi et al., 2002; Singh, 2005). The results show that both backscattering coefficients and copolarized phase difference at low frequencies are sensitive to the roughness of subsurface interfaces and deep soil moisture. Also, much larger depth sensitivity can be achieved using copolarized phase difference than scattering coefficients (Kuo and Moghaddam, 2007). The VV-pol is found to be more sensitive to soil moisture at higher look-angles whereas the HH-pol is found to be more sensitive at lower look-angles (Singh, 2005). The best suitable configuration of scatterometer system for the observation of soil moisture content has been reported at 5 GHz and 10° of incidence angle (Ulaby et al., 1978). However, only a limited number of bistatic experiments have been conducted (Singh et al., 1996; Khadhra et al., 2012; Gupta et al., 2014; Kumar et al., 2019; Yadav et al., 2020). Thus, it will be very interesting and worthy of pursuit to significantly increase the experience with the knowledge of bistatic remote sensing measurements. The bistatic measurement sets are composed of soils with different well-known statistical roughness scales and different moisture contents. The bistatic measurement facility (BMF) that has been calibrated using the Isolated Antenna Calibration Technique (IACT) demonstrates a nonlinear relationship between them (Khadhra, 2007). The scattering coefficient was found to increase with the increase in soil moisture content.

The modelling of the scattering mechanism of microwave from the soil surface is very difficult and complex. Several researchers have developed different modelling approaches based on linear regression and physical bases for the retrieval of soil moisture. Now, soft computing techniques are widely used for modelling different types of algorithms in different areas (Oh et al., 1992; Singh et al., 1996; Singh, 2005; Saleh et al., 2006). These models have advantages to provide reasonable results in most of the cases. Various researchers have used artificial neural network (ANN) for the estimation of soil surface parameters (Del Frate et al., 2003; Jiang and Cotton, 2004; Chai et al., 2009; Dharanibai and Alex, 2009).

The foremost objective of this chapter is to estimate the soil surface parameters using bistatic scatterometer data at X-band. The followed structure of this present chapter contains the Material and Methods in Section 2, Results and Discussions in Section 3. The last Section 4 contains the final remark and conclusion of this work.

2. Materials and methods

2.1 Bistatic scatterometer measurement and computation of bistatic scattering coefficients

Test beds of bare soil surfaces having dimensions $4m^2$ were prepared to carry out bistatic measurements at X-band besides the Department of Physics, Indian Institute of Technology (B.H.U.), Varanasi. The bistatic scatterometer system was designed for the soil moisture estimation at HH- and VV-polarisations in the angular range of incidence angle 20° to 70°. The calibration of the system was checked during the experiment to ensure the system integrity. The surface roughness was taken constant during entire observations to study the microwave response of soil moisture content only. The bistatic scatterometer system can be categorised into two sections. The first section was called transmitter and second section was called receiver. The transmitter sends the electromagnetic waves while the receiver receives the electromagnetic waves after interaction from scatter. In this study the transmitter part consists of a pyramidal dual polarised X-band horn antenna, a waveguide to N-female coaxial adaptor and PSG high power signal generator (E8257D, 10 MHz to 20 GHz). The receiver part consists of a pyramidal dual polarised X-band horn antenna, a waveguide to N-female coaxial adaptor, EPM-P series power meter (E4416A) and peak and average power sensor (E9327A, 50 MHz−18 GHz). The gains of these antennas were approximately 20 dB, whereas, their half power beam width were found to be 18° and 20° for E and H plane, respectively. The 90° E-H twisters were used to change the polarisation HH- to VV- and vice versa. The antennas were placed in far field region from the centre of the target to minimize the near-field interactions. Fig. 14.1 shows the photograph of bistatic scatterometer system used. The height and distance from the centre of target can also vary to adjust the focus of both antennas at the centre of the target. Table 14.1 shows the schematic specifications of scatterometer system used for bistatic measurements. The system was calibrated by noting the signals returned from an aluminum plate placed on the top of the target. Then, the plate was removed and angular measurements for target were

FIGURE 14.1 Photograph of bistatic scatterometer system used in the measurements beside department of Physics, IIT(BHU), Varanasi.

TABLE 14.1 Specification of bistatic scatterometer system.

RF generator		E8257D, PSG high power signal generator, 10 MHz to 20 GHz (agilent technologies)
Power meter		E4416A, EPM-P series power meter, 10 MHz to 20 GHz (agilent technologies)
Power sensor		Peak and average power sensor (E9327A, 50 MHz -18 GHz) (Agilent Technologies)
Frequency (GHz)		10 ± 0.05 (X-band)
Beam width	E plane (°)	17.3118
	H plane (°)	19.5982
Band width (GHz)		0.8
Antenna gain (dB)		20
Cross-polarisation isolation(dB)		40
Polarisation modes		Horizontal transmit—Horizontal receive (HH) Vertical transmit—Vertical receive (VV)
Antenna type		Dual-polarised pyramidal horn
Calibration accuracy (dB)		1
Platform height (meter)		3
Incidence angle (°)		20° (nadir)–70°
Measurement interval		20 min

carried out. The calibration of system was done regularly during the experiment to ensure the integrity of the system.

The scattered power in angular range of incident angle 20° to 70° in the step of 5° at both the polarisations HH and VV were measured by using bistatic scatterometer system. The height of antenna was taken as 1.50 m from the base and both were placed at same height. The calibration of the scatterometer was done with the help of aluminum sheet, which was used as the perfect reflector. The system is capable to measure the reflected and transmitted power at X-band (10.045 GHz) at both HH and VV polarisation.

If λ is the wavelength of the microwave, P_t is the transmitted power, G_r and G_t are the receiving and the transmitting antenna gains, R_1 and R_2 are the distances of the transmitting and receiving antenna from the centre of illuminated area, then the power received at the receiver due to the perfectly conducting flat aluminum sheet as the reflecting surface is given by Ulaby et al. (1982).

$$P_r^{Al} = \frac{P_t G_r G_t \lambda^2}{(4\pi)^2} (R_1 + R_2)^2 \tag{14.1}$$

If we use a reflecting target with (R_0) as its reflection coefficient, then the received power can be expressed as

$$P_r = \frac{P_t G_r G_t \lambda^2 |R_0|^2}{(4\pi)^2} (R_1 + R_2)^2 \tag{14.2}$$

The reflectivity(r) of the target may be obtained by the equation

$$r = |R_0|^2 = \frac{P_r}{P_r^{Al}} \tag{14.3}$$

$$\sigma^0(dB) = 10 log_{10} \left[\frac{2|R_0|^2 \cot\left(\frac{\phi_{az}}{2}\right) cosec\left(\frac{\phi_{el}}{2}\right)}{sec\left(\theta - \frac{\phi_{el}}{2}\right) + sec\left(\theta + \frac{\phi_{el}}{2}\right)} \right] \tag{14.4}$$

Therefore, knowing the value of the elevation, azimuth, look angle of an antenna and the reflectivity of the target, we can compute the scattering coefficient ($\sigma^0(dB)$) of the target by the equation 14.4 (Bertuzzi et al., 1992).

2.2 Soil moisture measurement

There are two methods, which are generally used for the measurement of soil moisture namely gravimetric and volumetric. The soil moisture content may be expressed by weight as the ratio of the mass of water present to the dry weight of the soil sample, or by volume as ratio of volume of water to the total volume of the soil sample. To determine any of these ratios for a particular soil sample, the water mass must be determined by drying the soil sample and weighted it before and after drying. The oven temperature has been taken as 105°C for drying the soil samples. It seems that this temperature range has been based on water boiling temperature and does not affect the soil physical and chemical characteristics.

2.2.1 Materials

(i) Oven with 100−110°C temperature.
(ii) A balance of precision of ± 0.001 g
(iii) Auger or tool to collect soil samples.

2.2.2 Procedure

(i) Place a soil sample of about 10 g in the oven weight as (wet soil).
(ii) Place the sample in the oven at 105°C, and dry it for 24 h or overnight.
(iii) Weigh the soil samples, and record this weight as weight of dry soil.
(iv) Return the sample to the oven and dry for several hours, and determine the weight of dry soil.
(v) Repeat step (iv) until there is no difference between any two consecutive measurements of the weight of dry soil.

2.2.3 Computation of soil moisture

The moisture content in dry weight basis may be calculated using the following formula:

$$m_d = \frac{(wt.of\ wet\ soil) - (wt.of\ dry\ soil)}{(wt.of\ dry\ soil)} \tag{14.5}$$

In some literature, the moisture content is expressed in wet weight basis that is defined as the ratio between water mass and the mass of wet soil (m_w). The conversion from m_d *to* m_w or vice versa can be carried out as follows:

$$m_d = \frac{wt.of\ wet\ soil}{wt.of\ dry\ soil} - 1 \tag{14.6}$$

$$m_d = \frac{m_d}{m_w} - 1 \tag{14.7}$$

But

$$\frac{m_d}{m_w} = \frac{(wt.\ of\ wet\ soil\ )}{(wt.\ of\ dry\ soil)} \tag{14.8}$$

Eqs. (14.4) and (14.5) can be used for the conversion of m_d *to* m_w and conversion of m_w to m_d, respectively,

$$m_w = \frac{m_d}{m_d + 1} \tag{14.9}$$

$$m_d = \frac{m_w}{1 - m_w}$$

(14.10)

The volumetric soil moisture computation is as follows,

$$m_v = \frac{volume\ of\ water}{volume\ of\ soil} \tag{14.11}$$

$$Volume\ of\ water = \frac{wt.of\ water}{water\ density} \tag{14.12}$$

$$Volume\ of\ soil = \frac{wt.of\ dry\ soil}{bulk\ density} \tag{14.13}$$

$$m_v = \frac{wt.of\ water}{wt.of\ dry\ soil} \times \frac{bulk\ density}{water\ density} \tag{14.14}$$

$$m_v = m_d \times \frac{d_b}{d_w} \tag{14.15}$$

2.3 Soil surface roughness measurement

Roughness of the soil surfaces from decimetre to millimetre range affects the microwave remote sensing observations greatly. Two important parameters of surface roughness are the standard deviation of height (σ) and correlation length(l).

Here 'σ' corresponds to vertical scale roughness while 'l' corresponds to horizontal scale roughness. The ratio of these two parameters is called the roughness parameter (h) which is proportional to RMS slope of surface roughness. The geometrical properties of the soil surface roughness parameter also depend on the polarisation and frequency of the microwave used and view angle of the sensor (Dobson and Ulaby, 1981).

2.3.1 RMS surface height

The root mean square height tells the surface height variation above an arbitrary plane. If the heights are spread over a larger value of heights then the value of RMS height will be higher. It represents the standard deviation of the surface height distribution.

For one-dimensional surface roughness profile consisting of N points with surface height z, the RMS height (σ) is given by the formula

$$\sigma = \sqrt{\frac{1}{N}\left[\sum_{j=1}^{N} \tau_i^2 - N\overline{z}^2\right]} \tag{14.16}$$

where,

$$\overline{z} = \frac{1}{N}\sum_{j=1}^{N} \tau_i^2 \tag{14.17}$$

The RMS height (σ) of a surface can vary from 0.25 cm to 4 cm. It is generally measured by using roughness profilometer.

2.3.2 Autocorrelation function

If the surface roughness is independent of the view direction, the correlation coefficient is said to be isotropic depending on a single parameter (ξ). It is related to the spatial resolution of the profile and is given by the equation

$$p(\xi) = \frac{\sum_{i=1}^{N-j} Z_i Z_{i+j}}{\sum_{i=1}^{N} Z_i^2} \quad (14.18)$$

where ξ represents a roughness character of a step of surface. For isotropic surfaces, ξ is a function of a single horizontal parameter x or y.

Thus

$$\xi = \mathrm{j}\Delta \mathrm{x} \quad (14.19)$$

where Δx is the spatial resolution of the profile and j is the particular number of profiles of the extended surface.

2.3.3 Correlation length

Correlation length describes the uniformity of the soil height over a certain distance along the surface. The maximum distance over which a significant change in the height occurs is called correlation length (l). For normal surfaces, the increase in distance results in decrease in the autocorrelation length.

In remote sensing applications, the roughness of the surface is defined by the ratio of RMS height (σ) to correlation length (l) and this ratio gives the effective slop of the surface. Smoother surfaces generally have higher correlation lengths while rougher surfaces have lower values of correlation lengths.

Two main autocorrelation functions are mostly used for describing the surface roughness and computation of correlation length, namely Exponential and Gaussian functions. The Gaussian and Exponential autocorrelation functions are used in the present study. The Exponential autocorrelation function describes the smooth natural surfaces while the Gaussian autocorrelation function correlates to the rough surfaces. The exponential function is given by the formula

$$\rho(\xi) = \exp\left(-\left(\frac{\xi}{1}\right)\right) \quad (14.20)$$

while the Gaussian function is given by the formula

$$\rho(\xi) = \exp\left(-\left(\frac{\xi}{1}\right)^2\right) \quad (14.21)$$

where ξ is the correlation function, l is the autocorrelation length.

The real surfaces cannot be defined by such simple isotropic autocorrelation functions and the spectrum may be computed numerically for the real surfaces. Hence, in the present study surfaces defined by different functions have been considered. Actually the angle of incidence depends on the slop of the scattering surface. Inclined surfaces imply an elevated horizon with variation depending on the azimuth. The local angle of incidence of an inclined surface depends on the orientation of the surface with respect to the view direction of the sensor.

2.4 Statistical performance indices

This study used some performance indices for comparing ANN models and linear regression model results for the estimation of soil moisture content from bare and rough soil surfaces using bistatic scatterometer data in specular direction. The three performance indices namely % Bias, root mean squared error (RMSE) and Nash-Sutcliffe Efficiency (NSE) were used.

The percentage bias (%Bias) measures the average tendency of the estimated values to be larger or smaller than their observed values. The optimum value of %Bias is 0.0 and the smaller value of %Bias indicates that accurate model prediction.

$$\%Bias = 100 * \left[\frac{\sum(y_i - x_i)}{\sum x_i}\right] \tag{14.22}$$

RMSE is frequently used to measure the differences between estimated values by a model or an estimator and the observed values.

$$RMSE = \sqrt{\frac{1}{n}\sum_{i=1}^{n}(y_i - x_i)^2} \tag{14.23}$$

where n is the number of observations.

The NSE is based on the sum of absolute squared differences between the estimated and observed values normalised by the variance of the observed values during the study. The NSE was calculated using the given formula, using the observed values during the study.

$$NSE = 1 - \frac{\sum_{i=1}^{n}(y_i - x_i)^2}{\sum_{i=1}^{n}(x_i - \bar{x}_i)^2} \tag{14.24}$$

where n is the number of observations; x_i is the observed and y_i is the simulated variable.

3. Result and discussions

3.1 Soil surface roughness analysis

Roughness of soil surface depends on wavelength of electromagnetic wave incident on the surface and surface characteristics. The term *surface roughness* in microwave region refers to the micro relief of the soil surfaces representing a scale range between millimetres and decimetres. For smooth surfaces, the surface irregularities and volume discontinuities are lesser in comparison to wavelength of microwaves. Five soil surfaces with different roughness were prepared besides the Department of Physics, Indian Institute of Technology (BHU), Varanasi. The pin profilometer was used for the measurement of soil surface roughness. The length of pin profilometer was 1 m. The pin profilometer had 100 spokes and the distance between any two spokes of the pin profilometer was 1 cm. The pin profilometer is put on the rough soil surface at the level of maximum depth of the soil surface roughness. All the spokes of pin profilometer created spectra according to the roughness of soil surfaces. The vertical height from the mean of roughness spectra is noted for all the 100 spokes. The 100 data points are generated according to the vertical and horizontal length of the surface roughness. The generated datasets are simulated in the computer to find the RMS height, correlation length and autocorrelation function. The plots of surface roughness, autocorrelation function for various plots are shown in Figs. 14.2−14.6. In the present study, five RMS heights are used: 1.61, 1.68, 1.78, 1.57 and 1.98 cm, respectively. The corresponding correlation lengths used are 11.69, 7.34, 5.96, 4.77 and 3.88 cm, respectively.

Fig. 14.2 shows that the autocorrelation function is more at par with the Gaussian function than Exponential function. It means the plot 1 can be considered as a Gaussian surface. Fig. 14.3 shows that the autocorrelation function is more on par with the Exponential function than Gaussian function. It means the plot 2 can be considered as Exponential surface. In Figs. 14.4−14.6, the autocorrelation function did not match either with Gaussian or Exponential functions.

3.2 Angular response of bistatic scattering coefficient

Figs. 14.7−14.11 show the angular variation of scattering coefficient at different surface roughness and soil moisture conditions at HH-polarisation while Figs. 14.12−14.16 show the angular variation of bistatic scattering coefficient at different surface roughness and soil moisture conditions at VV-polarisation. The soil moisture and roughness are varied to study the microwave response of the bistatic scattering coefficient during the entire observations. The RMS height of soil surfaces were 1.61, 1.68, 1.78, 1.57, and 1.98 cm. Their correlation lengths were found to be 11.7, 7.34, 5.96, 4.77 and 3.88 cm, respectively.

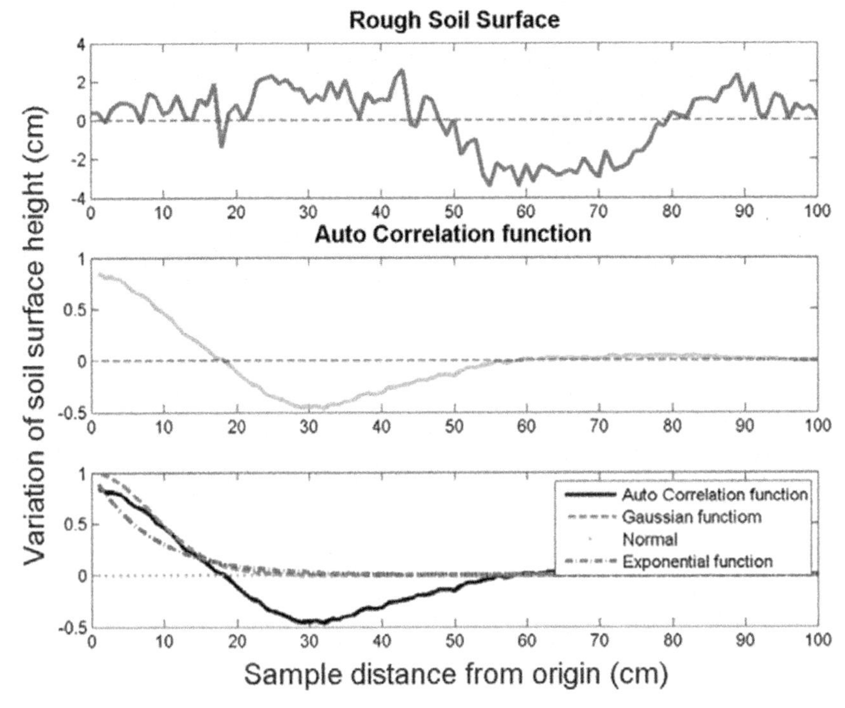

FIGURE 14.2 The plot between surface roughness and autocorrelation function for plot 1 (RMS height (σ) = 1.6 cm and correlation length (l) = 11.69 cm).

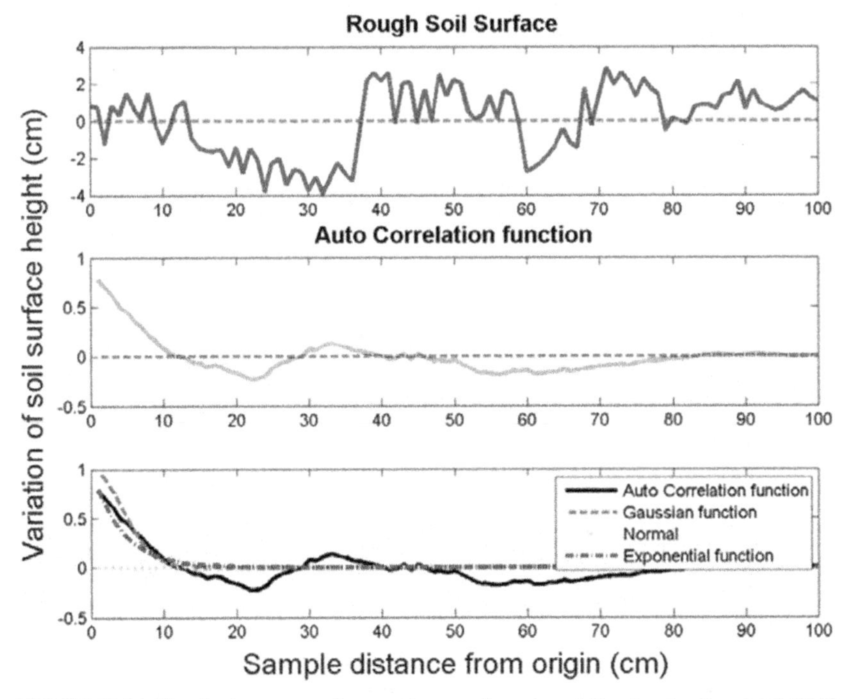

FIGURE 14.3 The plot between surface roughness and autocorrelation function for plot 2 (RMS height (σ) = 1.68 cm and correlation length (l) = 7.34 cm).

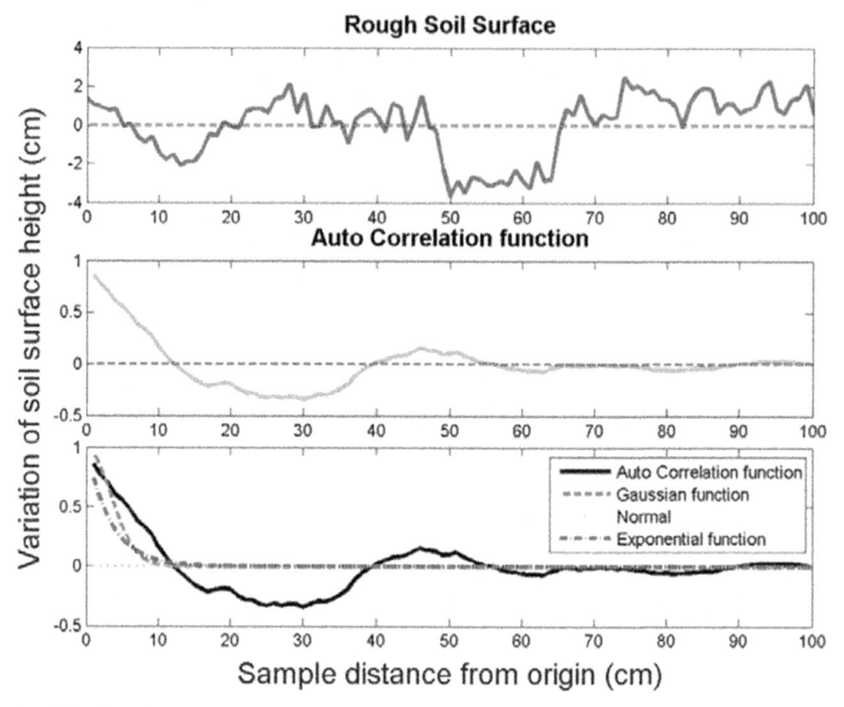

FIGURE 14.4 The plot between surface roughness and autocorrelation function for plot 3 (RMS height (σ) = 1.78 cm and correlation length (l) = 5.96 cm).

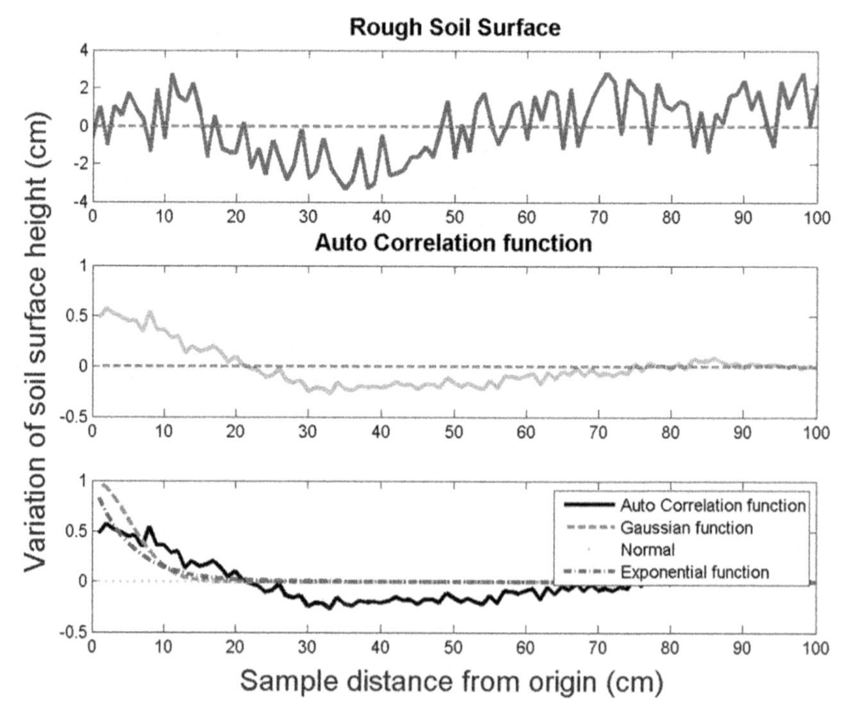

FIGURE 14.5 The plot between surface roughness and autocorrelation function for plot 4 (RMS height (σ) = 1.57 cm and correlation length (l) = 4.77 cm).

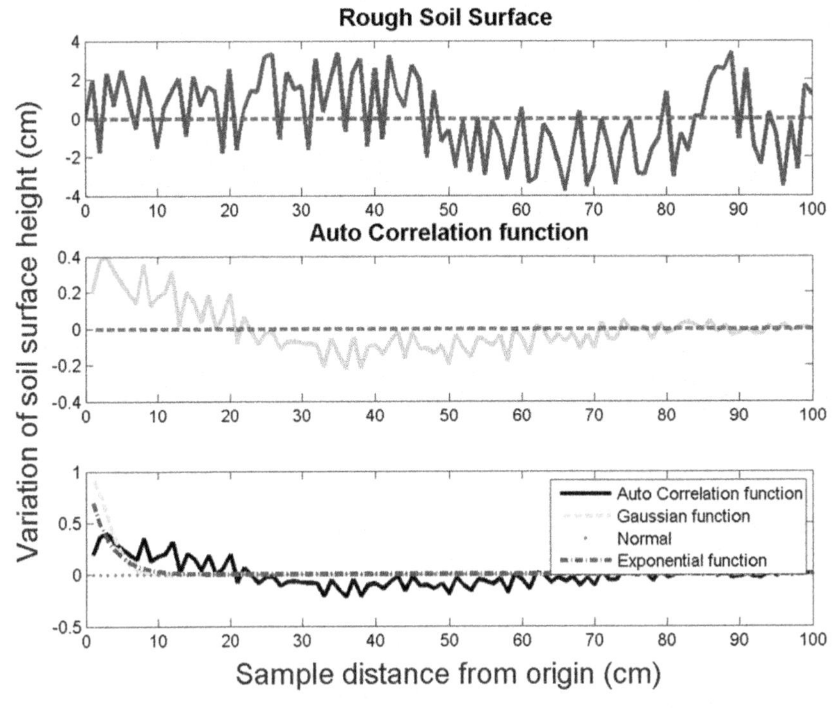

FIGURE 14.6 The plot between surface roughness and autocorrelation function for plot 5 (RMS height (σ) = 1.98 cm and correlation length (l) = 3.88 cm).

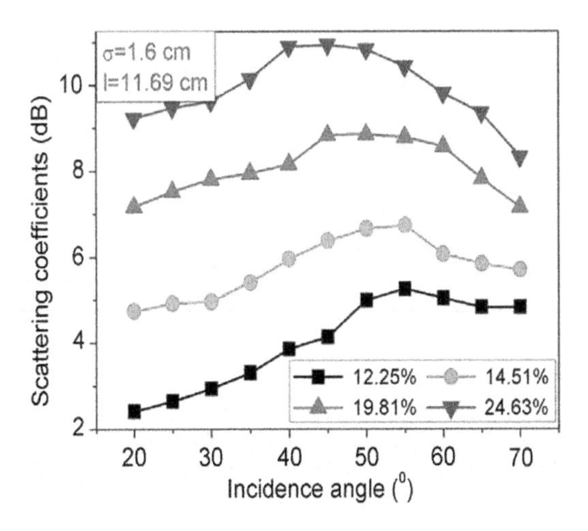

FIGURE 14.7 Angular variation of scattering coefficient with soil moisture content of soil plot 1 at HH-polarisation.

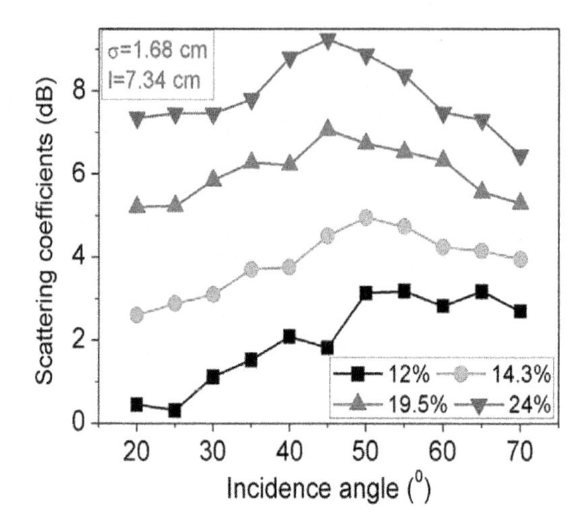

FIGURE 14.8 Angular variation of scattering coefficient with soil moisture content of soil plot 2 at HH-polarisation.

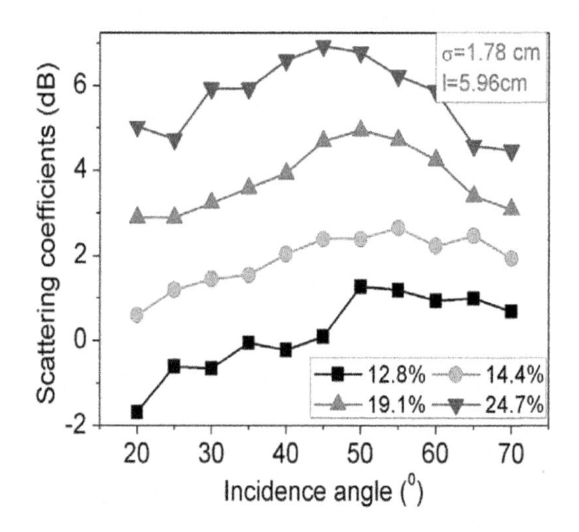

FIGURE 14.9 Angular variation of scattering coefficient with soil moisture content of soil plot 3 at HH-polarisation.

The bistatic scattering coefficient was found to increase with the soil moisture content and decrease with the increase in soil surface roughness. The maximum bistatic scattering coefficient was found at highest percentage of soil moisture (25%). The maximum bistatic scattering coefficient in terms of roughness was found for the smooth surface whose RMS height (σ) was 1.61 cm with correlation length (l) 11.69 cm. The minimum bistatic scattering coefficient was found at lowest percentage of soil moisture (12%). The

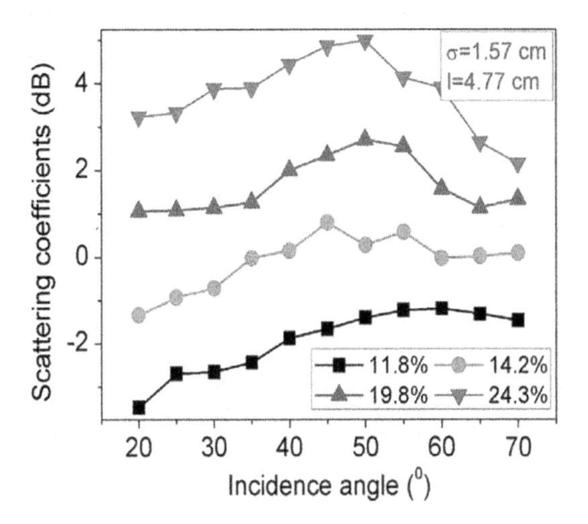

FIGURE 14.10 Angular variation of scattering coefficient with soil moisture content of soil plot 4 at HH-polarisation.

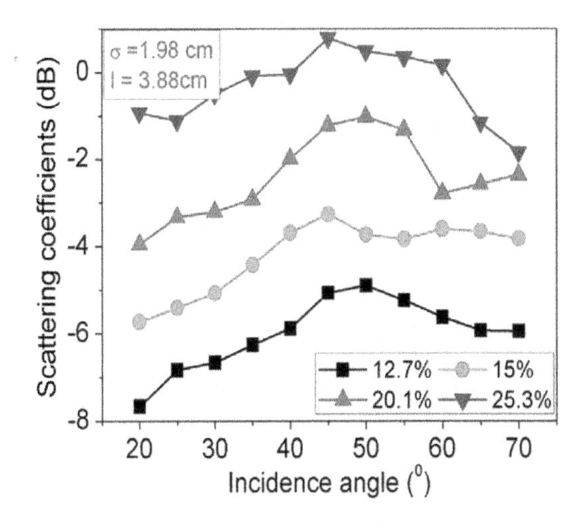

FIGURE 14.11 Angular variation of scattering coefficient with soil moisture content of soil plot 5 at HH-polarisation.

minimum bistatic scattering coefficient in terms of roughness was found for the smooth surface whose RMS height (σ) was 1.98 cm with correlation length (l) 3.88 cm.

All the bistatic scattering coefficient curves were found to decrease monotonically with the incidence angles. It was due to increase of dominant contribution of volume scattering at higher incidence angles. Thus, the separation between the bistatic scattering coefficients at higher incidence angles decreases.

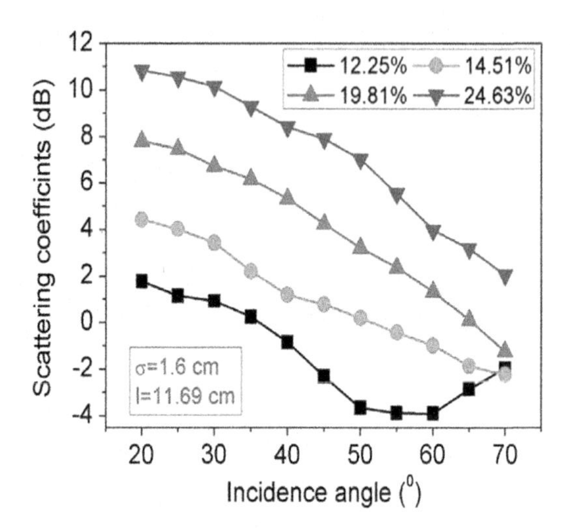

FIGURE 14.12 Angular variation of scattering coefficient with soil moisture content of soil plot 1 at VV-polarisation.

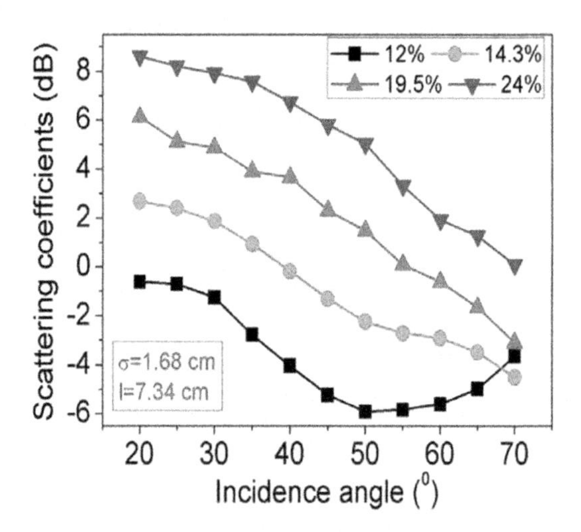

FIGURE 14.13 Angular variation of scattering coefficient with soil moisture content of soil plot 2 at VV-polarisation.

3.3 Estimation of soil moisture using ANN

A multilayer back propagation artificial neural network having simple processing units (neurons) is arranged in different layers as input, hidden and output layers. These layers are fully interconnected. The input to one node is the weighted sum of the outputs of the previous layer nodes. This sum is passed through an activation function to produce the final output. The

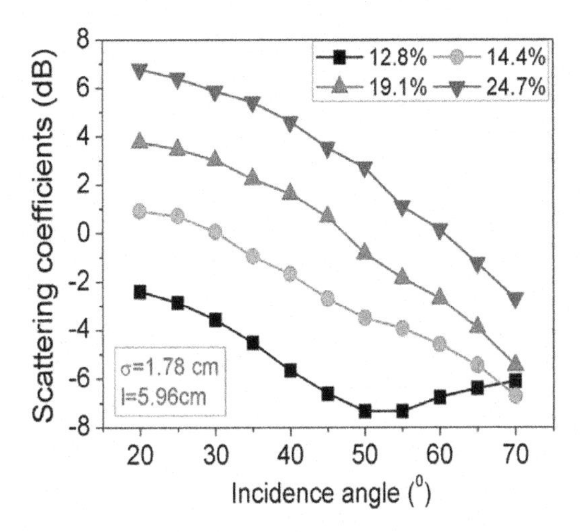

FIGURE 14.14 Angular variation of scattering coefficient with soil moisture content of soil plot 3 at VV-polarisation.

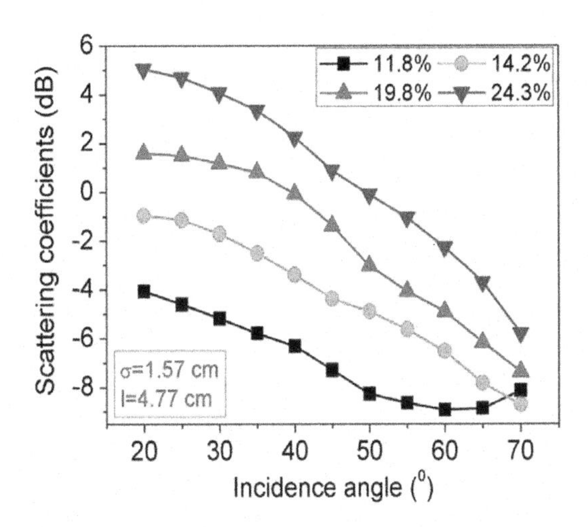

FIGURE 14.15 Angular variation of scattering coefficient with soil moisture content of soil plot 4 at VV-polarisation.

activation function is usually a sigmoid or hyperbolic tangent, which is a nonlinear function that has asymptotic behaviour. The calibration stage consists of adjusting the connection weights (randomly initialized) in order to decrease the difference between the network output and the desired outputs. The calibration data were presented to the input layer and propagated through the hidden layer to the output layer. The differences between the computed and the desired outputs were computed and fed backwards to adjust the network

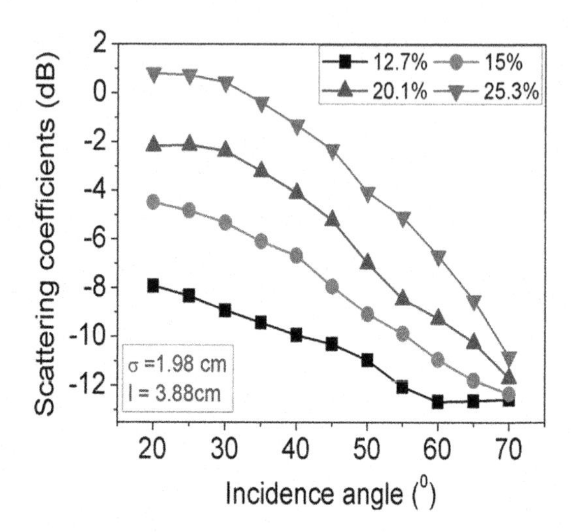

FIGURE 14.16 Angular variation of scattering coefficient with soil moisture content of soil plot 5 at VV-polarisation.

connections. This iterative process continued until the root mean square error reached a desired level. When the root mean square error reached an optimum level, the calibration is stopped and the weight and bias values saved. The trained network was used for the estimation of soil surface parameters of bare soil surfaces.

In this study, the linear regression analysis was made between scattering coefficient and soil moisture contents at different incidence angles for each plot under different roughness conditions. Tables 14.2 and 14.3 show the linear regression analysis between bistatic scattering coefficients and soil moisture of different plots at different angles of incidence for both HH- and VV- polarisation, respectively. This linear regression analysis was done to select the suitable scatterometer configuration. The higher value of R and Adj_R^2 were found at incidence angles 25°, 30°, 35°, 25°, 25° for Plot 1, Plot 2, Plot 3, Plot 4, Plot 5, respectively, at HH-polarisation. The higher value of R and Adj_R^2 were found at incidence angles 35°, 25°, 30°, 25°, 25° for Plot 1, Plot 2, Plot 3, Plot 4, Plot five5, respectively, at VV-polarisation. The values of the datasets whose adjusted R^2 found highest among the incident angles were taken as the plot angles of incidence along with its scattering coefficient at that angle of incidence.

The observed dataset (bistatic scattering coefficient and soil moisture) interpolated to 60 datasets for both like polarisations (HH- and VV-). The 2/3rd datasets are used for calibration and remaining 1/3rd datasets are used for the validation of BPANN.

In this study, MATLAB software was used for developing a BPANN consisting of one neuron at input layer, five neurons at hidden layer and one

TABLE 14.2 Calculated values of R and adjusted R^2 for soil plots at HH-polarisation.

	Plot 1		Plot 2		Plot 3		Plot 4		Plot 5	
Incident angle (°)	R	Adj_R^2	R	Adj_R^2	R	Adj_R^2	R	Adj_R^2	R	Adj_R^2
20	0.997	0.9903	0.997	0.9907	0.972	0.9177	0.992	0.9754	0.991	0.9741
25	0.998	0.9944	0.989	0.9663	0.975	0.9272	0.997	0.9919	0.998	0.9932
30	0.988	0.9647	0.998	0.9938	0.982	0.9465	0.994	0.9809	0.996	0.9867
35	0.991	0.9736	0.982	0.9479	0.99	0.9697	0.972	0.9182	0.989	0.9665
40	0.992	0.9769	0.988	0.9634	0.979	0.9376	0.989	0.9667	0.979	0.9375
45	0.987	0.9625	0.983	0.9506	0.974	0.923	0.978	0.9334	0.991	0.9719
50	0.994	0.9824	0.991	0.9732	0.99	0.9691	0.989	0.9683	0.991	0.9721
55	0.993	0.9793	0.994	0.9818	0.978	0.9333	0.988	0.9657	0.992	0.9764
60	0.991	0.9742	0.988	0.9641	0.985	0.9557	0.992	0.9767	0.97	0.9112
65	0.986	0.9572	0.992	0.9756	0.955	0.8693	0.986	0.9597	0.95	0.8541
70	0.986	0.9575	0.984	0.9538	0.978	0.9345	0.967	0.9013	0.918	0.7649

TABLE 14.3 Calculated values of R and adjusted R^2 for soil plots at VV-polarisation.

	Plot1		Plot2		Plot3		Plot4		Plot5	
Incident angle (°)	R	Adj_R^2	R	Adj_R^2	R	Adj_R^2	R	Adj_R^2	R	Adj_R^2
20	0.993	0.9798	0.983	0.9496	0.969	0.9072	0.985	0.9565	0.993	0.9781
25	0.992	0.9747	0.992	0.9756	0.962	0.8869	0.991	0.9738	1	0.999
30	0.996	0.9874	0.988	0.9647	0.997	0.9898	0.982	0.9452	0.995	0.9863
35	0.999	0.9961	0.984	0.9533	0.969	0.9072	0.984	0.9527	0.979	0.9389
40	0.998	0.9941	0.984	0.9529	0.958	0.8768	0.987	0.9607	0.978	0.9352
45	0.992	0.9773	0.986	0.9569	0.957	0.8748	0.984	0.9536	0.993	0.9788
50	0.983	0.948	0.989	0.9662	0.965	0.8969	0.97	0.9126	0.995	0.9845
55	0.982	0.9453	0.988	0.9631	0.959	0.8796	0.971	0.914	0.983	0.9492
60	0.981	0.9423	0.985	0.9543	0.983	0.9483	0.98	0.9407	0.975	0.9274
65	0.992	0.976	0.986	0.9597	0.96	0.8829	0.982	0.9452	0.976	0.9298
70	0.905	0.7291	0.876	0.6499	0.935	0.8113	0.917	0.76	0.976	0.9279

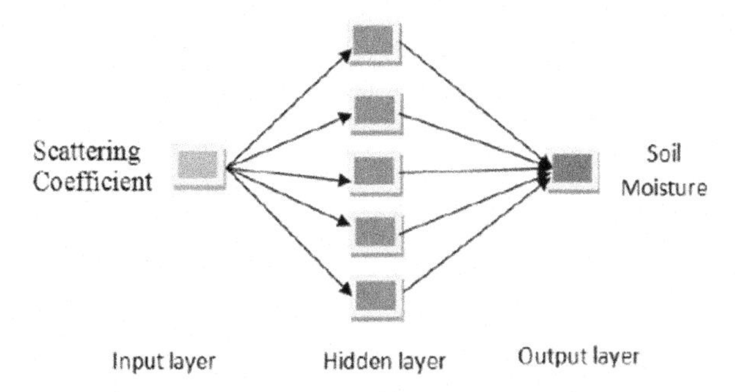

FIGURE 14.17 Architecture of BPANN used in study.

neuron at output layer. Sometimes, it is called as [1:5:1] architecture of BPANN as shown in Fig. 14.17. The hyperbolic tangent sigmoid transfer function (tansig) was used for input and hidden layer whereas linear transfer function (purelin) was used for the output layer. The BPANN was calibrated using input data (bistatic scattering coefficient) and target data (soil moisture content). The developed BPANN was used for validation. Remaining 20 datasets out of 60 datasets were used for validation of the developed and trained BPANN.

Figs. 14.18–14.23 show the scatter plot with 1:1 equal line between estimated soil surface parameters and observed soil surface parameters during validation of ANN model for HH- and VV- polarizations, respectively. These figures show the closeness between the estimated soil surface parameters and the observed soil surface parameters. The performances during the validation of ANN model for the estimation of soil surface parameters were evaluated in terms of %Bias, RMSE, NSE. Table 14.4 shows the performance indices during validation for the estimated soil surface parameters by ANN model.

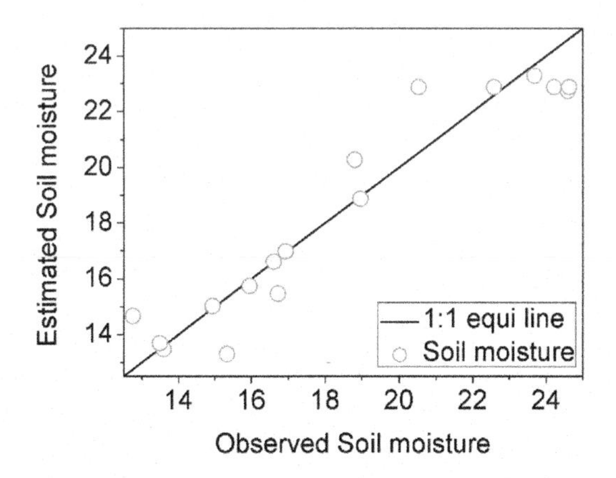

FIGURE 14.18 Scatter plot between observed and estimated soil moisture (%) by ANN at HH-polarisation.

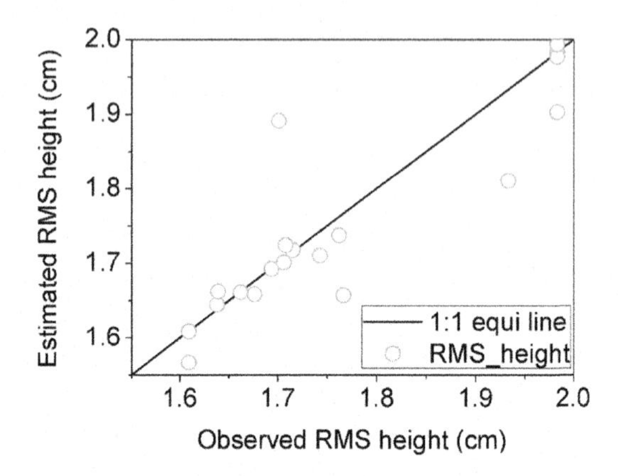

FIGURE 14.19 Scatter plot between observed and estimated RMS height (cm) by ANN at HH-polarisation.

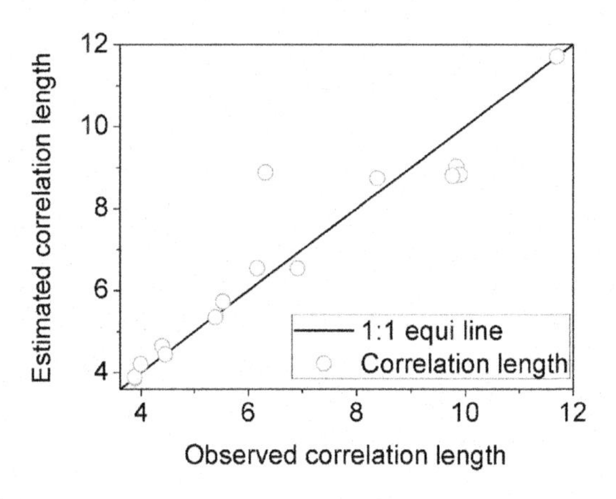

FIGURE 14.20 Scatter plot between observed and estimated correlation length (cm) by ANN at HH-polarisation.

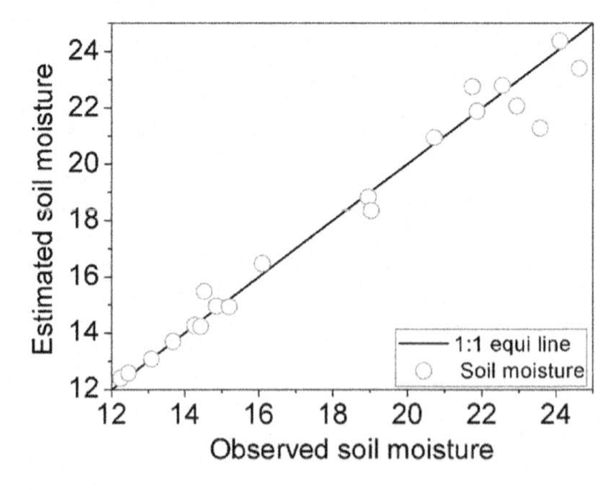

FIGURE 14.21 Scatter plot between observed and estimated soil moisture (%) by ANN at VV-polarisation.

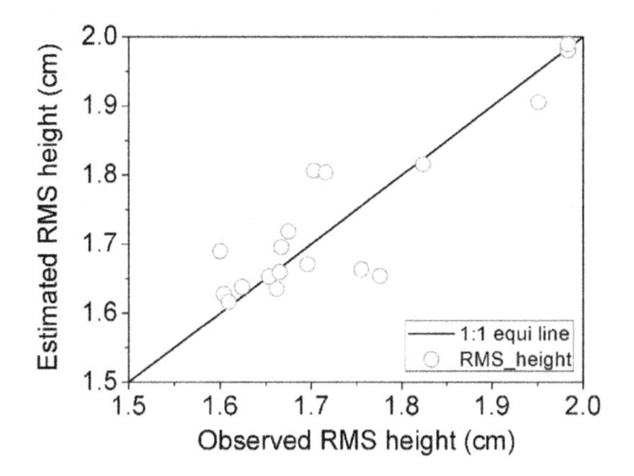

FIGURE 14.22 Scatter plot between observed and estimated RMS height (cm) by ANN at VV-polarisation.

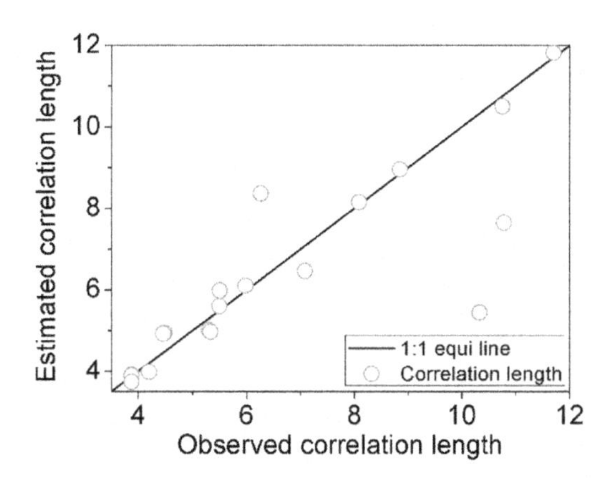

FIGURE 14.23 Scatter plot between observed and estimated Correlation length (cm) by ANN at VV-polarisation.

TABLE 14.4 Performance indices of ANN model during validation for the estimation of soil surface parameters using scatterometer data.

	HH-polarisation			VV-polarisation		
	RMSE	%Bias	NSE	RMSE	%Bias	NSE
Soil moisture	1.6392	−3.0693	0.8375	1.6436	−1.1188	0.8367
Correlation length	0.855	−1.0394	0.9082	0.5119	0.5878	0.9692
RMS height	0.0611	−0.5649	0.7922	0.0511	−0.8715	0.8815

The %Bias tells about the percentage of over/underestimated values by the models. The RMSE measures the closeness of the estimated values to the observed values. The value of RMSE closer to zero indicates high reliability of estimated and observed values.

For the HH-polarisation, the values of RMSE were found to be 1.64, 0.86 and 0.06 for the retrieval of soil moisture content, correlation length and RMS height, respectively. The values of percentage of bias were found to be −3.07, −1.04 and −0.57 for the retrieval of soil moisture content, correlation length and RMS height, respectively. The values of NSE were found to be 0.84, 0.91 and 0.79 for the retrieval of soil moisture content, correlation length and RMS height, respectively.

For the VV-polarisation, the values of RMSE were found to be 1.64, 0.51 and 0.05 for the retrieval of soil moisture content, correlation length and RMS height, respectively. The values of percentage of bias were found to be −1.12, 0.59 and −0.87 for the retrieval of soil moisture content, correlation length and RMS height, respectively. The values of NSE were found to be 0.84, 0.97 and 0.88 for the retrieval of soil moisture content, correlation length and RMS height, respectively.

For the retrieval of soil moisture, the values of RMSE were found to be higher at VV-polarisation than HH-polarisation, the values of NSE were found slightly higher at HH-polarisation than VV-polarisation and the percentage of bias was found higher at HH-polarisation than VV-polarisation. On behalf of these three performance indices, the VV-polarisation may be better than HH-polarisation for the estimation of soil surface parameters by bistatic scatterometer measurements. For the retrieval of correlation length of the soil surface, the values of RMSE were found to be higher at HH-polarisation than VV-polarisation, the values of NSE were found to be higher at VV-polarisation than HH-polarisation and the percentage of bias was found higher at HH-polarisation than VV-polarisation. On the basis of these three performance indices, the VV-polarisation may be better than HH-polarisation for the retrieval of correlation length of the soil surface.

For the retrieval of RMS height of the soil surface, the values of RMSE were found to be higher at HH-polarisation than VV-polarisation, the values of NSE were found to be higher at VV-polarisation than HH-polarisation and the percentage of bias was found higher at HH-polarisation than VV-polarisation. On the basis of these three derived performance indices, the VV-polarisation may be better than HH-polarisation. The retrieval of soil surface parameters by scatterometer data using ANN model was found to be more accurate at VV-polarisation than HH-polarisation.

4. Conclusion

The bistatic microwave scattering coefficient was found to increase with the soil moisture content and decrease with increase in soil roughness. The dynamic range of scattering coefficient was found more at VV- polarisation than HH-polarisation. The linear regression analyses were performed between scattering coefficients and soil moisture content at different soil surface roughness conditions for selecting the suitable incidence angle to generate the datasets for the calibration and validation of the ANN model. 1 hidden layer with 5 neurons may be sufficient to retrieve the soil surface parameters using ANN. The scattering coefficients at lower incidence angle were found more suitable to generate the datasets for the calibration and validation of ANN models. The calculated values of soil moisture, RMS height and correlation length were found to be nearly matching with the values estimated by ANN for both HH- and VV-polarisations. The retrieval of soil surface parameters using **bistatic** scatterometer data with the ANN model was found to be more accurate at VV-polarisation than HH-polarisation.

References

Baghdadi, N., King, C., Bourguignon, A., Remond, A., 2002. Potential of ERS and Radarsat data for surface roughness monitoring over bare agricultural fields: application to catchments in Northern France. Int. J. Rem. Sens. 23 (17), 3427–3442.

Bertuzzi, P., Chaànzy, A., Vidal-Madjar, D., Autret, M., 1992. The use of a microwave backscatter model for retrieving soil moisture over bare soil. Int. J. Rem. Sens. 13 (14), 2653–2668.

Chai, S.-S., Walker, J.P., Makarynskyy, O., Kuhn, M., Veenendaal, B., West, G., 2009. Use of soil moisture variability in artificial neural network retrieval of soil moisture. Rem. Sens. 2 (1), 166–190.

Chaube, N.R., Chaurasia, S., Tripathy, R., Pandey, D.K., Misra, A., Bhattacharya, B., Srivastava, P.K., 2019. Crop phenology and soil moisture applications of SCATSAT-1. Curr. Sci. 117 (6), 10.

Chauhan, N.S., 1997. Soil moisture estimation under a vegetation cover: combined active passive microwave remote sensing approach. Int. J. Rem. Sens. 18 (5), 1079–1097.

Del Frate, F., Ferrazzoli, P., Schiavon, G., 2003. Retrieving soil moisture and agricultural variables by microwave radiometry using neural networks. Rem. Sens. Environ. 84 (2), 174–183.

Deng, K.A.K., Lamine, S., Pavlides, A., Petropoulos, G.P., Srivastava, P.K., Bao, Y., Anagnostopoulos, V., 2019. Operational soil moisture from ASCAT in support of water resources management. Remote Sens. 11 (5), 579.

Dharanibai, G., Alex, Z., 2009. ANN technique for the evaluation of soil moisture over bare and vegetated fields from microwave radiometer data. Indian J. Radio Space Phys. 38, 283–288.

Dobson, M.C., Ulaby, F., 1981. Microwave backscatter dependence on surface roughness, soil moisture, and soil texture: Part III-soil tension. IEEE Trans. Geosci. Rem. Sens. (1), 51–61.

Dobson, M.C., Ulaby, F.T., Hallikainen, M.T., El-Rayes, M.A., 1985. Microwave dielectric behavior of wet soil-Part II: dielectric mixing models. IEEE Trans. Geosci. Rem. Sens. GE- 23 (1), 35–46.

Du, Y., Ulaby, F.T., Dobson, M.C., 2000. Sensitivity to soil moisture by active and passive microwave sensors. IEEE Trans. Geosci. Rem. Sens. 38 (1), 105–114.

Engman, E.T., Chauhan, N., 1995. Status of microwave soil moisture measurements with remote sensing. Rem. Sens. Environ. 51 (1), 189–198.

Gupta, D., Kumar, P., Mishra, V., Prasad, R., 2014. Soil moisture estimation by ANN using bistatic scatterometer data. ISPRS Ann. Photogram., Rem. Sens. Spatial Inform. Sci. 2 (8), 97.

Gupta, D.K., Prasad, R., Kumar, P., Vishwakarma, A.K., 2017. Soil moisture retrieval using ground based bistatic scatterometer data at X-band. Adv. Space Res. 59 (4), 996–1007.

Hallikainen, M.T., Ulaby, F.T., Dobson, M.C., El-Rayes, M.A., Lil-Kun, W., 1985. Microwave dielectric behavior of wet soil-Part 1: empirical models and experimental observations. IEEE Trans. Geosci. Rem. Sens. GE- 23 (1), 25–34.

Jiang, H., Cotton, W.R., 2004. Soil moisture estimation using an artificial neural network: a feasibility study. Can. J. Rem. Sens. 30 (5), 827–839.

Khadhra, K.B., 2007. Surface Parameter Estimation Using Bistatic Polarimetric X-Band Measurements.

Khadhra, K.B., Boerner, T., Hounam, D., Chandra, M., 2012. Surface parameter estimation using bistatic polarimetric X-band measurements. Prog. Electromag. Res. B 39, 197–223.

Kumar, P., Prasad, R., Choudhary, A., Gupta, D.K., Mishra, V.N., Vishwakarma, A.K., Singh, A.K., Srivastava, P.K., 2019. Comprehensive evaluation of soil moisture retrieval models under different crop cover types using C-band synthetic aperture radar data. Geocarto Int. 34 (9), 1022–1041.

Kuo, C.-h., Moghaddam, M., 2007. Electromagnetic scattering from multilayer rough surfaces with arbitrary dielectric profiles for remote sensing of subsurface soil moisture. IEEE Trans. Geosci. Rem. Sens. 45 (2), 349–366.

Njoku, E.G., Li, L., 1999. Retrieval of land surface parameters using passive microwave measurements at 6-18 GHz. IEEE Trans. Geosci. Rem. Sens. 37 (1), 79–93.

Oh, Y., Sarabandi, K., Ulaby, F.T., 1992. An empirical model and an inversion technique for radar scattering from bare soil surfaces. IEEE Trans. Geosci. Rem. Sens. 30 (2), 370–381.

Petropoulos, G.P., Srivastava, P.K., Piles, M., Pearson, S., 2018. Earth observation-based operational estimation of soil moisture and evapotranspiration for agricultural crops in support of sustainable water management. Sustainability 10 (1), 181.

Saleh, K., Wigneron, J.-P., de Rosnay, P., Calvet, J.-C., Kerr, Y., 2006. Semi-empirical regressions at L-band applied to surface soil moisture retrievals over grass. Rem. Sens. Environ. 101 (3), 415–426.

Schmugge, T.J., 1983. Remote sensing of soil moisture: recent advances. IEEE Trans. Geosci. Remote Sens. 3, 336–344.

Singh, D., 2005. A simplistic incidence angle approach to retrieve the soil moisture and surface roughness at X-band. IEEE Trans. Geosci. Rem. Sens. 43 (11), 2606–2611.

Singh, D., Mukherjee, P., Sharma, S., Singh, K., 1996. Effect of soil moisture and crop cover in remote sensing. Adv. Space Res. 18 (7), 63–66.

Srivastava, P.K., Han, D., Ramirez, M.R., Islam, T., 2013a. Appraisal of SMOS soil moisture at a catchment scale in a temperate maritime climate. J. Hydrol. 498, 292–304.

Srivastava, P.K., Han, D., Rico-Ramirez, M.A., O'Neill, P., Islam, T., Gupta, M., 2014a. Assessment of SMOS soil moisture retrieval parameters using tau–omega algorithms for soil moisture deficit estimation. J. Hydrol. 519, 574–587.

Srivastava, P.K., O'Neill, P., Cosh, M., Kurum, M., Lang, R., Joseph, A., 2014b. Evaluation of dielectric mixing models for passive microwave soil moisture retrieval using data from ComRAD ground-based SMAP simulator. IEEE J. Sel. Top. Appl. Earth Observ. Remote Sens. 8 (9), 4345–4354.

Srivastava, P.K., Han, D., Rico-Ramirez, M.A., O'Neill, P., Islam, T., Gupta, M., Dai, Q., 2015. Performance evaluation of WRF-Noah Land surface model estimated soil moisture for hydrological application: synergistic evaluation using SMOS retrieved soil moisture. J. Hydrol. 529, 200–212.

Srivastava, P.K., 2017a. Satellite soil moisture: review of theory and applications in water resources. Water Resour. Manag. 31 (10), 3161–3176.

Srivastava, P.K., Han, D., Yaduvanshi, A., Petropoulos, G.P., Singh, S.K., Mall, R.K., Prasad, R., 2017b. Reference evapotranspiration retrievals from a mesoscale model based weather variables for soil moisture deficit estimation. Sustainability 9 (11), 1971.

Srivastava, P.K., Han, D., Ramirez, M.R., Islam, T., 2013b. Machine learning techniques for downscaling SMOS satellite soil moisture using MODIS land surface temperature for hydrological application. Water Resour. Manag. 27 (8), 3127–3144.

Srivastava, P.K., Han, D., Rico-Ramirez, M.A., Al-Shrafany, D., Islam, T., 2013c. Data fusion techniques for improving soil moisture deficit using SMOS satellite and WRF-NOAH land surface model. Water Resour. Manag. 27 (15), 5069–5087.

Suman, S., Srivastava, P.K., Petropoulos, G.P., Pandey, D.K., O'Neill, P.E., 2020. Appraisal of SMAP operational soil moisture product from a global perspective. Remote Sens. 12 (12), 1977.

Ulaby, F.T., Batlivala, P.P., Dobson, M.C., 1978. Microwave backscatter dependence on surface roughness, soil moisture, and soil texture: Part I-bare soil. IEEE Trans. Geosci. Electron. 16 (4), 286–295.

Ulaby, F.T., Moore, R.K., Fung, A.K., 1982. Microwave Remote Sensing Active and Passive-Volume II: Radar Remote Sensing and Surface Scattering and Enission Theory. Artech House, Norwood, MA, US.

Wang, J.R., O'Neill, P.E., Jackson, T.J., Engman, E.T., 1983. Multifrequency measurements of the effects of soil moisture, soil texture, and surface roughness. IEEE Trans. Geosci. Rem. Sens. (1), 44–51.

Wang, J.R., Schmugge, T.J., 1980. An empirical model for the complex dielectric permittivity of soils as a function of water content. IEEE Trans. Geosci. Rem. Sens. GE- 18 (4), 288–295.

Yadav, V.P., Rajendra, P., Ruchi, B., Vishwakarma, A.K., 2020. An improved inversion algorithm for spatio-temporal retrieval of soil moisture through modified water cloud model using C-band Sentinel-1A SAR data. Comput. Electron. Agr. 173, 105447.

Zhuo, L., Han, D., Dai, Q., Islam, T., Srivastava, P.K., 2015. Appraisal of NLDAS-2 multi-model simulated soil moistures for hydrological modelling. Water Resour. Manag. 29 (10), 3503–3517.

Chapter 15

Soil water content influence on pesticide persistence and mobility

Manika Gupta[1], N.K. Garg[2], Prashant K. Srivastava[3,4]
[1]*Department of Geology, University of Delhi, New Delhi, India;* [2]*Department of Civil Engineering, Indian Institute of Technology, Delhi, New Delhi, India;* [3]*Remote Sensing Laboratory, Institute of Environment and Sustainable Development, Banaras Hindu University, Varanasi, Uttar Pradesh, India;* [4]*DST-Mahamana Centre of Excellence in Climate Change Research, Institute of Environment and Sustainable Development, Banaras Hindu University, Varanasi, Uttar Pradesh, India*

1. Introduction

In India, 70% of the population depends on agriculture and hence to increase the crop production, pesticide application is required to protect the crops from diseases and pests (Aggarwal et al., 2000; Joshi et al., 2007). However, several researchers have identified the hazardous nature of pesticides for environment as well as flora and fauna, and hence pesticide usage has become an issue of serious concern for the environmental safety (Forget, 1991; van der Werf, 1996; Gupta et al., 2012). According to Struthers et al. (1998) an understanding of pesticide mobility is essential for transport modelling and the rational design of remedial measures against pollution. Several factors may influence the persistence and mobility of the pesticides like full or partial irrigation, rainfall, and time of pesticide application. The pesticide dissipation is influenced by environmental factors like soil moisture. For many pesticides there is a 1.5- to 2.5-fold increase in pesticide half-life if soil moisture is reduced by a factor of two (Briggs and Lord, 1983; Mackay et al., 1997). The response of degradation rates to changes in temperature is less variable than responses to changes in soil moisture (Walker, 1987; Wauchope et al., 2002). Walker (1974) reported half-lives of 54, 63 and 90 days with soil moisture contents of 10.0%, 7.5% and 3.5%, respectively, for napropamide. Similarly, Choi et al. (1988) determined half-life values for DCPA (dimethyl tetrachloroterephthalate) to be 49, 33 and 31 days for low (0.1 kg H_2O kg^{-1} soil), medium (0.2 kg H_2O kg^{-1} soil) and high-soil moisture levels (0.4 kg H_2O kg^{-1} soil), respectively, thus suggesting that soil moisture was positively

Agricultural Water Management. https://doi.org/10.1016/B978-0-12-812362-1.00015-1

correlated with degradation rate. In another soil column study under laboratory conditions, mobility of pendimethalin in sandy loam soil was studied at two application rates, 1.0 and 2.0 kg/ha, with simulated rainfall of 300 mm (Chopra et al., 2010). The herbicide was distributed in soil at all depths at both doses with maximum concentration in top 10 cm. Harman et al. (2011) and Vereecken et al. (2011) also found that travel time of pesticides increases with increase in pore water velocity. Pesticide degradation rates in the above studies were measured on soil core samples incubated under laboratory conditions and do not consider the variations in environmental conditions that may affect in-situ degradation rates. In general, the presence of water favours pesticide diffusion.

The term 'pesticide' is a composite term that includes all chemicals that are used to kill or control pests. In agriculture, this includes herbicides (weeds), insecticides (insects) and fungicides (fungi). Pesticides can be classified in many different ways: according to the target pest, the chemical structure of the compound used, the action mode or the degree or type of health hazard involved. Ever since pesticides have been available, society has been concerned about the risks associated with their use. Agricultural use of pesticides is a subset of the larger spectrum of industrial chemicals used in modern society. To increase the agricultural production, pesticides have thus found a great usage in agricultural sector. The degradation of soil quality owing to overuse of pesticides is becoming a matter of concern. Some of the pesticides, which have been frequently found in use for protecting wheat crops, include Thiram, 2,4-D, etc. Thiram, a dithiocarbamate acts as a seed protectant and prevents fungal diseases in crops. These organo-sulphur compounds have extensively been used as pesticides in agriculture for decades (Szolar, 2007). Thiram presents very high chemical and biological activities as fungicide preventing crop damage in the field. Thus, wide use of Thiram has led to rise in concern for soil contamination (Filipe et al., 2007). On the other hand, 2,4-D, a chlorophenoxy herbicide, prevents the growth of broadleaf weeds. Over the past 50 years, 2,4-dichlorophenoxy acetic acid (2,4-D) is among the most utilised herbicides (Munro et al., 1992), which is ubiquitous in agriculturally dominated areas (Frank et al., 1992; Shen et al., 2005). 2,4-D is among various agrochemicals currently in use, being widely applied to control broad-leaved weeds in cereal crops (Powell et al., 1985) and may be quite mobile in aqueous systems because of its acidic carboxyl group (pKa = 2.8) (Aquino et al., 2007). The extensive use of 2,4-D in agriculture has become a serious soil-contamination problem (Sparks, 2003), and the residue of this herbicide is a source of severe surface and subsurface water pollution (Shukla et al., 2006; Wang et al., 2007). The contamination potential of any pesticide depends on its residential period in soil. Therefore, the movement and degradation properties of a pesticide in soil are very important for evaluating their environmental risk.

Pesticide degradation data are often obtained from simple laboratory experiments with the assumption that lab-based results can be extrapolated to field conditions. Nevertheless, the controlled studies like the lysimeter studies suffer from some of the limitations in presenting the real-field conditions, which are—the leachate collectors may interrupt capillary flow significantly in some soils, thus changing the upward or downward movement of mobile pesticides, normal root development in planted lysimeters might be hindered and also variation in structure of soil may affect the pesticide mobility studies (Winton and Weber, 1996). Also, the controlled studies may not be sufficient to adequately characterise soil contamination that can occur under actual field conditions. However, field monitoring and experimentation is both time-consuming and expensive. Thus, the numerical models can be utilised to predict pesticide concentration in unsaturated zone with the help of soil-water balance and hydro-meteorological parameters (Cisar and Snyder, 1996; Watanabe and Grismer, 2003). Once the models have been validated with field experimental data, these models could serve a useful purpose for regulating pesticide dosage. The studies related to the said pesticides for persistence and mobility are mostly limited to the lab and column studies and are not available under field conditions. Garg and Hassan (2007) showed that India is going to face alarming water scarcity, and it would be a real task to feed the growing population with limited water available for irrigation. In another study, Garg and Dadhich (2014) demonstrated that more area can be brought under irrigation using deficit irrigation along with increase in overall total crop production. Therefore, field studies related to the persistence and mobility of pesticides under deficit irrigation are required to determine its safe dosage to avoid soil contamination. In the present study, multiple depthwise temporal changes in the persistence of pesticides were determined experimentally under real-field conditions with varying irrigation treatments, which was simulated using HYDRUS-1D.

2. Methods

i. Laboratory equipment and field study

The pesticide standard (99% purity) was purchased from Sigma-Aldrich. The analytical grade reagents used in the study were obtained from Qualigens Fine Chemicals and Ranbaxy Fine Chemicals. Locally available pesticides used by the farmers have been applied in the experimental fields in the present study. A Hitachi UV-Visible Spectrophotometer Model, U -1500, with 1 cm matched quartz cells was used for determination of pesticide residue in the soil.

For this case study, all the experiments were performed in an agricultural field located at Roorkee (29° 51′ 0″ N latitude and 77° 53′ 0″ E longitude), India, shown in Fig. 15.1. The details of the same are presented in the earlier research communication (Gupta et al., 2012) with the mention here only of relevant information. The agricultural field was planted with wheat, and the experiments were conducted for 136 days, that is, till the wheat harvesting,

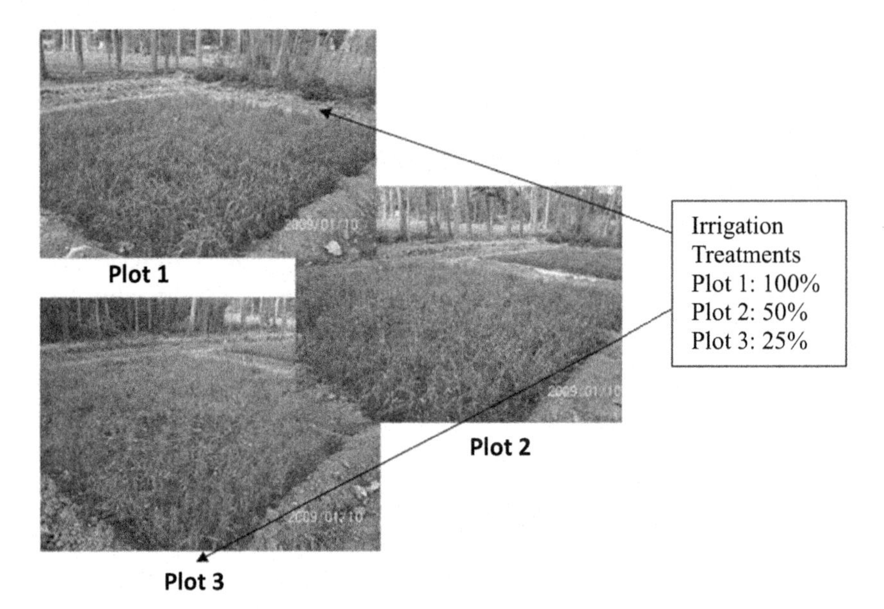

FIGURE 15.1 Photographs of experimental field plots utilised in case study.

with wheat being sown on 2nd Dec 2008 and harvested on 16th April 2009. The pesticides and fertilisers were applied to the fields as given in Table 15.1 and the properties of applied pesticides are provided in Table 15.2.

Meteorological data was obtained from the Department of Hydrology, IIT Roorkee, India, as shown in the following Fig. 15.2A–E.

Soil samples for pesticide residue analysis were collected before and after pesticide application from all the plots for the depth 0–60 cm at 15 cm interval. The 60 cm depth criterion was considered as the wheat roots normally grow to a depth of 60 cm. To understand the effect of soil moisture variation in soil on pesticide persistence, the plots were maintained at different soil moisture amounts following a maximum allowable depletion criterion, i.e., 25%, 50% and 75% of available soil moisture. The available soil moisture in the soil is taken as the difference between root zone water storage at field capacity and permanent wilting point. The percentage depletion of available soil moisture in the effective root zone can be estimated by using the following equation (Martin, 1990):

$$Depletion\% = 100 \times \frac{1}{n}\sum_{j=1}^{n}\frac{\theta_{f_j} - \theta_j}{\theta_{f_j} - \theta_{w_j}} \quad (15.1)$$

where n is the number of layers of the effective rooting depth used for the soil moisture sampling, θ_{f_j} is the volumetric moisture content in soil at field capacity of the jth layer (cm^3/cm^3), θ_j is the volumetric moisture content in jth

TABLE 15.1 Pesticide and fertiliser application details.

Properties	Thiram[a]	2,4-D[b]
Chemical name	Bis(dimethylthiocarbamoyl) disulphide	2,4-Dichlorophenoxyacetic acid
Pesticide class	Dithiocarbamate fungicide	Chlorophenoxy herbicide
Molecular formula	$C_6H_{12}N_2S_4$	$C_8H_6Cl_2O_3$
Molecular weight	240.4	221.04
Chemical structure	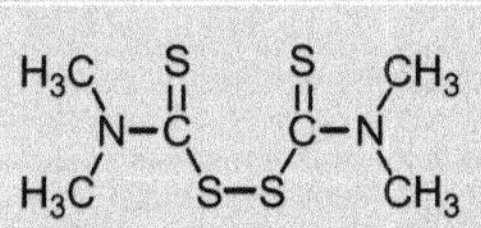 H_3C, S, S, CH_3, N–C, C–N, H_3C, S–S, CH_3	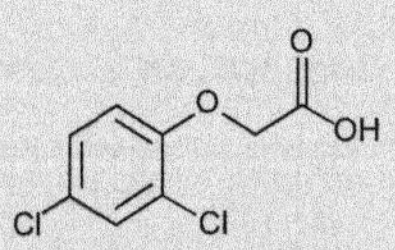 O, O, OH, Cl, Cl
Melting point	155–156°C	140.5°C
Adsorption coefficient, K_{oc}	670	67×10^{-3} cm^3/gm
Vapour pressure	2.3×10^{-3} Pa (25°C).	1×10^{-5} Pa (20°C).
Partition coefficient, Log K_{ow}	1.73	2.81 (octanol/water)
Dissociation constant, pKa	Nonionised	2.73
Solubility	In water 18 mg/L, ethanol less than 10 g/L, acetone 80 g/L, chloroform 230 g/L, hexane 0.04 g/L, dichloromethane 170 g/L toluene 18 g/L and isopropanol 0.7 g/L at 20°C	In water 890 mg/L, ethanol and acetone, 9.5 g/100 g; benzene, 1.07 g/100 g
Stability	Decomposed in acidic media, deteriorates on prolonged exposure to heat, air or moisture	

[a]*Walker and Keith, 1992.*
[b]*Walters, 2004.*

TABLE 15.2 Properties of pesticides applied in the case study.

S. No.		Application rate (kg/ha)	Application date
1.	Wheat sowing	–	2nd Dec'08
2.	Diammonium phosphate (DAP)	40.0	2nd Dec'08
3.	Potassium sulphate (potash)	08.0	2nd Dec'08
4.	Thiram	0.8	2nd Dec'08
5.	2,4-D	0.5	4th Jan'09
6.	Fertiliser, urea	40.0	10th Jan'09
7.	Fertiliser, urea	40.0	23rd Feb'09
8.	Wheat harvesting	–	16th April'09

layer before irrigation (cm^3/cm^3) and θ_{w_j} is the volumetric moisture content at wilting point of the *j*th layer (cm^3/cm^3). The θ can be obtained either through automatic sensors or the less expensive gravimetric method. This method involves gravimetrically determining the moisture content in the soil using Eq. (15.2):

$$\theta = \theta_g \frac{\rho_b}{\rho_w} \tag{15.2}$$

where ρ_w is the water density approximately 1 gm/cm^3 at normal temperatures and ρ_b is the soil bulk density of the undisturbed soil core samples (gm/cm^3) and θ_g is the gravimetric soil water content (mg/kg). The θ_f and θ_w are obtained from the field capacity and wilting point defined for the four depths using the pressure plate apparatus.

The depth of water to be applied at particular *MAD* was calculated by the following expression (Michael, 1978):

$$d_{irr} = \frac{1}{IRE} \times \sum_{j=1}^{n} \frac{MAD\left(\theta_{f_j} - \theta_{w_j}\right) \times Rz_j}{100} \tag{15.3}$$

where d_{irr} is the depth of irrigation water to be applied (cm), Rz_j is the depth of *j*th soil layer within the effective rooting depth (cm) and *IRE* is the irrigation efficiency (%).

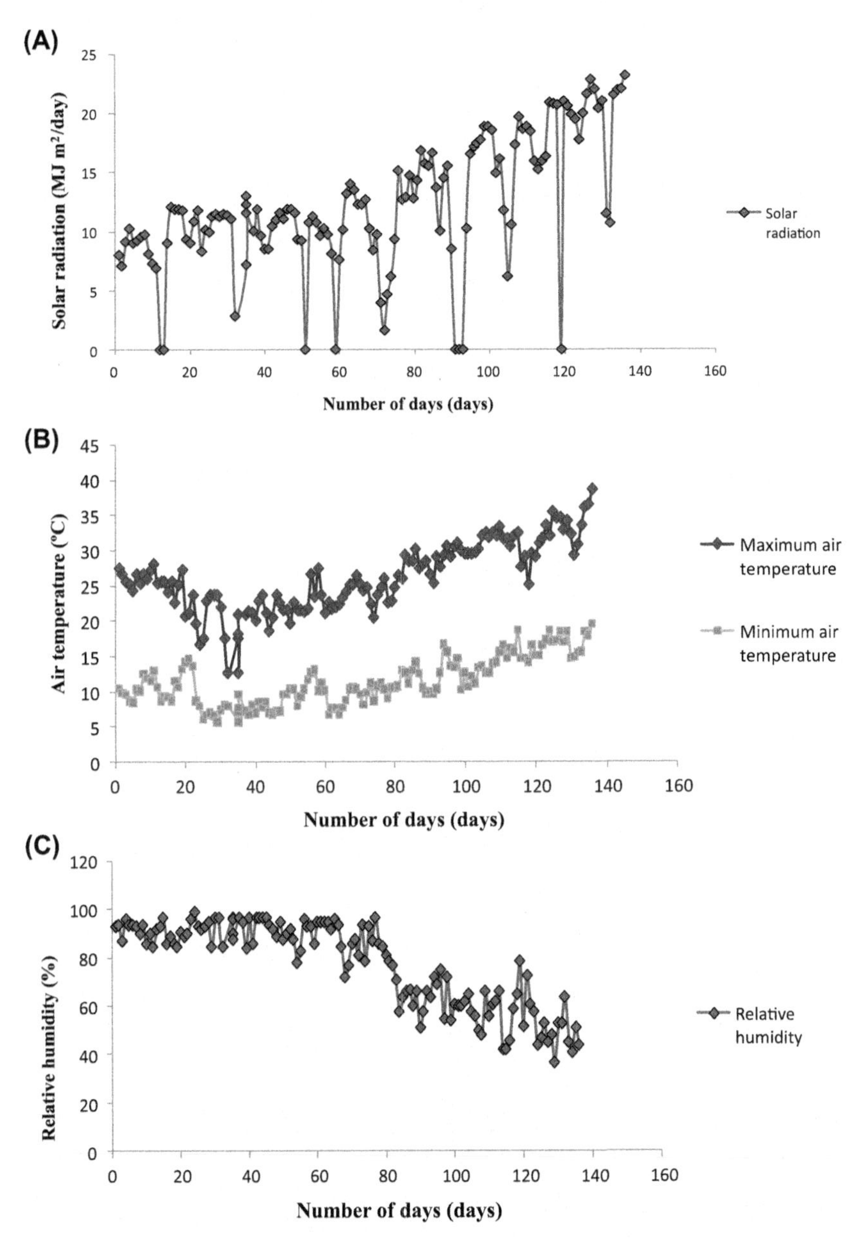

FIGURE 15.2 (A) Variation of daily average solar radiation for the study period. (B) Variation of daily maximum and minimum air temperature for the study period. (C) Variation of daily average relative humidity for the study period. (D) Variation of daily average wind speed for the study period. (E) Variation of daily rainfall for the study period.

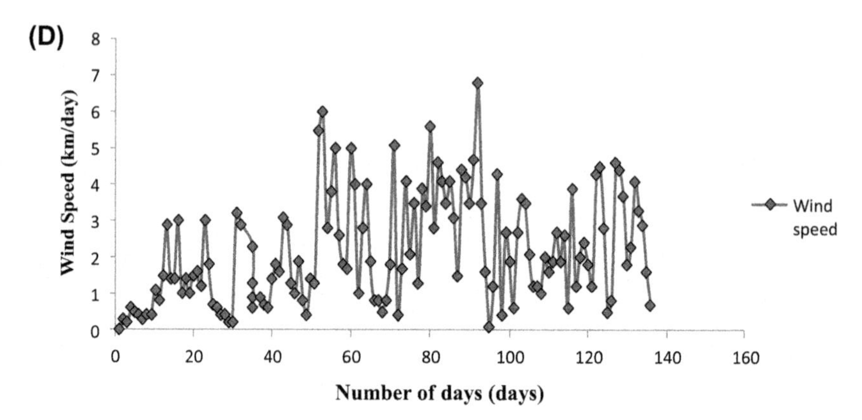

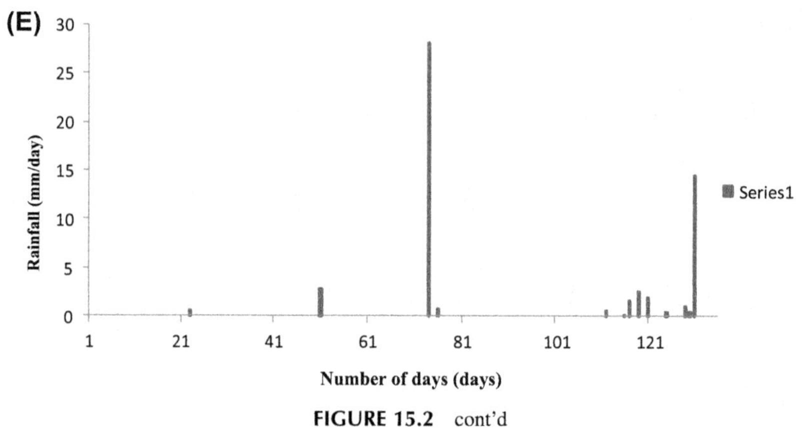

FIGURE 15.2 cont'd

ii. Laboratory extraction analysis method for pesticides

The initial concentration of pesticides was determined in the soil before sowing of wheat. After the pesticide application, the soil samples are collected based on crop growth stages. The samples in the current case study were taken for a depth of 0–60 cm at an interval of 15 cm. Standard fungicide and herbicide solutions were prepared in benzene and methanol, respectively, to obtain a stock solution of 100 ppm, which was diluted further to be used as working solutions. The wavelength scan was performed and the maxima peak was obtained at 284 and 564 nm for Thiram and 2,4-D, respectively, which was used for estimation of pesticide residue in the soil samples. The method followed for extracting Thiram and 2,4-D from the soil samples has been taken from Filipe et al. (2007) and Devi et al. (2001), which were modified for the current study with details already explained in Gupta et al. (2012) and Gupta et al. (2014). In short, Thiram extraction involved centrifugation of soil sample with acetonitrile followed by silica gel column cleanup, while 2,4 D extraction involved acetonitrile and carbon tetrachloride followed by cleanup, forming a complex with chromotropic acid. The samples were filtered through a 0.2 μm filter prior to analysis. Spectrophotometric analysis was performed and absorbance was at 284 nm for Thiram and at 564 nm for 2,4-D against a reagent blank and quantitated by comparison with external standards prepared from analytical-grade standards.

iii. Pesticide simulations

The temporal and spatial variation of the studied pesticides can be simulated by the coupling of the one-dimensional modified form of Richard's Eq. (15.4) with the one dimensional solute transport Eq. (15.5)

$$\frac{\partial \theta}{\partial t} = \frac{\partial}{\partial x}\left[K(h)\left(\frac{\partial h}{\partial x} + \cos\alpha\right)\right] - S \tag{15.4}$$

where h is the water pressure head (cm), θ is the volumetric soil water content (cm^3/cm^3), t is time (days), x is the spatial coordinate (cm) (positive upward), S is the sink term (cm^3/cm^3/day), α is the angle between the flow direction and the vertical axis (i.e., $\alpha = 0$ degree for vertical flow, 90 degrees for horizontal flow, and $0° < \alpha < 90°$ for inclined flow) and K is the unsaturated soil hydraulic conductivity function (cm/day).

$$\frac{\partial C_T}{\partial t} = \frac{\partial \theta c}{\partial t} + \frac{\partial \rho_b s}{\partial t} = \frac{\partial\left(\theta D^w \frac{\partial c}{\partial x}\right)}{\partial x} - \frac{\partial qc}{\partial x} - \mu_w \theta c - \mu_{so} \rho_b s \tag{15.5}$$

where,

$$C_T = \theta c + \rho_b s \tag{15.6}$$

$$s = K_d c \tag{15.7}$$

C_T is the total solute concentration in units of mass per volume of soil (mg/cm^3), c and s are solute concentrations in the liquid (mg/cm^3) and the solid (mg/mg) phases, respectively; q is the volumetric flux density (cm/day), μ_w, μ_{so} are first-order rate constants for solutes in the liquid and solid phases (day^{-1}), respectively; ρ_b is the soil bulk density (g/cm^3), D^w is the effective dispersion coefficient (cm^2/day) for the liquid phase, K_d is the distribution coefficient (cm^3/mg). The subscripts w and so correspond with the liquid and solid phases, respectively. Some of the solute transport parameters required for solving Eq. (15.5) can be obtained from the available literature. The parameters include the longitudinal dispersivity, partitioning coefficient and first-order decay constant or determination can be achieved for the specific field conditions like tortuosity factor and organic carbon fraction. The effective dispersion coefficient (D^w) in soil matrix for one dimensional solute transport, Eq. (15.5) has been taken as (Bear, 1972)

$$\theta D^w = D_L|q| + \theta D_w \tau_w \tag{15.8}$$

where D_w is the molecular diffusion coefficient in free water (cm^2/day), τ_w is a tortuosity factor in the liquid phase [−], $|q|$ is the absolute value of the Darcian fluid flux density (cm/day), and D_L is the longitudinal dispersivity (cm).

The tortuosity factor (τ_w) in Eq. (15.8) is evaluated using following relationship (Millington and Quirk, 1961),

$$\tau_w = \frac{\theta^{7/3}}{\theta^2} \tag{15.9}$$

The molecular diffusion coefficient in free water (D_w) in Eq. (15.8) can be calculated based on the method provided by EPA (Tucker and Nelken, 1990). The longitudinal dispersivities can be assumed within the ranges given in the literature (Perfect and Haszler, 2002; Vanderborght and Vereecken, 2007). The solid-phase concentration(s) in Eq. (15.5) is the adsorbed concentration given by Eq. (15.7). The adsorption isotherm can be described by the following equation:

$$s = \frac{K_d c^N}{1 + \eta c^N} \tag{15.10}$$

Where, N and η are the empirical coefficients. For, $N = 1$, Eq. (15.8) represents Langmuir isotherm while when $\eta = 0$, the equation represents Freundlich isotherm. However, in the shown case study linear adsorption has been assumed for all the pesticides, for which $N = 1$ and $\eta = 0$. Under linear adsorption, K_d can be determined using the empirical formula given by Eq. (15.11) (Karickhoff, 1981),

$$K_d = K_{oc} \times f_{oc} \tag{15.11}$$

where K_{oc} is partitioning coefficient (cm^3/gm), f_{oc} is soil's organic carbon fraction.

The first-order decay rate constant both in liquid (μ_w) and solid (μ_{so}) phases can be assumed within the specified ranges (Candela and Mariño, 2004).

Crank Nicholson's implicit finite difference scheme is utilised in HYDRUS-1D for the temporal approximation while the finite element scheme is utilised for the spatial variation using Galerkin method. The soil hydraulic properties can be determined either with the in-built ROSETTA module or be pedotransfer based or be laboratory based using pressure plate apparatus. The pedotransfer methods are usually locally based methods and should carefully be adopted for other regions (Garg and Gupta, 2015). In the current case study, the parameters for the four soil depths were obtained from laboratory-based soil water retention curves following the nonlinear least-square optimization approach with details mentioned in Garg and Gupta (2015). The initial soil moisture condition is taken to be the initial soil moisture measured and initial pesticide concentration profiles correspond to the pesticide concentration observed in the unsaturated zone at the start of the study at time, $t = 0$. The upper water flow boundary represents the soil-air interface, where water flux depends on the precipitation, evaporation and the moisture conditions in the soil (Simunek et al., 2005) given as

$$-K(h)\left(\frac{\partial h}{\partial z}+1\right)\bigg|_{z=0} = q_0(t) \quad \text{for } t>0 \tag{15.12}$$

where q_0 (cm/day) is the prescribed value of the soil water flux at the upper boundary. The upper boundary condition for the pesticide transport at the time when pesticide was applied is flux type given as

$$J_w c(0,t) - \theta D^w \frac{\partial c(0,t)}{\partial z} = J_{w0} c_0(t) \tag{15.13}$$

where J_w is the water flux and D^w is the effective dispersion coefficient.

Free drainage is set as the bottom boundary condition for water flow while zero concentration gradient is set for pesticide transport given by Eqs. (15.14) and (15.15), respectively.

$$\frac{\partial h}{\partial z} = 0, \quad z = -L \tag{15.14}$$

$$\frac{\partial c_i}{\partial z}\bigg|_{z \to L^+} = 0 \tag{15.15}$$

The sink term, S in Eq. (15.4), denotes the root water extraction rate, which depends on soil water availability and root distribution (Feddes et al., 1976). To determine the root water uptake under three irrigation treatments, Feddes et al. 1978, function is used, defined as

$$S(h,z) = \alpha_1(h) S_{\max}(h,z) \tag{15.16}$$

where $S_{\max}$ (m^3/m^3s) is the maximum possible root water extraction rate when soil water is not limiting, z is the soil depth (cm) and α_1 is a dimensionless water stress reduction factor as a function of pressure head h (cm). α_1 can be expressed as

$$\alpha_1(h) = \begin{cases} 0 & h \geq h_w \text{ or } h \leq h_1 \\ \dfrac{h-h_1}{h_2-h_1} & h_1 \leq h \leq h_2 \\ 1 & h_2 \leq h \leq h_3 \\ \dfrac{h_w-h}{h_w-h_3} & h_3 \leq h \leq h_w \end{cases} \tag{15.17}$$

Where h_1 is pressure head below which roots start to extract water from the soil (cm), h_2 and h_3 are pressure heads between which optimal water uptake exists (cm) and h_w is the permanent wilting point pressure head, below which root water uptake ceases (cm). The values of the parameters are taken from database provided by (Wesseling, 1991). The maximum possible root water extraction rate when soil water is not limiting, that is, the nonuniform distribution of the potential water uptake rate over a root zone, was given by

$$S_{\max} = \beta_z \times T_p \tag{15.18}$$

where β_z is the normalised water uptake depth-dependent root distribution function (cm^{-1}) and T_p is the potential transpiration rate (cm/day^1).

Potential evapotranspiration (ET_0) obtained using Penman-Monteith relationship is used to obtain both T_p in Eq. (15.18) along with potential soil evaporation rate (E_p) according to Eqs. (15.19) and (15.20), respectively.

Potential soil evaporation rate, E_p, is computed as:

$$E_p = \exp(-\alpha LAI) ET_0 \tag{15.19}$$

where LAI is the leaf area index, and α is an extinction coefficient of radiation (Feddes et al., 1978). The potential transpiration rate is then computed as

$$T_p = [1 - \exp(-\alpha LAI)] ET_0 \tag{15.20}$$

The value of α used is 0.39, as used for common crops (Feddes et al., 1978). The root distribution function (β_z) in Eq. (15.18) has been used as defined by Van Genuchten (1987),

$$\beta(z) = \left\{ \begin{array}{cc} \dfrac{5}{3L_r} & z \leq 0.2\, L_r \\ \dfrac{25}{12L_r}\left(1 - \dfrac{z}{L_r}\right) & 0.2\, L_r < z \leq L_r \\ 0 & z > L_r \end{array} \right\} \tag{15.21}$$

where L_r is the root depth (cm).

The model also assumes the root depth (L_r) to be the product of maximum rooting depth (L_m) and the root growth function ($f_r(t)$) (Simunek and Suarez, 1993).

$$L_r(t) = L_m f_r(t) \tag{15.22}$$

The root growth in HYDRUS-1D was determined utilising the Verhulst-Pearl logistic growth function, Eq. (15.23). The maximum rooting depth was assumed to be 60 cm with maturity being achieved in 60 days since sowing (Bhagat and Acharya, 1988).

$$f_r(t) = \frac{L_0}{L_0 + (L_m - L_0)e^{-rt}} \tag{15.23}$$

where L_0 is the initial rooting depth at the beginning of growth period (cm) and r is the growth rate (day^{-1}). The crop parameters included in the numerical simulation are provided in the following Table 15.3.

TABLE 15.3 Crop parameters.

Parameter	Value
Initial root growth time	5 days since sowing
Harvest time	136 days since sowing
Maximum rooting depth	60 cm (Bhagat and Acharya, 1988)
Maturity achieved	60 days since sowing (Bhagat and Acharya, 1988)

iv. Statistical evaluation

The model validation is performed to evaluate the accuracy of model simulations. The three criteria can be undertaken that are simple linear regression analysis (R^2), relative root mean square error (RMSE_R) and Nash-Sutcliffe modelling efficiency (E).

1. Simple linear regression analysis

Simple linear regression analysis is performed on the simulation results of moisture content for each depth, obtained with usage of water flow Eq. (15.4), to evaluate predictive performance of the numerical simulations conducted. Thus, coefficient of determination (R^2) and square of Pearson correlation coefficient (r), are evaluated to determine the goodness-of-fit between measured and simulated values of moisture.

$$r = \frac{n\sum M_i P_i - (\sum M_i)(\sum P_i)}{\sqrt{\left[n(\sum M_i^2) - (\sum M_i)^2\right]\left[n(\sum P_i^2) - (\sum P_i)^2\right]}} \tag{15.24}$$

where M_i is the measured soil water content, P_i is simulated soil water content corresponding to the measured value and n is the number of data points. The closer values of R^2 to 1 indicate that the regression explains better relationship between simulated and measured moisture content.

2. Relative root mean square error (RMSE_R)

RMSE_R is calculated for a comparison between measured and simulated moisture content values as well as between measured and simulated pesticide concentrations. RMSE_R is defined as:

$$RMSE_R = \sqrt{MSE} / \overline{M} \tag{15.25}$$

$$MSE = \frac{1}{n}\sum_{i=1}^{n}(M_i - P_i)^2 \tag{15.26}$$

where MSE is the mean square error, $\overline{M}$ is the measured mean value, M_i is the measured soil moisture content or the measured pesticide concentration, P_i is simulated soil moisture content or the simulated pesticide concentration corresponding to the measured value and n is the number of data points. Smaller values of $RMSE_R$ indicate that simulated values match the measured values.

3. Nash-Sutcliffe modelling efficiency (NSE)

NSE (Nash and Sutcliffe, 1970) is also used to assess the level of agreement between the simulated and measured soil moisture contents, as well as for assessing the agreement of pesticide concentrations. NSE is defined as:

$$E = 1 - \frac{\sum_{i=1}^{n}(M_i - P_i)^2}{\sum_{i=1}^{n}\left(M_i - \overline{M}\right)^2} \tag{15.27}$$

Modelling efficiencies (NSE) can range from $-\infty$ to 1 where $NSE = 1$ corresponds to a perfect match between simulated values and measured data and $NSE = 0$ indicates that the model predictions are as accurate as the mean of the measured data, whereas an efficiency less than zero ($-\infty < NSE < 0$) occurs when the model simulations are worse than the measured mean.

3. Results and discussion

The soil moisture observed and simulated for the crop period in three plots is shown in Fig. 15.3A–L. Fig. 15.3A–L depict the movement in the four considered depths (0–15, 15–30, 30–45, 45–60 cm) of the three plots 1 (Fig. 15.3A–D), 2 (Fig. 15.3E–H) and 3 (Fig. 15.3I–L), respectively. It can be noticed that the model was able to simulate the water movement with precise accuracy in all the three plots. The model simulates the soil moisture agreeably in fully irrigated as well as in deficit irrigation conditions. Accurate soil moisture movements are important as the solute or pesticide movement occurs mostly through being soluble in water. Thus, if the model is able to simulate the soil moisture precisely then only one can look at the movement and effects of a pesticide.

In the shown case study, both field experiments and numerical simulations were conducted. It can be seen in Fig. 15.4A and B that for the same irrigation treatment behaviour of the two pesticides varies. Thiram was applied along with seed sowing and it could not be traced in any of the treatments after 84 days of crop cycle. Also, this pesticide is not found to be mobile as the concentration was mostly detected and degraded in the upper soil region of

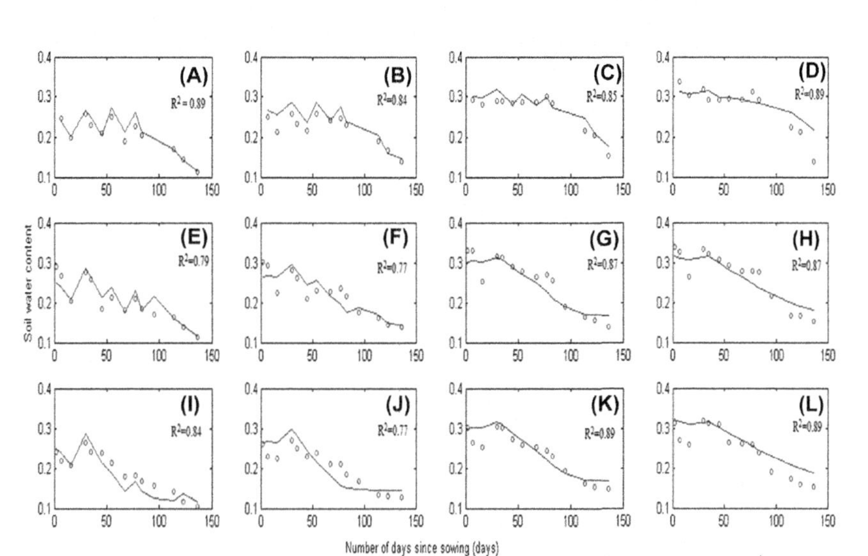

FIGURE 15.3 Soil moisture movement both observed and simulated in the three plots with varied irrigation amounts for the experiment period.

depth 0–15 cm. With 100% and 50% irrigation scenarios, the degradation of pesticide occurred a little quickly and the concentration was reduced to less than 50% within 20 days of pesticide application. In spite of higher irrigation for treatment 1, Thiram was found only in traceable limits in 15–30 cm soil depth. This behaviour of Thiram may be due to the low solubility as well as the high adsorption coefficient. Thus, concentration in 15–30 cm depth did not reach higher levels, even with two irrigations and one rainfall event. However, frequent irrigations or rainfall can lead to the slight mobility of the pesticide in the vertical profile before complete degradation occurs. The fate of this pesticide, both in experiments and numerical simulations, is seen not to be much influenced by soil moisture. Even the rate of decomposition is nearly similar for all three treatments, with a slightly slower rate in treatment 3 of 25% irrigation. So, the pesticide being a nonpolar compound is seen to be nonmobile and nonpersistent, which can be easily utilised for deficit irrigation conditions. The simulated Thiram concentrations show fine agreement with measured quantities. The efficiencies and index of agreement were found to be above 0.80 and 0.85, respectively, for the depth of 0–15 cm. However, the index of agreement was not more than 0.6 for the depth of 15–30 cm. On the other hand, irrigation had higher impact on mobility of 2,4-D. This pesticide was mobile in the 100% treatment plot with traceable quantities of pesticide reaching a depth of 30–45 cm. However, as the irrigation became deficit or in other words as the soil moisture was reduced with the other two treatments, no pesticide concentration could be detected below 15 cm. Moreover, for treatment 1 the pesticide was not detectable after 48 days of pesticide application (81 days after seed sowing) in any of the layers while in treatment 3 it

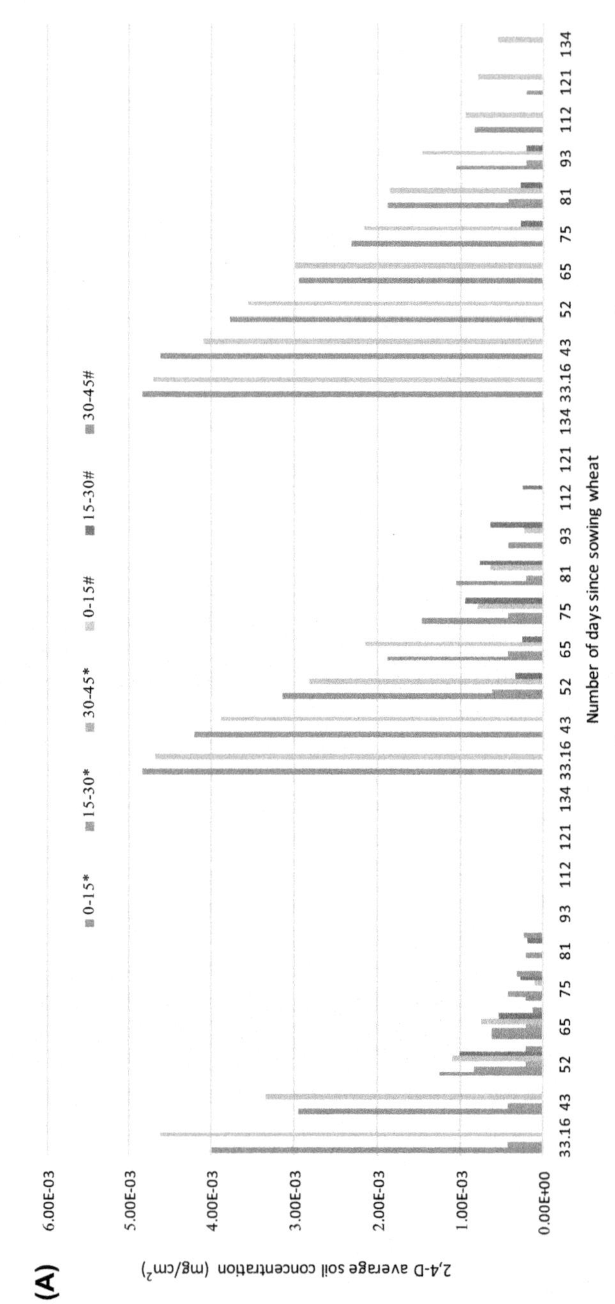

FIGURE 15.4 (A) 2,4-D average soil concentration (mg/cm^2). (B) Thiram average soil concentration (mg/cm^2): *Measured and #Simulated.

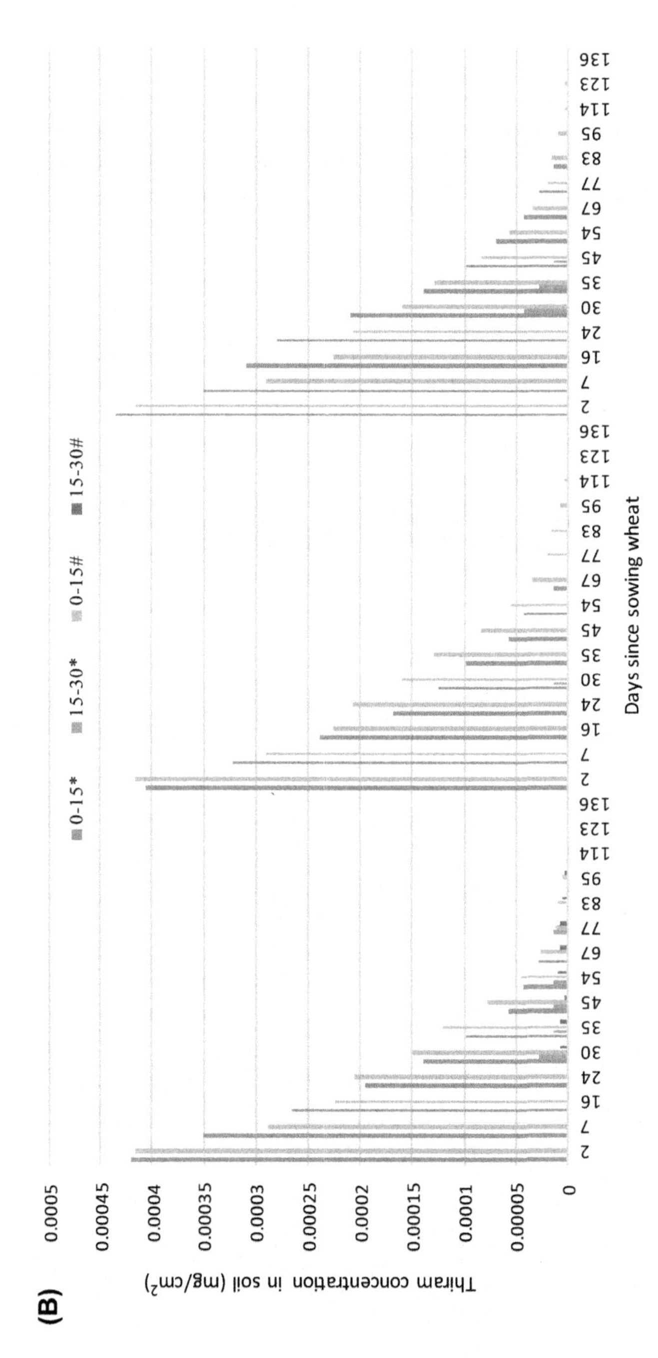

FIGURE 15.4 cont'd

remained up to 101 days after pesticide application (134 days after seed sowing), i.e., it could be traced up to the end of crop cycle under low soil moisture conditions. Thus, if proper soil moisture conditions are not maintained then this pesticide can sustain itself in soil for a long period. Also, subsequent application of the same pesticide in the soil will lead to concentration buildup in the soil. This could lead to further uptake by crop and be found in harvested food. The $RMSE_R$ was low and in the range of 0.3–1.6. Also, the efficiency was higher than 0.80. However, for treatment 1 the $RMSE_R$ values in the depth 30–45 cm were high, which could be attributed to measurement errors as concentration levels were near to detectable limits of instrument. The soil moisture plays a definite role in deciding the pathway of 2,4-D. It is seen that high soil moisture allows for fast degradation, as well as higher dispersion mobility. The travel time of concentration fronts has a reciprocal relationship with soil moisture. Significant changes in travel times can be observed for solute fronts for difference in application of water quantities. In plot 1, 239.91 mm of combined irrigation depth and rainfall lead to a faster dissipation of 2,4-D. However, 150.51 and 71.91 mm of applied water in treatment 2 and treatment 3, respectively, lead to increase in the travel time of the concentration front. Thus, the mobility of pesticide is quite restricted under the low water flux conditions (Gupta et al., 2012).

4. Conclusion

In this study, pesticides were applied to wheat plots under three different irrigation treatments and the behaviour (persistence and movement) of the pesticide concentrations in soil profile at different depths was studied both numerically using open source software-HYDRUS-1D and experimentally. The model efficiencies for the total solute concentration were found to be in the range of 80% for all the irrigation treatments. The study depicts that HYDRUS-1D can be utilised in simulation of various pesticides as has been done here for two pesticides. The Thiram concentrations were also found to be primarily limited to 15 cm depth of soil while 2,4-D was found to be a mobile pesticide in high flux moisture conditions. However, 2,4-D was restrictive under low flux conditions and stayed in the soil. Thus, the study indicated that the current dosage of pesticide applications in the agricultural fields need to be based on the irrigation conditions prevalent as to have sustainable development as agriculture is moving towards deficit irrigation. The model simulations can provide guidelines for policy makers and provide dosage recommendations based on water conditions like deficit irrigated or rain-fed fields.

Acknowledgements

The financial support for this research from University Grant Commission, Government of India, to Manika Gupta is highly acknowledged. The authors would like to thank Prof M. P. Sharma for providing facilities for conducting experiments in AHEC lab, IIT, Roorkee, and National Institute of Hydrology, Roorkee, for providing facilities for soil physical analysis.

References

Aggarwal, P., Bandyopadhyay, S., Pathak, H., Kalra, N., Chander, S., Kumar, S., 2000. Analysis of yield trends of the rice-wheat system in north-western India. Outlook Agric. 29, 259–268.

Aquino, A., Tunega, D., Haberhauer, G., Gerzabek, M., Lischka, H., 2007. Interaction of the 2, 4 dichlorophenoxyacetic acid herbicide with soil organic matter moieties: a theoretical study. Eur. J. Soil Sci. 58, 889–899.

Bear, J., 1972. Dynamics of fluids in porous media. American Elsevier.

Bhagat, R., Acharya, C., 1988. Soil water dynamics during wheat growth under different soil management practices. J. Indian Soc. Soil Sci. 36, 389–396.

Briggs, G.G., Lord, K.A., 1983. The distribution of aldicarb and its metabolites between *Lumbricus terrestris*, water and soil. Pestic. Sci. 14, 412–416.

Candela, L., Mariño, M.A., 2004. Simulation of 2, 4-D herbicide transport through the unsaturated zone using an analytical model. Int. J. Environ. Anal. Chem. 84, 123–131.

Choi, J., Fermanian, T., Wehner, D.J., Spomer, L., 1988. Effect of Temperature, Moisture, and Soil Texture on DCPA Degradation. Office of the Dean (CAFES), 13.

Chopra, I., Kumari, B., Sharma, S., 2010. Evaluation of leaching behavior of pendimethalin in sandy loam soil. Environ. Monit. Assess. 160, 123–126.

Cisar, J., Snyder, G., 1996. Mobility and persistence of pesticides applied to a USGA green. III: organophosphate recovery in clippings, thatch, soil, and percolate: pesticide and nutrient fate under turfgrass golf course conditions. Crop Sci. 36, 1433–1438.

Devi, K.M.D., Abraham, C., Chinnamma, N., 2001. Standardization of procedure for residue analysis of 2, 4-D in soil. J. Trop. Agric. 39, 175–177.

Feddes, R., Kowalik, P., Zarangy, H., 1978. Simulation of field water use and crop yield.

Feddes, R.A., Kowalik, P., Kolinskamalinka, K., Zaradny, H., 1976. Simulation of field water-uptake by plants using a soil-water dependent root extraction function. J. Hydrol. 31, 13–26.

Filipe, O., Vidal, M., Duarte, A., Santos, E., 2007. A solid-phase extraction procedure for the clean-up of thiram from aqueous solutions containing high concentrations of humic substances. Talanta 72, 1235–1238.

Forget, G., 1991. Pesticides and the third world. J. Toxicol. Environ. Health A Curr. Issues 32, 11–31.

Frank, H., Scholl, H., Sutinen, S., Norokorpi, Y., 1992. Trichloroacetic acid, a ubiquitous herbicide in finnish forest trees. Arctic Centre Publications, pp. 259–261.

Garg, N.K., Dadhich, Sushmita M., 2014. Integrated non-linear model for optimal cropping pattern and irrigation scheduling under deficit irrigation. Agric. Water Manag. 140, 1–13.

Garg, N.K., Gupta, M., 2015. Assessment of improved soil hydraulic parameters for soil water content simulation and irrigation scheduling. Irrig. Sci. 33 (4), 247–264.

Garg, N.K., Hassan, Q., 2007. Alarming scarcity of water in India. Curr. Sci 932–941.

Gupta, M., Garg, N., Joshi, H., Sharma, M., 2012. Persistence and mobility of 2, 4-D in unsaturated soil zone under winter wheat crop in sub-tropical region of India. Agric. Ecosyst. Environ. 146, 60–72.

Gupta, M., Garg, N.K., Joshi, H., Sharma, M.P., 2014. Assessing the impact of irrigation treatments on thiram residual trends: correspondence with numerical modelling and field-scale experiments. Environ. Monit. Assess 186 (3), 1639–1654.

Harman, C., Rao, P., Basu, N., McGrath, G., Kumar, P., Sivapalan, M., 2011. Climate, soil, and vegetation controls on the temporal variability of vadose zone transport. Water Resour. Res. 47, W00J13.

Joshi, A., Mishra, B., Chatrath, R., Ortiz Ferrara, G., Singh, R.P., 2007. Wheat improvement in India: present status, emerging challenges and future prospects. Euphytica 157, 431–446.

Karickhoff, S.W., 1981. Semiempirical estimation of sorption of hydrophobic pollutants on natural sediments and soils. Chemosphere 10, 833–846.

Mackay, D., Shiu, W.Y., Ma, K.C., 1997. Illustrated handbook of physical-chemical properties and environmental fate for organic chemicals: pesticide chemicals. CRC Press.

Martin, D.L., Stegman, E.C., Freres, E., 1990. Irrigation scheduling principles. In: Hoffman, G.L., Howell, T.A., Solomon, K.H. (Eds.), Management of farm irrigation systems. ASAE Monograph, pp. 155–372.

Michael, A.M., 1978. Irrigation theory and practice. Vikash Publishing House Private Limited, New Delhi, pp. 448–584. Reprint:1999.

Millington, R., Quirk, J., 1961. Permeability of porous solids. Trans. Faraday Soc. 57, 1200–1207.

Munro, I.C., Carlo, G.L., Orr, J.C., Sund, K.G., Wilson, R.M., Kennepohl, E., Lynch, B.S., Jablinske, M., 1992. A comprehensive, integrated review and evaluation of the scientific evidence relating to the safety of the herbicide 2, 4-D. Int. J. Toxicol. 11, 559.

Nash, J., Sutcliffe, J., 1970. River flow forecasting through conceptual models part I–A discussion of principles. J. Hydrol. 10, 282–290.

Perfect, E.S., Haszler, M., 2002. Prediction of dispersivity for undisturbed soil columns from water retention parameters. Soil Sci. Soc. Am. J. 66, 696.

Powell, W., Dean, G., Dewar, A., 1985. The influence of weeds on polyphagous arthropod predators in winter wheat. Crop Protect. 4, 298–312.

Shen, L., Wania, F., Lei, Y.D., Teixeira, C., Muir, D.C.G., Bidleman, T.F., 2005. Atmospheric distribution and long-range transport behavior of organochlorine pesticides in North America. Environ. Sci. Technol. 39, 409–420.

Shukla, G., Kumar, A., Bhanti, M., Joseph, P., Taneja, A., 2006. Organochlorine pesticide contamination of ground water in the city of Hyderabad. Environ. Int. 32, 244–247.

Simunek, J., Sejna, M., Van Genuchten, M.T., 2005. The HYDRUS-1D software package for simulating the one-dimensional movement of water, heat, and multiple solutes in variably-saturated media. University of California, Riverside, Research reports, p. 240.

Simunek, J., Suarez, D.L., 1993. Modeling of carbon-dioxide transport and production in soil 1. Model development. Water Resour. Res. 29, 487–497.

Sparks, D.L., 2003. Environmental Soil Chemistry. Academic Pr.

Struthers, J., Jayachandran, K., Moorman, T., 1998. Biodegradation of atrazine by *Agrobacterium radiobacter* J14a and use of this strain in bioremediation of contaminated soil. Appl. Environ. Microbiol. 64, 3368–3375.

Szolar, O., 2007. Environmental and pharmaceutical analysis of dithiocarbamates. Anal. Chim. Acta 582, 191–200.

Tucker, W.A., Nelken, L.H., 1990. Diffusion coefficients in air and water. In: Handbook of Chemical Property Estimation Methods: Environmental Behavior of Organic Compounds. American Chemical Society, Washington, DC, 17. 1–17–25.

van der Werf, H.M.G., 1996. Assessing the impact of pesticides on the environment. Agric. Ecosyst. Environ. 60, 81–96.

Van Genuchten, M.T., 1987. A numerical model for water and solute movement in and below the root zone. United States Department of Agriculture Agricultural Research Service US Salinity Laboratory.

Vanderborght, J., Vereecken, H., 2007. Review of dispersivities for transport modeling in soils. Vadose Zone J. 6, 29.

Vereecken, H., Vanderborght, J., Kasteel, R., Spiteller, M., Schäffer, A., Close, M., 2011. Do lab-derived distribution coefficient values of pesticides match distribution coefficient values determined from column and field-scale experiments? A critical analysis of relevant literature. J. Environ. Qual. 40, 879.

Walker, A., 1974. A simulation model for prediction of herbicide persistence. J. Environ. Qual. 3, 396–401.

Walker, A., 1987. Evaluation of a simulation model for prediction of herbicide movement and persistence in soil. Weed Res. 27, 143–152.

Walker, M., Keith, L.H., 1992. EPA's pesticide fact sheet database. CRC Press.

Walters, J., 2004. Environmental fate of 2, 4-dichlorophenoxyacetic acid. environmental monitoring and plant management, california department of pesticide regulation. Sacramento, CA.

Wang, Y., Merkel, B.J., Li, Y., Ye, H., Fu, S., Ihm, D., 2007. Vulnerability of groundwater in Quaternary aquifers to organic contaminants: a case study in Wuhan City, China. Environ. Geol. 53, 479–484.

Watanabe, H., Grismer, M.E., 2003. Numerical modeling of diazinon transport through inter-row vegetative filter strips. J. Environ. Manag. 69, 157–168.

Wauchope, R.D., Yeh, S., Linders, J.B.H.J., Kloskowski, R., Tanaka, K., Rubin, B., Katayama, A., Kördel, W., Gerstl, Z., Lane, M., 2002. Pesticide soil sorption parameters: theory, measurement, uses, limitations and reliability. Pest Manag. Sci. 58, 419–445.

Wesseling, J.G., Elbers, J.A., Kabat, P., van den Broek, B.J., 1991. SWATRE: Instructions for Input, Internal Note. Winand Staring Centre, Wageningen, the Netherlands.

Winton, K., Weber, J.B., 1996. A Review of Field Lysimeter Studies to Describe the Environmental Fate of Pesticides. Weed technology, pp. 202–209.

Part V

Geospatial techniques

Chapter 16

Irrigation water demand estimation in Bundelkhand region using the variable infiltration capacity model

Varsha Pandey[1], Prashant K. Srivastava[1,2], Pulakesh Das[3,4], Mukunda Dev Behera[3]

[1]*Remote Sensing Laboratory, Institute of Environment and Sustainable Development, Banaras Hindu University, Varanasi, Uttar Pradesh, India;* [2]*DST-Mahamana Center of Excellence in Climate Change Research, Institute of Environment and Sustainable Development, Banaras Hindu University, Varanasi, Uttar Pradesh, India;* [3]*Centre for Oceans, Rivers, Atmosphere and Land Sciences, Indian Institute of Technology Kharagpur, Kharagpur, West Bengal, India;* [4]*Remote Sensing and GIS Department, Vidyasagar University, Midnapore, West Bengal, India*

1. Introduction

Nowadays, the irrigation water demand is a major socioeconomic challenge in most of the developing countries. According to a study by United States (2012), around 70% of the world's annual fresh water is used for irrigation in the agricultural field. However, India uses 85% of its available fresh water resources for irrigation, thus becoming the largest net irrigated country in the world. The situation is worse in recent years due to recurrent occurrences of severe drought events. Therefore, scientific management of fresh water storage and its proper utilisation require precise and sustainable planning and approaches in the era of climate change and growing water demands. Soil Moisture Content (SMC), i.e., the water content in the root zone of the plant depends on the precipitation level, soil properties, climatic conditions, underlying geological formation and terrain slope, which are regarded as the critical parameters for irrigation water management and other agricultural practices (Seneviratne et al., 2010; Srivastava et al., 2013a,b; Srivastava, 2017). SMC is also considered as an integral component for weather phenomenon prediction (van den Hurk et al., 1997), natural hazards mitigation (Jeyaseelan, 2003; Pandey and Srivastava, 2019) and designing water balance equations (Srivastava et al., 2014a,b). The SMC in the soil profile is highly

Agricultural Water Management. https://doi.org/10.1016/B978-0-12-812362-1.00016-3

regulated by two important determinants such as precipitation and evapotranspiration (ET). Accurate assessment of these key significant fluxes made possible water management and irrigation scheduling. ET is a critical parameter responsible for water loss from the land surface and vegetation into the atmosphere due to enhanced temperature in ambient. Moreover, the rate of ET depends on various climatological variables such as rainfall, temperature, solar radiation, wind speed, humidity, as well as soil moisture availability. Though the estimation of evapotranspiration is not easy, it has the utmost importance in water inventory issues and irrigation scheduling. Additionally, for irrigation water management, one of the most useful indicators is the soil moisture deficit index (SMDI) or depletion used to represent the required amount of irrigation to maintain the sufficient water availability for effective crop production. Thus, the SMDI also determines the relative dryness and wetness of an area.

An accurate and real-time assessment of SMC and ET is one of the main research focus of the hydrologists for sustainable water resource management. SMC can be estimated through various observations, such as ground-based techniques (e.g., from probe or gravimetric measurements) (Hupet and Vanclooster, 2002; Srivastava et al., 2016); remote sensing techniques (e.g., optical and thermal infrared, active and passive microwave) (Carlson et al., 1994; Engman and Chauhan, 1995; Gupta et al., 2014) and various hydrological models such as Variable Infiltration Capacity (VIC), Noah Land Surface Model (Noah LSM) and Soil and Water Assessment Tool (SWAT) model (Srivastava et al., 2014b; Devia et al., 2015; Srivastava et al., 2015; Zhuo et al., 2015). Remote sensing technique is advantageous over field measurements by collecting data over the large areas, continuous estimation, including remote and inaccessible areas, free of cost and consume less processing time; however, there is a lack of daily data availability because of fixed sensor time intervals, lack of long-term data availability, coarse resolution soil moisture data, etc., that limit many applications including climatology of soil moisture, real-time assessment and others. On the contrary, the hydrological models have benefits through simulation of long-term hydrological variables for required temporal and spatial resolution. The macro-scale Variable Infiltration Capacity (VIC) hydrological model is extensively used by hydrologists in agricultural practices, watershed management, climate change studies, hazard monitoring and various hydrological and meteorology event assessments (Nijssen et al., 2001; Haddeland et al., 2006; Demaria et al., 2007; Luo and Wood, 2007; Wu et al., 2007; Gao et al., 2010; Wen et al., 2011).

VIC is a macroscale grid-based hydrologic model that addresses both the energy and water balance equation, originally developed by Xu Liang at the University of Washington. The model computes the vertical energy and moisture flux at grid level considering soil and vegetation information. This model has been widely tested and applied at larger scale over a long period by Lohmann et al. (1998a,b), Gao et al. (2010) and many others. Nowadays this model has been applied in various hydrological and climatic phenomena predictions and land-cover changes in the study area. Further, for detail study of the model, the reader can refer Markert (2017). The VIC model uses Penman-Monteith equation and simulates a number of climate parameters such as surface runoff, soil moisture content at various soil layers, and evapotranspiration. Devia et al. (2015) found VIC as the best model in the water management for agricultural purposes in moist areas compared to other hydrological models. Issac et al. (2017) estimated field level irrigation water requirements and trends using the VIC model.

In the current study, we aimed to use VIC-derived soil moisture content (VIC-SMC) at one-fourth degree spatial resolution for assessment of irrigation demand in terms of soil moisture deficit index (SMDI). Irrigation water management represents the use of an appropriate quantity of water at the proper time, and it is usually pursued by combining measurements of soil moisture content with an optimised irrigation plan. By irrigation scheduling, we can chart the exact time and required quantity of irrigation for viable crop production. The study is done over the Bundelkhand region of Uttar Pradesh, known as one of the groundwater scarce area.

2. Study site

This study was carried out in Bundelkhand region of Uttar Pradesh (Bundelkhand-UP), extending between latitudes 24°18′ and 26°45′ N and 78°16′ and 81°56′ E longitude (Fig. 16.1). The region's altitude ranges between 58 to 619 m above sea level (masl). The region lies in central India, surrounded by the Indo-Gangetic plain in the north and the undulating Vindhya mountain range at the northwest to the south, and covers an area of around 29,485.34 square km. The main rivers of this region are the Sindh, Betwa, Ken, Bagahin, Tons, Pahuj, Dhasan and Chambal, which constitute the part of Ganga River basin. The topography of the region is highly undulating, with rocky outcrops and boulder-strewn plains in a rugged landscape. The major soils include alluvial, medium black and mixed red and black soils (Pandey and Srivastava, 2018).

3. Materials and methods

3.1 Datasets

3.1.1 VIC input datasets

The main inputs of the VIC model are meteorological forcing data or climate data, soil parameters and vegetation parameters, grid vector data and Land-Use Land-Cover (LULC) map. The VIC model was simulated at 0.25° grid resolution, which was prepared using the ArcGIS software. For LULC mapping, the Landsat satellite data of the premonsoon season of 2014 was used.

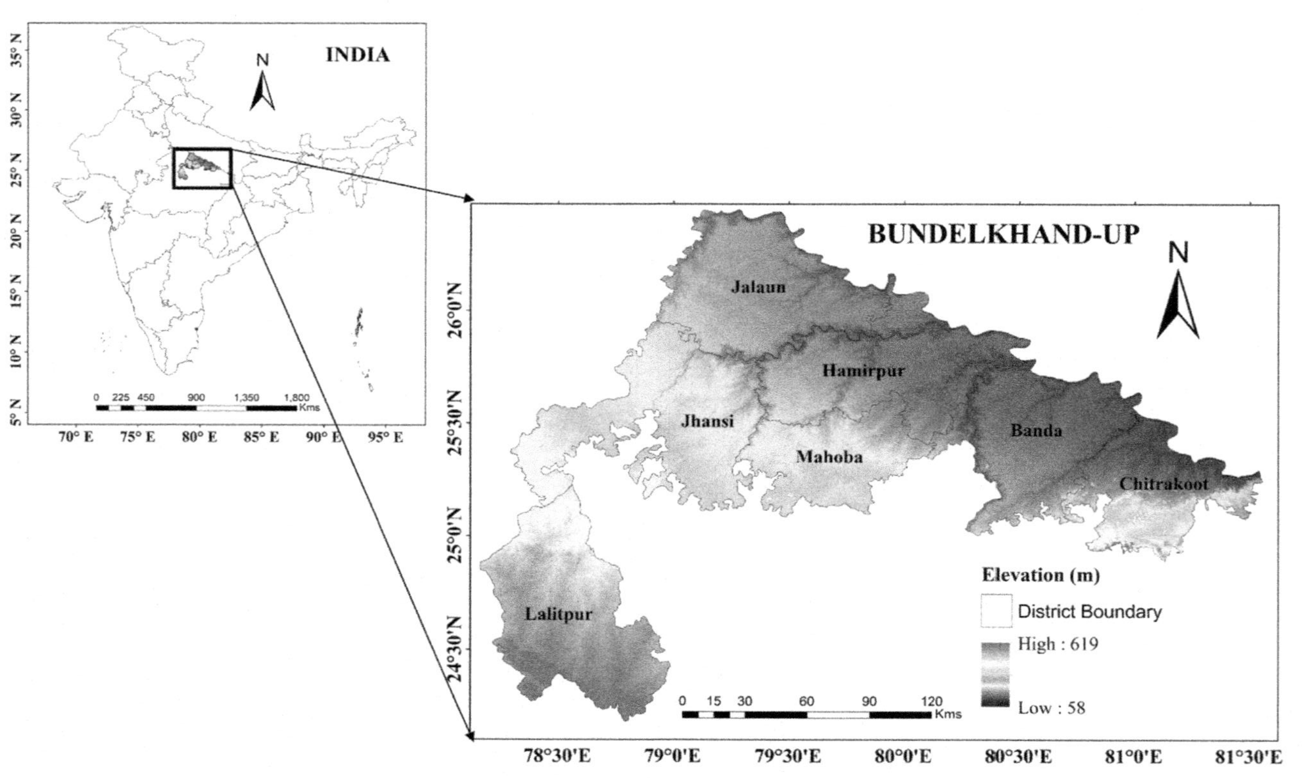

FIGURE 16.1 Study area map with digital elevation model (DEM).

The Landsat data was downloaded from the Earth explorer data portal (earthexplorer.usgs.gov/). Maximum likelihood supervised classification technique was applied for image classification, where the output LULC map was verified with high-resolution Google Earth imagery. The National Bureau of Soil Survey and Land Use Planning (NBSS and LUP) soil map was used to derive the soil type and depth information in the study area. The climate data (meteorological forcing data) such as daily precipitation, maximum and minimum temperature were accessed from the Indian Meteorological Department (IMD) at 0.25° and 1° grid scale, respectively.

3.1.2 ESA Climate Change Initiative soil moisture (CCI-SM)

The European Space Agency's (ESA) remote sensing product Climate Change Initiative soil moisture (CCI-SM) version 02.2 was used in this study for the validation of VIC-derived soil moisture content (VIC-SMC). The blended soil moisture product generated by fusing the active and passive data is used in the current study (Nicolai-Shaw et al., 2017). The CCI-SM product has a spatial resolution of 25 km and daily temporal interval available on a global scale. The CCI-SM represents the soil moisture contained in the upper few millimetres to centimetres from the soil surface.

3.2 Calibration and validation of VIC

For calibration, the soil parameters such as depths of first two soil layers (d1 and d2), infiltration curve parameter (binff), a fraction of maximum velocity of baseflow (Ds) and a fraction of maximum soil moisture where nonlinear baseflow begins (Ws) were optimised (Das et al., 2018). The obtained soil moisture was validated using ESA-derived CCI-SM for 5 years from 2010 to 14 at weekly time steps. The Coefficient of determination (R^2) and the root mean square error (RMSE) statistical measures were used to validate the VIC simulated SMC to the ESA CCI-SM, which can be expressed as:

$$r = \frac{\sum_{i=1}^{n}(C_i - \overline{C})(V_i - \overline{V})}{\sqrt{\sum_{i=1}^{n} C_i - \overline{C}} \cdot \sqrt{\sum_{i=1}^{n} V_i - \overline{V}}}$$

$$RMSE = \sqrt{\frac{1}{n}\sum_{i=1}^{n}(V_i - C_i)^2}$$

where n is the number of observations, V and C are the VIC simulated soil moisture and CCI soil moisture, respectively, and R^2 is calculated by the square of Pearson's correlation coefficient (r).

3.3 Soil moisture deficit index (SMDI)

The VIC derived daily soil moisture of the top soil layer was used to derive the weekly average for the Kharif season during the 5-year period (2010–14). The 5-year median, minimum and maximum were calculated for a particular week. Then the weekly percentage of soil moisture deficit (SDk,i) for each week *i* and year *k* is calculated using the following conditional expression:

$$SD_{k,i} = \frac{SM_{k,i} - SM_{median,i}}{SM_{msdian,i} - SM_{\min,i}} * 100 If SM_{k,i} \leq SM_{median,i}$$

$$SD_{k,i} = \frac{SM_{k,i} - SM_{median,i}}{SM_{\max,i} - SM_{median,i}} * 100 If SM_{k,i} \leq SM_{median,i}$$

where $SM_{k,i}$ is the VIC derived SM for the current week *i* and year *k* (2007–2014), and the $SM_{median,I}$, $SM_{max,i}$ and $SM_{min,i}$ denote the long-term median, maximum and minimum values of the current week *i* during the Kharif season (July–October), respectively. Then the weekly SMDI was computed for July 2010 to October 2014, as:

$$SMDI_t = 0.5 * SMDI_{i-1} + \frac{SD_i}{50}$$

where the $SMDI_{i-1}$ is the SMDI of the previous week and SD_i is the soil moisture deficit in percentage for a current week *i*. The SMDI was initialised as SMDI=SD_1/50 (Narasimhan and Srinivasan, 2005). The ranges of SMDI are given in Table 16.1.

4. Results and discussion

Six LULC categories were mapped employing the Landsat imagery for pre-monsoon season (Fig. 16.2), and the corresponding area statistics is shown as Table 16.2. The proportionate LULC area exhibits the dominance of agriculture fallow (53.25%), followed by scrubland (29.29%) and cultivated (12.37)

TABLE 16.1 Drought category according to the SMDI values.

Dynamic range	SMDI values
Wet land	0 or more
Mild dry land	−1 to −0.01
Moderate dry land	−2 to −1.01
Severe dry land	−3 to −2.01
Extreme dry land	−4 to −3.01

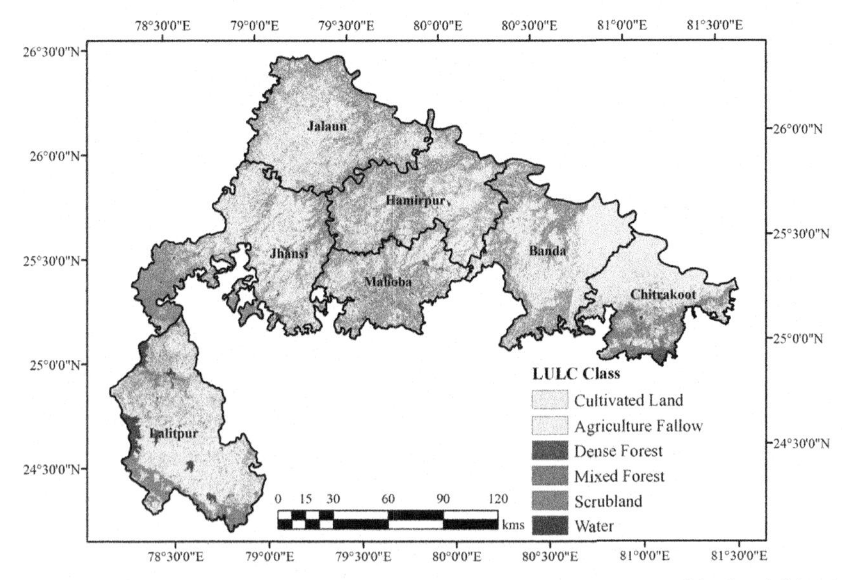

FIGURE 16.2 Land-use land-cover (LULC) map of the premonsoon season of the Bundelkhand region for the year 2014.

in the study area. Dense forests were only 0.3% of the total area, whereas, mixed forests were identified in 2.38% of the total land covers, and water body was observed in 2.41% area (Fig. 16.2). The proportionate land cover areas in different districts are shown in Table 16.3. Majority of the cultivated lands were identified in Banda (31.25%) followed by Chitrakoot (20.36%), Lalitpur (20.37%) and Jhansi (11.82%) districts (Table 16.3; Fig. 16.2). However, the

TABLE 16.2 Area statistics of LULC map of the Bundelkhand region for the year 2014.

LULC	Area	
	In km^2	In%
Cultivated land	3935.94	12.37
Agriculture fallow	16,939.36	53.25
Dense forest	96.44	0.30
Mixed forest	756.30	2.38
Scrubland	9318.03	29.29
Water	766.57	2.41

TABLE 16.3 Proportionate LULC areas across the districts.

	Cultivated land	Agriculture fallow	Dense forest	Mixed forest	Scrubland	Water
Banda	31.25	13.48	0.31	3.77	13.73	10.56
Chitrakoot	20.36	8.39	95.93	27.72	8.33	6.40
Hamirpur	4.79	16.43	0	7.46	16.13	11.55
Jalaun	7.32	18.29	0	7.74	15.60	5.23
Jhansi	11.82	16.80	0	13.28	21.27	10.71
Lalitpur	20.37	17.08	0	35.36	12.41	43.51
Mahoba	4.09	9.52	3.76	4.67	12.52	12.04

proportionate area under cultivated land was <10% in Jalaun, Hamirpur and Mahoba. The proportionate area of cultivated land and agriculture fallow clearly indicates the lack of water availability in the premonsoon season, which was the maximum in Jalaun, followed by Lalitpur, Jhansi and Hamirpur (Table 16.3).

Due to unavailability of river discharge data, the VIC simulated SMC was compared with the ESA CCI-SM to validate the model estimation. The VIC simulated daily SMC was used to generate the weekly average, which was compared with satellite-derived weekly SMC for 5 years. Statistical evaluation of VIC simulated weekly SMC shows good performance ($R^2 > 0.73$ and RMSE < 0.003) with ESA CCI-SM for all 5 years from 2010 to 2014 (Fig. 16.3). Chakravorty et al. (2016) also found better agreement between ESA CCI-SM and Modern Era Retrospective-analysis for Research and Applications-Land (MERRA-L) SMC simulation and with the IMD gridded rainfall over India and concluded that ESA CCI-SM perform well in India except in arid desert regions of western India. During 2010 to 2014, high coefficient of determinant (R^2) values 0.76, 0.90, 0.85, 0.80 and 0.73 with the corresponding low root mean square error (RMSE) values as 0.05, 0.05, 0.04, 0.05 and 0.03 m^3m^{-3} were obtained. However, VIC simulated SMC was observed slightly higher (overestimation) compared to the satellite observation (Fig. 16.3).

Fig. 16.4 shows the time-series plots of VIC simulated and ESA CCI satellite derived soil moisture content during 2010–14 on a weekly scale. Fig. 16.4 depicted that VIC simulated SMC trend shows similar pattern as CCI-SM. The VIC-SMC and CCI-SM ranges from 0.12 to 0.38 m^3m^{-3} and 0.09 to 0.32 m^3m^{-3}, respectively. The time-series shows the onset of monsoon in 2011 and 2013 was early (around 4 weeks in advance) in comparison to 2010, 2012 and 2014. However, during the months of January to March, the CCI-SM was observed to be much higher for the years 2010, 2013 and 2014 compared to the VIC-SMC. The temporal profile of soil moisture content identifies the favourable periods of cropping in the Bundelkhand region. In comparison to the rest of the period, extreme soil dryness (SMC $\approx$ 0.1 m^3m^{-3}) was mostly observed during the premonsoon season of April to June, which greatly affects the crop productivity in Kharif season.

In Fig. 16.5, the ET and SMC are shown, which depict a positive relationship, where the ET increased with an increase in SMC and vice-versa. Fig. 16.5 also shows that in the first 20 weeks (premonsoon season) for the year 2010, 2011 and 2012, the variation in ET was comparatively less as the SMC. On the contrary, in the 2013 and 2014, higher soil moisture in the premonsoon season induced higher ET, following the SMC. The seasonal variation is also very clearly seen in terms of ET and SMC. In monsoon, the ET was nearly equivalent as the SMC. However, in the postmonsoon season, the ET being higher compared to the SMC might be indicating the presence of subsurface SMC, which leads to higher ET as observed in the premonsoon season also.

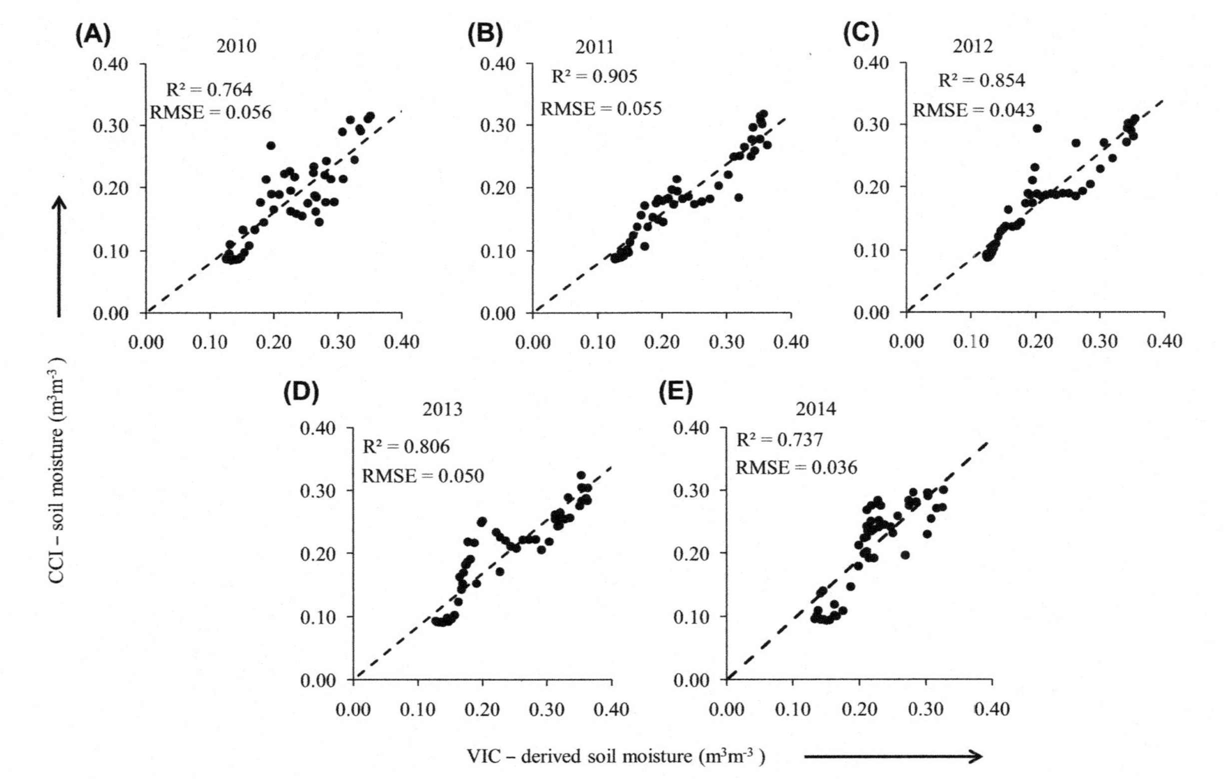

FIGURE 16.3 Scatter plots of weekly VIC-derived soil moisture content (VIC-SM) and ESA CCI satellite derived soil moisture (CCI-SM) in Bundelkhand-UP region for the period 2010–14.

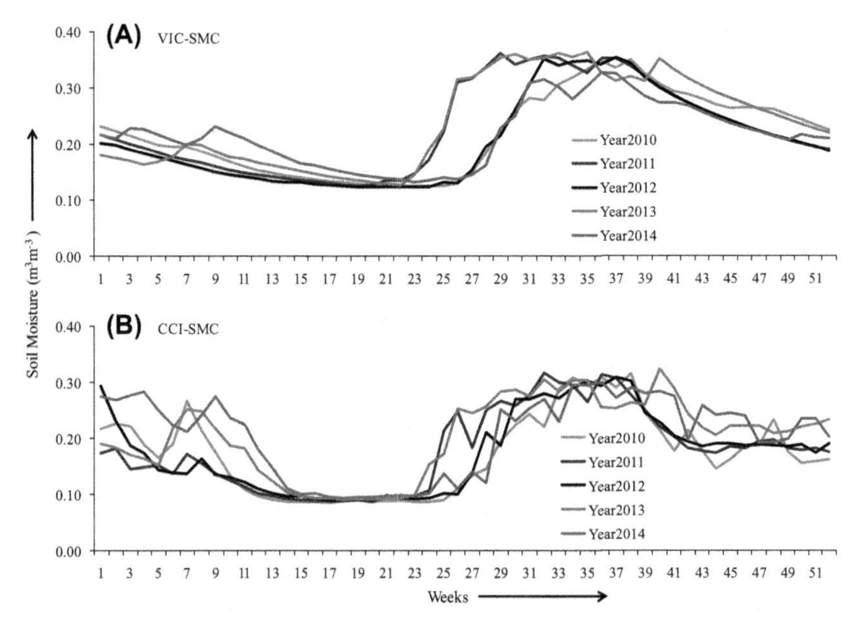

FIGURE 16.4 Time-series plot of weekly (A) VIC-simulated soil moisture content (VIC-SMC) and (B) ESA CCI satellite derived soil moisture (CCI-SM) in Bundelkhand-UP region for the period 2010–14.

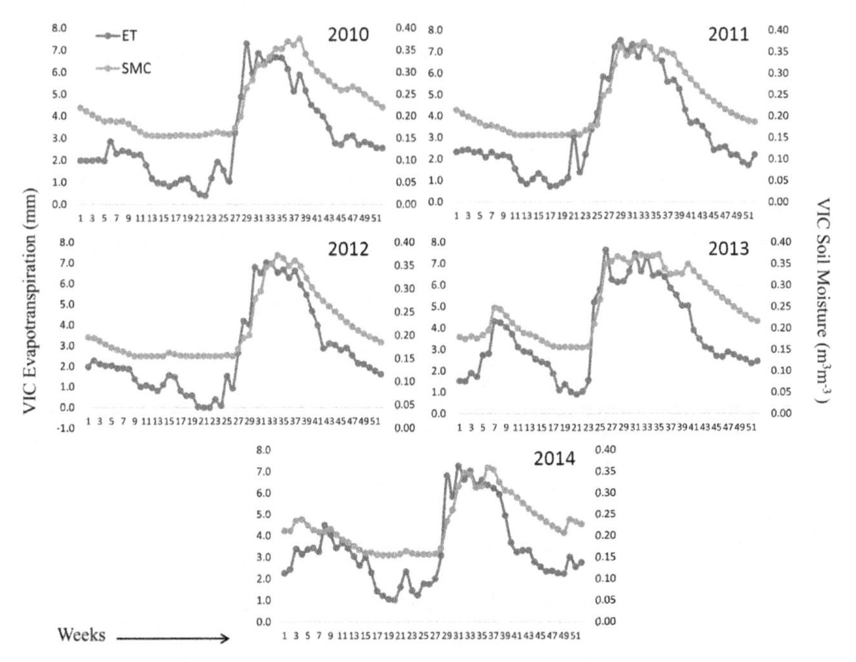

FIGURE 16.5 Variation of soil moisture content (SMC) with evapotranspiration (ET) for the period of 2010–15.

The Soil Moisture Deficit Index (SMDI), a measure of dryness, was computed to attribute the soil moisture deficit condition in the study area. For this, the VIC-derived soil moisture content was utilised for the Kharif period (July–October). The estimated SMDI value shows dry condition for the majority of the weeks in 2014; where the SMDI values ranged from 0.7 to −2.8. The zero and positive SMDI in 2014 was only estimated in 31st and 32nd week, compared to negative SMDI throughout the period. This indicates poor SMC condition whole of the year in 2014 and hence drought condition that required more water management options. The lowest SMDI in 2014 was estimated in 27th and 28th week. A similar trend was also observed in 2010 and 2012, where the lowest and negative SMDI was estimated in and around 27th and 28th week, i.e., the sowing period of Kharif crop. Moreover, in 2012, negative SMDI values (poor SMC) were estimated for the period of the 26th week to 30th week (5 weeks) followed by positive SMDI (good SMC) starting from the 31st week. The range of SMDI for the year 2012 varied between −2.2 and 2.4. A nearly similar pattern of SMC was estimated in 2010, where poor SMC was estimated during 26th to 34th week (9 weeks), followed by good SMC starting from the 35th week. The range of SMDI for the year 2010 varied between −2.4 and 2.0. On the contrary, positive SMDI values were estimated throughout the JJAS, which indicated good SMC condition in 2011 and 2013. Compared to the other years, the lowest SMDI value in 2013 was estimated as 0.2 during 37th to 39th week, which indicated less rainfall at the end of the monsoon season. Similarly, in 2011, the rainfall or the corresponding SMC was comparatively lower after the 35th week. However, the positive SMDI values in every week of 2011 and 2013 could be indicating water surplus condition in the soil. The results indicated both the natural SMC simulated by the VIC model incorporating the climate data. This result identified the dry periods and quantifies the required irrigation to retain sufficient SMC during the sowing period in 2010 and 2012. A number of governmental and private initiatives are adopted for irrigation in large areas of Bundelkhand region. Areas located near to reservoirs, canals and rivers were mostly benefitted by the irrigation facilities. In some regions, groundwater was also utilised for irrigation in dry seasons. However, the model simulation does not articulate the SMC due to irrigation, which plays a major role in regulating crop productivity in recent years. The number of poor SMC weeks was estimated maximum in 2014 followed by 2010.

The crop productivity data is accessed from the https://data.gov.in/ for comparison. Around 19 varieties of crops are cultivated in Kharif season including arhar/tur, bajra, dry chillies, groundnut, guar seed, jowar, maize, moong (green gram), rice, sannhamp, sesamum, small millets, soyabean, sugarcane, sweet potato, urad, dry ginger, cotton (lint), ginger. The total crop productivity from 2010 to 2014 was analysed with reference to the SMC in the Bundelkhand region. The total crop productivity in each year, the average of the period of 2010–2014 and the percentage deviation in productivity are tabulated in Table 16.4.

TABLE 16.4 Districtwise Kharif crop productivity and its percent deviation from 5-year average productivity.

	Total						Percentage deviation from average (in %)				
Districts	**2010**	**2011**	**2012**	**2013**	**2014**	**Average**	**2010**	**2011**	**2012**	**2013**	**2014**
Banda	122,365	165,265	172,541	176,927	138,742	155,168	−21.14	6.51	11.20	14.02	−10.59
Chitrakoot	53,312	80,532	73,714	62,802	46,296	63,331	−15.82	27.16	16.39	−0.84	−26.90
Hamirpur	149,087	191,573	182,892	267,324	250,992	208,374	−28.45	−8.06	−12.23	28.29	20.45
Jalaun	111,246	101,551	120,547	93,678	84,632	102,331	8.71	−0.76	17.80	−8.46	−17.30
Jhansi	81,627	130,627	110,427	89,588	110,333	104,520	−21.90	24.98	5.65	−14.29	5.56
Lalitpur	248,561	210,428	191,149	71,636	183,960	181,147	37.22	16.16	5.52	−60.45	1.55
Mahoba	115,880	159,479	140,091	199,393	130,029	148,974	−22.21	7.05	−5.96	33.84	−12.72
Total	882,078	1,039,455	991,361	961,348	944,984	963,845.2	−8.48	7.84	2.85	−0.26	−1.96

The overall maximum percentage deviation from the average crop productivity was negative or below the average productivity in the year 2010 (−8.48%), 2014 (−1.96) and 2013 (−0.26). In the year 2010 the maximum percentage deviation was estimated for Hamirpur (−28.45%), followed by Mahoba (−22.21%), Jhansi (−21.90%), Banda (−21.14%) and Chitrakoot (−15.82%); and positive deviation or more than average crop productivity was estimated in Lalitpur (37.22%) and Jalaun (8.71%). This could be attributed to the negative SMDI (poor SMC) in the development period of Kharif season (26th to 34th week), which reduced the crop productivity in all districts significantly, except Lalitpur. Similarly, overall lower than the average crop productivity was observed in 2014 (−1.96%), which was maximum in Chitrakoot (−26.90%) followed by Jalaun (−17.73%), Mahoba (−12.72%) and Banda (−10.59%); and positive values were observed in Hamirpur (20.45%), Jhansi (5.56%) and Lalitpur (1.55%). This also supports the negative SMDI (poor SMC) in whole Kharif period indicating reduced crop productivity. Again, below-average crop productivity was observed in 2013 (−0.26%); which could be attributed to the lower SMC condition during the peak crop production season (during 36th to 40th week) (Fig. 16.6). In 2013, the highest individual loss of crop was observed in Lalitpur (−60.45%). Moreover, negative crop productions were estimated in Jhansi (−14.29%), Jalaun (−8.46%) and Chitrakoot (−0.84%); and more than the average crop productivity was in Mahoba (33.84%), Hamirpur (28.29%) and Banda (14.02%). On the contrary, positive or more than average crop productivity was estimated in the year 2011 and 2012. However, below-average crop productivity was estimated in Hamirpur in both the years 2011 and 2012, and in Mahoba in 2012. Although there was poor SMC in 2011, the above-average crop productivity in the year could be attributed to the good SMC condition during the peak production periods (after 31st week). It can also be included that the artificial retention of SMC or irrigation management might enhance the crop productivity in this region.

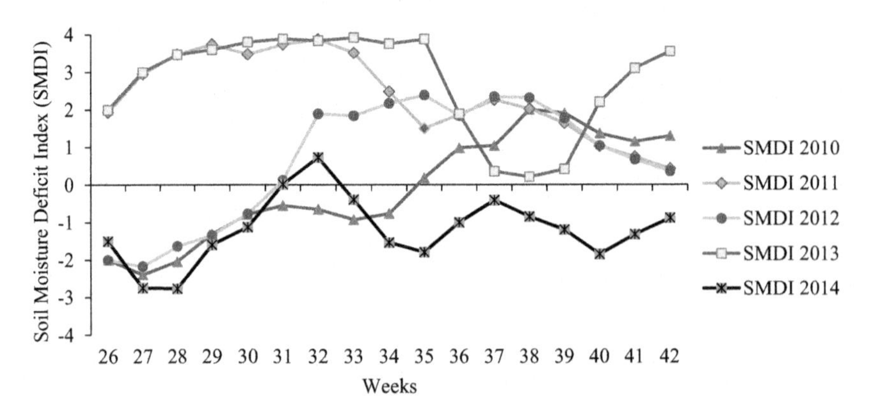

FIGURE 16.6 Soil moisture deficit index (SMDI) for Kharif season (July−October), where 26th week shows the first week of July and 42nd the last week of October.

5. Conclusions

Irrigation water demand is one of the challenging issue in the developing countries. To manage this demand precise irrigation is needed to check water wastage through irrigation sector. The Variable Infiltration Capacity Model–derived SMC is useful for adapting suitable measures to manage the irrigation water demand. Hydro-meteorological variables like ET and SMC infer the plant water use and deficit, thus recognized as a key indicators in scheduling irrigation properly. The weekly estimated soil moisture content could be useful for various applications including regular monitoring and assessment. In the current study, SMDI recognized as a robust approach for accurate irrigation scheduling at the high temporal resolution and would help in agriculture water demand management and planning. The result indicates the irrigation requirement and suggest suitable management options that are required in the Bundelkhand region.

References

Carlson, T.N., Gillies, R.R., Perry, E.M., 1994. A method to make use of thermal infrared temperature and NDVI measurements to infer surface soil water content and fractional vegetation cover. Remote Sens. Rev. 9, 161–173.

Chakravorty, A., Chahar, B.R., Sharma, O.P., Dhanya, C., 2016. A regional scale performance evaluation of SMOS and ESA-CCI soil moisture products over India with simulated soil moisture from MERRA-Land. Remote Sens. Environ. 186, 514–527.

Das, P., Behera, M.D., Patidar, N., Sahoo, B., Tripathi, P., Behera, P.R., Srivastava, S., Roy, P.S., Thakur, P., Agrawal, S., 2018. Impact of LULC change on the runoff, base flow and evapotranspiration dynamics in eastern Indian river basins during 1985–2005 using variable infiltration capacity approach. J. Earth Syst. Sci. 127, 19.

Demaria, E.M., Nijssen, B., Wagener, T., 2007. Monte Carlo sensitivity analysis of land surface parameters using the variable infiltration capacity model. J. Geophys. Res. Atmos. 112.

Devia, G.K., Ganasri, B., Dwarakish, G., 2015. A review on hydrological models. Aquat. Procedia 4, 1001–1007.

Engman, E.T., Chauhan, N., 1995. Status of microwave soil moisture measurements with remote sensing. Remote Sens. Environ. 51, 189–198.

Gao, H., Tang, Q., Shi, X., Zhu, C., Bohn, T., Su, F., Pan, M., Sheffield, J., Lettenmaier, D., Wood, E., 2010. Water Budget Record from Variable Infiltration Capacity (VIC) Model.

Gupta, M., Srivastava, P.K., Islam, T., Ishak, A.M.B., 2014. Evaluation of TRMM rainfall for soil moisture prediction in a subtropical climate. Environ. Earth Sci. 71, 4421–4431.

Haddeland, I., Lettenmaier, D.P., Skaugen, T., 2006. Effects of irrigation on the water and energy balances of the Colorado and Mekong river basins. J. Hydrol. 324, 210–223.

Hupet, F., Vanclooster, M., 2002. Intraseasonal dynamics of soil moisture variability within a small agricultural maize cropped field. J. Hydrol. 261, 86–101.

Issac, A.M., Raju, P., Joshi, S., Rao, V., 2017. Decadal trends in field level irrigation water requirement estimated by simulation of soil moisture deficit. Proc. Natl. Acad. Sci. India Sect. A Phys. Sci. 87, 901–910.

Jeyaseelan, A., 2003. Droughts & Floods Assessment and Monitoring Using Remote Sensing and GIS. Satellite Remote Sensing and GIS Applications in Agricultural Meteorology. World Meteorol. Org., Dehra Dun, India. Geneva, Switz.

Lohmann, D., Raschke, E., Nijssen, B., Lettenmaier, D., 1998a. Regional scale hydrology: I. Formulation of the VIC-2L model coupled to a routing model. Hydrol. Sci. J. 43, 131–141.
Lohmann, D., Raschke, E., Nijssen, B., Lettenmaier, D., 1998b. Regional scale hydrology: II. Application of the VIC-2L model to the Weser River, Germany. Hydrol. Sci. J. 43, 143–158.
Luo, L., Wood, E.F., 2007. Monitoring and predicting the 2007 US drought. Geophys. Res. Lett. 34.
Markert, K., 2017. VIC Model Overview.
Narasimhan, B., Srinivasan, R., 2005. Development and evaluation of soil moisture deficit index (SMDI) and evapotranspiration deficit index (ETDI) for agricultural drought monitoring. Agric. For. Meteorol. 133, 69–88.
Nicolai-Shaw, N., Zscheischler, J., Hirschi, M., Gudmundsson, L., Seneviratne, S.I., 2017. A drought event composite analysis using satellite remote-sensing based soil moisture. Remote Sens. Environ. 203, 216–225.
Nijssen, B., O'donnell, G.M., Hamlet, A.F., Lettenmaier, D.P., 2001. Hydrologic sensitivity of global rivers to climate change. Clim. Change 50, 143–175.
Pandey, V., Srivastava, K., 2018. Integration OF satellite, global reanalysis data and macroscale hydrological model for drought assessment IN SUB-tropical region OF India. Int. Arch. Photogram. Remote Sens. Spat. Inf. Sci. 42, 3.
Pandey, V., Srivastava, P.K., 2019. Integration of microwave and optical/infrared derived datasets for a drought hazard inventory in a sub-tropical region of India. Remote Sens. 11, 439.
Seneviratne, S.I., Corti, T., Davin, E.L., Hirschi, M., Jaeger, E.B., Lehner, I., Orlowsky, B., Teuling, A.J., 2010. Investigating soil moisture–climate interactions in a changing climate: a review. Earth Sci. Rev. 99, 125–161.
Srivastava, P.K., 2017. Satellite soil moisture: review of theory and applications in water resources. Water Resour. Manag. 31, 3161–3176.
Srivastava, P.K., Han, D., Ramirez, M.A.R., Islam, T., 2013a. Appraisal of SMOS soil moisture at a catchment scale in a temperate maritime climate. J. Hydrol. 498, 292–304.
Srivastava, P.K., Han, D., Ramirez, M.R., Islam, T., 2013b. Machine learning techniques for downscaling SMOS satellite soil moisture using MODIS land surface temperature for hydrological application. Water Resour. Manag. 27, 3127–3144.
Srivastava, P.K., Han, D., Rico-Ramirez, M.A., O'Neill, P., Islam, T., Gupta, M., 2014a. Assessment of SMOS soil moisture retrieval parameters using tau–omega algorithms for soil moisture deficit estimation. J. Hydrol. 519, 574–587.
Srivastava, P.K., O'Neill, P., Cosh, M., Kurum, M., Lang, R., Joseph, A., 2014b. Evaluation of dielectric mixing models for passive microwave soil moisture retrieval using data from ComRAD ground-based SMAP simulator. IEEE J. Sel. Top. Appl. Earth Obs. Remote Sens. 8, 4345–4354.
Srivastava, P.K., Han, D., Rico-Ramirez, M.A., O'Neill, P., Islam, T., Gupta, M., Dai, Q., 2015. Performance evaluation of WRF-Noah Land surface model estimated soil moisture for hydrological application: synergistic evaluation using SMOS retrieved soil moisture. J. Hydrol. 529, 200–212.
Srivastava, P., Pandey, V., Suman, S., Gupta, M., Islam, T., 2016. Available data sets and satellites for terrestrial soil moisture estimation. In: Satellite Soil Moisture Retrieval. Elsevier, pp. 29–44.
van den Hurk, B.J., Bastiaanssen, W.G., Pelgrum, H., van Meijgaard, E., 1997. A new methodology for assimilation of initial soil moisture fields in weather prediction models using Meteosat and NOAA data. J. Appl. Meteorol. 36, 1271–1283.

Wen, L., Lin, C.A., Wu, Z., Lu, G., Pomeroy, J., Zhu, Y., 2011. Reconstructing sixty year (1950-2009) daily soil moisture over the Canadian Prairies using the variable infiltration capacity model. Can. Water Resour. J. 36, 83–102.

Wu, Z., Lu, G., Wen, L., Lin, C.A., Zhang, J., Yang, Y., 2007. Thirty-five year (1971–2005) simulation of daily soil moisture using the variable infiltration capacity model over China. Atmos. Ocean 45, 37–45.

Zhuo, L., Han, D., Dai, Q., Islam, T., Srivastava, P.K., 2015. Appraisal of NLDAS-2 multi-model simulated soil moistures for hydrological modelling. Water Resour. Manag. 29, 3503–3517.

Chapter 17

Drainage network analysis to understand the morphotectonic significance in upper Tuirial watershed, Aizawl, Mizoram

Binoy Kumar Barman[1], Gautam Raj Bawri[1], K. Srinivasa Rao[1], Sudhir Kumar Singh[2], Dhruvesh Patel[3]

[1]*Department of Geology, Mizoram University, Aizawl, Mizoram, India;* [2]*K. Banerjee Centre of Atmospheric and Ocean Studies, IIDS, Nehru Science Centre, University of Allahabad, Prayagraj, Uttar Pradesh, India;* [3]*Department of Civil Engineering, School of Technology, PDPU, Raisan, Gandhinagar, Gujarat, India*

1. Introduction

Morphometric analysis mainly deals with the study of geomorphology that may be applied to a particular kind of landform or to drainage basin (Yadav et al., 2020). Drainage morphometry and morphotectonic parameters are generally studied to understand the drainage characteristics and also it reveals the prevailing geology, geomorphology and structure of the watershed (Yadav et al., 2020).

Morphometric analysis is a significant tool for prioritisation of sub-watersheds even without considering the soil map (Biswas et al., 1999). Morphometry is the measurement mathematical analysis of the configuration of Earth's surface shape and a dimension of its landform (Clarke, 1996). Morphometric analysis gives a quantitative description of drainage basin which is very useful in studies such as hydrologic modelling, watershed prioritisation, natural resources conservation and management (Kumar et al., 2018a,b; Yadav et al., 2020).

Tectonic geomorphology plays an important role in the field of Earth science, which deals with the various landforms formed by the tectonic activities of the Earth surface and is also helpful for understanding the landscape evolution. Mainly three types of morphometric parameters are classified, such

Agricultural Water Management. https://doi.org/10.1016/B978-0-12-812362-1.00017-5

as linear, aerial and relief aspects (Nag and Chakraborty, 2003; Patel et al., 2012). Stream order, stream length, mean stream length, stream length ratio and bifurcation ratio are linear aspects; relief ratio, basin length and total relief are the relief aspects; and the aerial aspects such as drainage density, texture ratio, stream frequency, form factor, circularity ratio, elongation ratio and length of overland flow (Patel et al., 2015). Morphotectonic parameters which are most commonly used for understanding the landscape evolution are elongation ratio, asymmetric factor, transverse topographic asymmetry, channel sinuosity and hypsometric integral.

Geographical information system (GIS) is a platform that provides advanced tools to obtain hypsometric information and to estimate the other associated parameters of landforms and do the spatial analysis (Kumar et al., 2018a,b; Rawat et al., 2018a; Singh et al., 2013). The satellite data have wide applicability in the fields of watershed (Kumar et al., 2018a,b), erosion (Rawat and Singh 2018b), groundwater (Singh et al., 2010), morphometric analysis (Yadav et al., 2014, 2016; Choudhari et al., 2018), landscape fragmentation (Kumar et al., 2018a,b; Singh et al., 2017, 2018; Shimrah et al., 2019), soil conservation (Rawat and Singh, 2017) and runoff estimation (Rawat and Singh, 2017).

In the present study, morphometric analysis and prioritisation of sub-watershed are carried out for Tuirial River of Mizoram, India. The parameters include bifurcation ratio, drainage density, stream frequency, stream length ratio, circulatory ratio, elongation ratio, etc. Prioritisation rating of five sub-watersheds of Tuirial River is carried out through ranking the computed morphological parameters. The sub-watershed with lowest ranking is given the highest priority in terms of soil erosion and conservation measure as well. The objectives of the current study are (i) to extract the morphometric parameters (bifurcation ratio, stream length, density of streams, number of streams, perimeter of study area, slope and elevation difference of drainage basin) from digital elevation model using GIS technique and (ii) to prioritise sub-watersheds primarily for soil erosion potentialities and conservation purposes.

2. Study area

The study area (Fig. 17.1) is situated at the southeastern part of the Tuirial river basin. The study area lies between 92° 44′ 40″–92° 51′ 40″ East longitude and 23°26′10″–23° 35′ 00″ North latitude in the Aizawl districts of Mizoram state in Northeast India. The topography varies significantly; elevation ranges from 210 m (River mouth) to 1619 m (River source) above mean sea level. The present work is covering an area of about 102.66 sq. km with an elongated shape of the watershed which falls in the parts of a survey of India toposheets number 84A/10, 84A/14 and 84A/15 at 1:50,000 scale. The maximum length of the study area is 16.25 km and the maximum width of the area is 5.96 km.

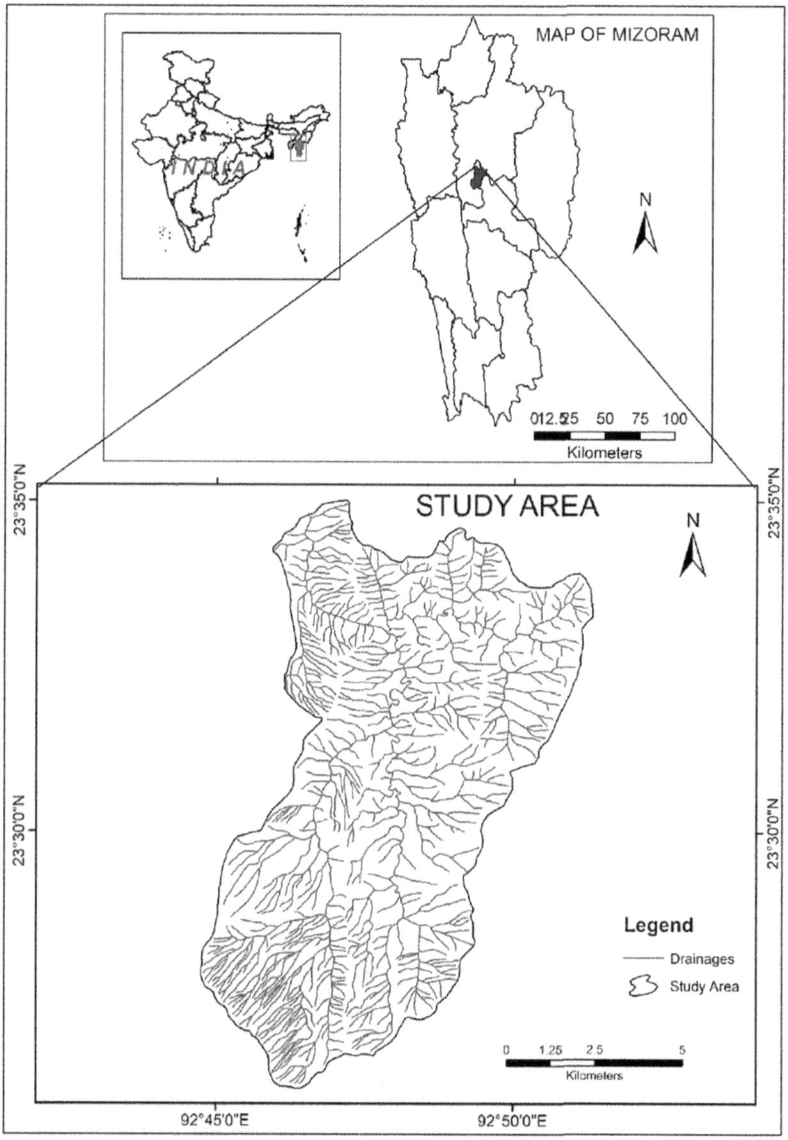

FIGURE 17.1 Location map of the study area.

3. Geology and geomorphology

Mizoram is entirely covered by a huge succession of Neogene which constitutes a part of the Assam-Arakan basin. Geomorphologically, the region is considered as the southern extension of Surma valley (Evans, 1964). The entire sedimentary column of the area is a repetitive succession of Paleogene and Neogene arenaceous and argillaceous rocks comprising of sandstone, siltstone, shale, silty-shale and their admixture in varying proportions along with few pockets of shell-limestone, calcareous sandstone and intraformational conglomerates. The tertiary rocks have been divided mainly into Barail, Surma and Tipam groups in ascending order of their ages, respectively (Sarkar and Nandy, 1976; Nandy et al., 1983; Karunakaran, 1974). The uppermost part of Tuirial watershed is the part of Tuirial basin which extends mainly in Aizawl and Serchhip district. The Tuirial River flows through the Tertiary rocks of Bhuban formation that belongs to the Surma group. The study area is mainly controlled by the folding and faulting activities and also occupied by a series of ridge and deep valleys. The ridges are elongated and narrow crested and valleys are mainly occupied by streams. The slope of the study area is gentle to moderate and some parts it showing very steep slope. Drainage pattern observed in the study sites are mostly dendritic type suggesting homogenous lithology and the southern part of the basin shows parallel drainage pattern which is the indication of tectonic tilting (Fig. 17.2).

4. Materials and Methods

Nowadays the usage of Remote Sensing and Geographical Information System (GIS) has become a standard technique (Table 17.1) to analyse the morphometric parameters and morphotectonic characteristics of a watershed. The following methodology has been adopted for attaining the objectives of the present study. Three types of datasets have been used in the present study (a) Survey of India (SOI) toposheets No. 84A/10, 84A/14 and 84A/15 on scale 1:50,000 and (b) SRTM Digital Elevation Model (DEM) data of 30 m resolution; IIRS LISS III imageand ArcGIS software. Fig. 17.3 expressed the applied methodology during the study.

The stream network and watershed boundaries were digitised from the SOI toposheets, after updating the same using satellite data (IRS LISS III) on 1:50,000 scale. The various morphometric and morphotectonic parameters of the watershed were computed as per the laws of Horton (1932, 1945), Strahler (1957, 1964), Schumm (1956), Bull and Mc Fadden (1977) and Cox (1994). The stream ordering was carried out by using the Strahler (1964) method. All the data are geometrically rectified and projected into Universal Transverse Mercator (UTM), WGS 1984 Datum; 46N zone projected system so that the error can be minimised in GIS platform.

5. Estimation of the hypsometric curve (HC) and hypsometric integral (HI)

The hypsometric analysis develops a relationship between the horizontal cross-sectional area of the river basin and its elevation in a dimensionless form

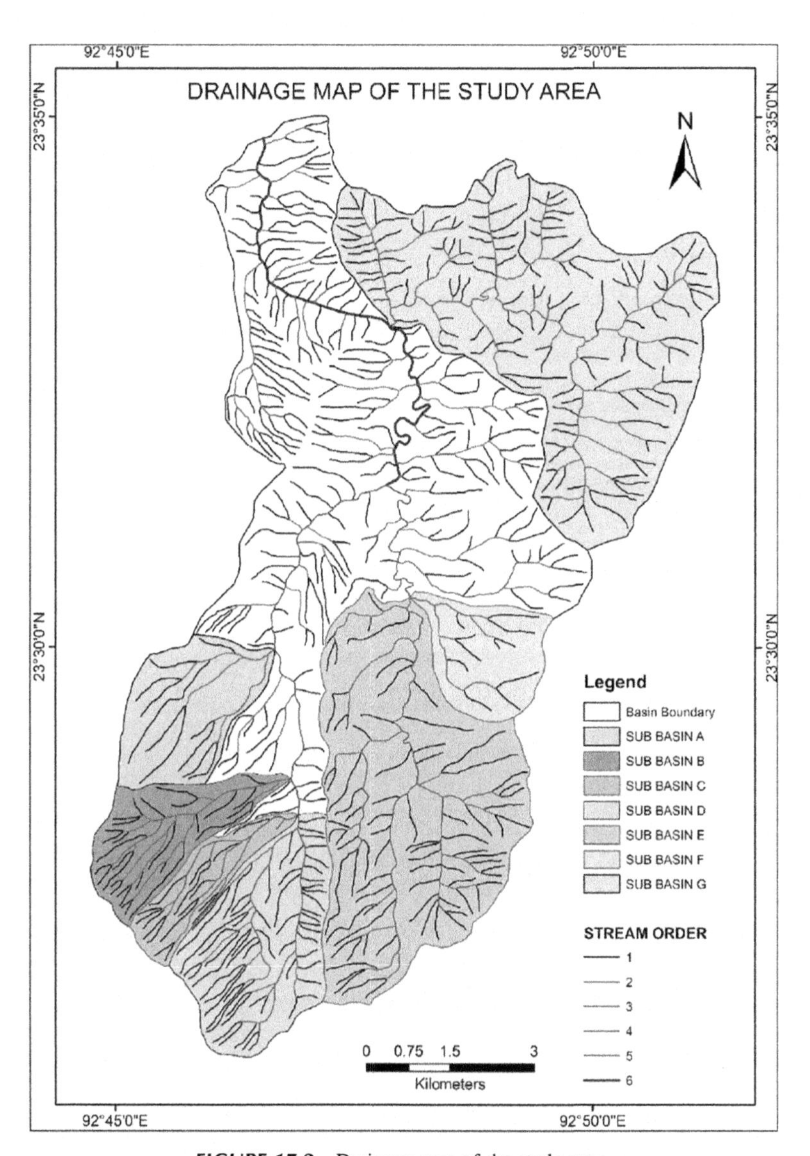

FIGURE 17.2 Drainage map of the study area.

(Singh and Singh, 2018; Yadav et al., 2020). The hypsometric curve is obtained by plotting the relative area (a/A) along the abscissa and relative elevation (h/H) along the ordinate. The hypsometric integral (HI) is obtained from the hypsometric curve (HC) and is equivalent to the ratio of the area. It is expressed in percentage units and is obtained from the percentage hypsometric curve by measuring the area under the curve. Strahler (1952) classified a

TABLE 17.1 Methods used for computing the morphometric and morphotectonic parameters.

Sl. No.	Morphometric parameter	Formula	References
Linear aspects			
1	Stream order (U)	Hierarchical rank	Strahler (1964)
2	Stream length (L_u)	Length of the stream	Horton (1945)
3	Mean stream length (L_{sm})	$L_{sm} = L_u/N_u$, where, L_u = Mean stream length of a given order (km), N_u = Number of stream segment.	Horton (1945)
4	Stream length ratio (R_l)	$R_l = L_u/L_{u-1}$, where, L_u = total stream length of order (u), L_{u-1} = The total stream length of its next lower order	Horton (1945)
5	Bifurcation ratio (R_b)	R_b=N_u/N_{u+1}, where, N_u = Number of stream segments present in the given order N_{u+1} = Number of segments of the next higher order	Schumn (1956)
6	Rho coefficient (ρ)	$P = R_l/R_b$, where, R_l = Stream length ratio, R_b = bifurcation ratio	Horton (1945)
Relief aspects			
7	Basin relief (R)	R=(H−h), where H= Highest elevation, h= Lowest elevation	Schumm (1956)
8	Relief ratio (R_r)	$R_r = R/L_b$, where, B_h=Basin relief, L_b = Basin length	Schumm (1956)
9	Relative relief (R_{hp})	R_{hp}= R/A, where R= Basin Relief, A= Basin Area	Smith (1935)
10	Dissection index (D_i)	Di=R_{hp}/R_a, where R_{hp}= Relative Relief, R_a= Absolute Relief	Dove Nir (1957); Singh and Dubey (1994)
11	Gradient ratio (G_r)	G_r= (a−b)/L_s, where, a= is the elevation at the source of the river, b= is the elevation at mouth of the river and L= is the length of main stream.	Sreedevi et al. (2005)
12	Ruggedness number (R_n)	$R_n = R \times D_d$, where, R = basin relief, D_d = Drainage density	Schumn (1956)

TABLE 17.1 Methods used for computing the morphometric and morphotectonic parameters.—cont'd

Sl. No.	Morphometric parameter	Formula	References
Aerial aspects			
13	Drainage density (D_d)	$D_d = L_u/A$, where, L_u = Total length of stream, A = area of basin	Horton (1945)
14	Stream frequency (F_s)	$F_s = N_u/A$, where, N_u = Total number of stream, A = Area of basin	Horton (1945)
15	Length of overland flow (L_o)	$L_o = D_d/2$, where, D_d = drainage density	Horton (1945)
16	Constant channel maintenance (C_m)	$C_m = 1/D_d$, where, D_d = drainage density	Schumn (1956)
17	Form factor (F_f)	$F_f = A/L_b^2$, where, A = Area of basin, L_b = Basin length	Horton (1932)
18	Circulatory ratio (R_c)	$R_c = 4\pi A/P^2$ where, A = area of basin; $\pi = 3.14$; P= perimeter of basin (Miller, 1953)	

watershed based on the hypsometric curve and HI as a youthful, mature and old stage. The threshold limits of HI as recommended by Strahler are (i) the watershed is at in-equilibrium (youthful) stage if the $HI \geq 0.6$, (ii) the watershed is at equilibrium stage if $0.35 \leq HI < 0.6$, and (iii) the watershed is at monadnock stage if $HI < 0.35$ helps in deciding the stage of the watershed.

6. Results and discussion

The different morphometric and morphotectonic parameters of the study area were calculated and summarised as below.

6.1 Analysis of morphometric parameters

The morphometric parameters of the upper Tuirial watershed have been classified into three categories namely linear aspects (Table 17.1), relief aspects and aerial aspects in the present study.

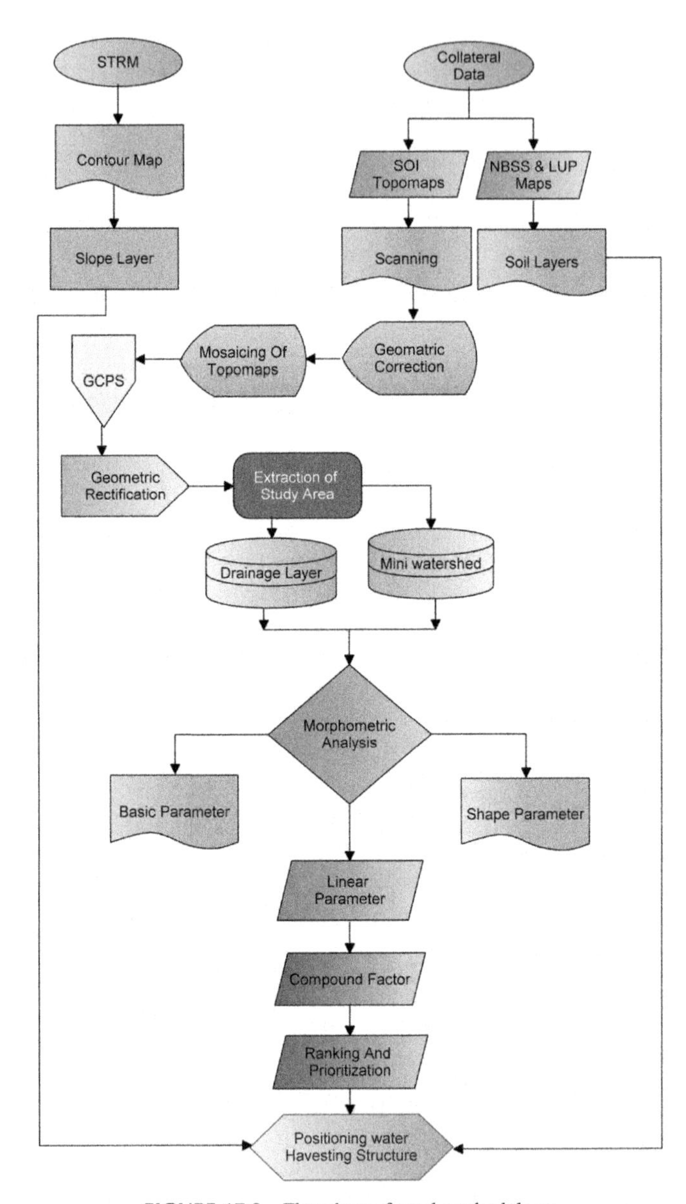

FIGURE 17.3 Flowchart of used methodology.

6.2 Linear parameters

Linear parameters evaluated for the present study include stream number, stream order (N_u), stream length (L_u), mean stream length (L_{sm}), stream length ratio (R_L) bifurcation ratio (R_b) and rho coefficient were determined and results have been presented (Tables 17.2 and 17.3).

TABLE 17.2 Computed linear morphometric aspects of upper Tuirial watershed.

Stream order	Number of streams (N_u)	Bifurcation ratio (R_b)	Mean bifurcation ratio	Total length of stream L_u (km)
1st	554	4.32	3.69	319.47
2nd	128	4.92		79.89
3rd	26	3.71		36.91
4th	7	3.5		14.15
5th	2	2		10.49
6th	1	–		9.75

TABLE 17.3 Linear morphometric parameters of the upper Tuirial watershed.

Sl. No	Linear parameters	Value
1	Stream number	718
2	Mean stream length (L_{sm})	0.65
3	Mean bifurcation ratio (R_b)	3.69
4	Rho coefficient (ρ)	0.14

6.2.1 Stream number

The number of stream segments in each order is known as stream number. Horton (1945) states that the number of stream segments of each order form an inverse geometric sequence with order number. The number of streams decreases as the stream order increases. Horton's law (1945) of stream numbers states that the number of stream segments of each order form an inverse geometric relationship with order number and it shows a straight line when plotted against order.

6.2.2 Stream order (N_u)

The first step in drainage basin analysis is the order designation where different streams are assigned with numeric values so that an universal counting pattern can be followed for any drainage basin called as stream ordering system. According to Strahler (1964) the smallest fingertips tributaries are designed as the first-order stream. Where two first-order stream channels join together then second-order stream is formed; where two second-order stream channels join together then the third-order stream is formed. In this way streams of two same lower order join together and they form a next higher order stream. The watershed comprises 554 first-order streams, 128 second-order streams, 26 third-order streams, 7 numbers of fourth-order streams, 3 numbers of fifth order and 1 number of sixth-order stream. Some parts of the watershed have dendritic drainages pattern which indicates the homogeneity in lithology and lack of structural control (Yadav et al., 2020).

6.2.3 Stream length (L_u)

Stream length is measured as the total length of streams in a particular order (L_u). The length of the streams has been calculated based on the law proposed by Horton (1945). It is the most important morphometric parameter which reveals surface runoff characteristics. The length of first-order stream is 319.47 km, second-order stream is 79.88 km, third-order stream is 36.91 km, fourth-order stream is 14.15 km, fifth-order stream is 10.49 km and sixth-order stream is 9.75 km. The L_u values for different sub watershed are given in table 17.2. It is observed that as the stream order increases the stream length of successive order decreases and this relation between stream order and stream length is called Horton's second law "**law of stream lengths**". It states that the total length of stream segment is the maximum in the first order stream and decreases as the stream order increases.

6.2.4 Mean stream length (L_{sm})

The mean stream length is a dimensional property, which is defined as the total stream length (L_u) divided by the total number of the streams in a given order (N_u). The mean stream length is a characteristic property related to the drainages network and its associated basin surface (Strahler, 1964). The river of fairly smaller length is characteristics of regions with steep slopes and less permeable rock formation and Rivers having longer lengths are commonly suggestive of smoother slope. The mean stream length of the watershed ranges from 0.58 km for first order stream to 9.75 km for sixth order stream (Table 17.2). These vast differences in mean length indicating that the lower order streams are originated from higher altitude with steep slopes and highly fractured topography (Yadav et al., 2020).

6.2.5 Stream length ratio (Rl)

Stream length ratio (R_l) is the ratio of the mean length of the one order to the next lower order (Horton, 1945). The stream length ratio is one of the most important hydrological features of the watershed which reveals the variations in slope and topography of that area.

6.2.6 Bifurcation ratio (R_b)

Bifurcation ratio is related to the branching pattern of a drainage network and is calculated by dividing the number of first order streams by the number of second order streams, then dividing the second order streams by the next higher order, and so on (Schumm, 1956). The higher values of bifurcation ratio (R_b) indicate a strong structural control in the drainage development whereas the lower values indicate that some of the areas in the basin are less affected by structural disturbances (Stahler, 1964; Nag, 1998; Vittala et al., 2004 and Chopra et al., 2005). According to Horton (1945, p. 290), the bifurcation ratio varies from a minimum of 2 in flat or rolling drainage basins to 3 or 4 in mountainous or highly dissected drainage basins. The mean bifurcation ratio of the study area is 3.69, which indicate that the watershed is largely controlled by the structure.

6.2.7 Rho coefficient

The rho coefficient is defined as the ratio between the stream length and the bifurcation ratio (Horton, 1945). The higher value of rho coefficient indicates high capacity for the storage of water and lower value indicates low capacity for the storage of water in the watershed. The value of rho coefficient in the study area is 0.5 which indicates that the area is having low water storage capacity.

6.3 Relief parameters

The relief parameters are the most important in drainage morphometry, which reveal the denudational characteristics of the watershed and also indicate the runoff condition of the area. The relief parameters such as basin relief, relief ratio, relative relief, gradient ratio, ruggedness number and dissection ratio are summarised below in the following sub-sections.

6.3.1 Basin relief

It is an important factor in understanding the denudational characteristics (the denudational landforms are formed as a result of active processes of weathering, mass wasting and erosion caused by different exogenetic geomorphic processes such as fluvial, glacier and aeolian) the difference in elevation between the highest point of a basin (H) and the lowest point on the valley floor (h) is called the basin relief (Strahler, 1957). The study area has the basin relief value of 1.40 km, which indicate that the area is in a low infiltration and high runoff condition.

6.3.2 Relief ratio

Relief ratio is the dimensionless height–length ratio equal to the tangent of the angles formed by two planes intersecting at the mouth of the basin, one representing the horizontal, the other passing through the height point of the basin (Schumm, 1963). It measures the overall steepness of a drainage basin and is an indicator of the intensity of erosion processes of operation on the slopes of the basin (Chopra et al., 2005). The relief ratio normally increases with decreasing drainage area and size of the given drainage basin (Gottschalk, 1994). The value of relief ratio is 0.086 indicating moderate to high relief with high rate of erosion.

6.3.3 Relative relief

Relative relief is determined as the ratio between the basins relief (R) to thebasin area of the watershed (Table 17.4). Its lower value indicate gentle topography whereas higher value suggests steeper slopes of the terrain. The relative relief of the study area is 0.027 and it is showing mature topography.

6.3.4 Gradient ratio

It is expressed as the ratio of the difference of source elevation (a) and mouth elevation (b) of major stream of the watershed to the length of the major channel (L_s) of that watershed. The gradient ratio of the watershed is 0.05 which indicate the area is moderate to high gradient of the terrain.

6.3.5 Ruggedness number

Ruggedness number is the product of the basin relief (R) and drainage density (D_d), where both parameters are in the same unit (Strahler, 1957). Low ruggedness value infers that the basin is less prone to soil erosion and the high ruggedness value of the basin implies that the basin is highly susceptible to

TABLE 17.4 The relief aspects of the upper Tuirial watershed.

Sl. No	Relief parameters	Value
1	Basin relief (R)	1.40
2	Relief ratio (R_r)	0.086
3	Relative relief (R_{hp})	0.027
4	Gradient ratio (G_r)	0.05
5	Ruggedness number (R_n)	6.47
6	Dissection ratio (D_i)	0.87
7	Basin area ($km^{2)}$	102.66
8	Basin length (km)	16.25
9	Basin perimeter(km)	50.06

erosion, due to structural complexity of the terrain. The study area has 6.28 ruggedness number value and this value indicate that the area is structurally complex with high relief and drainage density.

6.3.6 Dissection index

Dissection index is the ratio between basin relief to the absolute relief of an area and gives clue to stages of the landscape evolution and degree of dissection or vertical erosion of a region (Singh and Dubey, 1994). The value of dissection index ranges from 0 (complete absence of dissection) to 1 (vertical cliff). Dissection index value of the study area is 0.87 and suggest that the landforms of the study area are highly dissected mountainous terrains.

6.4 Aerial parameters

The aerial aspects determined include drainage density, stream frequency, drainage texture, form factor and length of the overland flow and the constant of channel maintenance that have been calculated, and their results (Table 17.5) are summarised below in the following sub-sections.

6.4.1 Drainage density (D_d)

Drainage density is the important aerial parameter which indicate the closeness of spacing between the channels (Horton, 1932). It is defined as the ratio of the total stream length of all orders within a basin to the area of the basin; according to Nag (1998), the low drainage density of a region indicate permeable subsurface material under dense vegetative cover with low relief. The drainage density of the study area varies between 2.82 and 7.64 km/km^2. The study area is divided into three classes: high drainage density (>4 km/km^2), moderate drainage density (2–4 km/km^2) and low drainage density

TABLE 17.5 The computed aerial morphometric parameters of the study area.

Sl. No.	Name of the parameter	Value
1	Drainage density (D_d)	4.62
2	Stream frequency (F_s)	7.05
3	Length of overland flow (L_o)	0.10
4	Constant channel maintenance (C_m)	0.21
5	Form factor (F_f)	0.38
6	Circulatory ratio (R_c)	0.50

(<2 km/km^2), where medium to high drainage density indicate that the region is characterized by impermeable subsurface material, high relief and sparse vegetation cover, and low drainage density indicate that region is characterized by permeable subsoil material and dense vegetation. Drainage density of the study area is high, the value of which is 4.62 km/km^2 indicating the region is characterised by sparse vegetation cover, impermeable subsurface material and high relief.

6.4.2 Stream frequency (F_S)

The stream frequency is defined as the ratio between the total number of stream segments of all the orders per unit area (Horton, 1932, 1945). It reveal the positive correlation with drainage density in the watershed that indicate an increase in stream population with respect to increase in drainage density. The study area has stream frequency (7.05) where higher stream frequency value indicate resistance of subsurface strata, sparse vegetation and high relief with low permeability of the rock formations.

6.4.3 Length of overland flow

The length of overland flow is one of the most important controlled variables which help hydrologic and physiographic development of drainage basins. Length over land flow (L_O) is defined as the length of the longest drainage path that water takes before its gets concentrated and is approximately equal to half of drainage density (Horton,1932, 1945). This factor depends on the rock types, permeability, climatic regime, vegetation cover and relief, as well as duration of erosion (Schumm, 1956). The study area having 0.10 length of over land flow indicates that the region is characterised by short flow path having less infiltration and area of high relief with steep slopes.

6.4.4 Constant of channel maintenance

Constant of channel maintenance is defined as the area of the watershed surface needed to maintain a unit length of stream channel and is expressed by the reciprocal of drainage density (Schumm, 1956, p. 607). The importance of this constant is that it provides a quantitative expression of the minimum, limiting area required for the development of length of channel. The constant of channel maintenance value for the study area is 0.21 sq. km/km, which means that 0.21 sq. km of surface area is required to maintain each km of the channel length and indicates that the area is characterised by high surface run off, low permeability and high drainage density.

6.4.5 Form factor

The form factor is a dimensionless number. It is defined as the ratio of basin area to the square of the basin length (Horton, 1945). The value of form factors would always be less than 0.75, which reveals a perfectly circular watershed.

Higher the value of form factor, more circular the shape of the watershed; whereas, long narrow watershed has low form factor value, which is close to zero. The form factor value of the area is 0.38. This indicate more or less elongated in nature with less side flow for longer duration.

6.4.6 Circularity ratio

Circularity ratio was introduced by Miller (1953), who states that it is a dimensionless quantity which is the ratio of the basin area to the area of a circle having the same circumference as the perimeter of the basin and expresses as the degree of circularity of the basin. The value of circulatory ratio of study area is 0.50. The high value of circularity ratio indicate that the region is more or less circular in shape and is characterized by high to moderate relief with structurally controlled drainage system.

6.5 Analysis of morphotectonic parameters

The most commonly used geomorphic indices for morphotectonic analysis such as basin elongation ratio, asymmetry factor, transverse topographic symmetry factor and channel sinuosity have been calculated and presented below in Table 17.6. The morphotectonic parameters are explained in the following sub-sections.

6.5.1 Basin elongation ratio (R_e)

Basin elongation ratio is the ratio of diameters of a circle of same area as the drainage basin and maximum length of the basin. Elongation ratio generally

TABLE 17.6 Methods used for computing the morphotectonic parameters.

Sl. No.	Morphotectonic parameter(s)	Formulae	Reference(s)
1	Basin elongation ratio (R_e)	$R_e = 2\ (\sqrt{A}/\sqrt{\pi})/L_b$, where A= basin area, π= 3.14 and L_b= basin length	Schumm (1956)
2	Asymmetry factor (A_F)	$A_F = (A_r/A_t) * 100$, where A_r = the area of basin and A_t = the total drainage basin area	Cox (1994)
3	Transverse topographic symmetry factor (T)	$T = D_a/D_d$, where D_a = distance from midline of the drainage basin to the midline of the active meander belt. D_d = distance from the basin midline to basin divide	Cox (1994)
4	Channel sinuosity (S)	S = CI/VI, where CI= Channel index and VI = valley index	Mueller (1968)

varies from 0.6 to 1.0 over a wide variety of climatic and geologic types and also indicate varying slopes of the watershed. The elongation ratio value of watershed is 0.70.

6.5.2 Asymmetry factor

The asymmetry factor is a morphotectonic variable, which is mainly used to find the presence or absence of regional tilt of a basin in regional scale (Keller and Pinter, 2002). Asymmetry factor (A_F) is a qualitative index, which helps in evaluating basin asymmetry. In a stable setting environment, A_F is 50. It is sensitive to tilting perpendicular to the main channel of the basin. The asymmetry factor may be defined as the ratio of the right hand area of drainage basin facing downstream of the trunk stream to the total area of the drainage basin. The study area asymmetry factor value is 73.82.

6.5.3 Transverse topographic symmetry factor (T)

The transverse topographic symmetry factor may be defined as the ratio between distance from the midline of the drainage basin to the midline of the meander belt (D_a) and distance from the basin midline to the basin divide (D_d). Perfectly symmetric basin has value of transverse topographic symmetry (T) as zero; as the asymmetry increases T also increases and approaches the value one. Transverse topographic symmetry factor is calculated for different segments of stream channels and indicates preferred migration of stream perpendicular to the drainage axis (Keller and Pinter, 2002). The estimated T value of the study area are 0.23 to 0.60, which indicate the asymmetric nature of the basin (Fig. 17.4).

6.5.4 Channel sinuosity (S)

Channel sinuosity, proposed by Mueller (1968), is used to understand the role of tectonism (Rhea, 1993). According to Mueller (1968), the channel sinuosity (S) is the ratio between the stream length (Sl) to the valley length (Vl), which is expressed as S=SL/Vl. The index value of 1.0 indicate straight river course. The values of 1.0 to 1.5 indicate the sinuous river course, whereas S values more than 1.5 represent the meandering course. The estimated S value of the study area range from 1.08 to 1.50. Table 17.7 indicate that sinuous nature of the river course.

7. Hypsometric curve (HC) and hypsometric integral (HI)

By examining the graphical diagram of the HC and HI value, it is revealed that (i) HC of Tuirial basin indicate that the overall basin is in mature or equilibrium stage while HC of sub-basins A, B, C, D, E, F, G and whole basin shows that these are in mature stage (Fig. 17.5). Hence, sub-basins are at the mature stage of geomorphic development (ii) due to variation in tectonic

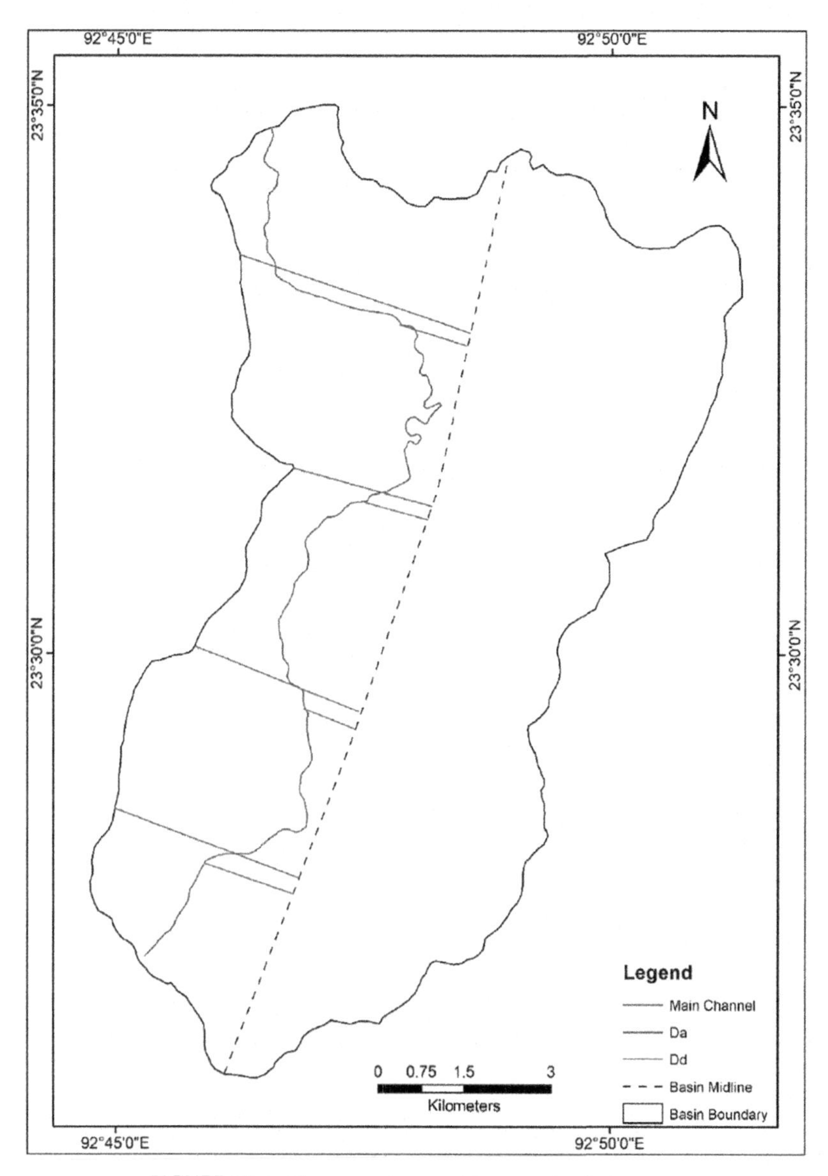

FIGURE 17.4 Transverse topographic symmetry factor map.

effect, lithology and rejuvenation processes there are some marginal differences which exist in the basin and the seven sub-basins. The entire basin suffers from serious geomorphic hazards like slope instability, floods, high sediment yield and severe soil erosion and the others suffer from anthropogenic effects such as deforestation, land-use/land-cover changes and poor

TABLE 17.7 Computed morphotectonic parameters of the study area.

Segment number	Transverse topographic symmetry factor (T) = D_a/D_d	Channel sinuosity (S) = Sl/Vl
1	1.40/3.23 = 0.43	3.07/2.82 = 1.08
2	1.00/3.28 = 0.30	4.32/3.48 = 1.24
3	1.69/2.81 = 0.60	4.07/3.54 = 1.14
4	0.95/4.06 = 0.23	9.34/6.20 = 1.50

conservation measures that maximise soil erosion. Soil erosion from the entire basin and its sub-basins is attributed to topography, rainfall and stage of river and is increased from the incision of channel beds, washout of top soil and undercut erosion of stream banks. The sub-basin wise hypsometric integral value are presented in Table 17.8.

8. Prioritisation of mini-watersheds/sub-basins and soil erosion hotspots

For prioritisation, the compound factors are figured by summing every one of the positions of direct parameters and in addition shape parameters and afterward separating by the quantity of parameters. From the gathering of these small-scale watersheds, most astounding position was allotted to the smaller than normal watershed having the least compound factor. Contingent on the estimation of compound factor, positioning to every less watershed is allotted as appeared in (Table 17.9). For Tuirial watersheds, watershed no. D is given position 1 with minimum compound factor esteem 3.1, and it is trailed by watersheds no. G and E, as second and third, respectively. The estimations of compound factor and particular position of every single smaller than normal watershed are shown in Table 17.9. The topical layers produced amid the GIS investigation are utilised to outline the potential regions for proposing check dam construction. The weighted overlay analysis in ArcGIS 10.0 was performed using spatial modeller. The following layers were considered example, soil type, waste thickness, incline from DEM and by keeping in view the organised position maps, appropriate fundamental check dam building locations are recognised. The need is given to the zones which have the blend of low compound factor, low seepage thickness, bring down incline and a higher invasion rate (Patel et al., 2012). In weighted overlay delineation, diverse zones are dispensed dependent on the mix and spoke to as low, medium/moderate, and high for check dam situating/situations/sites. The last conceivable check dams are proposed in the wake of directing escalated field

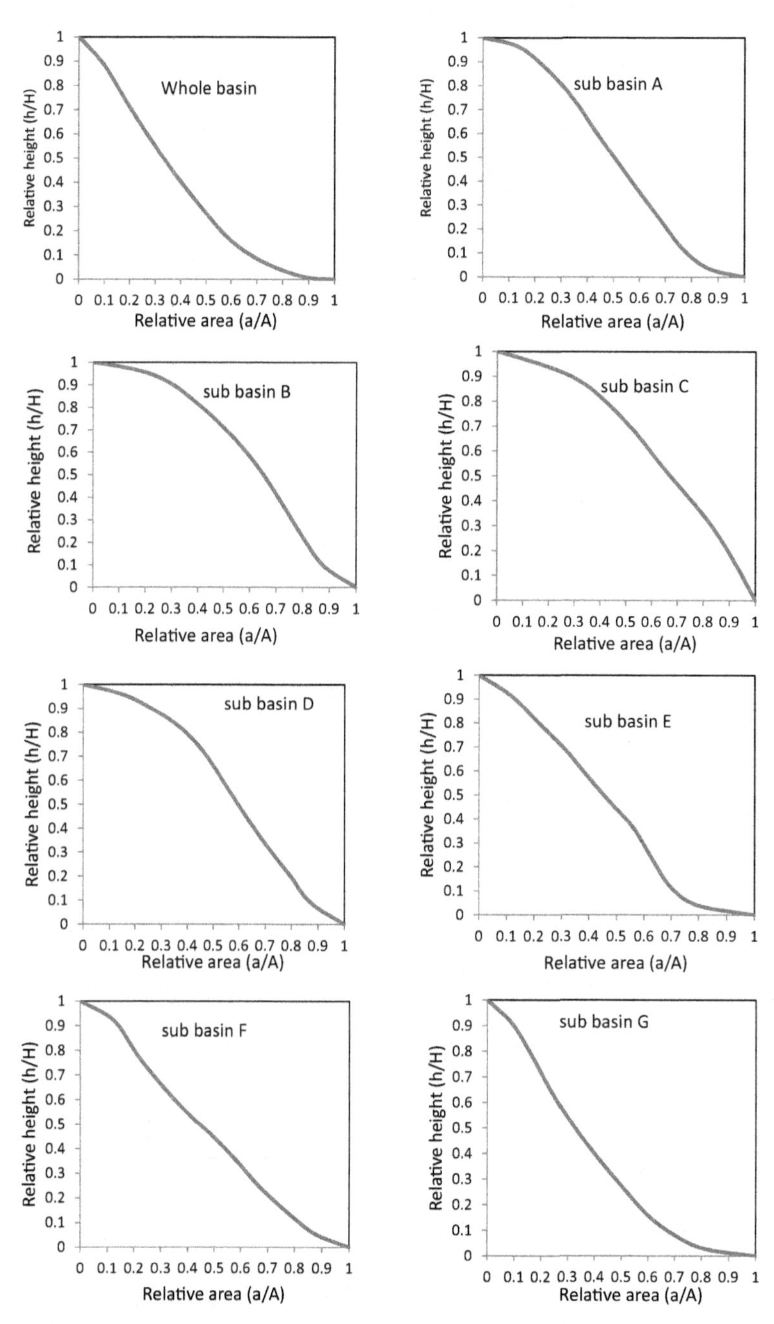

FIGURE 17.5 Hypsometric curves of the Tuirial River basin and seven sub-basins.

TABLE 17.8 The sub-basins wise Hypsometric integral values.

Subbasins	Area (km^2)	Max. elev.	Min. elev	Mean elev.	HI value	Geomorphic stage
A	4.455	1318	404	861	0.5	Mature stage
B	3.438	1494	601	1047.5	0.5	Mature stage
C	2.754	1494	660	1077	0.5	Mature stage
D	3.408	1567	662	1114.5	0.5	Mature stage
E	17.36	1392	334	863	0.5	Mature stage
F	3.592	1244	336	790	0.5	Mature stage
G	23.203	1403	265	834	0.5	Mature stage
Whole basin	101.38	1606	244	925	0.5	Mature stage

TABLE 17.9 Calculation of compound factor and prioritized ranks.

Watershed no.	R_b	D_d	F_u	T	Lo	R_f	B_s	R_e	C_c	R_c	Compound factor	Prioritization (ranking)
A	6	6	7	7	6	4	4	4	3	5	5.2	6
B	4	3	2	4	3	5	3	5	5	3	3.7	5
C	5	1	1	5	1	7	1	7	7	1	3.6	4
D	3	2	3	2	2	3	5	3	2	6	3.1	1
E	2	4	5	3	4	2	6	2	4	4	3.6	3
F	7	7	6	6	7	6	2	6	1	7	5.5	7
G	1	5	4	1	5	1	7	1	6	2	3.3	2

visits of the watershed. The check dams are recommended in smaller than normal watershed no. D, G, E, C, B, A and F, which have positions as 1, 2, 3, 4, 5, 6 and 7, respectively. The coordinated guide is shown in Table 17.9 with the proposed check dams.

9. Summary and conclusions

Geographical Information System (GIS) and Remote Sensing (RS) are one of the most important tools which are used for delineation of the different drainage basins. In the present study area the morphometric indices such as linear, relief and aerial parameters are analysed by using GIS and RS techniques and also help to establish the various terrain parameters such as nature of the tectonic activity in area bedrock condition, infiltration capacity and surface runoff. The upper Tuirial watershed which is mainly in Aizawl district and parts of Serchhip district proved that the dendritic drainage pattern. Bifurcation ratio, rho coefficient, stream length ratio of the watershed reveal that the study area is homogeneous in lithology and texture and lacks structural control; it also indicate that the area is having low water storage capacity, and low permeability rock formation in the area. Aerial aspects such as drainage density, stream frequency and length of the overland flow and the constant of channel maintenance have been calculated, which indicate that the region has sparse vegetation cover, impermeable subsurface material and high relief. Relief aspects such as basin relief, relief ratio, relative relief, dissection index, ruggedness number and gradient ratio have been carried out that reflects the outcome as the watershed is having high relief and steep slopes which show gentle topography. The analysis of morphotectonic parameters helps to understand that the watershed is more or less elongated in shape, and the area has a wide variety of climatic and geologic types. The calculated morphotectonic parameters like asymmetry factor ($AF = 73.82\%$) indicate basin is tectonic tilt and transverse topography symmetry factor is 0.23 to 0.60, which indicate an asymmetric nature of the basin and elongation ratio is 0.70, which indicate that the region is slightly tectonically active.

References

Biswas, S., Sudhakar, S., Desai, V.R., 1999. Prioritisation of subwatersheds based on morphometric analysis of drainage basin: a remote sensing and GIS approach. J. Indian Soc. Remote Sens. 27 (3), 155–166.

Bull W.B. and McFadden L.D., Tectonic geomorphology north and south of the Garlock fault, California. In: Doehring D.O. (Ed), Geomorphology in Arid Regions. In Proceedings of the Eighth Annual Geomorphology Symposium. Binghamton, NY: State University of New York at Binghamton, 115-138

Chopra, R., Dhiman, R.D., Sharma, P.K., 2005. Morphometric analysis of sub-watersheds in Gurdaspur district, Punjab using remote sensing and GIS techniques. J. Indian Soc. Remote Sens. 33 (4), 531.

Choudhari, P.P., Nigam, G.K., Singh, S.K., Thakur, S., 2018. Morphometric based prioritization of watershed for groundwater potential of Mula river basin, Maharashtra, India. Geol. Ecol. Landsc. 1–12 https://doi.org/10.1080/24749508.2018.1452482.

Clarke, J.I., 1996. Morphometry from maps. Essays in Geomorphology. Elsevier publication. Co., New York, pp. 235–274.

Cox, R.T., 1994. Analysis of drainage-basin symmetry as a rapid technique to identify areas of possible Quaternary tilt-block tectonics: An example from the Mississippi Embayment. Geol. Soc. Am. Bull. 106 (5), 571–581.

Dove Nir, 1957. The ratio of relative and absolute altitude of Mt. Camel. Geog. Rev. 47, 564–569.

Evans, P., 1964. The tectonic framework of Assam. Geol. Soc. India J. 5.

Gottschalk, L.C., 1964. Reservoir Sedimentation in Handbook of Applied Hydrology Ed. VT Cvhow Mc Graw Hill Book Company, New York section, 7-1.

Horton, R.E., 1932. Drainage-basin characteristics. TrAGU 13 (1), 350–361.

Horton, R.E., 1945. Erosional development of streams and their drainage basins; hydrophysical approach to quantitative morphology. Geol. Soc. Am. Bull. 56 (3), 275–370.

Karunakaran, 1974. Geology and Mineral Resources of the North Eastern States of India. Geol. Surv. India, Misc. Publ. 30, 93–101.

Keller, E.A., Pinter, N., 2002. Active tectonic earth quake—uplift and landscape, 2nd ed. Prentice hall Inc., New Jersey, Upper Saddle River, p. 337.

Kumar, N., Singh, S.K., Pandey, H.K., 2018a. Drainage morphometric analysis using open access earth observation datasets in a drought affected part of Bundelkhand, India. Appl. Geomat. org/10.1007/s12518-018-0218-2.

Kumar, M., Denis, D.M., Singh, S.K., Szabó, S., Suryavanshi, S., 2018b. Landscape metrics for assessment of land cover change and fragmentation of a heterogeneous watershed. Remote Sens. Appl. Soc. Environ. 10, 224–233. https://doi.org/10.1016/j.rsase.2018.04.002.

Miller, V.C., 1953. A quantitative geomorphic study of drainage basin characteristics in the clinch mountain area, Virginia and Tennessee. Department of Geology, Columbia University, New York, USA.

Mueller, J.E., 1968. An introduction to the hydraulic and topographic sinuosity indexes. Ann. Assoc. Am. Geogr. 58 (2), 371–385.

Nag, S.K., 1998. Morphometric analysis using remote sensing techniques in the Chaka sub-basin, Purulia district, West Bengal. J. Indian Soc. Remote Sens. 26 (1-2), 69–76.

Nag, S.K., Chakraborty, S., 2003. Influence of rock types and structures in the development of drainage network in hard rock area. J. Indian Soc. Remote Sens. 31 (1), 25–35.

Nandy, D.R., Gupta, S.D., Sarkar, S., Ganguly, 1983. A Tectonic Evolution of the Tripura- Mizoram fold belt, Surma basin, NE India. Quart. Jour. Geol. Min. Met. Soc. India 55 (4), 186–194.

Patel, D.P., Dholakia, M.B., Naresh, N., Srivastava, P.K., 2012. Water harvesting structure positioning by using geo-visualization concept and prioritization of mini-watersheds through morphometric analysis in the Lower Tapi Basin. J. Indian Soc. Remote Sens. 40, 299–312.

Patel, D.P., Srivastava, P.K., Gupta, M., Nandhakumar, N., 2015. Decision Support System integrated with Geographic Information System to target restoration actions in watersheds of arid environment: a case study of Hathmati watershed, Sabarkantha district. Guj. J. Earth Syst. Sci. 124, 71–86.

Rawat, K.S., Singh, S.K., 2017. Estimation of surface runoff from semi-arid ungauged agricultural watershed using SCS-CN method and earth observation data sets. Water Conserv. Sci. Eng. 1, 233–247.

Rawat, K.S., Singh, S.K., 2018b. Appraisal of soil conservation capacity using NDVI model-based C factor of RUSLE model for a semi arid ungauged watershed: a case study. Water Conserv. Sci. Eng. 3, 47–58. https://doi.org/10.1007/s41101-018-0042-x.

Rawat, K.S., Singh, S.K., Bala, A., 2018a. In: Singh, C.K. (Ed.), Estimation of Evapotranspiration through Open Access Earth Observation Data Sets and its Validation with Ground Observation Edited Book Geospatial Applications for Natural Resources Management. CRC Press, Boca Raton, FL, pp. 172–189, 33487–2742, (Chapter 11).

Rhea, S., 1993. Geomorphic observations of rivers in the Oregon Coast Range from a regional reconnaissance perspective. Geomorphology 6 (2), 135–150.

Sarkar, K., Nandy, D.R., 1976. Structures and tectonics of Tripura–Mizoram area, India. Geol. Surv. India, Miscellaneous Publication 34, 141–148.

Schumm, S.A., 1956. Evolution of drainage systems and slopes in badlands at Perth Amboy, New Jersey. Geol. Soc. Am. bull. 67 (5), 597–646.

Schumm, S.A., 1963. Sinuosity of Alluvial Rivers in the Great Plains. Bull. Geol. Soc. Am. 74, 1089–1100.

Shimrah, T., Sarma, K., Varga, O.G., Szilard, S., Singh, S.K., 2019. Quantitative assessment of landscape transformation using earth observation datasets in Shirui Hill of Manipur, India. Remote Sens. Appl. Soc. Environ. 15, 100237.

Singh, S., Dubey, A., 1994. Geoenvironmental planning of watershed in India. Chugh Publications, Allahabad, pp. 28–69.

Singh, S.K., Laari, P.B., Mustak, S.K., Srivastava, P.K., Szabó, S., 2018. Modelling of land use land cover change using earth observation data-sets of Tons River Basin, Madhya Pradesh, India. Geocarto Int. 33 (11), 1202–1222.

Singh, V., Singh, S.K., 2018. Hypsometric analysis using Microwave satellite data and GIS of Naina–gorma river basin (Rewa district, Madhya Pradesh, India). Water Conserv. Sci. & Eng. https://doi.org/10.1007/s41101-018-0053-7.

Singh, S.K., Singh, C.K., Mukherjee, S., 2010. Impact of land-use and land-cover change on groundwater quality in the Lower Shiwalik hills: a remote sensing and GIS based approach. Cent. Eur. J. Geosci. 2, 124–131.

Singh, S.K., Srivastava, P.K., Pandey, A.C., 2013. Fluoride contamination mapping of groundwater in Northern India integrated with geochemical indicators and GIS. Water Sci. Technol. Water Supply 13, 1513–1523.

Singh, S.K., Srivastava, P.K., Szabó, S., Petropoulos, G.P., Gupta, M., Islam, T., 2017. Landscape transform and spatial metrics for mapping spatiotemporal land cover dynamics using Earth Observation data-sets. Geocarto Int. 32 (2), 113–127.

Smith, G.H., 1935. The relative relief of Ohio. Geogr. Rev. 25, 272–284.

Sreedevi, P.D., Subrahmanyam, K., Ahmed, S., 2005. The significance of morphometric analysis for obtaining groundwater potential zones in a structurally controlled terrain. Environ. Geol. 47 (3), 412–420.

Strahler, A.N., 1952. Hypsometric (area-altitude) analysis of erosional topography. Geol. Soc. Am. Bull. 63, 1117–1141. https://doi.org/10.1130/0016- 7606, [1117:HAAOET]2.0.CO;2.

Strahler, A.N., 1957. Quantitative analysis of watershed Geomorphology. Trans. Am. Geophys. Union 38, 913–920.

Strahler, A.N., 1964. Quantitative geomorphology of drainage basins and channel networks. Section 4-II. In: Chow, V.T. (Ed.), Handbook of Applied Hydrology. McGraw Hill, New York, pp. 439–476.

Vittala, S.S., Govindaiah, S., Gowda, H.H., 2004. Morphometric analysis of sub-watersheds in the Pavagada area of Tumkur district, South India using remote sensing and GIS techniques. J. Indian Soc. Remote Sens. 32 (4), 351.

Yadav, S.K., Singh, S.K., Gupta, M., Srivastava, P.K., 2014. Morphometric analysis of upper tons basin from northern Foreland of Peninsular India using CARTOSAT satellite and GIS. Geocarto Int. 29 (8), 895–914.

Yadav, S.K., Dubey, A., Szilard, S., Singh, S.K., 2016. Prioritisation of sub-watersheds based on earth observation data of agricultural dominated northern river basin of India. Geocarto Int. 33 (4), 339–356.

Yadav, S.K., Dubey, A., Singh, S.K., Yadav, D., 2020. Spatial regionalisation of morphometric characteristics of mini watershed of Northern Foreland of Peninsular India. Arabian J. Geosci. 13, 435.

Chapter 18

Development of android application for visualisation of soil water demand

Prashant K. Srivastava[1,2], Prachi Singh[1], Varsha Pandey[1], Manika Gupta[3]
[1]*Remote Sensing Laboratory, Institute of Environment and Sustainable Development, Banaras Hindu University, Varanasi, Uttar Pradesh, India;* [2]*DST-Mahamana Centre of Excellence in Climate Change Research, Institute of Environment and Sustainable Development, Banaras Hindu University, Varanasi, Uttar Pradesh, India;* [3]*Department of Geology, University of Delhi, New Delhi, India*

1. Introduction

Now-a-days the problem of optimised irrigation system is recognised as a major challenge due to fresh water crisis all over the world. Agriculture sector is the primary consumer of the fresh water resource, which itself uses 70% of total fresh water consumption, i.e., 1500 billion cubic metres of water per year (Frenken and Gillet, 2012). FAO estimated that by 2050 the world's population will increase 34% from today that will reach around 9.1 billion. This huge population's food demand must be met by elevating the irrigation withdrawal significantly by 70% (Summary, n.d.). The nonoptimal usage of irrigation water is the main problem for an agricultural country like India. In India, water crisis is severe and agriculture water consumption is about 80% of the total fresh water consumption alone (Guidelines et al., 2005). Such growing water consumption demand has raised threats in many parts of the country to the future irrigated agriculture. In most of the Indian regions farmers are still using manually control irrigation techniques such as flooding and channel irrigation systems in which they irrigate the crop land frequently without knowing the water demand of crops that sometime consume excess water, which generally drains out and causes soil erosion and water resource contamination. Moreover, these traditional irrigations also provide an uneven distribution of irrigational water by providing unnecessary irrigation to one side while leading to dearth of irrigation water on the other side of a crop field. Whereas, the drip irrigation system is comparatively new and efficient technology that gives drips of water to each plant either in soil surface or their root zone directly, it is

Agricultural Water Management. https://doi.org/10.1016/B978-0-12-812362-1.00018-7

expensive and is not feasible for most of the major Indian crops like rice, wheat and barley. Scarcity of fresh water in India seeks optimal usage of water that should be provided to only that area where it is required and in appropriate quantity.

Despite scarcity, the unmanaged agricultural irrigation activity causes degradation of fresh water resources by overexploitation, nutrient contamination and salinization. It also depletes and degrades the ground water resources by over consumption. The effective way for improving the utilisation rate of irrigation water is to develop a precise controlled information system on the basis of crops' water demand data. Towards this, smart and automatic irrigation scheduling has become popular and proven its usefulness in irrigation water-use efficiency appropriately based on direct soil water measurements. Irrigation water management represents the use of a suitable quantity of water at the proper time, and it is usually pursued by combining measurements of soil moisture with an optimized irrigation plan. It will help in the conservation of water and nutrients leaching and contamination of groundwater (Gupta et al., 2013; Garg and Gupta, 2015).

The accurate estimation of soil moisture is important for understanding the ecosystem services such as agriculture, watershed management, climate change, hydrology and meteorology. (Srivastava et al., 2013b, 2013c, 2015, 2016b). In the agriculture sector, the estimation of Soil Moisture Content (SMC) will help greatly in irrigation water management (Srivastava et al., 2013a; Srivastava and Singh, 2016; Srivastava, 2017). Real-time monitoring of soil moisture will provide detailed information about soil moisture profile and irrigation requirement and indicates when crops are at risk (Srivastava et al., 2016a; Pandey and Srivastava, 2018). Soil moisture quantification could improve irrigation scheduling efficiency and can lead to improved quality, yield and profit and help in the conservation of water and nutrients leaching and contamination of groundwater. Soil moisture estimation can be achieved by in-situ measurements, remote sensing observation and simulation from different models (Petropoulos et al., 2014, 2015, 2016). In-situ measurements include ground-based techniques such as various probe or gravimetric measurements, which provide accurate measurements of soil moisture. The near-surface soil moisture content can be estimated by optical, thermal infrared, active microwave (Advanced Scatterometer (ASCAT), European Remote Sensing Satellite Scatterometer (ERS SCAT) etc.), passive microwave (Soil Moisture and Ocean Salinity (SMOS), Advanced Microwave Scanning Radiometer for Earth Observing (AMSR-E), Scanning Multichannel Microwave Radiometer (SMMR), etc.) and both active-passive microwave onboard remote sensing platforms (Carlson et al., 1994; Engman and Chauhan, 1995; Gupta et al., 2016; Srivastava, 2017; Petropoulos et al., 2018). The various hydrological models are also being actively used in estimation of soil moisture such as land surface models, global reanalysis datasets and Variable Infiltration Capacity (VIC) model (Liang et al., 1996).

In agriculture, application of quantitative irrigation is estimated precisely by the soil moisture indices known as soil moisture deficit/depletion (SMD), which is assessed using soil moisture and soil physical properties. SMD is an important indicator of the degree of soil saturation that is inversely related to the soil moisture in the soil layer. SMD provides the estimation of irrigation scheduling by quantifying the amount of water required to raise the soil moisture content of the plant root zone to field capacity (Srivastava et al., 2014). SMD is the key variable that will provide the estimation of irrigation scheduling by quantifying the amount of water required to raise the soil moisture content of the plant root zone to field capacity (Srivastava et al., 2013a, 2014). In a saturated soil, all of the available soil pores are full of water, but water will drain out of large pores under the force of gravity. The irrigation requires that when SMD condition is higher (positive) it indicates soil moisture is below field capacity of soil, whereas low SMD (negative) values represent low soil moisture.

For an effective management of irrigation water, Migliaccio et al. (2015) developed evapotranspiration (ET) and water balance–based smartphone tool for irrigation scheduling for some selected crops. The app estimates soil type, irrigation depth and duration and the output may vary for different crop types. An automated smartphone irrigation sensor was developed by Jaguey et al. (2015) that gives the digital picture of the root zone soil of the crop and also optically estimates the water content of the same area. Vellidis et al. (2016a,b) also developed a Cotton Smart Irrigation App using ET-based soil water balance model that uses meteorological data, soil parameters, crop phenology, crop coefficients and irrigation applications as input parameters and responds the root zone soil water deficits (RZSWD) in percentage/inches.

Thus, current work aims to estimate SMD by developing a smart and easy-to-use irrigation app that monitors and displays irrigation water demand to manage the applied irrigation and conserve water resources on an Android smartphone platform and provide an irrigation water demand of Varanasi district of Uttar Pradesh, India. The database file of Irrigation Scheduler App contains ground measured soil moisture and measured soil physical properties such as field capacity and texture that are visualised in a user-friendly Android app interface in terms of soil moisture amount in percentage, soil type and SMD. This app is easy to operate and gives all the contained information by selecting the geographical location of the crop site or village.

2. Materials and methods

2.1 Test site

Varanasi district extends between the latitude of 25°10′N to 25°37′ N and longitude of 82°39′ E to 83°10′ E in eastern Uttar Pradesh. It is a beautiful pilgrim city situated on the bank of River Ganga and spread over 1454.11 sq.

km of area (Fig. 18.1). The district showed that agriculture classes dominate the major area of the study area. Major urban agglomeration was found around Varanasi city block and northeastern part of Kashi Vidyapeeth block along the stretches of River Ganga while remaining blocks of Harhua, Pindra, Cholapur, Arajiline, Baragaon, Chiagigaon and Sewapuri having fewer urban parts except few central market locations. The overall nature of urbanisation was found to be spurious rather than planned. In general, elevation showed a declining trend from east to west mainly due to its geological characteristics and topographical feature of Vindhya plateau on the west. On an average, southwestern and northwestern parts exhibited more elevation above mean sea level than the rest of the areas ranging from 200 to 260 m. With the evident fact of Ganga flowing from northeast to southwest of the region, the corresponding area was marked with lower elevation more likely due to lateral erosion work of the river covering the blocks Chiagigaon, Harhua, Varanasi city, KVP and southern parts of Arajiline.

2.2 Soil sampling and generation of soil texture of Varanasi

To get an overview of soil types of Varanasi, soil samples from various blocks of the district were collected to assess different soil parameters such as soil moisture, soil texture, porosity, bulk density, organic matter, etc. Soil moisture at various sampling locations was measured using soil moisture probe at soil depths 5 and 10 cm. After soil sampling, to know moisture status in the Varanasi district, Inverse Distance Weighting (IDW) method was employed for map generation. This method estimates the values of an attribute at blank points using a linear combination of values at sampled points and weighted by an inverse function of the distance from the point of observation to the sampled points. The method follows an assumption that sampled points closer to the blank point are more similar to it than those that are farther away in their values. The variation of soil moisture can be seen in the plots provided.

2.3 Soil moisture deficit (SMD) computation

SMD is directly related to soil moisture and can be computed when linked with soil hydraulic parameters such as field capacity (FC). For calculation of FC, pedotransfer functions such as those available with ROSSETA model can be used. ROSETTA model is a computer program developed by Schaap et al. (2001) and useful for the estimation of water retention curve of the soil. It has five hierarchical pedotransfer functions (PTFs) to predict soil hydraulic properties (such as water retention parameters, saturated and unsaturated hydraulic conductivities) using easily obtainable input information (such as soil texture, bulk density and particle-size distribution) (Srivastava et al., 2013a). The PTFs calculate water retention curve based on empirical relationship between basic soil properties and parameters at a predefined potential.

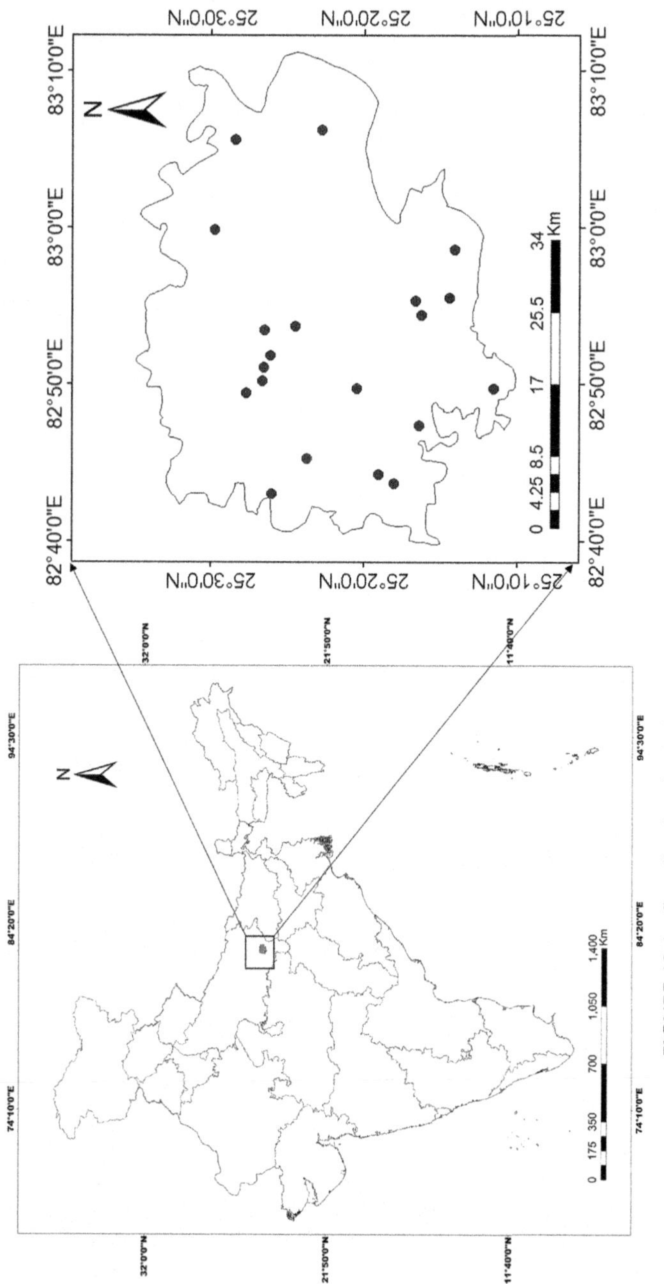

FIGURE 18.1 Sampling location in Varanasi district with geographical locations of villages.

For India, a pressure condition of −33 kPa (corresponding to the field capacity) and −1500 kPa (for the permanent wilting point) are generally used. Advanced form of ROSETTA model is equipped with neural network and bootstrap method. The model uses the most widely used van Genuchten equation (Van Genuchten, 1980), as given below:

$$\theta(h) = \theta_r + \frac{\theta_s - \theta_r}{(1 + (\alpha h)^n)^m}$$

where θ_r and θ_s are the residual and saturated water content, α is the scaling parameter, n is the curve shape factor and m is an empirical constant, which can be related to n by $m = 1 - 1/n$.

It is the difference of moisture content available in the plant root zone and the soil water that remains in soil against gravity, i.e., field capacity. We have developed the thematic layer of soil moisture content over Varanasi district, India, using software Arc GIS (version 10.1). The field capacity values are estimated from the ROSETTA model using soil hydraulic parameters and the field capacity map is generated. SMD is calculated by the following equation (Srivastava et al., 2014):

$$SMD = FC - SMC$$

2.4 Development of android application for irrigation scheduling

2.4.1 Irrigation Scheduler App development

The overall methodology of Irrigation Scheduler App development is represented by schematic diagram as Fig. 18.2.

This Irrigation Scheduler App is pioneer for Indian subcontinent. This is designed as a simple, user-friendly and cost-effective app through which we can provide SMD of the soil. For its proper calibration and validation many researchers are using this app. Irrigation Scheduler was designed to provide real-time irrigation schedules equally for all types of Rabi and Kharif crops. The user just needs to select the area (by geographical latitude and longitude) of crop site for knowing the moisture status of crop for irrigation. It readily displays useful information such as moisture status and SMD of the field. Future efforts will focus on inclusion of more study areas into the Irrigation Scheduler App.

Android Studio is the official Integrated Development Environment (IDE) for Android app development, based on IntelliJIDEA. IntelliJ is a powerful code editor and development tool. Android Studio is designed specifically for Android development. All the required tools to develop Android applications are open source and can be downloaded from the web along with some software that is needed before starting Android application programming, i.e., Java JDK5 or later version (Java Runtime Environment (JRE) 6).

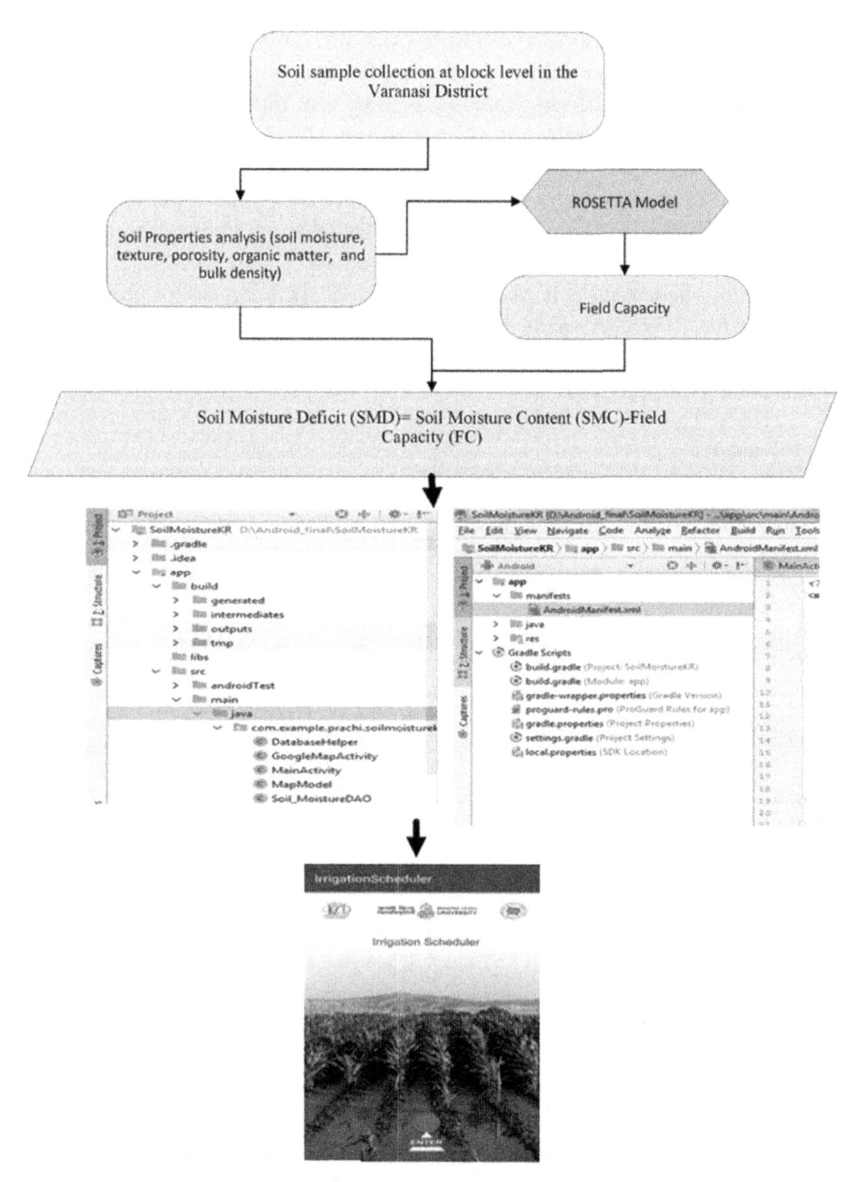

FIGURE 18.2 Schematic diagram.

Android Studio offers useful features for building apps having: a flexible Gradle-based build system, fast and feature-rich emulator, a unified environment (where you can develop for all Android devices), instant run to push changes (for running app without building a new APK), code templates and GitHub integration (for building common app features and importing sample code), extensive testing tools and frameworks, lint tools to catch performance, usability, version compatibility, C++ and NDK support and built-in support for Google Cloud Platform, making it easy to integrate Google Cloud Messaging and App Engine.

Irrigation Scheduler App development was provided using the Android Studio software using programming language of JAVA and tools of Android studio SDK (Software Development Kit). It follows the architecture of MVC (Model view controller). Database was created on MySql workbench, which returns data when request is made by app in JSON (Java Script Object Notation) format. AWS (Amazon web service) server was used to store MySql database for the app. Interaction between app, server and database can be shown clearly in Fig. 18.3.

2.4.2 Module

Android Studio contains one or more modules with source code files and resource files. Types of modules include Android app modules, Library modules and Google App Engine modules (Fig. 18.4).

- ➢ manifests: Contains the AndroidManifest.xml file.
- ➢ java: Contains the Java source code files, including JUnit test code.
- ➢ res: Contains all noncode resources, such as XML layouts, UI strings and bitmap images.

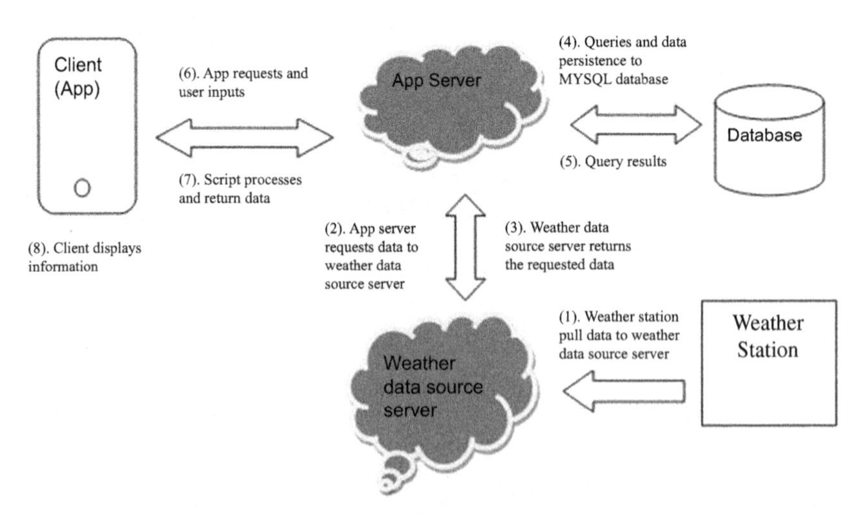

FIGURE 18.3 Diagram of interaction among client, server and automated weather stations (Migliaccio et al., 2016).

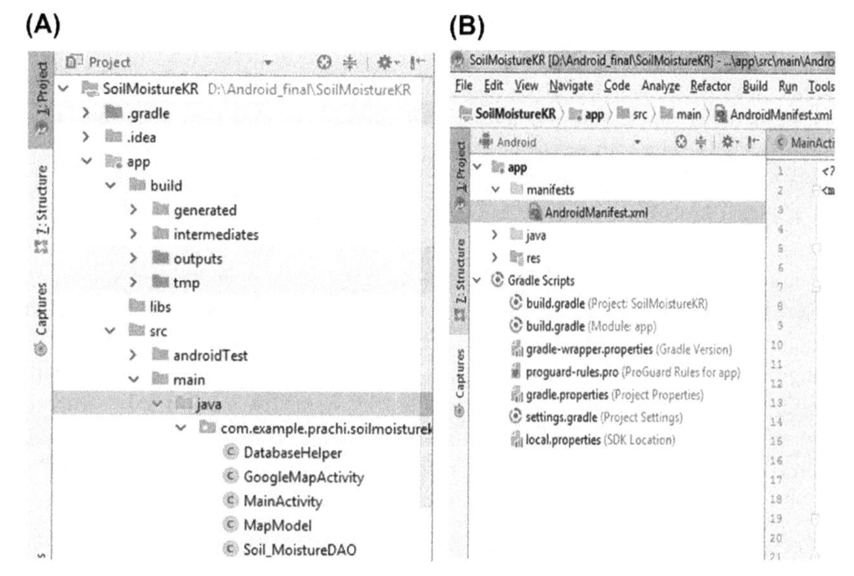

FIGURE 18.4 (A) Library module; (B) Interface of the app.

2.4.3 Android version

Currently, the developed app can be run in some specific mobile version given below in Table 18.1.

2.4.3.1 Geo-visualisation and testing

The analysis of soil moisture including all sample points showed that the eastern stretch from north to south covering Varanasi city block, parts of KVP, eastern Chiagigaon, central Sewapuri and northeastern Baragaon block contained high soil moisture ranging from 25% to 50%. It became evident from the field visit that these trends were mainly urban area mainly dominated by built up and concrete structures. Thus, it can be inferred that area of high urban

TABLE 18.1 This tabe shows the version supported by each app.

Code name	Version number	Release date	API level	Support status
Kitkat	4.4–4.4.4	31 Oct 2013	19–20	Supported
Lolipop	5.0–5.1.1	12 Nov 2104	21–22	Supported
Marshmallow	6.0–6.0.1	5 Oct 2015	23	Supported
Nougat	7.0–7.0.1	22 Aug 2016	24–25	Supported

agglomerations have a higher amount of soil moisture in comparison to the agricultural lands as plants utilise SM for their growth. Lands near to the water bodies and riverbeds were also found to be in line with the same trend of high soil moisture. From the map of DEM of Varanasi district, it was visualised that the area having high altitude has a negative bearing on soil moisture and vice versa.

The soil moisture and field capacity of the soil of the Varanasi district are shown in Fig. 18.5A and 18.5B respectively. A soil moisture map was generated for understanding the soil moisture variation over the test site using the Inverse Distance Weighting (IDW) interpolation technique by including all the sampling points. It is evident from the soil moisture map that soil moisture is significantly decreasing from east to west. Most of the eastern parts contain a high soil moisture that ranges between 25% and 50%, are mainly situated near to the water bodies and river beds while the majority of North-west and central part of the test site shows the medium moisture level ranging from 14% to 25%. It is evident from field visits that these medium moisture-holding sites are mainly the agricultural or fallow lands. The field capacity map of the soil of the Varanasi district according to the figure of Fig. 18.5B illustrates three types of soil categorised as sandy loamy, silt loamy and silt clay. The field capacity in terms of percentage was measured as 20% for sandy loamy soil, silt loamy 22% and silt clay 35%. Field capacity represents the water holding capacity of the soil, the result of analysis inferred majority of soil in Varanasi to be the sandy loamy type.

2.4.3.2 Irrigation Scheduler Android application

AVD (Android Virtual Device) was used to configure Irrigation Scheduler App than can be run on Android Emulator. This device is mostly useful to test any Android application, and also it is part of SDK (Software development kit) manager. Irrigation Scheduler App designed using Android Studio 3.1.4 software and JRE 1.8.0. Fig. 18.6A shows the main layout of the app whereas the next layout, i.e., Fig. 18.6B shows the locations of sampling points for Varanasi district on Google map. In database helper file of Irrigation Scheduler App, all the sampling databases are stored. After clicking on any sampling points, one can see the detailed information about the soil moisture that includes village name, soil type, SMD, latitude, longitude and soil moisture values as shown in Fig. 18.6C. This app will be helpful to know about the soil moisture–related information for the defined location from anywhere around the world at an almost real-time basis.

3. Conclusion

We have estimated soil moisture through Hydra probe, which was then analysed using advance data processing technique to assess the soil moisture

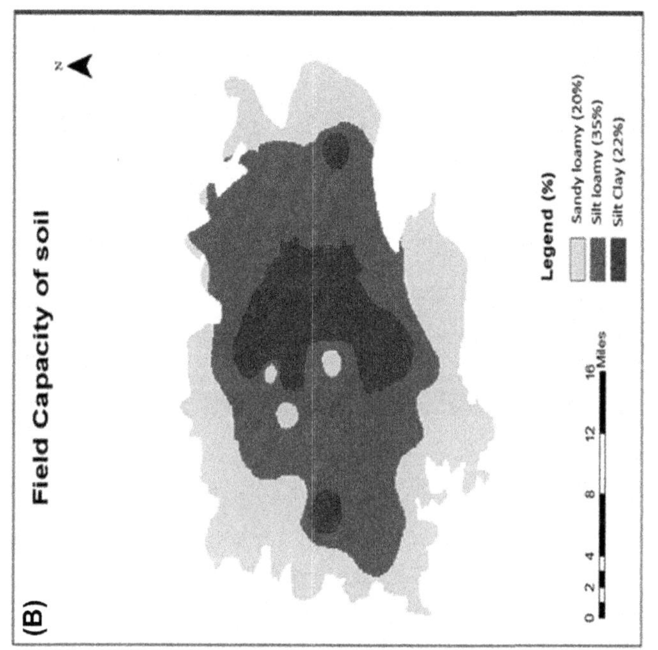

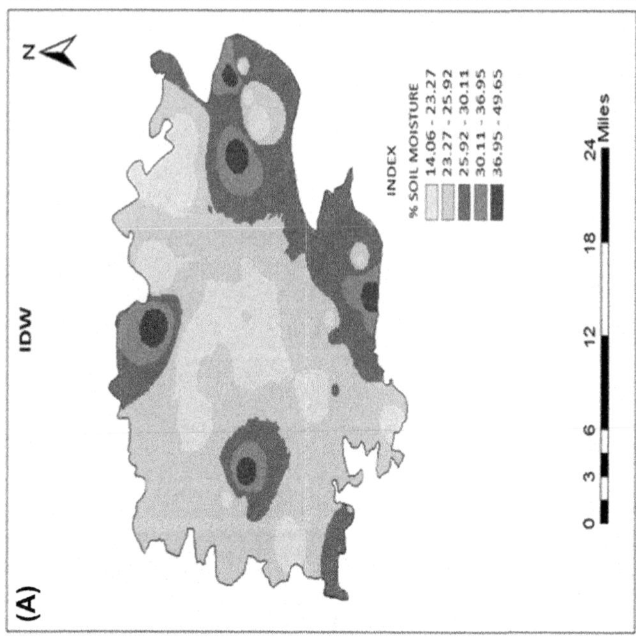

FIGURE 18.5 (A) Soil moisture variation in Varanasi district; (B) Field capacity variation in Varanasi district.

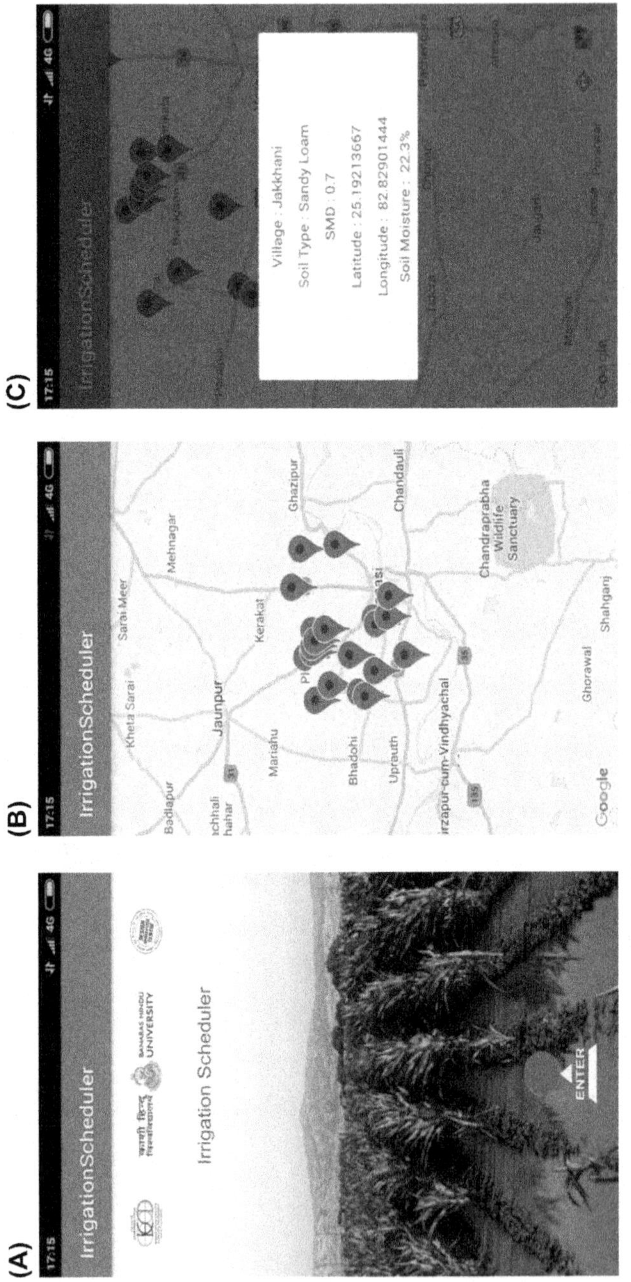

FIGURE 18.6 (A) App layout; (B) Google map location; and (C) SMD information.

deficit (SMD). The spatiotemporal database on SMD is the key information that will provide the accurate estimation of irrigation water requirement. The Irrigation Scheduler App provided in this chapter gives the quantitative irrigation water demand in terms of soil texture, SM that will help estimating the SMD.

The app can be used to generate real-time data, which will compute the required soil moisture content for optimum moisture condition for maximum crop yield. Moreover, it also assists in scheduling the irrigation with the required water, which will reduce the time and cost, and best optimize the water resources. Thus, it will help in minimising the adverse effects from droughts to secure the food and water. The developed methodology and app have potential as a prototype can be suitably used to prescribe water harvesting structures in this region, and can be linked with satellite-measured soil moisture content in future research for regular monitoring and providing the information of SMD.

The developed android mobile app would be useful for water resource managers, irrigation planners, agriculture practitioners, crop management team, central and state government agencies, and more appropriately for farmers. It will help in increasing crop productivity, minimising water loss, securing food and safeguarding future needs of the population. The android app will help effectively in irrigation water management practices by saving time and cost. Moreover, the agricultural community consisting of farmers, irrigation planners, central and state government agencies can use the outcome of the project. It will help in food and water security and safeguard future needs of the population.

Acknowledgement

We are thankful to Design and Innovation Centre, Banaras Hindu University, India for providing funding for this research.

References

Carlson, T.N., Gillies, R.R., Perry, E.M., 1994. A method to make use of thermal infrared temperature and NDVI measurements to infer surface soil water content and fractional vegetation cover. Remote Sens. Rev. 9, 161–173.

Engman, E.T., Chauhan, N., 1995. Status of microwave soil moisture measurements with remote sensing. Remote Sens. Environ. 51, 189–198.

Frenken, K., Gillet, V., 2012. Irrigation water requirement and water withdrawal by country. Food Agri. Org. U.N. 264.

Garg, N., Gupta, M., 2015. Assessment of improved soil hydraulic parameters for soil water content simulation and irrigation scheduling. Irrigat. Sci. 33, 247–264.

Guidelines, G., Audit, W., Conservation, W., 2005. Government of India ministry of water resources. General Guidelines for Water Audit & Water Conservation Central Water Commission, December 2005. New Delhi. Benchmarking.

Gupta, D., Prasad, R., Srivastava, P.K., Islam, T., 2016. Nonparametric model for the retrieval of soil moisture by microwave remote sensing. In: Satellite Soil Moisture Retrieval. Elsevier, pp. 159–168.

Gupta, M., Srivastava, P.K., Islam, T., Ishak, A.M.B., 2013. Evaluation of TRMM rainfall for soil moisture prediction in a subtropical climate. Environ. Earth Sci. https://doi.org/10.1007/s12665-013-2837-6.

Jaguey, J.G., Villa-Medina, J.F., Lopez-Guzman, A., Porta-Gandara, M.A., 2015. Smartphone irrigation sensor. IEEE Sens. J. 15 (9), 5122–5127. https://doi.org/10.1109/JSEN.2015.2435516.

Liang, X., Wood, E.F., Lettenmaier, D.P., 1996. Surface soil moisture parameterization of the VIC-2L model: Evaluation and modification. Glob. Planet Change. https://doi.org/10.1016/0921-8181(95)00046-1.

Migliaccio, K., Morgan, K., Fraisse, C., Vellidis, G., Andreis, J., 2015. Performance evaluation of urban turf irrigation smartphone app. Comput. Electron. Agric. 118, 136–142.

Migliaccio, K.W., Morgan, K.T., Vellidis, G., Zotarelli, L., Fraisse, C., Zurweller, B.A., Andreis, J.H., Crane, J.H., Rowland, D.L., 2016. Smartphone apps for irrigation scheduling. Trans. ASABE 59, 291–301.

Pandey, V., Srivastava, P.K., 2018. Integration of satellite, global reanalysis data and macroscale hydrological model for drought assessment in sub-tropical region of India. International Archives of the Photogrammetry. Remote Sens. & Spat. Inf. Sci. 42.

Petropoulos, G., Ireland, G., Griffiths, H., Islam, T., Kalivas, D., Anagnostopoulos, V., Hodges, C., Srivastava, P.K., 2016. Spatiotemporal estimates of surface soil moisture from space using the Ts/VI feature space. In: Satellite Soil Moisture Retrieval. Elsevier, pp. 91–108.

Petropoulos, G.P., Ireland, G., Srivastava, P.K., 2015. Evaluation of the soil moisture operational estimates from SMOS in Europe: results over diverse ecosystems. IEEE Sensor. J. 15, 5243–5251.

Petropoulos, G.P., Ireland, G., Srivastava, P.K., Ioannou-Katidis, P., 2014. An appraisal of the accuracy of operational soil moisture estimates from SMOS MIRAS using validated in situ observations acquired in a Mediterranean environment. Int. J. Remote Sens. 35, 5239–5250.

Petropoulos, G.P., Srivastava, P.K., Piles, M., Pearson, S., 2018. Earth observation-based operational estimation of soil moisture and evapotranspiration for agricultural crops in support of sustainable water management. Sustainability 10, 181.

Schaap, M.G., Leij, F.J., Van Genuchten, M.T., 2001. Rosetta: a computer program for estimating soil hydraulic parameters with hierarchical pedotransfer functions. J. Hydrol. 251, 163–176.

Srivastava, P.K., Pandey, V., Suman, S., Gupta, M., Islam, T., 2016a. Available data sets and satellites for terrestrial soil moisture estimation. In: Satellite Soil Moisture Retrieval. Elsevier, pp. 29–44.

Srivastava, P.K., Singh, R.M., 2016. GIS based integrated modelling framework for agricultural canal system simulation and management in Indo-Gangetic plains of India. Agric. Water Manag. 163, 37–47.

Srivastava, P.K., 2017. Satellite soil moisture: review of theory and applications in water resources. Water Resour. Manag. 31, 3161–3176.

Srivastava, P.K., Han, D., Ramirez, M.A.R., Islam, T., 2013a. Appraisal of SMOS soil moisture at a catchment scale in a temperate maritime climate. J. Hydrol. 498, 292–304.

Srivastava, P.K., Han, D., Ramirez, M.R., Islam, T., 2013b. Machine learning techniques for downscaling SMOS satellite soil moisture using MODIS land surface temperature for hydrological application. Water Resour. Manag. 27, 3127–3144.

Srivastava, P.K., Han, D., Rico-Ramirez, M.A., Al-Shrafany, D., Islam, T., 2013c. Data fusion techniques for improving soil moisture deficit using SMOS satellite and WRF-NOAH land surface model. Water Resour. Manag. 27, 5069–5087.

Srivastava, P.K., Han, D., Rico-Ramirez, M.A., O'Neill, P., Islam, T., Gupta, M., 2014. Assessment of SMOS soil moisture retrieval parameters using tau–omega algorithms for soil moisture deficit estimation. J. Hydrol. 519, 574–587.

Srivastava, P.K., Han, D., Rico-Ramirez, M.A., O'Neill, P., Islam, T., Gupta, M., Dai, Q., 2015. Performance evaluation of WRF-Noah Land surface model estimated soil moisture for hydrological application: synergistic evaluation using SMOS retrieved soil moisture. J. Hydrol. 529, 200–212.

Srivastava, P.K., Pandey, P.C., Kumar, P., Raghubanshi, A.S., Han, D., 2016b. Geospatial Technology for Water Resource Applications. CRC Press.

Summary, E., n.d., How to Feed the World in 2050. 2050 (1), 1–35.

Van Genuchten, M.T., 1980. A closed-form equation for predicting the hydraulic conductivity of unsaturated soils. Soil Sci. Soc. Am. J. 44, 892–898.

Vellidis, G., Liakos, V., Andreis, J., Perry, C., Porter, W., Barnes, E., Morgan, K., Fraisse, C., Migliaccio, K., 2016a. Development and assessment of a smartphone application for irrigation scheduling in cotton. Comput. Electron. Agric. 127, 249–259.

Vellidis, G., Liakos, V., Perry, C., Porter, W., Tucker, M., Boyd, S., Huffman, M., Robertson, B., 2016b. Irrigation scheduling for cotton using soil moisture sensors, smartphone apps, and traditional methods. In: Proceedings of the 2016 Beltwide Cotton Conference, New Orleans, LA. National Cotton Council Memphis, TN.

Chapter 19

GIS-based analysis for soil moisture estimation via kriging with external drift

Akash Anand[1], Prachi Singh[1], Prashant K. Srivastava[1,2], Manika Gupta[3]
[1]*Remote Sensing Laboratory, Institute of Environment and Sustainable Development, Banaras Hindu University, Varanasi, Uttar Pradesh, India;* [2]*DST-Mahamana Centre of Excellence in Climate Change Research, Institute of Environment and Sustainable Development, Banaras Hindu University, Varanasi, Uttar Pradesh, India;* [3]*Department of Geology, University of Delhi, New Delhi, India*

1. Introduction

Soil moisture plays a vital role in the modelling and monitoring of ecosystem variations with space and time (Falkenmark et al., 2004; Kerr et al., 2016). The amount of soil moisture in a particular area depends on several factors including topography, land cover, land surface temperature and other climatic parameters, whereas it has major implications for agriculture, wildlife, ecology and most importantly the hydrological cycle. Soil moisture is an integral part of plant growth and its continuous monitoring can provide information about drought-prone areas as well. Numerous studies have shown the impact of soil moisture on the interaction between earth surface and the atmosphere that has a direct influence on weather and global climate. Studies by Zhang and Anthes (1982) and Chang and Wetzel (1991) have shown the influence of variation in soil moisture and vegetation on intensity and development of a severe storm. For a large-scale modelling, soil temperature and soil moisture are key parameters to identify planetary boundary layers and wind patterns (Srivastava et al., 2015). Moisture present in the organic layer of the soil directly affects the energy fluxes coming from the land surface interfaces (Srivastava et al., 2013; Chaube et al., 2019; Srivastava et al., 2014a,b; Suman et al., 2020).

Spatial distribution analysis of soil moisture is quite an important prerequisite for modelling many land surface hydrological processes, which include soil erosion, infiltration, runoff, etc. Soil moisture is directly related to the hydrological cycle and can be affected by the regional land-use land-cover

Agricultural Water Management. https://doi.org/10.1016/B978-0-12-812362-1.00019-9

due to change in porosity and alteration in soil profile (Srivastava, 2017; Cambardella and Elliott, 1992; Srivastava et al., 2017). Regular monitoring of both soil moisture and soil temperature can be very effective for agriculture and farming purposes, as it can be used in the early monitoring of droughts and provides information about health of the soil. It can also help to improve the irrigational facility of any area and can result in better crop productivity, whereas regular monitoring also enables us to gather information about floods before it happens. Water-holding capacity of soil is a key parameter in modelling flood forecasting, at saturation, the capacity of soil to hold the water and percolation reaches its limit and results in flooding (Snepvangers et al., 2003; Kourgialas and Karatzas, 2014, 2016). Similarly, soil temperature influences the plant growth. A continuous monitoring of soil temperature is necessary for the different agricultural purposes (Junlong and Jianquan, 2009; Skierucha et al., 2012).

Spatial interpolation is one of the key components of GIS, which is used to predict the nearby value with a known central value (Lam, 1983; Mitas and Mitasova, 1999; Li et al., 2000; Li and Heap, 2008). It is widely used for the analysis of larger and complex datasets and is also a very time-consuming task that requires satisfactory knowledge of essential methods and their implementation. Interpolation is an interesting statistical method that is used in many disciplines like mathematics, earth science, geography and engineering, because as in many environmental applications, the remote measurements can be time-consuming, expensive and laborious. Interpolation methods are useful to predict values at a location for which there is no recorded observation or unsampled sites (Algarni, 2001). Broadly, interpolation is classified into two types, that is, nongeostatistical and geostatistical. Nearest neighbor (Dunlop, 1980), triangular irregular network (Wu et al., 2011), spline (Cox, 1975), etc., come under nongeostatistical interpolation; whereas simple kriging, ordinary kriging (Saito et al., 2005), etc. are univariate geostatistical interpolation; and universal kriging (Brus and Heuvelink, 2007), kriging with external drift (Snepvangers et al., 2003), ordinary cokriging (Diodato and Ceccarelli, 2005), etc., are multivariate geostatistical interpolation (Li and Heap, 2008). Spatial interpolation is necessary when (1) the discrete surface has a different level of resolution, cell size or orientation than another; (2) a continuous surface is represented by a data model that is different from that required; (3) and if the sample data do not cover the domain of interest completely (Burrough and McDonnell, 1998). In such instances, spatial interpolation methods provide tools to fulfil such task by estimating the values of an environmental variable at unsampled sites using data from point observations within the same region. Predicting the values of a variable at points outside the region covered by existing observations is called extrapolation (Burrough and McDonnell, 1998).

The presence of a spatial structure where observations close to each other are more alike than those that are far apart (i.e., spatial autocorrelation) is a prerequisite to the application of geostatistics (Goovaerts, 1999; Petropoulos et al., 2018). Different interpolation approaches have been used in different studies, but the multivariate interpolation approach shows higher accuracy for interpolation (Kravchenko and Bullock, 1999). In present study, KED approach is used as input for the semi-variogram analysis. The KED approach is one of most useful methods for the assumption of the values of an attribute of unsampled points. It uses linear combination of values at sampled points, weighted by an inverse function of the distance from the point of interest to the sampled points. The statement is that sampled points within a local neighborhood surrounding the unsampled point are more similar to it than those further away in their values (Isaaks and Srivastava, 1989). Whereas, another geospatial method known as semi-variogram is used which measures the average degree of dissimilarity between unsampled values and a nearby data value (Kupfersberger et al., 1998) and thus can depict autocorrelation at various distances. The value of the semi-variogram for a separation distance of h (referred to as the lag) is half the average squared difference between two nearby points (Western et al., 2002).

As geographic information systems (GIS) and modelling techniques are becoming powerful tools in natural resource management and biological conservation, spatial distribution of various climatic and environmental variables are increasingly required (Hartkamp et al., 1999; Pandey et al., 2019; Anand et al., 2020). The values of an attribute at unsampled points need to be estimated, meaning that spatial interpolation from point data to spatial continuous data is necessary (Li and Heap, 2008).

2. Study area

For the present study, Varanasi district was chosen as the study area. Varanasi district is located at 25°10′30′ and 25°35′15′ north latitude and 82°40′50′ and 83°12′18′ east longitude with a total area of 1535.00 sq. km and subtropical climate. The mean annual rainfall recorded is about 1056 mm/year (+172 SD) (Rao et al., 1971). The River Ganga flows south to north having the highest flood level at 73.90 m (1978) and the lowest river water level is approximately 58 m (Ranjan et al., 2017; Anand et al., 2018). It is a part of Indo-Gangetic Plain that supports good agricultural productivity and the land is composed of very fertile alluvial soil deposited by the Rivers Ganga and Varuna (Raju et al., 2009). In this study, 35 villages were used as a sample location for soil sampling using 5 cm diameter core pipe with depths of 5 and 20 cm from the surface. Almost all sample points are from agricultural regions. Analysis of study areas from total 50 locations data has been used for calibration and 15 locations data has been used for validation. The study area with sampling points is shown in Fig. 19.1.

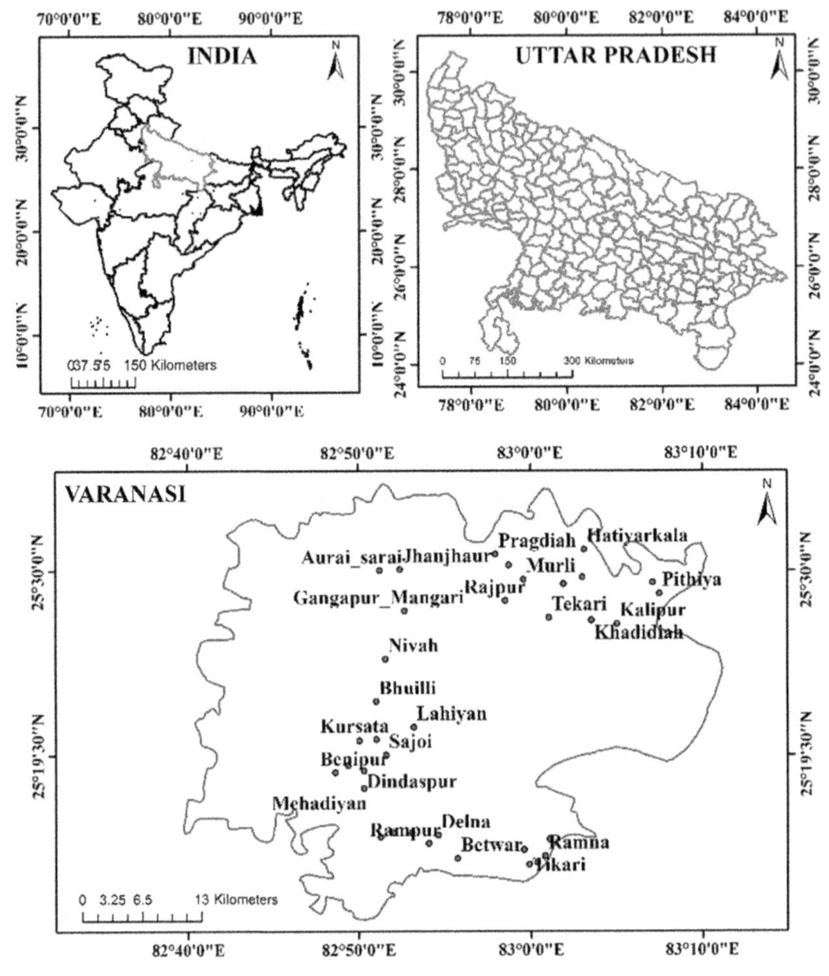

FIGURE 19.1 Location map of the study area.

3. Materials and methods

The in-situ soil moisture and soil temperature data are recorded using Stevens' Hydra probe at 50 different points within the study area, in which 35 points are used for generating the interpolation and remaining 15 are used for validation. The points are then interpolated using kriging with external drift method to increase the accuracy of the interpolation.

3.1 Kriging

Kriging is a geostatistical interpolation technique in which the interpolated values are simulated by a Gaussian Process using prior covariance

(Cressie, 1990; Knotters et al., 1995). Kriging is a linear model, which uses linear weights that depend not only on the distance between points but also on the direction and orientation of the local points.

Basic kriging equation is explained as,

$$\hat{Z}(x_0) - \mu = \sum_{i=1}^{n} \lambda_1 [Z(x_i) - \mu(x_0)] \tag{19.1}$$

where μ is a known stationary mean, assumed to be constant over the whole domain and calculated as the average of the data (Wackernagel, 2013). The parameter λ_I is kriging weight, n is the number of sampled points used to make the estimation and depends on the size of the sample points and $\mu(x_O)$ is the mean of samples within the sample points.

Kriging takes nonexhaustive secondary information to spatially correlate the primary and secondary variables (Goovaerts, 1997). So the secondary information is incorporated and shown in Eq. (19.2),

$$\widehat{Z_1}(x_0) - \mu_1 = \sum_{i_1=1}^{n_1} \lambda_1 [Z_1(x_{i_1}) - \mu_1(x_{i_1})] + \sum_{j=2}^{n_v} \sum_{i_j=1}^{n_j} \lambda_{i_j} [Z_j(x_{i_j}) - \mu_j(x_{i_j})] \tag{19.2}$$

where μ_1 is a known stationary mean of the primary variable, $z_{1(x_{i_1})}$is the data of the primary variable at point i_1, $\mu_1(x_{i_1})$ is the mean of samples within the sample point, n_1 is the number of sampled points within the sample points for point x_0 used to make the estimation, $(\lambda i_1 \cdot\ )$ is the weight selected to minimise the estimation variance of the primary variable, n_v is the number of secondary variables, n_j is the number of jth Secondary variable within the sample points, λ_{ij} is the weight assigned to i_j^{th} point of jth secondary variable, and is the data at i_j^{th} point of jth secondary variable and $\mu_j(x_{i_j})$ is the mean of samples of jth secondary variable within the sample points.

3.2 Semi-variogram

In few cases the interpolation can be aided by incorporating some additional variables, which help in increasing the sensitivity of the final product, which is termed as covariate. Whereas, all the geostatistical methods are related to a theory of assumptions, those values at close locations are more similar. Among geostatistical methods, semi-variogram is known as the most important tool for geostatistical analysis to measure the change in correlation with increasing distance (Curran, 1988; Zimmerman and Zimmerman, 1991). The semi-variogram is defined as half the variance of the increment in the random function (Bivand et al., 2008).

$$\gamma(h) = \frac{1}{2} E(Z(X) - Z(X + h))^2 \tag{19.3}$$

Calculation of and interpretation of $X + h$ distance and (h) value gets more difficult. The estimation of the semi-variogram can be calculated from this equation that is obtained from the sample.

$$\overline{\gamma}\left(\overline{h_j}\right) = \frac{1}{2n_h} \sum_{i=1}^{n_h} (Z(X_i) - Z(X_{i+h}))^2 \tag{19.4}$$

where $N(h)$ shows the number of sample pairs which have the distance h away from each other, whilst $Z(Xi)$ and $Z(Xi + h)$ show the value of variables at Xi and $Xi + h$ points (Isaaks and Srivastava, 1989; Ebdon, 1996; Bivand et al., 2008; Hengl, 2009).

3.3 Kriging with external drift (KED)

The kriging with external drift (KED) assimilates the local drift within the local sample points as a linear function (Goovaerts, 1997). The drift of the primary variable must be linearly associated to that of the secondary variable. And secondary variable should measure all primary data points. KED is also known as Universal Kriging (UK) or external drift kriging in (Pebesma, 2004). The KED estimator is mathematically explained as

$$z^*_{ced(\mu)} = \sum_{\alpha=1}^{n(\mu)} \lambda_\alpha^{ced}(\mu) z\,(\mu\alpha) \tag{19.5}$$

where $z^*_{ced}\,(\mu)$ is the KED estimator at location μ, $\lambda\alpha$ KED and (μ) are the KED weights corresponding to the n samples at location μ, and z $(\mu\ )$ are the sample values within the locations.

KED is preferred in present study, as it allows the incorporation of more than one covariate, as well as it requires a less complex variogram analysis compared to cokriging which requires variogram for each covariate. Goovaerts (2000) in his comparative study also found that kriging with external drift performs better than other kriging approaches because of its flexibility towards covariates. Araghinejad et al. (2006) used kriging with varying local means and kriging with external drift multivariate algorithms to model spatial variation of precipitation; his results show that kriging with external drift is showing higher accuracy. Different geostatistical methods were used by Musio et al. (2004) to model the spatial distribution of silver fir and Norway spruce in Black forest. She compared model with independent error, ordinary kriging, kriging with external drift and co-kriging, kriging with external drift performed better than others and it also halved the mean square prediction error than other models.

3.4 Geostatistical parameters

Few geostatistical parameters are calculated to see the accuracy of the analysis. Coefficient of Determination (R^2) can be calculated using this Eq. (19.6)

$$R^2 = 1 - \frac{\sum i(\hat{z}(x_i) - z(x_i))^2}{\sum_i (\hat{z}(x_i) - \overline{z(x_i)})^2} \tag{19.6}$$

Root Mean Square Error (RMSE) has been used as a standard statistical method and useful to measure model performance in meteorology, air quality and climate research studies (McKeen et al., 2005; Savage et al., 2013). RMSE can be calculated using this Eq. (19.7)

$$\text{RMSE} = \sqrt{\frac{1}{n}\sum_{i=1}^{n} (\hat{z}(x_i) - z(x_i))^2} \tag{19.7}$$

where N is the number of sample points used in the validation sets, $Z(X_i)$ are the observed values, $\hat{z}(x_i)$ are the estimated values and $\overline{z(x_i)}$ is the spatial average of the observed values.

The sum of all sample data divided by the number of entries is called sample mean values.

$$\overline{x} = \frac{\sum x}{n} \tag{19.8}$$

The standard deviation measures variability and steadiness of the sample. In statistical data analysis, less variation is often better. Standard deviation can be explained as

$$s.d. = \sqrt{\frac{\sum (x - \overline{x})^2}{n - 1}} \tag{19.9}$$

Skewness is a measure of symmetry, or more precisely, the lack of symmetry. A distribution, or dataset, is symmetric if it looks the same to the left and right of the centre point, whereas, kurtosis defines the relative sharpness of the normal frequency distribution. Positive kurtosis represents the higher peaks in the frequency distribution. Mathematical expression for skewness and kurtosis are explained as

$$Skewness = \frac{\sum_{i=1}^{n} (x_i - \overline{x})^2 / n}{\left[\sum_{i=1}^{n} (x_i - \overline{x})^2 / n\right]^{1.5}} \tag{19.10}$$

$$Kurtosis = \frac{\sum_{i=1}^{n}(x_i - \bar{x})^4/n}{\left[\sum_{i=1}^{n}(x_i - \bar{x})^2/n\right]^2} \quad (19.11)$$

where Σ is the summation for all observations (x_i) in a sample, x bar is the sample mean and n is the sample size.

3.5 Hydra probe

Hydra probe is an in-situ measurement tool used to measure soil moisture and soil temperature of different soil profiles. In this study, the in-situ data of soil temperature and soil moisture has been collected from the Steven's Hydra probe. This instrument has three sensors which are plugged in the soil at three different depths—5, 20 and 50 cm, respectively. The sensors work on the principle of Coaxial Impedance Dielectric Reflectometry (also known as 'Radiometric Coaxial Impedance Dielectric Reflectometry'). In this method soil moisture measurement employs an oscillator to generate electromagnetic signal by using metal tines into the soil. Soil reflects the propagated signal and sensor measures amplitude of incident signal and reflected signal in voltage (Seyfried et al., 2005; Burns et al., 2014). The ratio of these voltages; reflected signal and incident signal, are further used in Maxwell's equations, to calculate the imaginary and real dielectric permittivity that is further used to finding the soil water content. The unique advantage of this sensor is not requiring to calibrate soil moisture content for most of the soil. In this principle soil moisture calibration is less influenced by temperature, soil variability, soil salinity and intersensor variability (Bellingham, 2007).

4. Result and discussions

4.1 Interpolated soil moisture and soil temperature

Kriging is performed using the in-situ soil moisture and soil temperature points for the study area, which has given 79,250 different pixels. From the 79,250 different pixels, frequency plot is generated for both the parameters which has been shown in Fig. 19.2. Frequency plot of soil moisture has shown that the values are varying from 12.425% to 50.822% with a mean and median of 24.384% and 24.107%, respectively. Whereas, skewness and kurtosis are 1.941 and 15.304, kurtosis is showing a higher value due to the sudden sharpness of frequency plot. Frequency plot of soil temperature is showing higher variation throughout; it is varying between 15.638°C to 20.252°C with a mean and median of 18.083°C and 18.037°C, respectively. Skewness and kurtosis are showing a value of −0.103°C and 2.228°C; here the negative kurtosis is due to the presence of large number of peaks. First quartile value defines the middle number between lower limit and median of the dataset which is observed to be 22.908% in soil moisture dataset whereas for soil

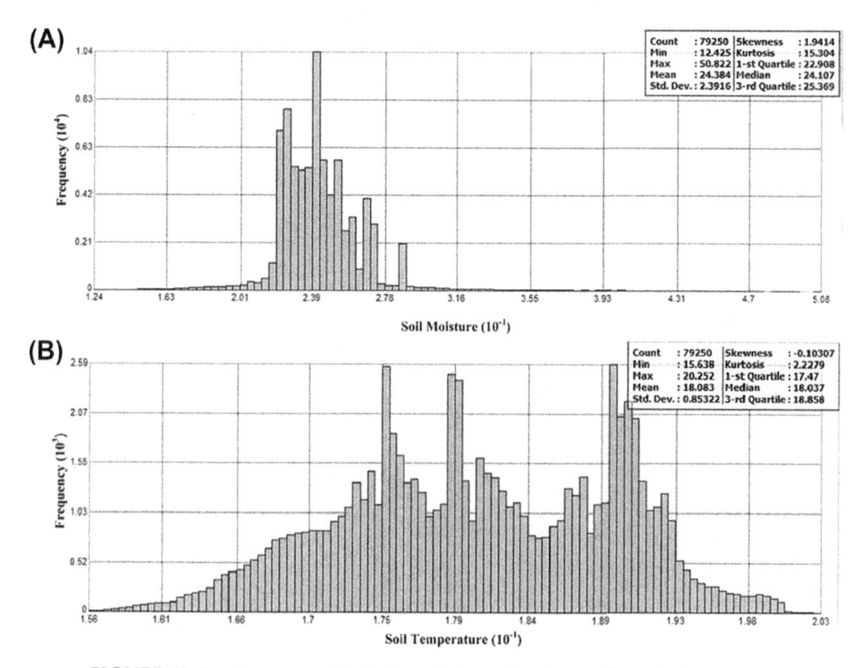

FIGURE 19.2 Frequency distribution of (A) soil moisture (B) soil temperature.

temperature the value is 17.47°C. Furthermore, third quartile is also calculated which shows the middle value between upper limit and median of the dataset, the value of third quartile for soil moisture and soil temperature are 25.369% and 18.858°C, respectively. So through this analysis, it is estimated that the majority pixels are having a soil moisture between 22.908% and 25.369% and soil temperature between 17.47°C and 18.858°C.

Interpolated soil moisture and soil temperature are shown in Fig. 19.3. Soil moisture and soil temperature are classified into five classes. Soil moisture is equally distributed from 14% to 51% into five classes whereas the soil temperature is also equally distributed into five classes with values ranging from 15°C to 20°C. As kriging is a univariate geostatistical approach that predicts values of nearby location using a known point, an external drift is also calculated to provide another dimension to make the prediction more accurate.

4.1.1 Semi-variogram analysis

Semi-variogram is a graph used to relate the variation in the difference of values of the variable pairs to the separation distance between them also the direction of the variation is considered. This geostatistical approach is widely used for establishing spatial correlation for sampled data and spatial prediction for nearby unsampled locations.

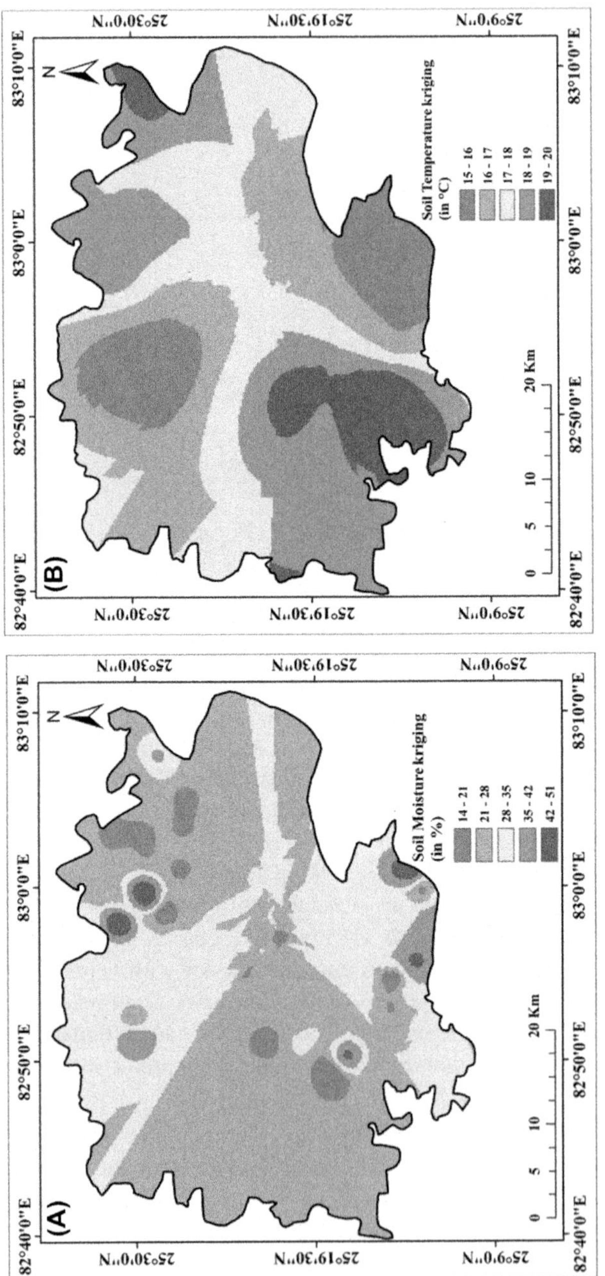

FIGURE 19.3 Kriging of (A) soil moisture and (B) soil temperature.

In present analysis, the semi-variogram is developed by assigning a bin size of 0.007 and total 12 lags. Parameter is a constant value in semi-variogram modelling, which defines the stability of a model, the value of parameter varies between 0 and 2 in which value 0 represents zero stability of the model whereas 1 represents an exponential model; 2 represents a Gaussian model. Presently, the value of parameter is observed to be 1.284, which states that the model is stable. Whereas the major range is 0.0632 and partial still is 0.0147, which represents that the model is following a similar trend throughout. Through the semi-variogram plot it is observed that the semi-variance is varying from 1.61×10^{-3} to 1.768×10^{-2}, whereas the upper limit of the distance is 8.608×10^{-2} decimal degrees (Fig. 19.4). According to the law of spatial geography, it is well known that the nearer objects have more similar values than farther objects. Similarly, here the low semi-variance value represents two input variables having very similar values or closer to each other, whereas high semi-variance value represents more variation between two variables. An average line and model line is plotted through the graph to show the variation of bins from these lines, it is observed that the bins having low values are more nearer to the model line whereas the higher values tend to scatter more.

After performing the semi-variogram analysis, it is verified that the model is stable and following a similar trend throughout. Thereafter weights are assigned to estimate the value at nearby locations with maximum neighbor possibility of 5 and minimum neighbor possibility of 2. A regression is performed between the observed and predicted soil moisture values for every

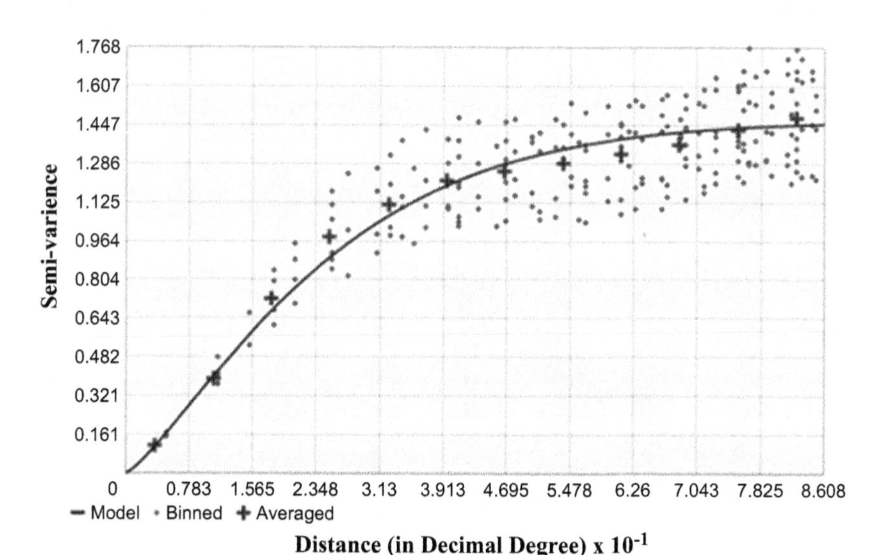

FIGURE 19.4 Semi-variogram plot.

pixel value which has shown an R^2 of 0.9896 with a regression equation of $Y = 0.939X + 1.507$ as shown in Fig. 19.5. The predicted values have shown a mean error of 0.029 and RMSE of 0.32 throughout. After the regression, the predicted values are used for generating the final interpolated product, which is shown in Fig. 19.6. Soil moisture kriging with external drift is generated as a final product, which is further classified into five equally distributed classes as 14%–21%, 21%–28%, 28%–35%, 35%–42% and 42%–48%. The predicted mean soil moisture value is 24.158%, whereas the mean of observed values is 24.384% and predicted standard deviation is 2.383% and observed standard deviation is 2.39%. Skewness and kurtosis is 1.867 and 12.949, respectively, for the predicted values, whereas for the observed values the kurtosis and skewness is 1.941 and 15.304 (Fig. 19.7).

The kriging of soil moisture with external drift has shown a significantly higher accuracy. Through present analysis it is observed that kriging with external drift is one of the most accurate multivariate tools for the analysis of spatial distribution and calculation of interpolation.

A similar analysis is done by Srivastava et al. (2019), in which they used four interpolation techniques namely Inverse Distance Weighting (IDW), spline, ordinary kriging and KED to model soil moisture based on 82 in-situ points. The results from that analysis show that KED is performing better than other models, KED has an RMSE of 8.69, percentage bias of −0.9 and Mean Absolute Error (MAE) of 5.52, which is very close to the performance statistics of ordinary kriging with RMSE of 8.73, percentage bias of −0.9 and MAE of 5.56. IDW has shown the lowest performance statistics than

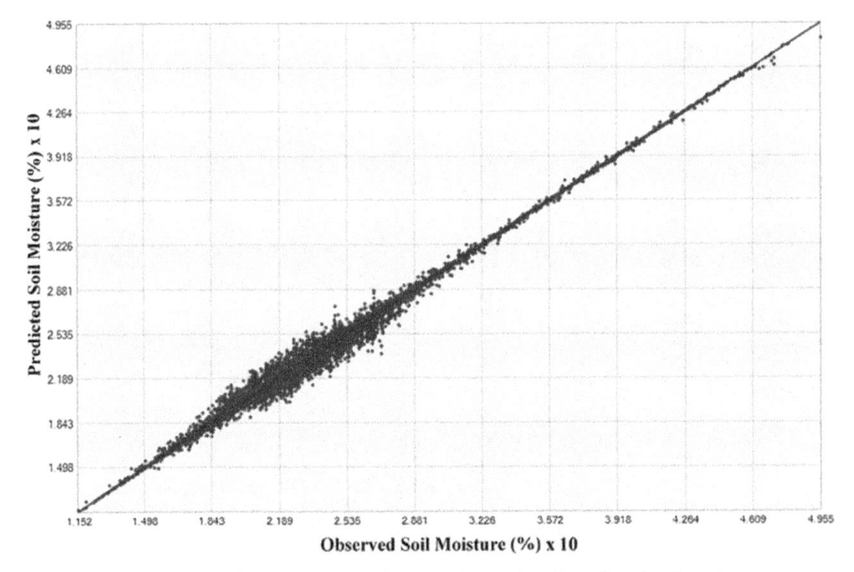

FIGURE 19.5 Correlation plot between observed and predicted soil moisture.

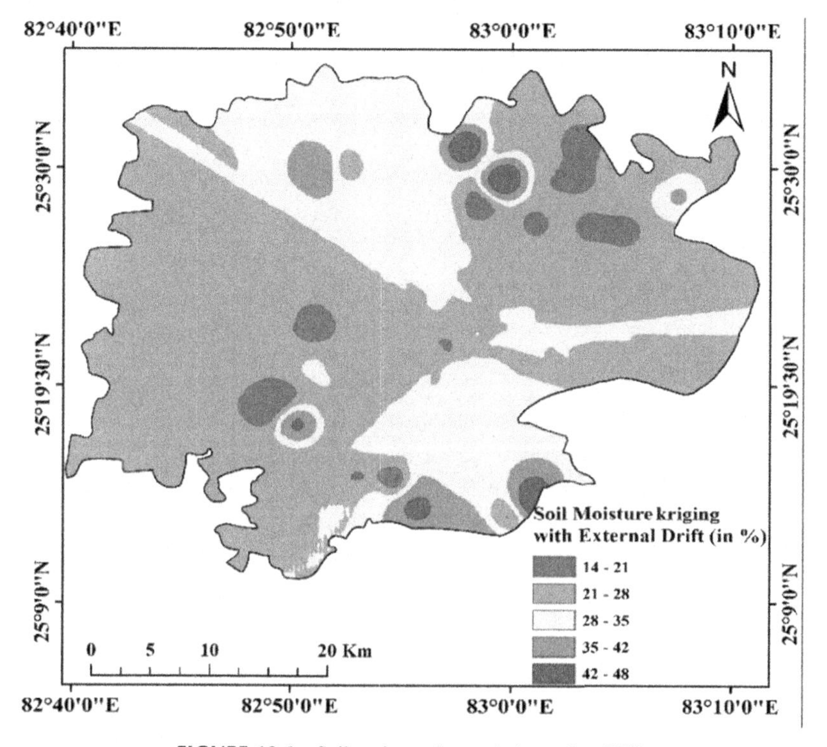

FIGURE 19.6 Soil moisture interpolation using KED.

other models having an RMSE value of 9.32, percentage bias of −0.3 and MAE value 6.14. Whereas in present study, semi-variogram analysis is performed for two covariates namely soil moisture and soil temperature, it has significantly improved the results. The performance of KED for 50,851 random sample pixels within the study has shown RMSE value of 0.32 and an MAE of 1.52.

Through this analysis it is observed that in order to integrate the datasets of different covariates, advanced geostatistical approach like KED can combine the covariates into a single layer. Although the results for KED are better than other models, there are some computational requirements, which need to be satisfied in order to run the model. Mathematically, KED follows a straightforward rule in which it requires the variogram coefficients as inputs. Using the covariance matrix and auxiliary parameters, kriging weights can be calculated in KED but the satisfactory variogram is required in order to get the output. Before executing the model, it is required that the relation between target and input parameters must be linear and the variables should vary smoothly within the spatial limit to avoid the data instability. One of the major challenges in performing multivariate geostatistical analysis is the heterogeneity in the data and its lower dependencies with auxiliary variables.

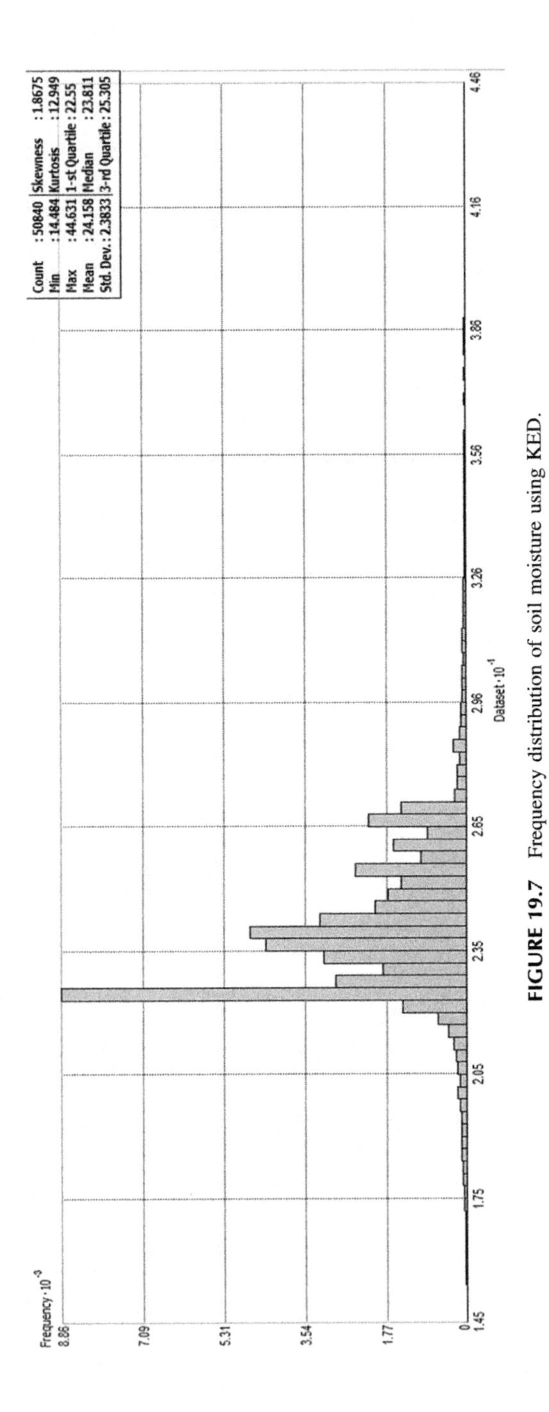

FIGURE 19.7 Frequency distribution of soil moisture using KED.

5. Conclusion

As soil moisture is a key element in regional as well as global water cycle, it has also a direct relation with global climate and agriculture practices. So, a regular monitoring at a high spatial and temporal resolution is needed to improve our understanding towards soil moisture variation and factors affecting it, which can help in predicting the future scenarios with better accuracy. The satellite data are a major source of information for mapping soil moisture but their spatial resolution is yet to be improved to benefit the regional scale, and most importantly satellite data lacks in dimensionality. But the geostatistical analysis can be used to incorporate regional factors to enhance the accuracy.

Present study stated the role of multivariate geostatistical approach for generating spatial distribution of soil moisture. Spatial distribution of in-situ soil moisture is very complex to model, as it varies with time and space very significantly. Kriging with external drift approach is used presently, in which soil moisture and soil temperature are considered as covariates and used as an input parameter in the model. The predicted soil moisture has shown an RMSE value of 0.32 and an MAE of 1.52. Semi-variogram analysis is done using interpolated soil moisture and soil temperature data by assigning the weights, which has shown a semi-covariance parameter value of 1.284, which means the model is stable and its trend is similar throughout. The results shown in this study illustrate the role of multivariate approach for interpolation, especially the kriging with external drift approach. Further the multivariate geostatistical approach can be used in future studies to model stochastic simulations specially to downscale coarse resolution data using fine-scale covariates.

References

Algarni, D.A., 2001. Comparison of thin plate spline, polynomial, CI—function and Shepard's interpolation techniques with GPS-derived DEM. Int. J. Appl. Earth Obs. Geoinf. 3, 155–161.

Anand, A., Kumar, A., Patil, R.G., 2018. Remote sensing based approach on recent changes in platform of River Ganga from Mirzapur to Ballia. i-Manag. J. Futur. Eng. Technol. 13, 19.

Anand, A., Malhi, R.K.M., Pandey, P.C., Petropoulos, G.P., Pavlides, A., Sharma, J.K., Srivastava, P.K., 2020. Use of hyperion for mangrove forest carbon stock assessment in Bhitarkanika forest reserve: a contribution towards blue carbon initiative. Rem. Sens. 12, 597.

Araghinejad, S., Burn, D.H., Karamouz, M., 2006. Long-lead probabilistic forecasting of streamflow using ocean-atmospheric and hydrological predictors. Water Resour. Res. 42.

Bellingham, K., 2007. The Stevens Hydra Probe Inorganic Soil Calibrations.

Bivand, R.S., Pebesma, E.J., Gómez-Rubio, V., Pebesma, E.J., 2008. Applied Spatial Data Analysis with R. Springer.

Brus, D.J., Heuvelink, G.B., 2007. Optimization of sample patterns for universal kriging of environmental variables. Geoderma 138, 86–95.

Burns, T.T., Adams, J.R., Berg, A.A., 2014. Laboratory calibration procedures of the Hydra Probe soil moisture sensor: infiltration wet-up vs. dry-down. Vadose Zone J. 13.

Burrough, P.A., McDonnell, R.A., 1998. Creating Continuous Surfaces from Point Data. Principles of Geographic Information Systems. Oxford University Press, Oxford, UK.

Cambardella, C., Elliott, E., 1992. Particulate soil organic-matter changes across a grassland cultivation sequence. Soil Sci. Soc. Am. J. 56, 777–783.

Chaube, N.R., Chaurasia, S., Tripathy, R., Pandey, D.K., Misra, A., Bhattacharya, B., Srivastava, P.K., 2019. Crop phenology and soil moisture applications of SCATSAT-1. Curr. Sci. 117, 10.

Chang, J.-T., Wetzel, P.J., 1991. Effects of spatial variations of soil moisture and vegetation on the evolution of a prestorm environment: a numerical case study. Mon. Weather Rev. 119, 1368–1390.

Cox, M.G., 1975. An algorithm for spline interpolation. IMA J. Appl. Math. 15, 95–108.

Cressie, N., 1990. The origins of kriging. Math. Geol. 22, 239–252.

Curran, P.J., 1988. The semivariogram in remote sensing: an introduction. Rem. Sens. Environ. 24, 493–507.

Diodato, N., Ceccarelli, M., 2005. Interpolation processes using multivariate geostatistics for mapping of climatological precipitation mean in the Sannio Mountains (southern Italy). Earth Surf. Process. Landforms: J. Brit. Geomorphol. Res. Group 30, 259–268.

Dunlop, G., 1980. A rapid computational method for improvements to nearest neighbour interpolation. Comput. Math. Appl. 6, 349–353.

Ebdon, D., 1996. A review of: "interactive spatial data analysis. By T. C. Bailey & A. C. Gatrell. Int. J. Geogr. Inf. Syst. 10, 511–512.

Falkenmark, M., Rockstrom, J., Rockström, J., 2004. Balancing Water for Humans and Nature: The New Approach in Ecohydrology. Earthscan.

Goovaerts, P., 1997. Geostatistics for Natural Resources Evaluation. Oxford University Press on Demand.

Goovaerts, P., 1999. Geostatistics in soil science: state-of-the-art and perspectives. Geoderma 89, 1–45.

Goovaerts, P., 2000. Geostatistical approaches for incorporating elevation into the spatial interpolation of rainfall. J. Hydrol. 228, 113–129.

Hartkamp, A.D., De Beurs, K., Stein, A., White, J.W., 1999. Interpolation Techniques for Climate Variables. Cimmyt.

Hengl, T., 2009. A Practical Guide to Geostatistical Mapping. Hengl Amsterdam.

Isaaks, E.H., Srivastava, R.M., 1989. An Introduction to Applied Geostatistics. Oxford university press.

Junlong, Z.X.Z.C.F., Jianquan, Y.X.L., 2009. Smart sensor nodes for wireless soil temperature monitoring systems in precision agriculture. Trans. Chin. Soc. Agric. Mach. S1.

Kerr, Y.H., Wigneron, J.P., Al Bitar, A., Mialon, A., Srivastava, P.K., 2016. Soil moisture from space: techniques and limitations. In: Satellite Soil Moisture Retrieval. Academic Press, USA, pp. 3–27. Elsevier.

Knotters, M., Brus, D., Voshaar, J.O., 1995. A comparison of kriging, co-kriging and kriging combined with regression for spatial interpolation of horizon depth with censored observations. Geoderma 67, 227–246.

Kourgialas, N., Karatzas, G., 2014. A hydro-sedimentary modeling system for flash flood propagation and hazard estimation under different agricultural practices. Nat. Hazards Earth Syst. Sci. 14, 625.

Kourgialas, N.N., Karatzas, G.P., 2016. A flood risk decision making approach for Mediterranean tree crops using GIS; climate change effects and flood-tolerant species. Environ. Sci. Pol. 63, 132–142.

Kravchenko, A., Bullock, D.G., 1999. A comparative study of interpolation methods for mapping soil properties. Agron. J. 91, 393–400.

Kupfersberger, H., Deutsch, C.V., Journel, A.G., 1998. Deriving constraints on small-scale variograms due to variograms of large-scale data. Math. Geol. 30, 837–852.

Lam, N.S.-N., 1983. Spatial interpolation methods: a review. Am. Cartogr. 10, 129–150.

Li, J., Heap, A.D., 2008. A Review of Spatial Interpolation Methods for Environmental Scientists.

Li, X., Cheng, G., Lu, L., 2000. Comparison of spatial interpolation methods [J]. Adv. Earth Sci. 3.

McKeen, S., Wilczak, J., Grell, G., Djalalova, I., Peckham, S., Hsie, E.Y., Gong, W., Bouchet, V., Menard, S., Moffet, R., 2005. Assessment of an ensemble of seven real-time ozone forecasts over eastern North America during the summer of 2004. J. Geophys. Res.: Atmosphere 110.

Mitas, L., Mitasova, H., 1999. Spatial Interpolation. Geographical Information Systems: Principles, Techniques, Management and Applications 1.

Musio, M., Augustin, N., Kahle, H.-P., Krall, A., Kublin, E., Unseld, R., Von Wilpert, K., 2004. Predicting magnesium concentration in needles of Silver fir and Norway spruce—a case study. Ecol. Model. 179, 307–316.

Pandey, P.C., Anand, A., Srivastava, P.K., 2019. Spatial distribution of mangrove forest species and biomass assessment using field inventory and earth observation hyperspectral data. Biodivers. Conserv. 28, 2143–2162.

Pebesma, E.J., 2004. Multivariable geostatistics in S: the gstat package. Comput. Geosci. 30, 683–691.

Petropoulos, G.P., Srivastava, P.K., Piles, M., Pearson, S., 2018. Earth observation-based operational estimation of soil moisture and evapotranspiration for agricultural crops in support of sustainable water management. Sustainability 10, 181.

Raju, N.J., Ram, P., Dey, S., 2009. Groundwater quality in the lower Varuna river basin, Varanasi district, Uttar Pradesh. J. Geol. Soc. India 73, 178.

Ranjan, A.K., Prakash, S., Anand, A., Verma, S.K., Murmu, L., Kumar, P.B., 2017. Ground water prospect variability analysis with spatio temporal changes in Ranchi City, Jharkhand India using Geospatial Technology. Int. J. Earth Sci. Eng. 10, 616–625.

Rao, K., George, C., Ramasastri, K., 1971. Potential evapotranspiration over India. India Met. Dept. Sci. Rep.

Saito, H., McKenna, S.A., Zimmerman, D., Coburn, T.C., 2005. Geostatistical interpolation of object counts collected from multiple strip transects: ordinary kriging versus finite domain kriging. Stoch. Environ. Res. Risk Assess. 19, 71–85.

Savage, N., Agnew, P., Davis, L., Ordóñez, C., Thorpe, R., Johnson, C., O'Connor, F., Dalvi, M., 2013. Air quality modelling using the Met Office Unified Model (AQUM OS24-26): model description and initial evaluation. Geosci. Model Dev. (GMD) 6, 353–372.

Seyfried, M., Grant, L., Du, E., Humes, K., 2005. Dielectric loss and calibration of the Hydra Probe soil water sensor. Vadose Zone J. 4, 1070–1079.

Skierucha, W., Wilczek, A., Szypłowska, A., Sławiński, C., Lamorski, K., 2012. A TDR-based soil moisture monitoring system with simultaneous measurement of soil temperature and electrical conductivity. Sensors 12, 13545–13566.

Snepvangers, J., Heuvelink, G., Huisman, J., 2003. Soil water content interpolation using spatio-temporal kriging with external drift. Geoderma 112, 253–271.

Srivastava, P.K., 2017. Satellite soil moisture: review of theory and applications in water resources. Water Res. Manag. 31, 3161–3176.

Srivastava, P.K., Han, D., Ramirez, M.A.R., Islam, T., 2013. Appraisal of SMOS soil moisture at a catchment scale in a temperate maritime climate. J. Hydrol. 498, 292–304.

Srivastava, P.K., Han, D., Rico-Ramirez, M.A., O'Neill, P., Islam, T., Gupta, M., 2014a. Assessment of SMOS soil moisture retrieval parameters using tau–omega algorithms for soil moisture deficit estimation. J. Hydrol. 519, 574–587.

Srivastava, P.K., Han, D., Rico-Ramirez, M.A., O'Neill, P., Islam, T., Gupta, M., Dai, Q., 2015. Performance evaluation of WRF-Noah Land surface model estimated soil moisture for hydrological application: synergistic evaluation using SMOS retrieved soil moisture. J. Hydrol. 529, 200–212.

Srivastava, P.K., Han, D., Yaduvanshi, A., Petropoulos, G.P., Singh, S.K., Mall, R.K., Prasad, R., 2017. Reference evapotranspiration retrievals from a mesoscale model based weather variables for soil moisture deficit estimation. Sustainability 9, 1971.

Srivastava, P.K., O'Neill, P., Cosh, M., Kurum, M., Lang, R., Joseph, A., 2014b. Evaluation of dielectric mixing models for passive microwave soil moisture retrieval using data from ComRAD ground-based SMAP simulator. IEEE J. Sel. Top. Appl. Earth Obs. Remote Sens. 8 (9), 4345–4354.

Srivastava, P.K., Pandey, P.C., Petropoulos, G.P., Kourgialas, N.N., Pandey, V., Singh, U., 2019. GIS and remote sensing aided information for soil moisture estimation: a comparative study of interpolation techniques. Resources 8, 70.

Suman, S., Srivastava, P.K., Petropoulos, G.P., Pandey, D.K., O'Neill, P.E., 2020. Appraisal of SMAP operational soil moisture product from a global perspective. Remote Sens 12 (12), 1977.

Wackernagel, H., 2013. Multivariate Geostatistics: An Introduction with Applications. Springer Science & Business Media.

Western, A.W., Grayson, R.B., Blöschl, G., 2002. Scaling of soil moisture: a hydrologic perspective. Annu. Rev. Earth Planet Sci. 30, 149–180.

Wu, C., Wu, J., Luo, Y., Zhang, H., Teng, Y., DeGloria, S.D., 2011. Spatial interpolation of severely skewed data with several peak values by the approach integrating kriging and triangular irregular network interpolation. Environ. Earth Sci. 63, 1093–1103.

Zhang, D., Anthes, R.A., 1982. A high-resolution model of the planetary boundary layer—sensitivity tests and comparisons with SESAME-79 data. J. Appl. Meteorol. 21, 1594–1609.

Zimmerman, D.L., Zimmerman, M.B., 1991. A comparison of spatial semivariogram estimators and corresponding ordinary kriging predictors. Technometrics 33, 77–91.

Chapter 20

Modelling key parameters characterising land surface using the SimSphere SVAT model

Swati Suman[1], Matthew R. North[2], George P. Petropoulos[2], Prashant K. Srivastava[1], Dionissios T. Hristopulos[2], Daniela Silva Fuzzo[3], Toby N. Carlson[3]

[1]*Remote Sensing Laboratory, Institute of Environment and Sustainable Development, Banaras Hindu University, Varanasi, Uttar Pradesh, India;* [2]*Geostatistics Laboratory, School of Mineral Resources Engineering, Technical University of Crete, Chania, Greece;* [3]*Pennsylvania State University, Department of Meteorology, University Park, State college, PA, United States*

1. Introduction

Accurate monitoring of water and vegetation stress is now of prominent global concern, and it is regarded as a high priority issue (Petropoulos et al., 2016). Much emphasis is placed on the accurate monitoring of the effects of climate change on water and vegetation, particularly for communities located in the Mediterranean region having water-scarce ecosystems (Amri et al., 2014). Thus, studies on the partitioning of incoming energy into heat and water fluxes are crucial in understanding the mechanism of climate change. The terrestrial boundary layer and its vegetation play a critical role in regulating the partitioning of incoming energy (into Latent (LE), Sensible (H) and Ground (G) heat fluxes), having an effect in photosynthesis and the energy and water vapour cycles (Prentice et al., 2014).

Research on improving our understanding of the representation of land atmosphere interactions has led to the development and exploration of a wide variety of different modelling schemes. A number of Land Surface Models (LSMs) for assessing the contribution of different variables associated with land surface interaction at various degrees of complexities have been developed since the 1970s. Since then, LSMs have evolved from simple bucket models without vegetation consideration (e.g., Manabe, 1969) into contemporary versions with credibly detailed representations of the exchanges of

Agricultural Water Management. https://doi.org/10.1016/B978-0-12-812362-1.00020-5

energy, water and CO_2 in the soil-vegetation-atmosphere continuum. Among various forms of LSMs, Soil Vegetation Atmosphere Transfer (SVAT) models are increasingly gaining recognition in land surface processes and Earth's system component studies (Ireland et al., 2014). SVATs are mathematical representations of vertical 'views' of the physical mechanisms controlling energy and mass transfers in the soil-vegetation-atmosphere continuum. Those models are able to provide deterministic estimates of the time course of soil and vegetation state variables at time-steps compatible with the dynamics of atmospheric processes. Fine temporal resolution (often <1 h) of SVAT models allows simulations to be in satisfactory agreement with the timescale of the physical process being simulated.

Developed by Carlson and Boland (1978), SimSphere is an SVAT model that simulates and enhances our understanding of boundary layer processes and is being extensively used as a research, educational and training tool in several universities worldwide. SimSphere these days has gained a lot of popularity as an extensive tool being synergistically used with Earth Observation (EO) data due to its ability to provide spatiotemporal estimates of evapotranspiration (ET) rates and surface soil moisture. Most of these investigations have been based around the implementation of a data assimilation technique termed the 'triangle' (Petropoulos and Carlson, 2011). Variants of this technique are currently investigated by different space agencies for developing related operational products (Chauhan et al., 2003; Piles et al., 2011, 2016). A series of SA experiments has already been conducted on SimSphere (Petropoulos et al., 2009b; Petropoulos et al., 2013a–c; 2014). Those studies provided for the first time independent evidence to enhance our understanding of the model's behaviour, coherence and correspondence to that it has been built to simulate (Petropoulos et al., 2009a; Petropoulos et al., 2013a–c; 2014). However, SimSphere validation has previously only been performed over a very small range of land-use/cover types (e.g., Todhunter and Terjung, 1988; Ross and Oke, 1988; Petropoulos et al., 2015). Given its current global expansion, such a comprehensive validation of it is both timely and fundamental importance to further establish the model's structure, coherence and representativeness in terms of its ability to realistically represent Earth's land surface interactions.

In light of the above, this study's objective has been to investigate the ability and applicability of SimSphere to simulate a series of significant variables characterising land surface interactions and specifically: Net Radiation (R_{net}), Latent Heat (LE) and Sensible Heat (H). For this purpose, in-situ measurements for a total of 70 days selected from seven model European ecosystems sites representative of different conditions have been used by the CarboEurope monitoring network in Europe to validate the model's output.

2. Model formulation

SimSphere simulates the land-atmosphere exchanges taking place in a vertical column that extends from the root zone below the soil surface up to a level well above the surface canopy, the top of the surface mixing layer. SimSphere was considerably modified to its current state by Gillies et al. (1997) and later by Petropoulos et al. (2013d) and Anagnostopoulos et al. (2017). It is currently maintained and freely distributed by Aberystwyth University, United Kingdom (http://www.aber.ac.uk/simsphere). A detailed description of its architecture can be found in Gillies (1993) and an overview on its use can be found in Petropoulos et al. (2009b).

Briefly, SimSphere is a unidimensional two-source SVAT model with a plant component (input parameters and simulated parameters shown in Table 20.1). The model structure is an integrated form of three major components namely the *physical, vertical* and *horizontal* layers. The *physical* component determines the microclimate in the model and primarily takes

TABLE 20.1 Summary of the main SimSphere inputs (top) and of its simulated outputs (bottom). The units are also provided in parentheses where applicable

Name of the model input	Process in which parameter is involved	Min value	Max value
Slope *(degrees)*	Time & location	0	45
Aspect *(degrees)*	Time & location	0	360
Station height *(meters)*	Time & location	0	4.92
Fractional vegetation cover *(%)*	Vegetation	0	100
LAI *(m^2m^{-2})*	Vegetation	0	10
Foliage emissivity *(unitless)*	Vegetation	0.951	0.990
[Ca] (external [CO_2] in the leaf) *(ppmv)*	Vegetation	250	710
[Ci] (internal [CO_2] in the leaf) *(ppmv)*	Vegetation	110	400
[03] (ozone concentration in the air) *(ppmv)*	Vegetation	0.0	0.25
Vegetation height *(meters)*	Vegetation	0.021	20.0
Leaf width *(meters)*	Vegetation	0.012	1.0
Minimum stomatal resistance *(sm^{-1})*	Plant	10	500

Continued

TABLE 20.1 Summary of the main SimSphere inputs (top) and of its simulated outputs (bottom). The units are also provided in parentheses where applicable—cont'd

Name of the model input	Process in which parameter is involved	Min value	Max value
Cuticle resistance *(sm^{-1})*	Plant	200	2000
Critical leaf water potential *(bar)*	Plant	−30	−5
Critical solar parameter *(Wm^{-2})*	Plant	25	300
Stem resistance *(sm^{-1})*	Plant	0.011	0.150
Surface moisture availability *(vol/vol)*	Hydrological	0	1
Root zone moisture availability *(vol/vol)*	Hydrological	0	1
Substrate Max. Volum. water content *(vol/vol)*	Hydrological	0.01	1
Substrate climatol. mean temperature *(°C)*	Surface	20	30
Thermal inertia *($Wm^{-2}K^{-1}$)*	Surface	3.5	30
Ground emissivity *(unitless)*	Surface	0.951	0.980
Atmospheric precipitable water *(cm)*	Meteorological	0.05	5
Surface roughness *(meters)*	Meteorological	0.02	2.0
Obstacle height *(meters)*	Meteorological	0.02	2.0
Fractional cloud cover *(%)*	Meteorological	1	10
RKS (satur. thermal conduct.) (Cosby et al., 1984)	Soil	0	10
Cosby B (see Cosby et al., 1984)	Soil	2.0	12.0
THM (satur.vol. water cont.) (Cosby et al., 1984)	Soil	0.3	0.5
PSI (satur. water potential) (Cosby et al., 1984)	Soil	1	7
Wind direction (*degrees*)	Wind sounding profile	0	360
Wind speed *(knots)*	Wind sounding profile	—	—

TABLE 20.1 Summary of the main SimSphere inputs (top) and of its simulated outputs (bottom). The units are also provided in parentheses where applicable—cont'd

Name of the model input	Process in which parameter is involved	Min value	Max value
Altitude *(1000's feet)*	Wind sounding profile	—	—
Pressure *(mBar)*	Moisture sounding profile	—	—
Temperature *(Celsius)*	Moisture sounding profile	—	—
Temperature-dewpoint temperature *(Celsius)*	Moisture sounding profile	—	—

SimSphere Simulated Outputs

Output Name	Units	Output Name	Units
Air temperature at 1.3m	°C	Radiometric temperature	°C
Air temperature at 50m	°C	Root zone moisture avail.	n/a
Air temperature at foliage	°C	Sensible heat flux	Wm^{-2}
Bowen ratio	n/a	Short-wave flux	Wm^{-2}
[CO_2] on canopy	ppmv	Specific humidity at 1.3 m	gKg^{-1}
[CO_2] flux	Micromoles m^2s^{-1}	Specific humidity at 50 m	gKg^{-1}
Epidermal water potential	Bars	Specific humidity at foliage	gKg^{-1}
Global O_3 flux	$Ugm^{-2}s^{-1}$	Stomatal resistance	sm^{-1}
Ground flux	Wm^{-2}	Surface moisture availability	n/a
Ground water potential	bars	Vapor pressure deficit	Mbar
Latent heat flux	Wm^{-2}	Water use efficiency	n/a
Leaf water potential	bars	Wind at 10 m	Kts
Net radiation	Wm^{-2}	Wind at 50 m	Kts
[O_3] canopy	ppmv	Wind in foliage	Kts
[O_3] flux plant	$Ugm^{-2}s^{-1}$		

account of the available radiant energy reaching the surface in clear sky condition or the plant canopy. The component is calculated as a function of Sun and Earth geometry, atmospheric transmission factors for scattering and absorption, the atmospheric and surface emissivity's and surface (including soil and plant) albedos. The *vertical structure* components (Fig. 20.1, *right*) effectively correspond to the components of the Planetary Boundary Layer (PBL) that is divided into three layers—a surface mixing layer, a surface of constant flux layer and a surface vegetation or bare soil layer. Vegetation and soil fluxes mix at the top of the vegetation canopy. Their relative weights depend on the fractional vegetation cover (FVC), specified as an input to the model. The soil hydraulic parameters are prescribed from the Clapp and Hornberger (1978) classification. The soil surface turbulent fluxes are determined following the Monin and Obukhov (1954) similarity theory, which takes into account atmospheric stability. The Atmospheric Boundary Layer (ABL) conditions are provided by a unidimensional ABL model.

SimSphere simulates the processes and the interaction between soil, plant and atmosphere layers over a 24-h cycle. The cycle runs at a chosen time step, starting generally from the early morning (at 06: 00 am local time) to monitor the continuously evolving interaction between the input layers. A number of input parameters are required to parameterise the model, categorised into seven defined groups (Table 20.1), and the model provides predictions as a function of time for a total of more than 30 variables (Table 20.1).

3. Materials and methods

Fig. 20.2 provides details of the methodology followed to parameterise and validate SimSphere targeted outputs, whereas the major steps involved in this process are outlined below.

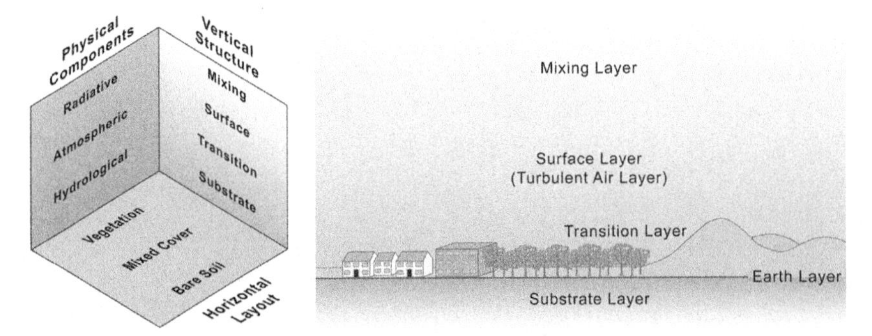

FIGURE 20.1 (*Left*) The three facets of SimSphere architecture, (*Right*) different layers represented within SimSphere's vertical domain.

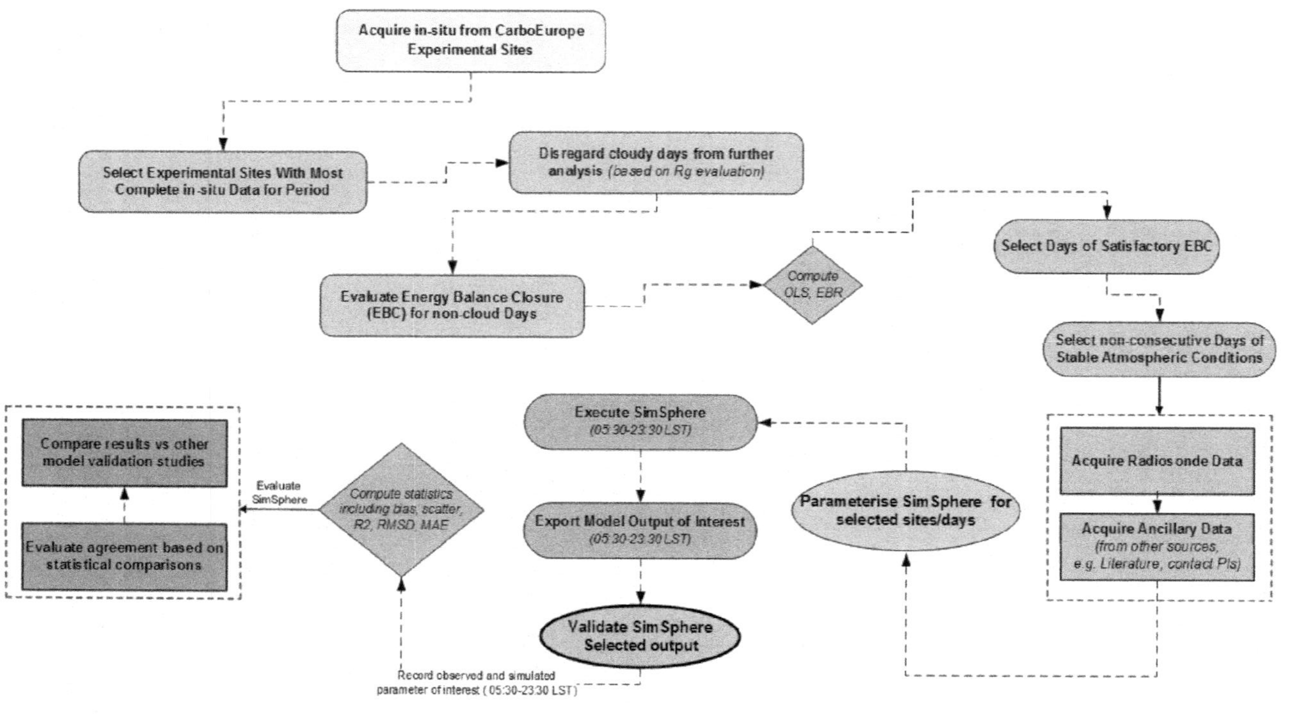

FIGURE 20.2 Overall methodology of SimSphere validation followed in this study.

3.1 In-situ datasets collection

This study evaluates the ability of SimSphere Soil Vegetation Atmosphere Transfer (SVAT) model in providing diurnal estimates of key variables characterising water and energy balance at seven CarboEurope sites, as part of a larger observational network, FLUXNET (Baldocchi et al., 2001). The sites used in our study were selected as representative of different ecosystem types (see Table 20.2). In-situ data for selected sites were acquired from the European Fluxes database Cluster (http://gaia.agraria.unitus.it/) for the year 2011. In particular, Level 2 data were obtained across all selected sites for consistency. This product includes the originally acquired in-situ measurements from which only the removal of erroneous data caused by obvious instrumentation error were undertaken. In addition, atmospheric profile (i.e., radiosonde) data as atmospheric temperature profile, dew point temperature, wind direction, wind speed and atmospheric pressure were obtained for each site/day from the University of Wyoming (http://weather.uwyo.ddeedu/upperair/sounding.html).

Initially, for each site, cloudy days were identified and were subsequently excluded from further analysis. Identification of cloudy days was carried out using diurnal incoming global solar radiation (Rg) observations. As cloud-free days were flagged as those having smoothly symmetrical Rg curves and cloudy days as those having asymmetrical ones (Carlson et al., 1991). Subsequently, energy balance closure (EBC) for those cloud-free days only was evaluated. EBC is believed to be the most relevant energy measurement tool, as its magnitude depends on more accurate entities such as Latent Heat (LE) and Sensible Heat (H) and not on other scaler fluxes such as CO_2 (Wilson et al., 2002; Foken et al., 2006). EBC was evaluated principally by calculating the linear regression coefficients (slope and intercept), as well as the coefficient of determination (R^2) from the ordinary least squares (OLS) relationship between the half-hourly estimates of the dependent flux variables (LE + H) and the independently derived available energy (R_{net}-G-S). In addition, the Energy Balance Ratio (EBR) was also computed by cumulatively summing R_{net}-G-S and LE + H from the 30-min mean average surface energy flux components, and then rationing each of the cumulative sums as follows (Liu et al., 2006):

$$EBR = \frac{\sum (LE + H)}{\sum (Rnet - G - S)} \tag{20.1}$$

where LE is the Latent Heat, H is the Sensible Heat, R_{net} is the net radiation, G is the heat flux into the soil and S is the rate of change of heat storage (air and biomass). This index ranges generally from zero to one, with values closer to one highlighting a satisfactory diurnal energy closure, indicating a good quality of in-situ measurements.

TABLE 20.2 Some of the main characteristics of the selected CarboEurope sites used for SimSphere validation

Site Name	Site Abbreviation	County	Geographic Location	PFT	Ecosystem Type	Dominant Species	Elevation	Climate
Llano de los Juanes	Es_Lju	SPAIN	36.9266/-2.1521	OLI	Olive Plantation	*Olea europea, Macchia*	1622 m	Warm temperate with dry, hot summer
Collelongo-SelvaPiana	It_Col	ITALY	41.8493/13.5881	DBF	Deciduous Broadleaf Forest	*Fagus sylvatica*	1645 m	Warm temperate fully humid with warm summer
Monte Modone	It_Mbo	ITALY	46.0296/11.0829	GRA	Grassland	Alpine meadow	1547 m	Snow, fully humid warm summer
Aguamarga	Es_Agu	SPAIN	36.8347/-2.2511	SHR	Annual Broadleaf Shrub	Sumac (*Rhus*), Toyon (*Heteromeles*), and coffeeberry (*Rhamnus*) species	195 m	Arid Steppe Cold

Continued

TABLE 20.2 Some of the main characteristics of the selected CarboEurope sites used for SimSphere validation—cont'd

Site Name	Site Abbreviation	County	Geographic Location	PFT	Ecosystem Type	Dominant Species	Elevation	Climate
Lavarone	It_Lav	ITALY	45.9553/ 11.2812	ENL	Evergreen Needle leaf forest	*Pinus sylvestris*	1353 m	Warm temperate fully humid with warm summer
Puechabon	Fr_Pue	FRANCE	43.7414/ 3.5958	EBF	Evergreen Broadleaf forest	*Quercus ilex*	211 m	Warm temperate with dry, hot summer
Roccarepampani	It_Ro3	ITALY	42.3753/ 11.9154	CRO	Cropland	Cereal crop	320 m	Warm temperate with dry, hot summer

All days with low EBC (i.e., EBR<0.750, slope <0.85, $R^2 < 0.930$) were excluded from further analysis. Further constraints were applied to calibrate the selected data quality with the in-situ data quality which was performed over several steps. Secondly, atmospherically stable conditions, such as low wind speeds and small available energy, were selected for the evaluation simulation days (Maayar et al., 2001). Such conditions were identified during evaluation of the in-situ dataset, where direct measurements of wind speed and energy flux amplitude and diurnal trend were used as indicators of atmospherically stable conditions. In total a set of 70 nonconsecutive days from the seven CarboEurope sites were identified as being suitable to include in the model verification.

3.2 SimSphere parameterisation and implementation

SimSphere parameterisation was carried out at the measurement scale of the flux tower observations, i.e., the area of the possible measurement fetch around which the tower is built and the footprint of the turbulent flux measurements, representing an area of $\sim 1\ km^2$ for the test sites as they are relatively homogeneous. On this basis, SimSphere was parameterised to the daily conditions existent at the flux tower for each of the selected days.

For each day the model was parameterised to the daily existing conditions at the flux tower up to a height of 54,000 ft. Initial conditions for air temperature, dew point temperature, atmospheric pressure, wind speed and direction were used within the 'Wind Sounding' and 'Water Vapour Sounding' components of the model. These data were acquired from the publically available University of Wyoming database, and were collected at 6:00 am GMT to correspond to the model's initialisation. Ancillary information on vegetation and soil parameters (leaf area index—LAI, FVC, vegetation height, soil type, etc.) was also used directly within the model's initialisation. Such information was acquired in most cases directly from communication with the principal investigators of each respective site, though in some cases it had to be acquired from standard literature sources (e.g., Mascart et al., 1991; Carlson et al., 1991). The soil type parameters were obtained using the soil texture data provided at each CarboEurope test site. Similarly, this was also the case for the topographical information that was required in model initialisation. Upon model initialisation, the model was executed for each site/day and the 30′ average value of each of the evaluated parameters per site for the period 0530–2330 h was subsequently exported in SPSS for comparisons against the corresponding in-situ data.

3.3 Validation

To analyse the correlation of the model's simulated values to the observed, a series of statistical approaches based on the results of many previous similar studies (e.g., Giertz et al., 2006; Marshall et al., 2013) were undertaken.

Those included were root mean square difference [RMSD], the linear regression fit model coefficient determination [R^2], the Bias or Mean Bias Error [MBE], the Scatter or mean standard deviation [MSD], the mean absolute error [MAE] and the NASH index, tabulated in Table 20.3. MSD was employed to express the model precision and ultimately for the correction of nonsystematic error. All statistical matrices were computed from the comparative analysis of the two datasets for each day of comparison at 30′ intervals. The same set of statistical metrics was performed on the dataset for each of the CarboEurope sites for each of the selected days.

4. Results

4.1 Net radiation (R_{net}) flux

The results of the analysis between SimSphere predicted and in-situ net radiation measurement are summarised in Table 20.4. Furthermore, Fig. 20.3A illustrates the agreement between the in-situ and the predicted R_{net} for all days of comparisons from all experimental sites. For most of the compared days diurnal variation of the simulated R_{net} in general was found in close correspondence with the observed R_{net} both in shape and magnitude (although results are not shown here for brevity).

Overall, R_{net} simulated by SimSphere was found to be reasonably accurate with an average RMSD of 64.65 Wm^{-2} and a correlation coefficient of 0.96. A minor underestimation of the in-situ data was evident for all sites and days combined (MBE = −2.07 Wm^{-2}), though overall R_{net} showed a significant range of agreement, with RMSD ranging from 24.38 to 98.26 Wm^{-2} between the validation days. Interestingly, a noticeable trend between extended observation time period and simulation accuracy was observed within a number of test sites. Also, notably, there were increased periods within a number of test sites where simulation accuracy was found increasing depending on the period in which the simulation days were located. Such trends were observed for the IT_Ro3 cropland site, where error ranges decreased for the period between late April (21/04/2011) and late August (28/08/2011), before increasing in early September (09/09/2011). However, the periods of increased accuracy varied on a per site basis and were only prevalent within the olive plantation (ES_Lju), grassland (IT_Mbo), cropland (IT_Ro3) and deciduous broadleaf forest (IT_Col) sites. Daily R^2 values exhibited less variance with generally more comparable ranges (0.909−0.998) between all the study days, suggesting a satisfactory agreement between both datasets, also illustrated by the distribution of the points around the 1:1 line in Fig. 20.3A. This was also reflected within the NASH index values reported (0.897−0.999).

When averaged per site, RMSD showed significantly less variance, exhibiting a range from 55.86 (IT_Lav) to 68.49 Wm^{-2} (FR_Pue). This trend

TABLE 20.3 An overview of the statistical measures implemented in this study to evaluate SimSphere's outputs against the corresponding in-situ data

Name	Description	Mathematical Definition
Bias/MBE	Bias (accuracy) or Mean Bias Error	$bias = \frac{1}{N}\sum_{i=1}^{N}(P_i - O_i)$
R^2	Linear Correlation Coefficient of Determination of P_i to O_i	$R^2 = \left[\sum_{i=1}^{N}(P_i - \overline{P})(O_i - \overline{O}) \Big/ \left[\sum_{i=1}^{N}(O_i - \overline{O})^2 \sum_{i=1}^{N}(P_i - \overline{O})^2\right]^{0.5}\right]^2$
Scatter/MSE	Scatter (precision) or Mean Standard Deviation	$scatter = \frac{1}{(N-1)}\sum_{i=1}^{N}\left(P_i - O_i - \overline{(P_i - O_i)}\right)^2$
RMSD	Root Mean Square Difference	$RMSD = \sqrt{bias^2 + scatter^2}$
MAE	Mean Absolute Error	$MAD = N^{-1}\sum_{i=1}^{N}\lvert P_i - O_i\rvert$
NASH	Nash Sutcliffe Efficiency	$NASH = 1 - \left[\frac{\sum_{i=1}^{N}(Oi-Si)^2}{\sum_{i=1}^{N}(Oi-\overline{O})^2}\right]$

TABLE 20.4 An overview of R_{net} simulation accuracy

			Statistical Test								Statistical Test				
Site	PFT	Day	Bias	Scatter	RMSD	MAE	NASH	Site	PFT	Day	Bias	Scatter	RMSD	MAE	NASH
ES_LJU	OLI	14/04/2011	−24.55	42.31	48.91	32.45	0.921	IT_RO3	CRO	09/04/2011	−8.20	85.76	86.16	76.40	0.912
		09/05/2011	−19.34	60.31	63.33	47.55	0.976			11/04/2011	−52.87	46.21	70.22	55.97	0.913
		24/06/2011	12.18	67.54	68.63	57.97	0.916			18/04/2011	13.74	80.88	82.03	72.17	0.990
		27/06/2011	6.06	66.98	67.25	47.26	0.978			21/04/2011	24.95	56.34	61.62	55.09	0.982
		19/07/2011	26.05	57.38	63.01	44.21	0.934			20/06/2011	−12.51	53.15	54.60	48.95	0.937
		28/07/2011	34.52	56.12	65.89	47.60	0.971			26/06/2011	−22.36	48.39	53.30	42.70	0.972
		04/08/2011	15.06	51.08	53.25	33.81	0.930			24/08/2011	13.94	54.53	56.28	41.84	0.961
		22/08/2011	8.26	57.55	58.14	47.33	0.899			28/08/2011	−8.98	59.95	60.62	51.20	0.899
		25/08/2011	10.23	59.03	59.91	49.44	0.978			09/09/2011	−19.92	67.62	70.49	62.77	0.897
		28/09/2011	−19.69	92.19	94.27	78.84	0.998			11/09/2011	2.40	68.15	68.19	55.23	0.971
		Average	**4.88**	**64.78**	**64.96**	**48.65**	**0.950**			**Average**	**−6.98**	**66.53**	**66.90**	**56.23**	**0.943**

IT_COL	DBF	26/06/2011	−29.91	67.82	74.12	52.94	0.969	IT_LAV	ENL	27/06/2011	−24.60	57.52	62.56	46.13	0.971
		08/07/2011	−23.15	46.34	51.80	41.84	0.978			03/07/2011	−60.69	39.12	72.21	63.35	0.986
		13/07/2011	−12.95	56.81	58.27	50.16	0.934			09/07/2011	−35.90	57.43	67.73	58.59	0.971
		18/07/2011	−23.69	54.99	59.87	48.72	0.978			11/08/2011	−16.51	31.22	35.32	30.06	0.998
		11/08/2011	−10.67	63.23	64.12	50.03	0.974			12/08/2011	−0.79	31.24	31.25	24.10	0.996
		23/08/2011	14.50	64.17	65.79	54.93	0.940			20/08/2011	3.59	31.32	31.53	21.85	0.975
		11/09/2011	40.85	53.96	67.67	47.63	0.899			21/08/2011	23.69	29.01	37.46	32.13	0.989
		15/09/2011	38.95	59.52	71.13	52.79	0.969			24/08/2011	47.45	25.99	54.10	47.45	0.990
		16/09/2011	18.84	70.23	72.71	50.39	0.999			09/09/2011	33.71	46.83	57.70	49.08	0.979
		17/09/2011	44.54	54.46	70.36	47.23	0.920			30/09/2011	58.84	78.66	98.26	78.02	0.954
		Average	**4.61**	**68.03**	**68.19**	**51.16**	**0.956**			**Average**	**−9.70**	**55.01**	**55.86**	**44.02**	**0.981**

Continued

TABLE 20.4 An overview of R_{net} simulation accuracy—cont'd

Site	PFT	Day	Statistical Test					Site	PFT	Day	Statistical Test				
			Bias	Scatter	RMSD	MAE	NASH				Bias	Scatter	RMSD	MAE	NASH
IT_MBO	GRA	10/04/2011	−45.49	54.34	70.87	47.71	0.979	FR_PUE	EBF	06/04/2011	−48.91	48.89	69.15	52.63	0.978
		10/05/2011	−22.05	41.00	46.56	37.14	0.936			09/04/2011	−39.03	51.27	64.43	50.03	0.913
		25/06/2011	−11.70	21.39	24.38	18.92	0.901			16/04/2011	−57.09	45.67	73.11	57.57	0.932
		03/07/2011	−12.38	66.20	67.35	56.63	0.978			17/05/2011	−27.98	49.22	56.62	46.95	0.946
		24/08/2011	40.61	55.84	69.04	46.81	0.925			28/05/2011	−38.36	48.14	61.55	50.92	0.961
		25/08/2011	41.22	61.04	73.66	50.97	0.978			19/06/2011	−58.10	49.41	76.27	64.97	0.947
		13/09/2011	−23.86	80.95	84.39	78.38	0.963			08/07/2011	−27.62	38.41	47.31	37.66	0.975
		21/09/2011	−21.12	75.19	78.10	69.16	0.910			26/09/2011	49.90	44.96	67.17	49.90	0.963
		26/09/2011	−3.44	67.29	67.38	59.95	0.912			14/09/2011	60.09	48.58	77.27	60.09	0.978
		30/09/2011	−5.05	49.55	49.81	43.63	0.978			20/09/2011	47.71	62.85	78.91	51.51	0.938
		Average	**−6.33**	**65.07**	**65.38**	**50.93**	**0.946**			**Average**	**−15.99**	**66.60**	**68.49**	**52.47**	**0.953**

ES_AGU	SHR	07/04/2011	−49.42	23.11	54.55	49.42	0.978								
		27/04/2011	−62.87	26.14	68.09	62.87	0.963								
		08/05/2011	−41.11	19.67	45.58	41.11	0.974								
		14/05/2011	−14.87	34.17	37.26	33.38	0.954								
		23/05/2011	−24.01	24.79	34.51	31.38	0.960								
		13/07/2011	27.95	26.78	38.71	32.17	0.980								
		29/07/2011	52.86	64.52	83.40	68.43	0.979								
		14/08/2011	55.68	50.21	74.97	67.51	0.968								
		26/08/2011	59.11	52.30	78.92	70.46	0.989								
		07/09/2011	41.81	48.79	64.25	59.21	0.972								
		Average	15.02	60.92	62.75	53.40	0.972								
All sites		**Average**	**−2.07**	**63.85**	**64.65**	**50.98**	**0.96**								

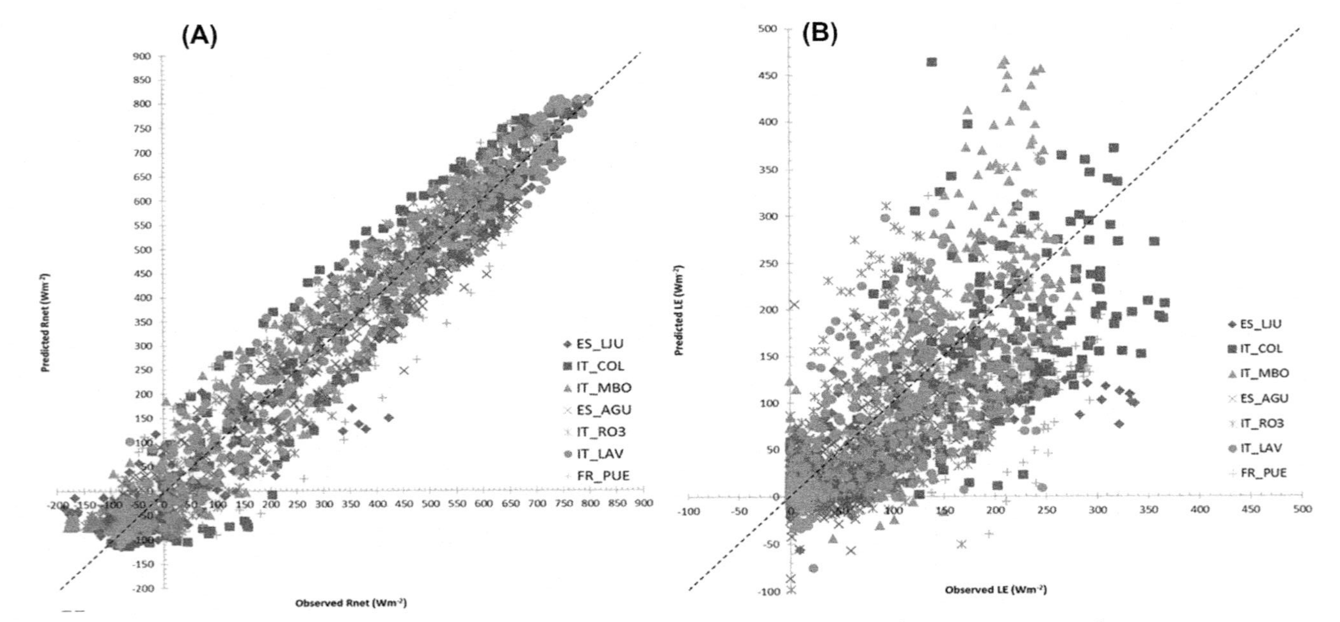

FIGURE 20.3 Comparisons of predicted and observed (A) R_{net} fluxes (Wm^{-2}), (B) LE fluxes (Wm^{-2}), (C) H fluxes (Wm^{-2}) and (D) Tair at 50 m (°C).

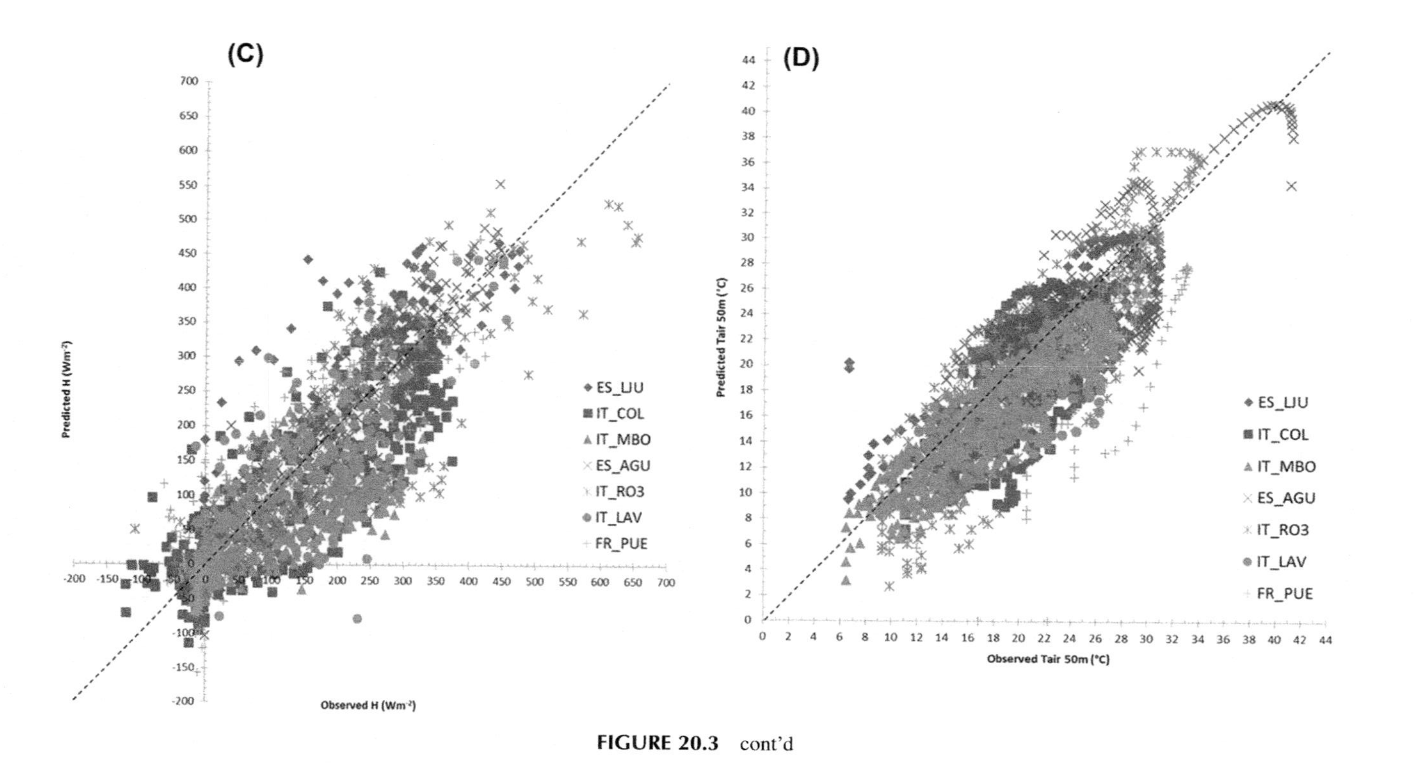

FIGURE 20.3 cont'd

was also reflected by lower variance in correlation coefficients ($R^2 = 0.936-0.970$) and NASH index values (0.943–0.981) for the per site averages. The evergreen needle-leaf forest site, IT_Lav, consistently demonstrated the highest model performance in simulating R_{net} with an RMSD value of 55.86 Wm^{-2}, that being 8.79 Wm^{-2} lower than the overall average. MBE between sites showed significant variability, ranging from a moderate underestimation of the in-situ measurements over the evergreen broadleaf forest site (−15.99 Wm^{-2}), to a moderate overestimation within the shrubland site (15.02 Wm^{-2}). All in all, SimSphere was able to reproduce R_{net} reasonably well in terms of both amplitude and trend. Indeed, this is reflected in the low MSD values of all sites (55.01–68.03 Wm^{-2}), particularly so at sites such as IT_Lav (55.01 Wm^{-2}) and ES_Agu (60.92 Wm^{-2}).

4.2 Latent Heat (LE) flux

SimSphere simulated LE flux and the CarboEurope LE measurement for all combined days exhibited an overall average RMSD error of 62.75 Wm^{-2} and a correlation coefficient value of 0.542, respectively (Table 20.5). Although RMSD for the LE output showed a better agreement in comparison to the R_{net} output (Section 4.1), R^2 was significantly lower (a decrease of 0.408). As can be seen from Fig. 20.3B, the distribution of points shows an increased dispersion from the 1:1 line in comparison to the R_{net} output. There was also an apparent overestimation of the in-situ measurements by the model for the LE flux (MBE = 15.78 Wm^{-2}). R^2 values varied significantly between all simulation days from 0.020–0.961, suggesting notable discrepancies between the predictions and observations. Additionally, daily RMSD values also varied significantly, reflecting the trends observed in the R^2 statistics. RMSD varied from 22.08 to 86.45 Wm^{-2} between all days of simulation. When analysed on a site-by-site basis, average RMSD exhibited comparable ranges to those reported for the individual simulation days, with RMSD varying from 37.25 Wm^{-2} (ES_Agu—shrubland) to 75.36 Wm^{-2} (IT_Col, deciduous broadleaf forest). On a per site basis, ES_Agu shrubland site consistently demonstrated above-average correlation to the in-situ measurements with the lowest RMSD and MAE values of all sites, 37.25 and 25.58 Wm^{-2}, respectively. Lowest agreement between the LE fluxes predicted from SimSphere and those from the in-situ measurements was in the IT_Col deciduous broadleaf forest site (RMSD = 75.36 Wm^{-2}, MAE = 55.86 Wm^{-2}) and IT_Mbo grasslands site (RMSD = 74.66 Wm^{-2}, MAE = 52.87 Wm^{-2}), respectively. On the whole, SimSphere was consistent in terms of its ability to reproduce in-situ LE fluxes, with low MSD values across most sites. Yet, the IT_Mbo (grassland) and IT_Ro3 (cropland) sites exhibited the largest MSD of 74.58 and 68.48 Wm^{-2}, respectively, an increase of 15.64 and 9.54 Wm^{-2} on the overall average suggesting a weaker systematic replication of LE fluxes over those sites (Table 20.5). There was a systematic overestimation of LE for

TABLE 20.5 An overview of LE simulation accuracy

Site	PFT	Day	Statistical Test					Site	PFT	Day	Statistical Test				
			Bias	Scatter	RMSD	MAE	NASH				Bias	Scatter	RMSD	MAE	NASH
ES_LJU	OLI	14/04/2011	13.10	43.69	45.62	34.00	0.987	IT_RO3	CRO	09/04/2011	−34.88	54.19	64.45	39.69	0.996
		09/05/2011	−8.48	37.57	38.51	26.45	0.993			11/04/2011	−39.35	43.02	58.30	41.49	0.997
		24/06/2011	42.62	62.22	75.42	63.34	0.977			18/04/2011	−17.47	21.90	28.02	20.97	0.998
		27/06/2011	46.98	59.15	75.53	60.96	0.968			21/04/2011	1.65	27.69	27.74	20.70	0.998
		19/07/2011	17.78	25.03	30.70	23.02	0.954			20/06/2011	51.85	54.15	74.97	55.86	0.954
		28/07/2011	26.35	23.88	35.57	30.00	0.961			26/06/2011	38.33	31.82	49.81	39.17	0.960
		04/08/2011	−13.97	24.09	27.85	21.57	0.966			24/08/2011	12.15	28.29	30.79	22.73	0.984
		22/08/2011	−3.40	38.77	38.92	28.53	0.987			28/08/2011	18.05	26.51	32.07	23.96	0.973
		25/08/2011	22.97	33.43	40.56	29.31	0.902			09/09/2011	46.93	45.17	65.14	47.73	0.972
		28/09/2011	22.00	28.76	36.21	26.91	0.903			11/09/2011	49.09	54.13	73.07	51.67	0.986
		Average	**21.09**	**51.49**	**55.64**	**37.22**	**0.983**			**Average**	**−0.87**	**68.48**	**68.48**	**47.51**	**0.982**

Continued

TABLE 20.5 An overview of LE simulation accuracy—cont'd

Site	PFT	Day	Statistical Test					Site	PFT	Day	Statistical Test				
			Bias	Scatter	RMSD	MAE	NASH				Bias	Scatter	RMSD	MAE	NASH
IT_COL	DBF	26/06/2011	26.53	30.72	40.59	30.21	0.915	IT_LAV	ENL	27/06/2011	−9.09	38.54	39.59	29.72	0.938
		08/07/2011	2.34	71.20	71.24	51.70	0.936			03/07/2011	23.40	41.88	47.97	38.47	0.973
		13/07/2011	33.33	53.23	62.81	47.75	0.976			09/07/2011	−16.39	55.28	57.66	41.60	0.912
		18/07/2011	35.85	70.07	78.71	62.73	0.935			11/08/2011	32.47	44.84	55.36	41.66	0.899
		11/08/2011	32.46	68.31	75.63	65.57	0.894			12/08/2011	29.70	67.43	73.68	59.10	0.937
		23/08/2011	−25.34	81.15	85.01	50.98	0.900			20/08/2011	31.48	80.52	86.45	63.16	0.936
		11/09/2011	56.10	42.26	70.23	56.10	0.986			21/08/2011	−12.13	45.44	47.04	33.46	0.938
		15/09/2011	60.69	49.42	78.27	61.47	0.984			24/08/2011	−21.87	57.06	61.11	46.97	0.989
		16/09/2011	50.25	47.72	69.30	53.45	0.987			09/09/2011	27.18	69.22	74.37	59.71	0.935
		17/09/2011	6.74	26.51	27.35	21.59	0.993			30/09/2011	9.78	40.27	55.69	48.69	0.913
		Average	**33.67**	**67.43**	**75.36**	**55.86**	**0.951**			**Average**	**8.47**	**58.32**	**58.93**	**41.39**	**0.937**

IT_MBO	GRA	10/04/2011	16.85	25.39	30.47	21.85	0.989	FR_PUE	EBF	06/04/2011	52.85	57.24	77.91	56.05	0.980
		10/05/2011	−35.35	42.72	55.45	40.52	0.913			09/04/2011	−17.44	39.39	43.08	25.79	0.996
		25/06/2011	6.87	59.93	60.33	49.33	0.976			16/04/2011	43.76	41.67	60.43	45.93	0.977
		03/07/2011	−26.51	73.75	78.37	56.20	0.911			17/05/2011	45.00	59.73	74.78	56.06	0.990
		24/08/2011	−19.29	51.79	55.27	37.79	0.978			28/05/2011	46.25	61.55	76.99	55.46	0.985
		25/08/2011	26.85	68.15	73.25	61.21	0.936			19/06/2011	28.64	43.41	52.01	39.13	0.993
		13/09/2011	−8.09	44.20	44.93	36.71	0.998			08/07/2011	22.05	38.52	44.38	33.47	0.983
		21/09/2011	14.93	53.34	55.39	34.19	0.936			26/09/2011	49.04	44.60	66.28	50.75	0.985
		26/09/2011	14.52	52.12	54.10	39.33	0.978			14/09/2011	62.28	39.97	74.00	62.28	0.954
		30/09/2011	26.21	37.65	45.88	33.52	0.980			20/09/2011	11.54	19.56	22.71	18.02	0.987
		Average	**−3.45**	**74.58**	**74.66**	**52.87**	**0.959**			**Average**	**37.56**	**57.77**	**68.91**	**47.46**	**0.988**

Continued

TABLE 20.5 An overview of LE simulation accuracy—cont'd

			Statistical Test								Statistical Test				
Site	**PFT**	**Day**	**Bias**	**Scatter**	**RMSD**	**MAE**	**NASH**	**Site**	**PFT**	**Day**	**Bias**	**Scatter**	**RMSD**	**MAE**	**NASH**
ES_AGU	SHR	07/04/2011	−20.76	30.09	36.55	25.02	0.990								
		27/04/2011	−21.86	29.03	36.34	28.04	0.994								
		08/05/2011	−9.68	21.12	23.23	16.54	0.996								
		14/05/2011	9.05	20.14	22.08	17.51	0.990								
		23/05/2011	10.84	25.10	27.35	19.64	0.986								
		13/07/2011	27.01	28.63	39.36	31.06	0.884								
		29/07/2011	34.47	25.94	43.14	34.81	0.754								
		14/08/2011	25.42	24.42	35.25	28.31	0.947								
		26/08/2011	28.00	52.61	59.60	40.41	0.975								
		07/09/2011	36.65	37.96	52.76	39.47	0.953								
		Average	**13.99**	**34.53**	**37.25**	**25.58**	**0.947**								
All sites		**Average**	**15.78**	**58.94**	**62.75**	**43.98**	**0.964**								

the majority of sites. Exceptions were only the IT_Mbo and IT_Ro3 sites, exhibiting a small average underestimation (MBE) of −3.45 and −0.87 Wm^{-2}, respectively. Interestingly, both broad-leaf forest sites, IT_Col (deciduous broad-leaf forest) and FR_Pue (evergreen broad-leaf forest), showed the highest overestimation of LE fluxes with moderately high MBE values of 33.67 and 37.56 Wm^{-2}, respectively.

4.3 Sensible Heat (H) flux

Concerning the H fluxes, results showed high performance of the model in simulating values for H fluxes with an average RMSD of 55.36 Wm^{-2} and an R^2 value of 0.83 (Fig. 20.3C, Table 20.6). A significant improvement in the accuracy of the simulation of the model output in comparison to both the R_{net} and LE was evident. H flux results exhibited a decrease in overall RMSD of 9.29 and 7.39 Wm^{-2}, respectively. Similar trends were also evident in both the MBE (−0.08 Wm^{-2}) and MSD (53.56 Wm^{-2}) results for this output, where model performance was better in comparison to both the R_{net} and LE outputs. Although with regards to R^2, the H flux output exhibited a minor decrease in correlation (0.83) compared to the R_{net} output, when examining the R^2 values for the individual simulation days, there was a significant variation in both correlation coefficients ($R^2 = 0.607{-}0.982$) and RMSD (RMSD = 20.03−91.07 Wm^{-2}). RMSD ranged from 35.50 (ES_Agu) to 71.93 Wm^{-2} (IT_Ro3) on a site-by-site basis. Similarly to LE flux, the ES_Agu site reported the highest simulation accuracy (RMSD = 35.50 Wm^{-2}, $R^2 = 0.944$, MBE = −7.01 Wm^{-2}, MSD = 34.80 Wm^{-2}). On the contrary, the cropland site IT_Ro3 consistently reported a less satisfactory agreement between model prediction and in-situ data for H flux. Generally, SimSphere was often unable to represent the peak of H flux across all sites diurnally; this is shown by a scatter of peak values as reported in Fig. 20.3C. However, the model did neither consistently overestimate nor underestimate H flux, but produced a range of bias values, with an average error of −0.08 Wm^{-2}. Both the FR_Pue and ES_Lju sites showed a predominant underestimation of H flux at −16.29 and −17.17 Wm^{-2}, respectively. Yet, for the IT_Mbo site, a moderate overestimation of 16.41 Wm^{-2} was reported, suggesting land-cover type may be related to simulation accuracy.

5. Discussion

This study presents an evaluation of the SimSphere SVAT model's ability in simulating key variables characterising Earth's land/surface interaction across a range of European ecosystems. The model was parameterised for seven sites where a total of 70 days (10 days per site) from the year 2011 were selected to validate the model's ability to predict Net Radiation (R_{net}), Latent Heat (LE) and Sensible Heat (H). The agreement between the two datasets was evaluated

TABLE 20.6 An overview of H simulation accuracy

Site	PFT	Day	Statistical Test					Site	PFT	Day	Statistical Test				
			Bias	Scatter	RMSD	MAE	NASH				Bias	Scatter	RMSD	MAE	NASH
ES_LJU	OLI	14/04/2011	−29.24	44.75	53.45	39.51	0.985	IT_RO3	CRO	09/04/2011	10.92	39.80	41.27	26.92	0.934
		09/05/2011	−11.76	32.57	34.63	30.29	0.963			11/04/2011	31.67	30.24	43.79	34.75	0.919
		24/06/2011	−47.07	39.11	61.20	48.54	0.945			18/04/2011	42.10	42.34	59.71	44.00	0.958
		27/06/2011	−28.81	38.98	48.47	37.58	0.948			21/04/2011	33.35	52.28	62.01	42.53	0.961
		19/07/2011	−27.46	38.74	47.48	35.77	0.978			20/06/2011	−9.57	73.29	73.91	52.42	0.958
		28/07/2011	−43.87	50.48	66.88	51.27	0.915			26/06/2011	17.25	89.42	91.07	70.44	0.983
		04/08/2011	18.95	38.42	42.84	31.95	0.934			24/08/2011	16.30	43.62	46.56	36.97	0.917
		22/08/2011	−3.39	51.14	51.25	39.75	0.964			28/08/2011	−17.29	48.32	51.32	30.11	0.913
		25/08/2011	17.21	52.08	54.85	44.13	0.964			09/09/2011	−15.89	39.23	42.32	28.03	0.978
		28/09/2011	13.23	41.60	43.65	29.29	0.978			11/09/2011	−22.61	61.45	65.48	44.20	0.928
		Average	**−17.17**	**60.22**	**62.62**	**43.97**	**0.957**			**Average**	**15.53**	**70.23**	**71.93**	**47.95**	**0.945**

IT_COL	DBF	26/06/2011	1.74	46.77	46.80	33.26	0.899	IT_LAV	ENL	27/06/2011	−22.70	68.75	72.40	51.93	0.968
		08/07/2011	18.13	64.78	67.27	51.57	0.924			03/07/2011	−35.97	64.90	74.20	54.32	0.974
		13/07/2011	9.77	44.49	45.55	41.51	0.970			09/07/2011	−25.35	48.49	54.72	40.30	0.913
		18/07/2011	12.29	57.20	58.50	51.31	0.941			11/08/2011	5.65	41.04	41.42	32.01	0.978
		11/08/2011	−3.40	37.51	37.66	29.44	0.991			12/08/2011	0.32	32.85	32.85	25.04	0.963
		23/08/2011	55.49	53.01	76.74	60.69	0.997			20/08/2011	7.77	56.67	57.20	38.05	0.918
		11/09/2011	32.16	37.20	49.17	36.64	0.969			21/08/2011	9.11	51.09	51.90	38.97	0.978
		15/09/2011	21.18	73.90	76.88	62.74	0.879			24/08/2011	18.93	56.46	59.55	46.52	0.899
		16/09/2011	23.20	43.50	49.30	41.64	0.969			09/09/2011	3.34	71.63	71.71	55.63	0.910
		17/09/2011	−0.51	59.69	59.69	45.19	0.914			30/09/2011	41.43	41.04	58.31	43.60	0.989
		Average	**14.72**	**58.78**	**60.59**	**46.84**	**0.945**			**Average**	**−6.72**	**56.95**	**57.34**	**39.18**	**0.949**

Continued

TABLE 20.6 An overview of H simulation accuracy—cont'd

Site	PFT	Day	Statistical Test					Site	PFT	Day	Statistical Test				
			Bias	Scatter	RMSD	MAE	NASH				Bias	Scatter	RMSD	MAE	NASH
IT_MBO	GRA	10/04/2011	−29.74	51.93	59.84	48.15	0.910	FR_PUE	EBF	06/04/2011	−36.45	36.93	51.89	38.72	0.978
		10/05/2011	0.29	20.03	20.03	16.50	0.971			09/04/2011	−4.73	61.85	62.03	46.98	0.995
		25/06/2011	4.97	32.86	33.23	25.14	0.896			16/04/2011	−42.22	50.00	65.44	49.12	0.914
		03/07/2011	15.82	67.80	69.62	42.00	0.941			17/05/2011	−50.66	49.10	70.55	53.69	0.968
		24/08/2011	36.06	22.46	42.48	37.55	0.879			28/05/2011	−4.18	60.90	61.04	49.30	0.978
		25/08/2011	32.11	22.49	39.20	32.69	0.986			19/06/2011	−37.85	59.70	70.69	64.09	0.925
		13/09/2011	15.15	26.73	30.73	22.44	0.976			08/07/2011	−14.58	40.37	42.93	35.78	0.946
		21/09/2011	31.57	24.50	39.96	32.22	0.936			26/09/2011	11.57	31.31	33.38	26.11	0.917
		26/09/2011	16.48	13.24	21.14	17.15	0.914			14/09/2011	23.07	42.11	48.01	38.77	0.913
		30/09/2011	41.43	41.04	58.31	43.60	0.989			20/09/2011	−6.86	28.55	29.36	20.38	0.979
		Average	**16.41**	**40.97**	**44.13**	**31.74**	**0.940**			**Average**	**−16.29**	**52.98**	**55.43**	**42.29**	**0.951**

ES_AGU	SHR	07/04/2011	−1.09	30.30	30.32	25.05	0.991								
		27/04/2011	−17.07	24.53	29.89	24.17	0.930								
		08/05/2011	−8.29	29.72	30.85	22.23	0.978								
		14/05/2011	−10.76	24.77	27.00	22.46	0.915								
		23/05/2011	−30.75	33.29	45.32	33.51	0.997								
		13/07/2011	−27.78	33.14	43.24	31.19	0.937								
		29/07/2011	−4.41	37.58	37.84	28.45	0.914								
		14/08/2011	20.68	35.58	41.16	31.22	0.989								
		26/08/2011	8.19	47.52	48.22	34.04	0.937								
		07/09/2011	0.07	30.02	30.02	22.99	0.993								
		Average	**−7.01**	**34.80**	**35.50**	**25.03**	**0.958**								
All sites		**Average**	**−0.08**	**53.56**	**55.36**	**39.57**	**0.95**								

based on a series of computed statistical metrics using, as reference, in-situ data acquired from selected sites belonging to the CarboEurope monitoring network.

In overall, results showed highest agreement of H fluxes to the measured in-situ values for all ecosystems, with an average RMSD of 55.36 Wm^{-2}. Predicted LE fluxes and R_{net} also agreed well with the corresponding in-situ data with RMSDs of 62.75 and 64.65 Wm^{-2}, respectively. Very high values of the Nash-Sutcliffe efficiency index were also reported for all of the model outputs evaluated, ranging from 0.720 to 0.998, suggesting a very good model representation of the observations.

Those findings are largely in accordance to previous analogous verification studies reported on the model. For example, Ross and Oke (1988) performed a validation of a previous version of SimSphere over an urban environment of Vancouver, Canada, and reported an acceptable agreement for H fluxes (average RMSD = 56 Wm^{-2}); however, significant average error ranges for LE fluxes (RMSD = 107 Wm^{-2}) were also reported in their study. Also, Ross and Oke (1988) noted that peak values of air temperature diurnal variability should be observed between 1030 and 1430 LST, this is in close correlation to this present study, further appraising SimSphere's representation of T_{air} at 50 m. Todhunter and Terjung (1988) further described in detail how earlier versions of SimSphere dissipated too much of Rnet as LE flux and too little to be lost to H; the latter correlates well to the Ross and Oke's findings (1988) but also the findings reported within; where average bias values indicate general net overestimations of LE flux in the order of 15.78 Wm^{-2}, compared to the slight average underestimation of H flux at −0.08 Wm^{-2}. Yet when compared with R_{net}, the simulated values of LE and H fluxes demonstrated improved model performance confirmed by the low average RMSD and high overall R^2. Petropoulos et al. (2015) in a verification of the model outputs at ecosystems located in the USA and Australia found a good agreement between the model predictions and the in-situ measurements (particularly so for the LE, H, with RMSDs of 39.47 and 55.06 Wm^{-2}, respectively).

Among all selected experiment sites, the shrubland located at ES_Agu consistently showed remarkably low average RMSD in all model outputs assessed, particularly so for LE and H fluxes. This is likely to be related to the site's characteristics, located within a water limited environment, where transpiration effects are much lower in amplitude and thus more predictable, especially given the site's relative homogeneity (Maayar et al., 2001). Akkermans et al. (2012) stated that underestimations of LE can largely be attributed to overestimations of H. Such effects were seen most prominently in our validation site ES_LJU, where a general underestimation of LE (MBE = −17.17 Wm^{-2}) partly contributed to the significant overestimation of H flux (MBE = 21.09 Wm^{-2}). Also, for example, Marshall et al. (2013) have suggested that ecosystems which exhibit increased stand complexity and heterogeneity, such as forested environments (particularly those with understorey vegetation), can have a profound effect on the overall exchange of mass and energy.

In the overall evaluation of the results concerning the model agreement to the in-situ instrumentation uncertainty in the measured variables themselves should also be partially taken into account when attempting to explain the disagreement between the simulated and observed variables (Bellocchi et al., 2010; Oncley et al., 2007; Verbeeck et al., 2008). Generally, R_{net} measurement accuracy error is in the order of 10%, although, an additional 10% instrumentation uncertainty should be added due to limited view angle/measuring volume (especially in the case of rugged terrains) (Baldocchi et al., 2001). Typical uncertainty in the LE and H estimation using the eddy covariance generally varies between 10% and 20% but can be much higher during periods of low flux magnitude and/or limited turbulent mixing such as at night (Petropoulos et al., 2013d). For example, Hollinger and Richardson (2005) showed that uncertainty in flux measurements is inversely proportional to magnitude, the smaller the flux the greater the relative uncertainty. Also, it should be noted that for some days included in our comparisons, a characteristic of the acquired in-situ data for those days was the presence of many spikes (indicative of very high or very low values). Possible reasons for those spikes could be instrumental errors, horizontal advection of H_2O and CO_2, footprint changes, as well as a nonstationarity of turbulent regime within the atmospheric surface layer (Papale et al., 2006). For those days, comparisons resulted in a somewhat lower accuracy of model predictions as such conditions cannot be replicated by the model which assumes homogeneity of vegetation canopy and ignores horizontal advection.

On the whole, despite the occasionally inferior performance of SimSphere in simulating the examined model outputs for some days/sites, model predictions were found significant in terms of the representation of the physical and dynamic processes involved in the interactions of the complex nature of the soil-land-atmosphere system. Moreover, it is important to recognise that uncertainty is inevitable in any model, which as a model will never be as complex as the reality it portrays. In this way, SimSphere fulfils its objective as a tool as it identifies the expected trends and patterns of change, if not always the magnitudes.

6. Concluding remarks

In this study, key findings from a large-scale validation of the SimSphere land biosphere model in numerous European environments were reported. In total, the model's ability to predict Net Radiation (R_{net}), Latent Heat (LE) and Sensible Heat (H) at seven ecosystems and for 70 cloud-free days in 2011 were examined. A systematic statistical analysis was employed to assess the agreement between model predictions and corresponding in-situ measurements. To our knowledge, this is the first study reporting results on the validation of SimSphere's ability to accurately simulate key variables characterising land surface processes, particularly so in European ecosystems.

In overall, SimSphere accurately predicted the evaluated parameters for most of the experimental sites. The evaluation and analysis of a model performance allowed for an increased understanding of the model's capabilities. The results of this study provide further independent evidence that SimSphere has a high capability of simulating variables associated with the Earth's energy and water balance. As noted by Verbeeck et al. (2008), discrepancies found in any validation study should be regarded as a positive step when evaluating model performance. Such studies can also advance our understanding on the amount of complexity required for adequate representation of land surface processes and interactions between different components of our Earth system. Further efforts should be directed towards validating SimSphere at other ecosystems globally, as this will allow assessing its applicability as a universally applied SVAT model. Moreover, as use of the model is currently being explored synergistically with EO data, including its possible expansion to a 2D model, it would be of utmost interest to evaluate the overriding effects of SimSphere predictions to the overall prediction error derived from such synergistic methods.

Acknowledgements

Dr. Petropoulos acknowledges the financial support provided by the European Commission under the Marie Curie Individual Fellowship project 'ENViSIoN-EO' for the completion of this work. We thank as well the PIs of the CarboEurope network for sharing the acquired data of their experimental sites.

References

Akkermans, T., Lauwaet, D., Demuzere, M., Vogel, G., Nouvellon, Y., Ardö, J., Van Lipzig, N., 2012. Validation and comparison of two soil-vegetation-atmosphere transfer models for tropical Africa. J. Geophys. Res. Biogeo. 117, 2005–2012.

Amri, R., Zribi, M., Lili-Chabaane, Z., Szczypta, C., Calvet, J.C., Boulet, G., 2014. FAO-56 dual model combined with multi-sensor remote sensing for regional evapotranspiration estimations. Remote Sens 6, 5387–5406.

Anagnostopoulos, V., Petropoulos, G.P., 2017. A modernized version of a 1D soil vegetation atmosphere transfer model for use in land surface interactions studies. Environ. Mod. Soft. 90, 147–156.

Baldocchi, D., Falge, E., Gu, L., Olson, R., Hollinger, D., Running, S., Wofsy, S., 2001. FLUXNET: a new tool to study the temporal and spatial variability of ecosystem-scale carbon dioxide, water vapour, and energy flux densities. Bull. Am. Meteorol. Soc. 82, 2415–2434.

Bellocchi, G., Rivington, M., Donatelli, M., Matthews, K., 2010. Validation of biophysical models: issues and methodologies. A review. Agron. Sustain. Dev. 30, 109–113.

Carlson, T.N., Boland, F.E., 1978. Analysis of urban-rural canopy using a surface heat flux/temperature model. J. Appl. Meteorol. 17, 998–1014.

Carlson, T.N., Belles, J.E., Gillies, R.R., 1991. Transient water stress in a vegetation canopy: simulations and measurements. Remote Sens. Environ. 35, 175–186.

Chauhan, N.S., Miller, S., Ardanuy, P., 2003. Spaceborne soil moisture estimation at high resolution: a microwave-optical/IR synergistic approach. Int. J. Remote Sens. 22, 4599–4646.

Clapp, R.B., Hornberger, G.M., 1978. Empirical equations for some soil hydraulic-properties. Water Resour. Res. 14, 601–604.

Foken, T., Wimmer, F., Mauder, M., Thomas, C., Liebethal, C., 2006. Some intercepts of energy balance closure. Atmos. Chem. Phys. 6, 4395–4402.

Giertz, S., Diekkrüger, B., Steup, G., 2006. Physically-based modelling of hydrological processes in a tropical headwater catchment (West Africa)–process representation and multi-criteria validation. Hydrol. Earth Syst. Sci. 10, 829–847.

Gillies, R.R., Carlson, T.N., Cui, J., Kustas, W.P., Humes, K.S., 1997. Verification of the "triangle" method for obtaining surface soil water content and energy fluxes from remote measurements of the Normalized Difference Vegetation Index NDVI and surface radiant temperature. Int. J. Remote Sens. 18, 3145–3166.

Gillies, R.R., 1993. A Physically-Based Land Sue Classification Scheme Using Remote Solar and Thermal Infrared Measurements Suitable for Describing Urbanisation. PhD Thesis. University of Newcastle, UK, 121pp.

Hollinger, D., Richardson, A., 2005. Uncertainty in eddy covariance measurements and its application to physiological models. Tree Physiol. 25, 873–885.

Ireland, G., Volpi, M., Petropoulos, G.P., 2014. Semi-automated flooded area cartography from EO single image analysis and machine learning: a case study from a Mediterranean river flood. Remote Sensing MDPI 7, 3372–3399.

Liu, Y., Hiyama, T., Yamaguchi, Y., 2006. Scaling of land surface temperature using satellite data: a case examination on ASTER and MODIS products over a heterogeneous terrain area. Remote Sens. Environ. 105, 115–128.

Maayar, M., Price, D.T., Delire, C., Foley, J.A., Black, T.A., Bessemoulin, P., 2001. Validation of the Integrated Biosphere Simulator over Canadian deciduous and coniferous boreal forest stands. J. Geophys. Res. Atmos. 106, 14339–14355, 1984–2012.

Manabe, S., 1969. Climate and the ocean circulation 1. The atmospheric circulation and the hydrology of the earth's surface. Mon. Weather Rev. 97, 739–774.

Marshall, M., Tu, K., Funk, C., Michaelsen, J., Williams, P., Williams, C., Kutsch, W., 2013. Improving operational land surface model canopy evapotranspiration in Africa using a direct remote sensing approach. Hydrol. Earth Syst. Sci. 17, 1079–1091.

Mascart, P., Taconet, O., Pinty, J.P., Mehrez, M.B., 1991. Canopy resistance formulation and its effect in mesoscale models: a HAPEX perspective. Agric. For. Meteorol. 54, 319–351.

Monin, A.S., Obukhov, A., 1954. Basic laws of turbulent mixing in the surface layer of the atmosphere. Contrib. Geophys. Inst. Acad. Sci. USSR 151, 163–187.

Oncley, S.P., Foken, T., Vogt, R., Kohsiek, W., DeBruin, H.A.R., Berhofer, C., Christen, A., Van Gorsel, E., Grantz, D., Feigenwinter, C., Lehner, I., Liebethal, C., Liu, H., Mauder, M., Pitacco, A., Ribeiro, L., Weidinger, T., 2007. The energy balance experiment EBEX-2000. Part I: overview and energy balance. Bound. Lay. Meteoro. 123 (1–28).

Papale, D., Reichstein, M., Aubinet, M., et al., 2006. Towards a standardized processing of net ecosystem exchange measured with eddy covariance technique: algorithms and uncertainty estimation. Biogeosciences 3 (4), 571–583.

Petropoulos, G.P., Carlson, T.N., 2011. Retrievals of turbulent heat fluxes and soil moisture content by remote sensing. Chapter 19. In: Advances in Environmental Remote Sensing: Sensors, Algorithms, and Applications, vol. 556. Taylor and Francis, 667–502.

Petropoulos, G.P., Griffiths, H.M., Tarantola, S., 2013a. Towards operational products development from earth observation: exploration of SimSphere land surface process model sensitivity using a GSA approach. In: 7th International Conference on Sensitivity Analysis of 25 Model Output, 1–4 July 2013, Nice, France.

Petropoulos, G., Griffiths, H.M., Ioannou-Katidis, P., 2013b. Sensitivity exploration of SimSphere land surface model towards its use for operational products development from earth observation data. chapter 14. In: Mukherjee, S., Gupta, M., Srivastava, P.K., Islam, T. (Eds.), Advancement in Remote Sensing for Environmental Applications. Springer.

Petropoulos, G.P., Griffiths, H., Tarantola, S., 2013c. Sensitivity analysis of the SimSphere SVAT model in the context of EO-based operational products development. Environ. Model. Software 49, 166–179.

Petropoulos, G.P., Konstas, I., Carlson, T.N., 2013d. Automation of SimSphere Land Surface Model Use as a Standalone Application and Integration with EO Data for Deriving Key Land Surface Parameters. European Geosciences Union, Vienna, Austria, pp. 7–12. April 2013.

Petropoulos, G., Wooster, M.J., Kennedy, K., Carlson, T.N., Scholze, M., 2009a. A global sensitivity analysis study of the 1d SimSphere SVAT model using the GEM SA software. Ecol. Model. 220, 2427–2440.

Petropoulos, G., Carlson, T., Wooster, M.J., 2009b. An overview of the use of the SimSphere soil vegetation atmospheric transfer (SVAT) model for the study of land atmosphere interactions. Sensors 9, 4286–4308.

Petropoulos, G.P., Ireland, G., Lamine, S., Ghilain, N., Anagnostopoulos, V., North, M.R., Srivastava, P.K., Georgopoulou, H., 2016. Evapotranspiration estimates from SEVIRI to support sustainable water management. J. Appl. Earth Observ. & Geoinf. 49, 175–187.

Petropoulos, G.P., Griffiths, H.M., Carlson, T.N., Ioannou-Katidis, P., Holt, T., 2014. SimSphere model sensitivity analysis towards establishing its use for deriving key parameters characterising land surface interactions. Geosci. Model Dev. Discuss. 7, 1873–1887.

Petropoulos, G.P., North, M.R., Ireland, G., Srivastava, P.K., Rendall, D.V., 2015. quantifying the prediction accuracy of a 1-D SVAT model at a range of ecosystems in the USA and Australia: evidence towards its use as a tool to study Earth's system interactions. Geosci. Model Dev. 8, 3257–3284.

Piles, A., Camps, A., Vall-Llossera, M., Corbella, I., Panciera, R., Rüdiger, C., Walker, J., 2011. Downscaling SMOS-derived soil moisture using MODIS visible/infrared data. Geosci. Rem. Sen. IEEE Trans. 49 (9), 3156–3166.

Prentice, I.C., Liang, X., Medlyn, B.E., Wang, Y.P., 2014. Reliable, robust and realistic: the three R's of next-generation land surface modelling. Atmos. Chem. Phys. 14, 24811–24861.

Ross, S.L., Oke, T.R., 1988. Tests of three urban energy balance models. Bound. Lay. Meteorol. 44, 73–96.

Todhunter, P.E., Terjung, W.H., 1988. Intercomparison of three urban climate models. Bound. Lay. Meteorol. 42, 181–205.

Verbeeck, H., Samson, R., Granier, A., Montpied, P., Lemeur, R., 2008. Multi-year model analysis of GPP in a temperate beech forest in France. Ecol. Model. 210, 85–109.

Wilson, K., Goldstein, A., Falge, E., Aubinet, M., Baldocchi, D., Berbigier, P., Verma, S., 2002. Energy balance closure at FLUXNET sites. Agric. For. Meteorol. 113, 223–243.

Part VI

Future challenges in agricultural water management

Chapter 21

Future challenges in agricultural water management

Sumit Kumar Chaudhary[1], Prashant K. Srivastava[1,2]
[1]*Remote Sensing Laboratory, Institute of Environment and Sustainable Development, Banaras Hindu University, Varanasi, Uttar Pradesh, India;* [2]*DST-Mahamana Centre of Excellence in Climate Change Research, Institute of Environment and Sustainable Development, Banaras Hindu University, Varanasi, Uttar Pradesh, India*

1. Introduction

Agriculture is the prime exercise that directly impacts human survival. With growing population, pressures on agricultural production are increasing day by day. In agriculture, water is one of the key players to increase production. Agriculture is the major consumer of water with 65%–75% of freshwater being presently used for irrigation (Bennett, 2000). In some cases, it draws about 90% of the total water (Allan, 1997). Most of the agricultural areas are irrigated with groundwater, which is the main source of freshwater and makes up a very small fraction of all water on the Earth. Only 0.007% of the Earth's water is available to fuel and feed the entire population on the Earth and most of this water is concentrated in some specific regions leaving other regions water-deficit (Shiklomanov, 1993; Pimentel et al., 1999). Due to this uneven distribution of water resources, nearly 80 densely populated countries (more than 40% population of the world) worldwide are facing deficit of freshwater supply (Bennett, 2000).

As per the study of Postel (1996), the minimum average annual amount of water required per capita for food production is approximately 42 L. For human health, the minimum daily per capita water requirement is 50 L, which is eight-times less than the water being used in United States (Gleick, 1996). Between 1960 and 1997, availability of per capita freshwater worldwide has dropped by about 60%. Another 50% weakening in per capita water supply is expected by the year 2025 (Hinrichsen, 1998).

The contribution of the precipitation to groundwater primarily depends on the climatic conditions of the region. For example, the precipitation contribution to groundwater for temperate and humid climates is about 30%–50%, whereas only 2% or even less than this contributes for hot and dry climates

Agricultural Water Management. https://doi.org/10.1016/B978-0-12-812362-1.00021-7

(Tyler et al., 1996; Bouwer, 2002). However, it is very difficult to predict the actual rate of recharge; these estimates are used to assess the pumpable amount of groundwater without depleting the groundwater resources (Stone et al., 2001). The groundwater is the main source of water for agriculture and daily human needs in various arid and semi-arid regions. In these areas, more pumping than recharge of ground water is common, which results in decrease in ground water levels with frightening rates (Pimentel et al., 1999). Asia's two major breadbaskets namely the Punjab of Indian subcontinent and the North China Plains are on the high alert for declining water table (Seckler et al., 1999).

Thus, water management in agriculture is a very important aspect in order to maintain balance between the use of water and its impact on our environment. There is heavy demand of agricultural water management to produce more food, create revenues and prosperity among rural people and contribute to the sustainability of Earth's ecology (World Bank, 2005).

Irrigation, one of the forms of agricultural water management (AWM), along with floodplain farming, drainage and flood control, watershed restoration, recycled water use and rainwater harvesting, has increased the importance of AWM for agricultural production and has enabled the farming of more than 60% of the world's agricultural land (Maslov, 2009). However, increase in irrigation to increase the production is an unsustainable process and leads to degradation of environment (Gleick, 1998; Hajkowicz and Young, 2005). So, water management practices in agriculture might be a noble possibility for sustainable development. Various governments are trying to promote the adoption of efficient water-saving technologies to reduce groundwater discharge (Dhawan, 2000).

Due to the lack of monetary incentives for saving water, this approach becomes potentially less effective (Narayanamoorthy, 2004). With these subsidy programmes and promotions by government authorities, minor successes have been achieved in these directions. For example, according to India's third Minor Irrigation Census 2001, India's some 8.5 million tubewell owners (i.e., only 3%) used drip or sprinkler irrigation, while about 88% used water for irrigation by direct flooding. In India, groundwater is rarely controlled or even priced. However, heavy subsidy on electricity to pump groundwater has been given by various governments (Badiani et al., 2012; Fishman et al., 2015).

The main sources of the irrigation in most of the countries are rivers, canals and groundwater, which are directly influenced by rainfall in their geographical regions (Pimentel et al., 2004). Better management of storing rain water in open reservoirs or by recharging the groundwater should require to enhance the irrigation resources (Bhan, 2009). In arid regions, use of groundwater is not practical due to various reasons. As an alternative, water treatment or transportation of water from other regions has been adopted, but still people living in arid regions face problems due to economic limitations and because the alternatives are not practical in most of the cases. Increased cropping intensities, both vertically and horizontally, require more water per unit area cultivated and simultaneously result in degradation of the land and associated water resources at some places.

Near dense urban populations, industrial and domestic use of good-quality water has increased drastically (Loucks and van Beek, 2017). So treatment of used water and its transportation have been adopted to provide water to nearby rural areas for irrigation (Pedrero et al., 2010). The contamination of surface and groundwater resources by a variety of point and nonpoint pollution sources is possible due to the extensive use of fertilisers, pesticides, etc. The drained water from agriculture fields can pollute the rivers and finally our ecology will be at risk (Foster et al., 2018). Therefore, it is apparent that agriculture water management is crucial for future growth of both developed and developing nations.

2. Future challenges

Challenges in evolving sustainable, equitable and efficient management methods for conservation of water resources in agriculture, in order to meet the growing demand of food, are discussed as follows:

2.1 Irrigation scheduling

Irrigation scheduling is a very important way to use water efficiently in crop production. For irrigation scheduling, monitoring soil moisture during the crop growth is very demanding and needs to be accurately measured on ground (Srivastava et al., 2013a; 2015; Srivastava 2017a). The problem in the measurement of soil moisture is the availability of cheap, efficient and automatic devices that need to be developed.

For large scale, the remote sensing–based soil moisture measurement is required, which can provide spatially and temporally rich information during growth period of the crop. This will definitely benefit the exact time and amount of water required for crop. There is a need of efficient algorithms and techniques for remote sensing measurement of soil moisture to overcome spatial variability problems (Srivastava et al., 2013b, 2014a,b). Various studies have been performed to effectively retrieve the soil moisture by remote sensing techniques (Bai and He, 2015; Baghdadi et al., 2006; Sikdar and Cumming, 2018; Lawrence et al., 2007; Cui et al., 2016; Baghdadi et al., 2016; Kim et al., 2018; Li et al., 2018; Van Rompaey et al., 2002; Said et al., 2012; Alfa et al., 2018; Henderson-Sellers et al., 1993), but still various factors need to be studied in future.

Early prediction of rainfall is necessary for the better management of water for irrigation and thus an effective rainfall forecasting system needs to be developed in order to manage irrigation scheduling in agricultural field.

Rainfall prediction has been one of the most challenging issues around the world in the past year. Widely used techniques for prediction are regression analysis, clustering and Artificial Neural Network (ANN) etc. (Islam, 2012a,b,c). These methods are suitable for predicting rainfall very effectively, but still have some limitations (Islam, 2012d). Those limitations need to be observed closely and rectified accordingly. The available rainfall forecasting at regional levels needs to be enhanced to overcome the variability of rainfall as well as the water needs for irrigation.

Remote sensing can also be applied in thematic mapping of available water capacity per metre depth of soil in the area (Srivastava, 2017b). To know the local convection effect on water consumption of irrigated fields surrounded by arid zones, in-depth study will be required in future research directions. Since large part of the agricultural field is covered by ground water discharge, hence continuous data of water tables with allowable depletion levels during whole year with adequate temporal gap need to be provided for farmers. And proper training regarding the use of these data and maps should be provided in order to ensure the better use of water in irrigation. Plant water stress indices need to be regularly estimated qualitatively and quantitatively. This will create simple and low-cost warning system to alert the farmers about the water status of their crops.

The information technologies have played a very important role in agricultural water management and simulating real-world problems is very easy and economical to carry out research and testing. So, development of irrigation simulators by taking into account hydraulic or socio-economic network constraints is highly required, which will help in strategic planning and real-time irrigation scheduling.

2.2 Irrigation water need

For the calculation of irrigation water need, two important factors are taken into consideration:

1. The total water need of the various crops,
2. The amount of rain water which is available to the crops.

The difference between crop water need and amount of available rainwater is referred to as the irrigation water need for a crop. Thus the irrigation water need is basically dependent on the climate condition of the crop area. For humid climates, the amount of rainfall is excessive to the crop water needs, so drainage of excessive water is required to save crops from destruction. In sub-humid and semi-arid climates, the amount of rainfall plays a significant role but is not sufficient in many cases to cover the crop water need. However, in arid climates without irrigation it is not possible to maintain the crop water needs at all. The calculation of the irrigation water needs of a crop is very essential to manage the water in agriculture. In order to do so, first a database of various crops, their water need and rainfall available in that area are required.

The depth (or amount) of water required to balance the water loss through evapotranspiration is known as crop water need (ET crop). In other words, it is the quantity of water required by the various crops to grow favourably. For hot, dry, windy and sunny areas, crop water needs are high, whereas for cool, humid and cloudy areas with little or no wind, it is lower.

The climatic impact on crop water needs is characterised by the reference crop evapotranspiration (ET_O). The ET_O is usually expressed in millimetres per unit of time, e.g., mm/day, mm/month or mm/season. Grass has been taken as the reference crop. ET_O is referred as rate of evapotranspiration from a large area which is covered by 8–15 cm tall green grass that grows actively in plenty of water, and completely shades the ground.

To measure the evapotranspiration, experimental (using an evaporation pan) and theoretical (using measured climatic data e.g., the Blaney-Criddle method) methods have been adopted by the researchers (Srivastava et al., 2017a,b; Petropoulos et al., 2018). Evapotranspiration also plays a significant role in deciding the irrigation requirement and amount of water needs for crop growth. Evapotranspiration is intrinsically difficult to quantify and predict exclusively at large spatial level. Remote sensing can provide a cost-effective approach for estimation of evapotranspiration at regional and global levels.

In the past few decades, significant number of studies on remote sensing–based evapotranspiration estimation have been observed in literature (Anderson et al., 2016; Cristóbal et al., 2011; Lawrence et al., 2007; Gong et al., 2017; Zhang et al., 2016; Gowda et al., 2008). Several research problems need to be discussed in future including identification of uncertainty sources in remote sensing–based evapotranspiration models, assimilation of various remote sensing data and techniques in order to get robust evapotranspiration estimates. These advances in the mapping of evapotranspiration through remote sensing will help in deciding the irrigation requirement of the crop with spatial variability and leads to better agriculture water management practices.

2.3 Farmers knowledge limitation

Farmers' knowledge can play a significant role in increasing water-use efficiency, which directly influences the agricultural water management schemes. Farmers should be aware of technologies that help to manage the water efficiently and to optimise the crop production without harming the environment. For different crop types, different amounts of water are required. Farmers should also be aware of the requirement of the water for various crop types. Finally, there is strong need of developing detailed guidelines in the form of booklets in accordance with farmers' perspectives and requirements for agricultural water management (Hill and Allen, 1996). The guidelines should be simple and understandable for high, medium and low-input farmers.

However, knowledge of various water management techniques is required for farmers, but the expenses to implement those techniques are still very high. So, either we need to develop low-cost techniques of water management or increase the incomes of farmers. In current decades, the genetic underpinnings of drought-resistant crops have been acknowledged by researchers, which could grow and even thrive in dry conditions. Knowledge of these crops and proper planning can produce optimum yields even in dry conditions and save water for future generations. To spread this knowledge and practice the managing authority should hold awareness programmes in rural areas.

2.4 Policy adaptation, integration and coherence between agriculture, water, energy and environmental policies

The adaptation, integration and coherence of policies need to be mainstreamed across multiple sectors (between agriculture, water, energy and environment) and greater policy coherence is essential. This will definitely create better prospects for the water management practices in agriculture and other streams. First, the various contradictions and vulnerability between policies should be analysed and then proper solutions should be incorporated with the consultancy of various departments. This can be achieved by improving institutional structures and adopting advances in cross-sectoral planning (England et al., 2018).

Adaptation is referred as "adjustment in natural or human systems in response to actual or expected climatic stimuli or their effects, which moderates harm or exploits beneficial opportunities" (IPCC, 2007:869). For sustainable development, planned adaptation, which is defined as "adaptation that is the result of a deliberate policy decision, based on an awareness that conditions have changed or are about to change and that action is required to return to, maintain, or achieve a desired state" (IPCC, 2007:869) is needed across all divisions to diagnose the complexities in sectoral method of policy-making and highlight the significance of policy coherence in addressing future challenges (Conway and Mustelin, 2014; Conway et al., 2015; Nilsson et al., 2016). It also ensured the link between different national and international policies (Cochrane et al., 2017). This link is definitely helpful in mitigation of the various complexities associated with agriculture water management and finally helpful in sustainable development.

Organisation for Economic Co-operation and Development (OECD) defined policy coherence as "the systematic promotion of mutually reinforcing policy actions across government departments and agencies creating synergies towards achieving the agreed objectives" (OECD, 2004:3). Coherence among sectoral policies ensure that more agreement and less conflict among various organisations and governments policies can lead to better management of budgets and resources for sustainable development (Akhtar-Schuster et al., 2011).

To mitigate the climate change–related problems many countries have been working on such policies by establishing interministerial climate change committees and task forces (Stringer et al., 2014). This types of practice is also needed in agriculture water management community and organisations.

The major challenges in implementation of such practices is due to the lack of relative accessibility (i.e., communication, information sharing and collaboration) between government ministries and departments (Stringer et al., 2012). In order to encourage institutional support for cross-cutting policies, practices and partnerships, many approaches have been endorsed (Stringer et al., 2014).

a. Reinforcing national level coordination and defining clearer roles across sectors;
b. Development of partnership to deal with different stakeholders across sectors;
c. Facilitate learning and knowledge-sharing approaches;
d. The development of more equitable and transparent distribution systems to optimise costs and benefits;
e. Adaptation to climate change.

3. Conclusion

Scarcity of water posed a threat to the food production for the rising population, economic growth, and social security and environmental protection. Since agriculture is the prime user of water, agriculture water management practices need to evolve for better future of living beings on the Earth. In earlier days, the farmers used conventional methods to manage the rainfall water for irrigation as well as other purposes at local levels, which was destroyed by the introduction of large-scale irrigation systems after Western colonisation.

Due to growth in irrigated agriculture, and increasing demand in drinking water supplies, nonsustainable development of water resources and their uncontrolled uses are very common in many areas, which leads to various social and environmental problems such as groundwater depletion, groundwater quality deterioration, decreasing supplies. Hence, water management challenges are increasing day by day due to inadequate practices in agriculture sector. There are wide gaps between understanding physical problems and management solutions. Technically and economically feasible, socially and politically acceptable management solutions need be adopted in future and policies and programmes should be in line of water management rather than development of new water resources. The lack of adequate scientific data regarding the availability of good-quality water and their demand in different sectors are also the main difficulties in developing strategies for agriculture water management.

The situations also become severe when there is a lack of coordination among different agencies for data collections, processing and retrieval, and the lack of integration of social, economic and environmental factors in assessment of resource condition. Technology and its reach to the local level are also few challenges which significantly affect the agriculture water management. The available technologies for water conservation and management are either less effective or out of reach of the local farmers and communities. The advancement in sustainable water management techniques is also very slow and needs to accelerate in near future. After all, the lack of policies coherence between agriculture, energy and environment sectors is needed to remove conflicts arising across sectors that lead to efficient water management practices in agriculture.

References

Akhtar-Schuster, M., Thomas, R.J., Stringer, L.C., Chasek, P., Seely, M., 2011. Improving the Enabling Environment to Combat Land Degradation: Institutional, Financial, Legal and Science-Policy Challenges and Solutions. Land Degradation and Development. https://doi.org/10.1002/ldr.1058.

Alfa, M.I., Ajibike, M.A., Donatus, B.A., Mudiare, O.J., 2018. Assessment of the effect of land use/land cover changes on total runoff from Ofu river catchment in Nigeria. J. Degrad. & Min. Lands Manag. 5 (3), 1161–1169. https://doi.org/10.15243/jdmlm.2018.053.1161.

Allan, J., 1997. 'Virtual Water': A Long Term Solution for Water Short Middle Eastern Economies? London: School of Oriental and African Studies, University of London. https://doi.org/10.1016/S0921-8009(02)00031-9.

Anderson, M., Hain, C., Gao, F., Kustas, W., Yang, Y., Sun, L., Yang, Y., Holmes, T., Dulaney, W., 2016. Mapping evapotranspiration at multiple scales using multi-sensor data fusion. In: 2016 IEEE International Geoscience and Remote Sensing Symposium (IGARSS), pp. 226–229. https://doi.org/10.1109/IGARSS.2016.7729050.

Badiani, R., Jessoe, K.K., Plant, S., 2012. Development and the environment: the implications of agricultural electricity subsidies in India. J. Environ. Dev. https://doi.org/10.1177/1070496512442507.

Baghdadi, N., Holah, N., Zribi, M., 2006. Soil moisture estimation using multi-incidence and multi-polarization ASAR data. Int. J. Remote Sens. 27 (10), 1907–1920. https://doi.org/10.1080/01431160500239032.

Baghdadi, N., El Hajj, M., Zribi, M., 2016. Coupling SAR C-band and optical data for soil moisture and leaf area index retrieval over irrigated Grasslands. In: International Geoscience and Remote Sensing Symposium (IGARSS) 2016–Novem, pp. 3551–3554. https://doi.org/10.1109/IGARSS.2016.7729919, 3.

Bai, X., He, B., 2015. Potential of Dubois model for soil moisture retrieval in prairie areas using SAR and optical data. Int. J. Remote Sens. 36 (22), 5737–5753. https://doi.org/10.1080/01431161.2015.1103920. Taylor & Francis.

Bennett, A.J., 2000. Environmental consequences of increasing production: some current perspectives. Agric. Ecosyst. Environ. https://doi.org/10.1016/S0167-8809(00)00218-8.

Bhan, S., 2009. Rainwater management. J. Soil Water Conserv. 8 (1), 42–48.

Bouwer, H., 2002. Artificial recharge of groundwater: hydrogeology and engineering. Hydrogeol. J. https://doi.org/10.1007/s10040-001-0182-4.

Cochrane, L., Cundill, G., Ludi, E., New, M., Nicholls, R.J., Wester, P., Cantin, B., et al., 2017. A reflection on collaborative adaptation research in Africa and Asia. Reg. Environ. Change. https://doi.org/10.1007/s10113-017-1140-6.

Conway, D., Mustelin, J., 2014. Strategies for improving adaptation practice in developing countries. Nat. Clim. Change. https://doi.org/10.1038/nclimate2199.

Conway, D., Van Garderen, E.A., Deryng, D., Steve, D., Krueger, T., Landman, W., Lankford, B., et al., 2015. Climate and southern Africa's water-energy-food Nexus. Nat. Clim. Change. https://doi.org/10.1038/nclimate2735.

Cristóbal, J., Poyatos, R., Ninyerola, M., Llorens, P., Pons, X., 2011. Combining remote sensing and GIS climate modelling to estimate daily forest evapotranspiration in a mediterranean mountain area. Hydrol. Earth Syst. Sci. 15 (5), 1563–1575. https://doi.org/10.5194/hess-15-1563-2011.

Cui, Q., Dong, X., Shi, J., Zhao, T., Xiong, C., 2016. An algorithm for retrieving soil moisture using L-band H-polarized multiangular brightness temperature data. IEEE Geosci. Rem. Sens. Lett. 13 (9), 1295–1299.

Dhawan, B.D., 2000. Drip irrigation evaluating returns. Econ. Polit. Wkly. 35 (42), 3775–3780. https://doi.org/10.1016/j.solmat.2011.02.016.

England, M.I., Dougill, A.J., Stringer, L.C., Vincent, K.E., Pardoe, J., Kalaba, F.K., Mkwambisi, D.D., Namaganda, E., Afionis, S., Afionis safionis, S., 2018. Climate Change Adaptation and Cross-Sectoral Policy Coherence in Southern Africa, vol. 18, pp. 2059–2071. https://doi.org/10.1007/s10113-018-1283-0.

Fishman, R., Devineni, N., Raman, S., 2015. Can improved agricultural water use efficiency save India's groundwater? Environ. Res. Lett. 10 (8) https://doi.org/10.1088/1748-9326/10/8/084022. IOP Publishing.

Foster, S., Pulido-Bosch, A., Vallejos, Á., Molina, L., Llop, A., MacDonald, A.M., 2018. Impact of irrigated agriculture on groundwater-recharge salinity: a major sustainability concern in semi-arid regions. Hydrogeol. J. 26 (8), 2781–2791, 2018.

Gleick, P.H., 1996. Basic water requirements for human activities: meeting basic needs. Water Int. https://doi.org/10.1080/02508069608686494.

Gleick, P.H., 1998. The World's Water 1998-1999: The Biennial Report on Freshwater Resources. Island Press, Washington, DC.

Gong, T., Lei, H., Yang, D., Yang, J., Yang, H., 2017. Monitoring the variations of evapotranspiration due to land use/cover change in a semiarid shrubland. Hydrol. Earth Syst. Sci. 21 (2), 863–877. https://doi.org/10.5194/hess-21-863-2017.

Gowda, P.H., Jose, L.C., Colaizzi, P.D., Evett, S.R., Howell, T.A., Tolk, J.A., 2008. ET mapping for agricultural water management: present status and challenges. Irrigat. Sci. 26 (3), 223–237. https://doi.org/10.1007/s00271-007-0088-6.

Hajkowicz, S., Young, M., 2005. Costing yield loss from acidity, sodicity and dryland salinity to Australian agriculture. Land Degrad. Dev. https://doi.org/10.1002/ldr.670.

Henderson-Sellers, A., Dickinson, R.E., Durbidge, T.B., Kennedy, P.J., McGuffie, K., Pitman, A.J., 1993. Tropical deforestation: modeling local- to regional-scale climate change. J. Geophys. Res. Atmos. 98 (D4), 7289–7315. https://doi.org/10.1029/92JD02830.

Hill, R.W., Allen, R.G., 1996. Simple irrigation calendars: a foundation for water management. In: Irrigation Scheduling: From Theory to Practice, Proceedings ICID/FAO Workshop, Sept. 1995, Rome. Water Report No. 8, FAO, Rome.

Hinrichsen, D., 1998. "Feeding a Future World." People Planet (Article).

IPCC, 2007. Appendix I: glossary. In: Parry, M.L., Canziani, O.F., Palutikof, J.P., Van der Linden, P.J., Hanson, C.E. (Eds.), Climate Change 2007: Impacts, Adaptation and Vulnerability. Contribution of Working Group II to the Fourth Assessment Report of the Intergovernmental Panel on Climate Change. Cambridge University press, Cambridge.

Islam, T., Rico-Ramirez, M.A., Han, D., Srivastava, P.K., 2012a. Using S-band dual polarized radar for convective/stratiform rain indexing and the correspondence with AMSR-E GSFC profiling algorithm. Adv. Space Res. 50, 1383–1390.

Islam, T., Rico-Ramirez, M.A., Han, D., Srivastava, P.K., Ishak, A.M., 2012b. Performance evaluation of the TRMM precipitation estimation using ground-based radars from the GPM validation network. J. Atmosph. Solar-Terr. Phys. 77, 194–208.

Islam, T., Rico-Ramirez, M.A., Han, D., Srivastava, P.K., 2012c. A Joss–Waldvogel disdrometer derived rainfall estimation study by collocated tipping bucket and rapid response rain gauges. Atmosph. Sci. Lett. 13 (2), 139–150.

Islam, T., Rico-Ramirez, M.A., Han, D., Srivastava, P.K., 2012d. Artificial intelligence techniques for clutter identification with polarimetric radar signatures. Atmosph. Res. 109, 95–113.

Kim, S.-B., Huang, H., Liao, T.-H., Colliander, A., 2018. Estimating vegetation water content and soil surface roughness using physical models of L-band radar scattering for soil moisture retrieval. Remote Sens. 10 (4), 556. https://doi.org/10.3390/rs10040556.

Lawrence, D.M., Peter, E., Thornton, Oleson, K.W., Bonan, G.B., 2007. The partitioning of evapotranspiration into transpiration, soil evaporation, and canopy evaporation in a GCM: impacts on land–atmosphere interaction. J. Hydrometeorol. 8 (4), 862–880. https://doi.org/10.1175/JHM596.1.

Li, J., Wang, S., Gunn, G., Joosse, P., Hazen, A., Russell, J., May 2018. A model for downscaling SMOS soil moisture using sentinel-1 SAR data. Int. J. Appl. Earth Obs. Geoinf. 72, 109–121. https://doi.org/10.1016/J.JAG.2018.07.012. Elsevier.

Loucks, D.P., van Beek, E., 2017. Urban water systems. Water Resource Systems Planning and Management. Springer, Cham, pp. 527–565.

Maslov, B.S., 2009. Drainage of Farmlands:, vol. 1. Eolss Publishers Co. Ltd, Oxford, United Kingdom.

Narayanamoorthy, A., 2004. Drip Irrigation in India: Can it Solve Water Scarcity? IWA Publishing.

Shiklomanov, I.A., 1993. World fresh water resources. In: Gleick, P. (Ed.), Water in Crisis: A Guide to the World's Fresh Water Resources. Oxford University Press, Oxford, pp. 13–24.

Nilsson, M., Griggs, D., Visbeck, M., 2016. Policy: map the interactions between sustainable development Goals. Nature. https://doi.org/10.1038/534320a.

OECD, 41–19 May 2004. Institutional Approaches to Policy Coherence for Develop- Ment. A Comparative Analysis of Institutional Mechanisms to Promote Policy Coherence for Development. OECD Policy Workshop. Available at: http://www.oecd.org/development/pcd/31659769.pdf.

Pedrero, F., Kalavrouziotis, I., Alarcón, J.J., Koukoulakis, P., Asano, T., 2010. Use of treated municipal wastewater in irrigated agriculture—Review of some practices in Spain and Greece. Agri. Water Manag. 97 (9), 1233–1241.

Petropoulos, G.P., Srivastava, P.K., Piles, M., Pearson, S., 2018. Earth observation-based operational estimation of soil moisture and evapotranspiration for agricultural crops in support of sustainable water management. Sustainability 10 (1), 181.

Pimentel, D., Bailey, O., Kim, P., Mullaney, E., Calabrese, J., Walman, L., Nelson, F., Yao, X., 1999. Will limits of the Earth's resources control human numbers? Environ. Dev. Sustain. https://doi.org/10.1023/A:1010008112119.

Pimentel, D., Berger, B., Filiberto, D., Newton, M., Wolfe, B., Karabinakis, E., Clark, S., Poon, E., Abbett, E., Nandagopal, S., 2004. Water resources: agricultural and environmental issues. BioScience 54 (10), 909–918.

Postel, S., 1996. Dividing the Waters: Food Security, Ecosystem Health, and the New Politics of Scarcity. Worldwatch Paper.

Said, S., Kothyari, U.C., Arora, M.K., 2012. Vegetation effects on soil moisture estimation from ERS-2 SAR images. Hydrol. Sci. J. 57 (3), 517–534. https://doi.org/10.1080/02626667.2012.665608.

Seckler, D., Barker, R., Amarasinghe, U., 1999. Water scarcity in the twenty-first century. Int. J. Water Resour. Dev. https://doi.org/10.1080/07900629948916.

Sikdar, M., Cumming, I., 2018. A Modified Empirical Model for Soil Moisture Estimation in Vegetated Areas Using SAR Data. Accessed September 26. https://sar.ece.ubc.ca/papers/Sikdar_IGARSS_2004.pdf.

Srivastava, P.K., 2017b. Satellite soil moisture: Review of theory and applications in water resources. Water Res. Manag. 31 (10), 3161–3176.

Srivastava, P.K., Han, D., Rico-Ramirez, M.A., Al-Shrafany, D., Islam, T., 2013a. Data fusion techniques for improving soil moisture deficit using SMOS satellite and WRF-NOAH land surface model. Water Res. Manag. 27 (15), 5069–5087.

Srivastava, P.K., Han, D., Rico-Ramirez, M.A., O'Neill, P., Islam, T., Gupta, M., Dai, Q., 2015. Performance evaluation of WRF-Noah Land surface model estimated soil moisture for hydrological application: Synergistic evaluation using SMOS retrieved soil moisture. J. Hydrol. 529, 200–212.

Srivastava, P.K., Han, D., Ramirez, M.R., Islam, T., 2013b. Machine learning techniques for downscaling SMOS satellite soil moisture using MODIS land surface temperature for hydrological application. Water Res. Manag. 27 (8), 3127–3144.

Srivastava, P.K., Han, D., Rico-Ramirez, M.A., O'Neill, P., Islam, T., Gupta, M., 2014a. Assessment of SMOS soil moisture retrieval parameters using tau–omega algorithms for soil moisture deficit estimation. J. Hydrol. 519, 574–587.

Srivastava, P.K., O'Neill, P., Cosh, M., Kurum, M., Lang, R., Joseph, A., 2014b. Evaluation of dielectric mixing models for passive microwave soil moisture retrieval using data from ComRAD ground-based SMAP simulator. IEEE J. Select. Top. App. Earth Observ. Rem. Sens. 8 (9), 4345–4354.

Srivastava, P.K., Han, D., Yaduvanshi, A., Petropoulos, G.P., Singh, S.K., Mall, R.K., Prasad, R., 2017a. Reference evapotranspiration retrievals from a mesoscale model based weather variables for soil moisture deficit estimation. Sustainability 9 (11), 1971.

Stone, D.B., Moomaw, C.L., Davis, A., 2001. Estimating recharge distribution by incorporating runoff from mountainous areas in an alluvial basin in the Great basin region of the southwestern United States. Gr. Water. https://doi.org/10.1111/j.1745-6584.2001.tb02469.x.

Stringer, L.C., Dougill, A.J., Mkwambisi, D.D., Dyer, J.C., Kalaba, F.K., Mngoli, M., 2012. Challenges and opportunities for carbon management in Malawi and Zambia. Carbon Manag. https://doi.org/10.4155/cmt.12.14.

Stringer, L.C., Dougill, A.J., Dyer, J.C., Vincent, K., Fritzsche, F., Leventon, J., Falcão, M.P., et al., 2014. Advancing climate compatible development: lessons from southern Africa. Reg. Environ. Change. https://doi.org/10.1007/s10113-013-0533-4.

Tyler, S.W., Chapman, J.B., Conrad, S.H., Hammermeister, D.P., Blout, D.O., Miller, J.J., Sully, M.J., Ginanni, J.M., 1996. Soil-water flux in the southern Great basin, United States: temporal and spatial variations over the last 120,000 years. Water Resour. Res. https://doi.org/10.1029/96WR00564.

Van Rompaey, Anton, J.J., Govers, G., Puttemans, C., 2002. Modelling land use changes and their impact on soil erosion and sediment supply to rivers. Earth Surf. Process. Landf. 27 (5), 481–494. https://doi.org/10.1002/esp.335.

World Bank, 2005. The World Bank Annual Report 2005: Year in Review, Volume 1. Washington, DC.

Zhang, K., Kimball, J.S., Running, S.W., 2016. A review of remote sensing based actual evapotranspiration estimation. Wiley Interdiscip. Rev. Water 3 (6), 834–853. https://doi.org/10.1002/wat2.1168.

Index

Note: 'Page numbers followed by "*f*" indicate figures and "t" indicate tables.'

E

F

G

N

O

P

R

S

T

Printed in the United States
By Bookmasters